Individual Differences in Hemispheric Specialization

NATO ASI Series

Advanced Science Institutes Series

A series presenting the results of activities sponsored by the NATO Science Committee, which aims at the dissemination of advanced scientific and technological knowledge, with a view to strengthening links between scientific communities.

The series is published by an international board of publishers in conjunction with the NATO Scientific Affairs Division

A Life Sciences **B Physics**	Plenum Publishing Corporation New York and London
C Mathematical and Physical Sciences	D. Reidel Publishing Company Dordrecht, Boston, and Lancaster
D Behavioral and Social Sciences **E Engineering and Materials Sciences**	Martinus Nijhoff Publishers The Hague, Boston, Dordrecht, and Lancaster
F Computer and Systems Sciences **G Ecological Sciences** **H Cell Biology**	Springer-Verlag Berlin, Heidelberg, New York, London, Paris, and Tokyo

Recent Volumes in this Series

Volume 127—The Organization of Cell Metabolism
edited by G. Rickey Welch and James S. Clegg

Volume 128—Perspectives in Biotechnology
edited by J. M. Cardoso Duarte, L. J. Archer, A. T. Bull, and G. Holt

Volume 129—Cellular and Humoral Components of Cerebrospinal Fluid in Multiple Sclerosis
edited by A. Lowenthal and J. Raus

Volume 130—Individual Differences in Hemispheric Specialization
edited by A. Glass

Volume 131—Fat Production and Consumption: Technologies and Nutritional Implications
edited by C. Galli and E. Fedeli

Volume 132—Biomechanics of Cell Division
edited by Nuri Akkas

Volume 133—Membrane Receptors, Dynamics, and Energetics
edited by K. W. A. Wirtz

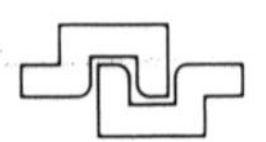

Series A: Life Sciences

Individual Differences in Hemispheric Specialization

Edited by

A. Glass

The Medical School
University of Birmingham
Birmingham, United Kingdom

Plenum Press
New York and London
Published in cooperation with NATO Scientific Affairs Division

Proceedings of a NATO Advanced Research Workshop on
Individual Differences in Hemispheric Specialization,
held October 15-19, 1984,
in Maratea, Italy

Library of Congress Cataloging in Publication Data

NATO Advanced Research Workshop on Individual Differences in Hemispheric Specialization (1984: Maratea, Italy)
Individual differences in hemispheric specialization.

(NATO ASI series. Series A, Life sciences; v. 130)
"Proceedings of a NATO Advanced Research Workshop on Individual Differences in Hemispheric Specialization, held October 15-19, 1984, in Maratea, Italy"—T.p. verso.
"Published in cooperation with NATO Scientific Affairs Division."
Includes bibliographies and index.
1. Cerebral dominance—Congresses. 2. Laterality—Congresses. 3. Individuality—Congresses. I. Glass, A. (Alan) II. North Atlantic Treaty Organization. Scientific Affairs Division. III. Title. IV. Series. [DNLM: 1. Dominance, Cerebral—congresses. 2. Individuality—congresses. WL 335 N279i 1984]
QP385.5.N37 1984 152 87-14055
ISBN 0-306-42586-6

A Division of Plenum Publishing Corporation
233 Spring Street, New York, N.Y. 10013

Printed in the United States of America

PREFACE

This volume originates from a NATO Advanced Research Workshop held in Maratea, Italy from 8th-15th October 1984. Aims and contributions are described at greater length in the Introduction and the following chapters. It is hoped that this volume will provide a critical overview of hemispheric specialization in relation to individual differences, but one that is not intended to be comprehensive. Three contributions on this theme are made by authors who were invited to the Workshop but were unable to participate in it.

The volume contains a critical appraisal of the differentially specialized functions of left and right human cerebral hemispheres in verbal and visuospatial domains respectively (formerly cerebral dominance), in relation to individual variation due, for example, to gender and handedness.

Critical cross-comparison of several methods of assessing hemispheric specialization such as perceptual/behavioral, clinical/neurological, electrophysiological and "real time" methods of assessment of cerebral orientation have been made. Individual differences have been considered in relation to statistical concepts in the assessment of cerebral lateralization. Some emphasis has been placed on the application of these methods and concepts to psychopathology.

It is a consensus of the views expressed at the meeting (since considered and matured) and the different methods of investigation of cerebral orientation represented, that hemispheric specialization has re-emerged in the guise not of an absolute structural model of cerebral dominance, but as a dynamic process modulating the utilization by hemispherically differential strategies, activation and arousal of a relative structural specialization. The model must eventually incorporate a regional intrahemispheric specificity of function.

In addition to the thanks expressed in the Acknowledgements of the Introduction, I wish to thank Dr. S.R.Butler and Mr D.Symons for invaluable help in preparing the Index.

A.Glass

CONTENTS

INDIVIDUAL DIFFERENCES IN BEHAVIOURAL INDICES OF CEREBRAL ORIENTATION

ASYMMETRIES IN PSYCHOPATHOLOGY AND INDIVIDUAL DIFFERENCES

INTRODUCTION

PREAMBLE

The "Think-Tank"

At the outset of the Workshop (of which this volume is the record), it was pointed out that the Workshop itself was, in current parlance, a sort of transient "Think-tank", meaning an establishment or institute temporarily housing the "thinkers"-the participants. It is interesting that this expression evolved from the term "think-tank" signifying head or cranium. This quotation can be taken as an example of the earlier meaning:

"... a roscoe said: ´Whr-r-rang!´ and a lead pill split the ozone past my noggin... Neon lights exploded inside my think-tank ..Kane Fewster was on the floor. There was a bullet hole through his think-tank." (Perelman, 1951).

The original meaning could also be associated with the Workshop. The theme might be encapsulated by the question: what models of cerebral organization inside the "think-tank" can be inferred from measures of individual differences in hemispheric specialization?

Since the earlier usage evolved, much of relevance to the Workshop has occurred in the Neurosciences and related fields, but it could be argued that among these, together with a greatly increased knowledge of cerebral dominance, the development of the computer and the parallel appreciation that human cognitive processes may be mimicked by it, may have been among the most relevant. It could be speculated, perhaps, that the extension of the meaning of "Think-tank" has in some way, symbolised these developments.

Origin of the Workshop

This volume has its origin in a NATO Advanced Research Workshop on "Individual Differences in Hemispheric Specialization" held in Maratea, Italy in October, 1984.

The concept of the Workshop, in turn, sprang from a Round Table entitled "Hemispheric Specialization and Lateral Asymmetries in the EEG" given under the auspices of the Second European Winter Conference on Brain Research, in Chamonix, France, in March 1982. The papers emanating from the Round Table were later published as a Special Issue of Biological Psychology (Glass, 1984). The earlier Round Table was divided into two sections, one dealing with normal lateralization of cognitive function and psycho-physiological studies and the other concerned with EEG asymmetries as signs of disturbed laterality in psychiatric disorders. Insofar, as the plan of the Workshop followed that of the Round Table, although it broadened and extended the plan which the Round Table adopted, the contents of the Workshop and Round Table correspond. After an initial section in which the issues are defined, this volume is divided into three sections, dealing with the normal brain, and corresponding to the first part of the Round Table. The first of these three sections deals with individual differences in hemispheric specialization from the viewpoint of cerebral anatomical and circulatory asymmetries (although the emphasis is upon hemispheric specialization related to individual differences in the anatomy of the corpus callosum). The second section covers hemispheric specialization in respect of electrophysiological asymmetries largely, but not exclusively, in relation to asymmetries in the EEG alpha rhythm. The third section covers behavioural indices of cerebral orientation, in respect to lateral asymmetries in visual and tactile stimuli and divided field studies. The second part of this volume which covers psychopathological and psychiatric aspects of individual differences in hemispheric specialization, corresponds to the second section of the Round Table.

Theme of Workshop

The original theme of the Workshop was to have encompassed the theme of the Round Table, to include not only EEG alpha and evoked potential (VER, CNV, event-related potential studies) measures of lateral asymmetry in relation to hemispheric specialization but also the cross-comparison of these with other measures of assessment of hemispheric specialization including imaging asymmetry techniques such as Positron Emission Tomography (PET) (Raichle, 1985), blood flow studies such as those of Lassen, Ingvar and Skinhoj (1978) and Gur and Reivich (1980), and Gur and Gur (this volume) clinical and neurological evidence of lateralization, together with the behavioural perceptual techniques of dichotic listening (Kimura, 1967) and divided field studies (Beaumont, 1982). The aim, with a methodological emphasis was to cross-compare the results of assessment of laterality by these techniques in individuals in specific clusters such as gender and handedness groups and in clusters encountered in psychiatry and psychopathology.

However, valuable such a methodologically based comparison of different techniques might be, it would not have had the conceptual value appropriate for a NATO workshop, so a new theme was evolved. In the formulation of the new theme much was owed to the insight of my colleague Dr. S. R. Butler who drew attention to a concept formulated by Segalowitz. The problem of individual variation in lateralization had been highlighted by Segalowitz and Bryden (1983) who emphasized the importance of individual differences in cerebral lateralization for cognitive function. Evidence from brain imagery, cerebral blood flow patterns of distribution, divided field studies and dichotic listening studies together with clinical examination of the effects of lateralized lesions indicated that the pattern of functional specialization for language was not consistent across individuals. Some of the differences were correlated with handedness some with gender, although not to a sufficient extent to account for them all. Thus the same "ingredients" (i.e. methods of laterality assessment) could be used as for the earlier formulation of the Workshop theme, but with a significant shift of emphasis from the purely methodological to that of a fresh but related concept, that of individual differences in hemispheric specialization. The concept of individual differences in language lateralization or functional specialization was broadened to include together with the conventional cognitive functions of the "left" hemisphere, the cognitive functions of the "right" hemisphere (Bogen, 1969), encapsulated in the term "hemispheric specialization", that is, in individual differences in the cognitive specialization of both cerebral hemispheres, including its variation in gender and handedness groups among other factors and in effect comparing different methods of laterality assessment.

Thus, the proposal for the Workshop was redrafted extensively drawing on the original concept of individual differences in language lateralization (Segalowitz and Bryden, 1983, and Segalowitz, see this volume) to include not only language but other lateralizable functions of both left and right hemispheres. The basic idea in the original proposal of a comparison of the different methods of assessment of hemispheric specialization in clusters of individuals i.e. gender and handedness, familial laterality groups, developmental groups and, of course, psychiatric and psychopathological groups was retained and was implicit in the new context.

"Crossed Aphasia" and Associated Factors

The problem of individual differences in hemispheric specialization had emerged in the study of task-related EEG alpha asymmetries in different groups of normal subjects, specifically gender and familial handedness groups (Glass, Butler and Carter, 1984; Butler, 1984). But, to retrace the steps, it was an important and challenging problem, generally, potentially apparent since the time of Broca (1865). Indeed as early as the end of the nineteenth century Bramwell (1899) used the term "crossed aphasia" to describe individual patients who differed from or were exceptions to Broca's generalization that, uniformly, the left hemisphere was specialized for language and that aphasia or dysphasia rarely or never occurred after a lesion of the right hemisphere. "Crossed aphasia" in both dextrals and sinistrals (setting aside discussion of the aetiology) was defined as aphasia or dysphasia occurring as the result of a cerebral lesion which was ipsilateral not contralateral to the preferred hand. It was encountered in sinistral as well as dextral patients. Thus, in both right and especially in left handers there is long established clinical evidence of individual differences in language lateralization (reversed dominance) (Zangwill, 1982). This conclusion is supported by evidence derived from the use of other techniques for assessing lateralization. Right handers, in a sample of normal subjects, showed greater suppression of EEG alpha rhythm over the left hemisphere when performing mental arithmetic, but left handers almost no alpha asymmetry, suggesting that the verbal-symbolic cerebral function of right handers was more strongly lateralized than that of left handers (Butler and Glass, 1974) or that a greater proportion of left handed than right handers had reversal of dominance (see Butler and Glass, this volume; and Marshall this volume, in regard to graded versus discrete effects of lateralization). The conclusion regarding individual differences is supported by evidence derived from earlier behavioural assessment of lateralization. For example, in dichotic listening studies, Satz, Achenbach, Pattishall and Fennell (1965) showed that left and right ear difference scores were twice as large for dextrals as sinistrals. Sinistrals showed no LVF superiority for recognising dot patterns (Harcum and Dyer, 1962) and no differences between sides to a unilateral auditory stimulus (Provins and Jeeves 1975).

The effects of familial laterality on cerebral dominance have also been investigated. Recovery from aphasia, for example, was faster if there were left handed relatives (Zangwill, 1960). Sinistrals with left handed relatives developed speech disorders with equal frequency after either left or right sides lesions, sinistrals with only dextral relatives tended to develop aphasia after only left sided lesions (Hécaen and Sauguet, 1971). RVF superiority for verbal material was reduced in non-familial sinistral right handers. A left handed relative reduced the probability of a left ear superiority for the recognition of verbal material in right handers (Satz, Achenbach and Fennell, 1967) and left handers (Zurif and Bryden, 1969). Thus, a close sinistral relative increases the probability of weak or even reversed lateralisation of cerebral function. McKeever and VanDeventer (1977) had shown a familial laterality effect in left handed subjects in that left handers with left handed relatives produced a significant right visual field superiority for tachistoscopically presented verbal material, indicating a left hemisphere dominance for language, which left handed subjects without left handed relatives did not. No such familial laterality effect was found in right handers, however. Using dichotic listening studies, Lishman and McMeekan (1977) found that a family history of sinistrality among left handers was associated with smaller ear difference scores, indicating reduced lateralisation or bilateral representation of language, equally in left dominant and right cerebrally dominant left handers.

AIMS OF THE WORKSHOP

Therefore, it must be appreciated that, to attempt to define the aims of the Workshop in terms of the elucidation of the role of individual differences in hemispheric specialization, there would appear to be a single key issue. This appears to be that the lateral asymmetry of human cerebral function is describable at one extreme in terms of the variation in cognitive strategies employing one hemisphere rather than the other from one individual to another, or, from task to task performed by a single individual and the intrinsic "absolute" functional identity of the hemispheres, what is termed the dynamic process model versus the fixed structure model (see Gruzelier, this volume; Cohen, 1982). It is worth clarifying the distinction between a fixed structure model of cerebral asymmetry and a gross morphological feature of cerebral asymmetry, for example, that between left and right plana temporales. A gross morphological asymmetry can only be one component of the fixed structural model, as Marshall, (this volume) points out, the relationship between size and hemispheric specialization or local gross cerebral asymmetry is inferred from an assumptive relationship between size and proficiency.

The models of hemispheric specialization that we shall consider have been categorised (Cohen, 1982) as either absolute or relative fixed structure models where a given cognitive function is completely lateralized to one hemisphere and relative models in which a given function is performed preferentially, better, or more efficently, by one hemisphere than the other.

Until relatively recently, the possibility of individual variation in hemispheric specialization had been regarded as stumbling-block in the assessment of human cerebral functional organization. This has happened because, as Segalowitz has pointed out, (Segalowitz and Bryden, 1983; Segalowitz, see this volume), the analytical techniques for determining lateralizable functions have been limited.

Analyses currently employed have not in general permitted us to distinguish variation due to differences in individual hemispheric specialization, related to expected biological or experiential variability from the unavoidable errors of measurement made in the various types of assessment of laterality in these individual subjects. Taking a specific example, although one which is not necessarily the most familiar, in measuring EEG alpha asymmetry during mental arithmetic, suppression being greater over the more active hemisphere, the amplitude of the alpha rhythm then, is lower during calculation over the left side of the brain in right-handers. However, analyses currently employed have not allowed us to distinguish variation in asymmetry of the alpha rhythm, due to differences in individual hemispheric specialization or even individual differences in hemispheric usage or utilization from unavoidable errors of measurement made in the electrophysiological and other types of assessment of cerebral laterality. Bryden (1982), for example, has pointed to the existence of factors which may be unrelated to the lateralization of language function or even to hemispheric specialization and which may cause lateralization of the perceptual measures of asymmetry in dichotic listening studies and in divided field studies (split field tachistoscopy), factors such as variation in strategies used to perform the tasks and attentional biases which may contribute, it is suggested, to the variability in measurement of lateralization. One could add another source of variance which might simply be that due to asymmetries of function in the primary receptor organs or their centripetal connections, which cannot always be controlled for satisfactorily. These are valid points, for, as Cohen (1982) has pointed out, a model of hemisphere asymmetry is essentially a neurological model. This has the consequences that the model must incorporate all the relevant features of the actual brain, even if this merely entails attempting to design procedures for rejection of sensory and motor asymmetries, among other features, into the proposed experiments.

The inflation of the error of measurement by the biological variation of individuals has been an apparent stumbling block, because it has tended almost outweigh the main outline of the "generalised" case, that the left hemisphere is specialized for language and the right for visuospatial function. Now that these main outlines have, apparently by consensus, received clarification by the experimental efforts of the past decades and because of the relative measure of agreement reached by perceptual, clinical-neurological and "realtime" (including EEG and evoked response) indices of cerebral orientation, it may be profitable to take stock, as attempted in this volume and to study the factors affecting variation in hemispheric specialization between individuals. At the same time, it must be borne in mind that such individual differences or variation may well tend to undermine such generalisations concerning cerebral functional organisation that have cost so much to achieve. Although Popper (1972) has stated that the function of experimental test or trial can only result in the disproof, not the proof of hypotheses, it is a paradox that no prizes are offered for negative results. Yet it may be that in this volume we are approaching the point of dissolution of the paradigm where consensus give way to dissensus (Laudan, 1984) and a fresh paradigm may emerge. To quote Schofield (this volume) in the context of individual differences in bimanual response to lateralized stimuli and referring to gender differences:

"It is one thing to find that males and females choose, possibly as set by unnoticed situational influence, prior experience, or experimenter reactivity, different task strategies, either linguistic or non-linguistic, and quite another to propose that there are functional brain differences between males and females".

Factors such as handedness, familial laterality, gender, developmental experience, literacy, diversity in the cognitive strategies adopted in problem solving may each contribute to this individual variation. Indeed, it would not be surprising if one method of assessing specialization is more appropriate to a particular factor than to another. It was, at the outset of the Workshop, the aim (or hope) that by the cross-comparison of the various asymmetry measures, it should have been possible to develop guidelines and future strategies for teasing-out the relative contributions of those factors affecting cerebral orientation. The reader will be able to ascertain how far this hope has been fulfilled, bearing in mind some of the caveats that have already been put forward. Nevertheless, to quote Segalowitz and Bryden (1982): "The field of experimental neuropsychology has not advanced beyond the stage where error of measurement and individual differences in brain organisation can remain confounded". The reverse side of the coin is that the extent of non-pathological individual differences in cerebral orientation may be far greater than was once suspected.

This need to find out at a fundamental level more about variation in the functional asymmetry of the human brain must obviously be relevant to understanding how the brain "works", which in turn is of importance in fields as diverse as computer science and psychiatry. In computer science, for example, accounts of lateralized function have sparked off an important and fruitful area of investigation in current artificial intelligence studies of visual perception.

However, it is to be hoped that these fields have, in turn, and will continue themselves to contribute to the understanding of cerebral function. Psychiatry and psychopathology are dealt with in the last Section of this volume. Computer science, for example, through artificial intelligence (or cognition) may have much to contribute. Marr (1982) has recently provoked us to ask through his modular concept of localization, what is it exactly that we consider is being lateralized? For example, the "shading" of a three-dimensional object gives rise to the perception of its solidarity. Each of these factors is what Marr terms a module, that is, "tree recognition" and three-dimensional viewing of the tree itself are each, in cerebral terms, a separate module. Does this provide us with a clue as to the way in which lateralization or hemispheric specialization proceeds?

BRIEF OVERVIEW

It will be helpful, in view of the foregoing discussion of the aims of the Workshop, to consider, in the light of this discussion, the contributions to this volume, to ascertain whether in the time intervening since the Workshop there has been a development of views towards a change in the theoretical stance of the contributors.

However, although three of the twenty three chapters were not in fact presented at the Workshop, these contributors were invited to attend but, in the event, were unable to participate at Maratea. Their chapters, are, however, relevant to the theme of the volume. The delay ensuing in publishing the volume after the Workshop is to some extent valuable in that the views expressed at the Workshop have thereby been considered and matured. On the other hand, the chapters are, in some cases less up-to-date than otherwise they might have been. In some instances, the delays have been due to the tardiness of a few contributors and editorial flexibility which did not disallow this! On the other hand, the delays have also been due to the decision taken at the Workshop itself to referee the chapters independently. This has been done in nearly all chapters and, hopefully, has improved the quality of the contributions, but has delayed their publication.

Defining the Issues

An attempt has been made to define the issues in the discussion of the theme and the aims of the Workshop, in this Introduction, and this is the aim of the first Section.

In the first chapter, Segalowitz defines what appears to be a central problem for the Workshop, that is, the possible confounding of biological differences in brain asymmetry with variation in the measurement of asymmetry itself, whether by perceptual, neurological or clinical and "real time" measures of cerebral orientation. Consideration is given to the factors which contribute variance to lateralization scores, whether or not these factors relate to hemispheric specialization. The issue is whether these variables correspond to variables other than those of the commonly considered individual differences: for example, attentional bias and hemispheric activation are considered. To resolve some of these problems, a method of statistical analysis is developed which allows for variation in lateralization, and this is illustrated by a repeated measures study on visual half-field data and with single session EEG alpha asymmetries:

intersubject variation in asymmetry, whatever the source, can be isolated.

The issues considered by Marshall, initially, is concerned with what he terms the duality hypothesis. He draws an important distinction between hypotheses that propose a difference between the functional capacity of left and right hemispheres, and the concept of cerebral duality, that one hemisphere can function independently of the other, almost to the extent of mental duality. These two hypotheses are evaluated critically and separately and the concept of hemispheric specialization is contrasted with the concept of focal specialization, localization, between and within hemispheres, not a duality of functional, lateralization but the concept of a multiplicity of loci is emphasized.

The chapter by Annett summarises issues concerned with hemispheric specialization, related to her well-known Right Shift theory of handedness, based on analysis of her extensive left- and right- hand preference and skill data (Annett, 1985). Individual differences in hemispheric specialization are hypothesised to depend on a gene (the RS gene) which may be present in single or double doses and which may promote the early control of speech by the left hemisphere. A bias to right handedness is secondary. In the absence of the gene, there is a chance bias to either left or right hemisphere or either hand preference, but in this case slight cultural pressure and other factors, may cause a tendency towards right handedness. It follows that testable predictions concerning hemispheric specialization can be formulated in respect of sinistral tendencies among individuals and their relatives and in respect to gender, as well as hyptheses specifying the consequences for improved interhemispheric cooperation in the absence of the double RS gene.

Marshall's second chapter raises an issue germane to any discussion of individual differences in hemispheric specialization, that of whether lateralization can be considered a graded or a discrete characteristic. Consideration of this problem, is essential to the study of individual differences in lateralization. A metric is developed over which gradation of lateralization can be measured to correspond to, or be compared to, empirically derived measures of laterality, but, of course, it must be remembered that, overall, within a given group, the main metric will also be contributed to by individual variation, and this could in theory occur as a graded effect even if lateralization in the individual cases making up the group were discrete, not graded.

Individual differences in Cerebral Anatomical Asymmetry and Circulatory Asymmetry

Witelson and Kigar have addressed the problem of individual differences in anatomical asymmetry not directly in relation to hemispheric specialization, but to a related field, that of individual variation in the morphology of the corpus callosum, that broad and extensive band of myelinated fibres which link left and right hemispheres. A review of the considerable literature which includes methodological considerations, reveals variation in callosal morphology between handedness and age groups and which is consistent across diverse studies. Earlier studies which showed increased callosal thickness in schizophrenia are not supported by later work, although interactions with gender and handedness are possibilities. Such findings in regard to callosal topography may provide an anatomical substrate for individual differences in hemispheric specialization, although it is not yet clear, of course, how such differences could be incorporated in a fixed structure model of cerebral dominance. It must be remembered, as Marshall (this volume) has pointed out that it is an assumption, although a plausible one, that a larger cerebral neurostructure is evidence of greater functional proficiency.

Gur and Gur summarise their findings in relation to individual differences in the direction and degree of hemispheric specialization. Studies during cognitive function have been made in brain damaged and left- and right-handed normal subjects. Employing not only perceptual methods of laterality assessment, dichotic listening and split field tachistoscopic techniques, but also regional blood flow measures using 133-Xenon inhalation techniques and local cerebral glucose metabolism studies with Positron Emission Tomography (PET), both of the latter techniques showing lateralized activity during standardised lateralization tasks. Also, the neuro-imaging studies have been used in the study of emotion and affect and have shown individual difference effects in regard to handedness and gender. The main focus of the results is on regional blood flow studies, however.

Individual Differences in Electrophysiological Asymmetries

Butler and Glass, in their chapter, review data on individual differences in EEG alpha asymmetries during cognitive tasks designed to engage either left or right hemispheres. Such asymmetries have already been shown to be associated with cognitive processes and not due to asymmetric motor or sensory loading (Butler & Glass, 1985). The results are interpreted using the assumption that suppression of alpha rhythm over homotopic areas of one hemisphere compared with the other is an indication of the activation of that hemisphere. Thus, this indicates a relative increase in the utilization of that hemisphere during specific cognitive task performance. On this basis, individual differences in EEG alpha asymmetry involving handedness and gender groups are interpreted less in terms of a fixed structure model of hemispheric specializaton, a model for which the results cannot provide support, but a more labile, dynamic process of interhemispheric functional organisation, which even perhaps, in part, involves cognitive strategy differences between groups. An example of this lability, the possible influence of dynamic processes and of cognitive style, an investigation of the effect of a potential conversant on EEG alpha asymmetries is presented (Butler, Glass and Fisher, 1986).

Flor-Henry, Koles and Reddon, in their chapter, report on the analysis of the EEG from an eight-electrode montage in 57 males and 56 females, all right handed, whose ages ranged from 18 to 59. Recordings were made during various cognitive tasks. Measures were made of EEG power, phase and coherence. The details of this normative data are recorded in the appendices to the chapter. EEG differences were found between males and females, consistent with cognitive differences. The ageing brain in EEG asymmetry terms was found to approximate more to the female pattern of cerebral organization, than to the male, with EEG signs of a right hemisphere preponderance.

Ray, in his chapter, has summarised work in his laboratory which has been directed towards understanding how individual differences occur in psychophysiological measures (especially the EEG alpha rhythm) of cognitive and emotional processes. He points out the basic, pioneering approach of early EEG research into the alpha rhythm in his field. He then reviews task-related EEG alpha asymmetry measures of hemispheric lateralization, especially in relation to gender differences. However, the possibility of there being apparently unresolved methodological issues still complicating the interpretation of EEG alpha asymmetries has led him to propose other lines of research. The investigation of factors such as emotionality, direction of attention, intake and rejection tasks (the former requiring the use of external information, the latter not requiring it) on the asymmetry of EEG alpha activity are described. Finally, in regard to the possibility that individual differences in personality outside those of gender and handedness might influence hemispheric specialization, an EEG study in alpha rhythm and other frequency bands of the introversion/extraversion dimension is described. Among other effects, these preliminary findings showed the importance of the resting EEG "baseline" condition in this field.

Molfese and Molfese, in their chapter, describe the use of the auditory evoked response (AER) in studying the development of language and its interrelationship with hemispheric specializaiton at different ages. The technique is described in a study in which the brain's electrical response over left and right hemispheres to speech and non-speech syllables is recorded at birth. The results at birth especially from the left hemisphere can be used to predict accurately performance on two language tests in 3-year-olds. Although the individual differences here studied were those of age or development and speech proficiency, it is considered that the electrophysiological parameters do not indicate lateralization as the sole factor involved.

Rockstroh and Lutzenberger describe experiments with event-related slow brain potentials to provide evidence for hemisphere-specific processing in anhedonic subjects. These subjects are characterised by a marked defect in pleasure capacity, a personality trait which has for long been associated with schizophrenia. Results of Contingent Negative Variation (CNV) and post-imperative slow potential (SP) studies indicated significantly greater post-imperative negativity (PINV) in anhedonics than in controls, over the left precentral region. Although a similar effect can be seen normally (Butler, Glass and Heffner, 1981). This, together with other evidence suggests greater compensatory left hemisphere activation in the anhedonic group, to counteract for attentional and preparatory deficits in the group.

Elbert and Birbaumer describe experiments to investigate hemispheric interaction in smokers and non-smokers, using cigarettes with differing nicotine concentrations. They presented tactile stimuli to both hands, the left requiring pattern discrimination, the right simple enumeration. The slow potentials (CNV's) recorded during task-performance showed a task-dependent asymmetry which developed earlier with nicotine. This, and biofeedback evidence suggests that nicotine either interacts asymmetrically with right hemisphere arousal or facilitates interchange between hemispheres. Smokers with nicotine, were better adapted as a group, to switching between hemispheres, but without nicotine were less well adapted to switching.

Puente and Peacock describe, an EEG interval histogram analysis (using half-wave duration) from occipital sites (01 and 02) in brain damaged and non-brain damaged schizophrenics and patients with affective disorders, in EEG's recorded while subjects rested, and performed multiplications and solved geometric problems. Relatively slow right hemisphere activity is reported for the schizophrenic group during the geometrical task. This and other evidence suggests, it is emphasised, the dynamic nature of the hemispheric dysfunction in schizophrenia.

Individual Differences in Behavioral Indices of Cerebral Orientation

Young, Bion and McWeeny' report, on a series of experiments involving right-handed children, in which stimuli (chiefly face recognition and dot enumeration) were presented to left and right visual fields and left and right hands respectively, to determine age and gender differences in lateral asymmetries. No developmental changes in the size of asymmetries are reported but gender differences in asymmetry across age proved to be stable. However, as these gender differences could be induced or eliminated by small procedural changes, the model of variation in functional asymmetry in male or female brains is not well-supported, but, instead, the concept of a gender difference in subjects' reliance on, or use of, lateralized cognitive processes is affirmed. Possibly, these processes could either be under voluntary control, or, what is considered to be more likely, procedurally determined or material-specific.

Schofield has provided, an extensive critical review of divided field studies of cerebral lateralization using manual reaction time to dots and flashes of light, studies in which an attempt has been made to measure interhemispheric transmission times. Inconsistencies in the reported findings could be explained by failure to take into account individual differences in processing strategy in relation to hemispheric specialization. A simple reaction time experiment in children is reported and thoroughly analyzed, in which bimanual responses to bilateral stimuli is related to bimanual responses to unilateral stimuli. Gender differences in laterality to unilateral presentations are reduced to factors which may represent individual hemispheric differences in processing strategies. Hand differences (fasthand, slowhand) in reaction time and other measures are suggested to relate not to the hemisphere of initial visual projection but the hemisphere in which the response is initiated.

O'Connell, Tucker and Scott present, a self-report scale designed to measure accurately the constructs of emotional and cognitive self-control, set in the context of individual differences in lateralized cognitive style. The methodological and theoretical problems encountered are discussed in respect of hemispheric activation particularly in relation to direction of eye movement studies. Self-regulation and emotional factors either traits or states are involved in the dynamic determination of asymmetric brain activation and, therefore, of cognitive style which is not seen as synonymous with hemispheric specialization. Following discussion of the self-report scale, the proposition that an anterior-posterior dimension of cerebral organization may be at least as important when considering activation and arousal as a 'hemispheric' dimension is put forward.

Weber and Bradshaw's contribution is devoted to a critical examination of the evidence for Levy's revised hypothesis (Levy, 1982) and is a rejoinder to it. The hypothesis is, of course, that the preferred writing hand and, especially in left handed writers, hand posture; that is, whether the position of the hand in writing is inverted or normal, is an indication of the direction of lateralization of cerebral function or hemispheric specialization. This problem and the associated question of ipsilateral motor control has received a great deal of experimental investigation. Weber and Bradshaw make the point of general relevance, perhaps, based on Levy's assertion of possible differences in educational methods, that if hand posture is a "cultural" phenomenon (Levy, 1982) then it is less likely to be an "important correlate of neurological organization". In general, they claim to find little clear evidence to link writing posture with the lateralization of cerebral function.

Asymmetries in Psychopathology and Individual Differences

Gruzelier, in his first chapter, fully develops the concept of individual differences in lateralization of cerebral function applying to the normal and pathological brain. He extends its scope from the widely accepted forms of individual differences in hemispheric specialization, gender and handedness, to consideration of other differences. This he does by enlarging the concept of hemispheric specialization beyond an exclusively structural model of lateralization to a combined structural and dynamic process model (Cohen, 1982). Evidence is presented from a variety of conditions in both the normal and pathological brain. Assessment is mainly through asymmetries in the rate of habituation of electrodermal responses and non-specific electrodermal responses. Thus, a contribution is made towards specifying the role of dynamic process asymmetries, with reference to individual differences not only in gender and handedness but individual differences in cases where hemispheric activation, in particular, may be a factor.

Miran and Miran, gather evidence from a variety of sources, to place what appears to be an increasingly appreciated flexibility of lateralization of cerebral function in the context of an integrated, homeostatic brain model, which they would argue is more realistic than other models. They draw on supporting but circumstantial evidence from "hardwired" perceptual and motor systems; go on to consider several lateralized, including cognitive, functions in the same framework and adduce evolutionary and developmental dimensions to their theory. They regard the homeostatic brain model as providing a model, based on hemispheric specialization for understanding individual differences. They propose that psychopathological conditions may be caused by a disruption of homeostatic systems rather than site-specific deficits.

Cromwell, focuses provocatively on schizophrenia in relation to hemispheric specialization. He emphasises the role of hemispheric advantage, not of absolute hemispheric specialization, and a sequential, dynamic rather than a static concept of cerebral lateralization. A model of schizophrenia is put forward in which the processing of information is conceptualised as flowing from right to left hemisphere prior to verbal or other responses. Left hemisphere hyperarousal is seen as a secondary phenomenon, resulting from "faulty" information being transferred to it, derived from an earlier, "preattentional" stage dysfunction in the right hemisphere.

Gruzelier, in his second contribution, reviews the role of the interhemispheric disconnection hypothesis stemming from experimental disconnection (Sperry, 1964) and clinical callosotomy (Bogen and Vogel, 1962) in schizophrenia in the context of hemispheric specialization. Two initial influences on the concept of disturbed interhemispheric integration are described, that of left sided temporal lobe epileptic foci associated with schizophrenia (Flor-Henry, 1969) and that of the enlarged corpus callosum in schizophrenia (Rosenthal and Bigelow, 1972; however, see Witelson and Kigar, this volume). From this standpoint, the evidence for disordered interhemispheric transfer is reviewed not on the basis of callosal agenesis or callosal section, but of faulty transmission such as might be involved by a reduced signal-to-noise ratio (Butler, 1979). Evidence from auditory processing sources, haptic tasks involving interhemispheric transfer, visual processes (divided field studies) and somatosensory evoked potentials in schizophrenic patients are summarised. The conclusion appears to be that although a consistent pattern of results has emerged across studies, evidence is against the concept of frank, functional disconnection as an aetiological factor. For example, a lateralized deficit could also give rise to defective interhemispheric transfer. More questions than answers are said to emerge from the results and the heuristic value of the callosal theory in future investigation of the schizophrenic brain is emphasised.

Miran and Miran, in their second, contribution, focus on the role of intrahemispheric and interhemispheric communication in schizophrenia. They propose that typical schizophrenic deficits in cognition and perception can be interpreted as breakdowns in internal communication within a homeostatic brain. Studies of left and right hemispheric dysfunction and callosal dysfunction in schizophrenia are evaluated and frontal lobe dysfunction (which has produced negative findings) and deficits in parietal and occipital function are considered.

Schizophrenia is regarded as a dysfunction of the homeostatic brain and implications for its assessment and treatment are examined. In conclusion, a number of points are made in relation to the homeostatic brain model and hemispheric specialization, but it is emphasised that to consider merely the over- or under- activation of a dominant hemisphere is too simplistic. Schizophrenia, in terms of psychopathology, is considered to be a breakdown in both intra- and interhemispheric feedback systems.

Serafetinides, linking cerebral laterality and psychopathological disorders, has recalled his earlier evidence for agressive behaviour in young, male temporal lobe epileptics, the majority having a left dominant hemisphere focus, and he reviews subsequent supporting evidence that assaultive psychopathological disorders are associated with communication difficulties. It is concluded that the role of the verbal hemisphere in control of agressive impulses is considerable and that a similar sequence of dysfunction may hold for other hemisphere specific impairments. The need for constant redefinition of such testable formulations of brain-behaviour relations is emphasised.

CONCLUSIONS

It is hardly feasible to attain a complete synthesis of the contributions to this volume and it would certainly be foolhardy to claim to have done so. It is also difficult to represent a contributor's views with much semblance of accuracy, but in spite of this it was felt worth risking possible misrepresentation, if some form of synthesis were to be achieved. The reader must judge whether this goal has been gained.

A common thread can be discerned and appears to run through those chapters in which what appears the central problem of the Workshop is addressed. These chapters are to be found in each Section but perhaps especially in the Psychopathology Section. There appears to be a detectable gradient shift in the paradigm concerning hemispheric specialization. This shift was perhaps just apparent at the time of the Workshop and has since been consolidated in these contributions. There is a tendency for a move towards a greater lability in the concept of hemispheric specialization, particularly in the area of individual gender and handedness but also in other less well-studied areas of individual difference. The move is in a direction away from a fixed structure model towards one based on hemispheric usage and activation. There is a growing realisation that there are factors affecting the measurement of lateralization other than those purely of hemispheric or language lateralization, between those individuals making up the commonly acknowledged groups of gender and handedness. The measures of laterality are the perceptual/behavioural, and the "real-time" indices of orientation, the electrophysiological asymmetries and brain imagery. These measures may contrast with, and, have to be reconciled, with the data regarding individual differences in hemispheric specialization derived from clinical/neurological sources.

The paradigm has shifted, in the last decade, from perhaps a simple "black and white" model of cerebral dominance (which, of course, was never held in totality!) which had been essentially unchanged from the nineteenth century, but with the more recent acknowledgement of the importance of right hemisphere function. It was based on and invigorated by the split-brain work of Bogen and Vogel and Sperry and Gazzaniga (1970). It has now moved towards the formulation of a more complex model, less simplistic, perhaps, but hopefully nearer "reality" in which differences in strategy may play their role as part of a dynamic process in cerebral asymmetry. Strategy differences (Butler & Glass, this volume) as a basis for individual differences are not the whole component of the dynamic process (Gruzelier, this volume). Possibly, further insight be sought into dynamic processes of hemispheric specialization by following-up the study of performance indicators in relation to task-related EEG alpha asymmetries, as predictors of lateralized task performance (Furst, 1976, for right hemisphere tasks; Glass and Butler, 1977, for left hemisphere tasks). This change in the paradigm is reflected in the application of hemispheric specialization to psychopathology and psychiatry (See Gruzelier and Miran and Miran, this volume).

It has been emphasised, however, and this should be taken as a warning that the "best" hypothesis, a combined structural and dynamic model of hemispheric specialization (Cohen, 1982) is, perhaps paradoxically, least effective in forming predictions that can be tested experimentally. Also unpredictable results are too easily explained by it. Its heuristic value is limited by its underdetermination (See .Laudan, 1984).

An example of a less than fixed model of specialization is that involving differences of cognitive style, described as "a disposition to adopt a particular processing strategy which effectively changes the cognitive demands of the task" (Cohen). Fixed structure and dynamic models are not mutually exclusive, in fact they could be mutually dependent in a "real" brain. A functional asymmetry, that is, a dynamic process, must be combined with an underlying structural asymmetry (which is not necessarily the same as a gross anatomical asymmetry, of the type of the asymmetrical planum temporale). Unless, of course, the structure model is like a standard set of "pigeon holes", which are uniform in structure but have different objects or messages inside them. To extend this analogy, let us imagine two structurally identical microprocessors (PC's, or personal computers), side-by-side, representing left and right cerebral hemispheres, with cables interlinking inputs and outputs to represent the corpus callosum. It is possible for these computers to work together with different operating systems, CP/M, MS-DOS, for example. Perhaps, each microprocessor could support different languages, Fortran, Basic, for example and which in turn could also support quite different functional categories of software, let us say, a Wordprocessing program in one computer and a Graphics program, in the other, for example. It is possible, in fact, for the two microprocessors to have different structures and still be functionally linked by the cable if their software is compatible, or can be made compatible. It is difficult to know how such an information processing model or analogy, could be commensurate with those current theories of cerebral organization seeking to account for laterality findings and which form the theme underlying the Workshop and its proceedings. Would it be a fixed structure asymmetry? A dynamic process asymmetry? A function of cognitive style or strategy? A verbal or visuospatial dichotomy? A holistic or analytical process? It is interesting to speculate how, for example, a heterarchical concept of hemispheric specialization could be made to fit such a model. It is difficult to ascertain where in current models of hemispheric specialization such an analogy could be fitted or at what level. Intuitively, one feels that any such theories should be compatible with an information processing model of this type, however simplistic it may appear initially. Alternatively, we should consider how current models could be reconciled with such an analogy to give us more insight into hemispheric specialization in the "real" brain. There are difficulties here which have not yet been tackled "head-on", but if they are may represent a way forward into individual differences in hemispheric specialization.

The final question we must ask is, what we consider is being physically lateralized when we say, for example, that "language" is lateralized to the left hemisphere or visuospatial function is lateralized to the right hemisphere? Will this reduce to an asymmetry of neurotransmitter substances (Reynolds, 1984), for example? When we are in a position to answer this question we will be in a position to understand better the basis for individual differences in hemispheric specialization and to evaluate their significance.

ACKNOWLEDGEMENT

Thanks are due to the NATO Scientific Affairs Division for their generous support without which the Workshop would not have been possible, and also to the Wellcome Foundation for their additional help with the project.

My thanks must be expressed to the late Dr. M. di Lullo, formerly Director of the ARW Programme for his support and are also due to Dr. Tilo and Mrs Barbara Kester for help in organizing the Workshop. Mr Guzzardi and his staff of the Hotel Villa del Mare at Maratea, Italy ably looked after the participants during the meeting.

I am deeply indebted to my colleague Dr. Stuart R. Butler for stimulating and helpful discussion before, during and after the Workshop and also to Dr. Joseph E. Bogen for helpful discussion some years ago, and to Professor Dietrich Lehmann for earlier discussion leading to the initial suggestion of the Workshop and its later formulation. The support of Professor J.J.T. Owen for allowing the facilities of the Department of Anatomy, University of Birmingham to be used for preparation of the text is greatly appreciated. Mr Roland A. Hill is thanked for help in organization, preparation and support. Also the editorial support of Madeleine Carter of Plenum Publishing Company Limited is gratefully acknowledged. I am particularly grateful to Sue Burton and Claire Hundley for their patient and unremitting work in processing the entire text of the volume.

The following are thanked for refereeing the chapters or for advice and assistance with refereeing: M. Annett, S.R. Butler, D. Carroll, J.H. Gruzelier, L.J. Harris, D.H. Ingvar, M.A. Jeeves, J.C. Marshall, M. Moscovitch, R. Obrecht, M.A.J. O'Callaghan, K.P.O'Connor, D.J. Parker, P. Prior, S. Segalowitz, J.C. Shaw and D.M. Tucker.

Especial thanks are due to the Organizing Committee, Professor Pierre Flor-Henry and Dr. Michael Miran and I am deeply grateful to the participants for their stimulating contributions.

Preparation for publication of the chapters in this volume has been made possible by Contract Number DA7 A45-85-M0184 from the U.S. Army Research Institute for the Behavioral and Social Sciences through its European Science Coordination Office in London, England. The opinions expressed are those of the authors and do not necessarily represent those of the U.S. Army. In this connection, Dr. Michael Kaplan, Chief, European Science Coordination Office, Army Research. Institute is thanked for his patient support.

A. Glass,
University of Birmingham

REFERENCES

Annett, M. (1985). Left, Right, Hand and Brain: The Right Shift Theory. London: Lawrence Erlbaum.

Beaumont, J.G. (1982). (Ed). Divided Visual Field Studies of Cerebral Organization. London: Academic Press.

Bogen, J.E. (1969). The other side of the brain II: An appositional mind. Bulletin of the Los Angeles Neurological Society, 34, 191-220.

Bogen, J.E. and Vogel, P.J. (1962). Cerebral commissurotomy in man. Bulletin of the Los Angeles Neurological Society, 27, 169-172.

Bramwell, B. (1899). On "crossed" aphasia. Lancet, 1, 1473-1479.

Broca, P. (1865). Sur le siege de la faculté du langage articulé. Bulletin de la Societe d'Anthropologie, 6, 337-393.

Bryden, M.P. (1982). The behavioral assessment of lateral asymmetry: problems, pitfalls, and partial solutions. In R.N. Malatesha and L.C.Hartlage (Eds). Neuropsychology and Cognition - Volume II. The Hague: Martinus Nijhoff Publishers, 44-54.

Butler, S.R. (1979). Interhemispheric relations in schizophrenia. In J. Gruzelier and P. Flor-Henry (Eds). Hemisphere Asymmetries of Function in Psychopatholgy, Amsterdam: Elsevier/North Holland Biomedical Press, 47-63.

Butler, S.R. (1984). Sex differences in human cerebral function. In G.D.De Vries et al. (Eds). Progress in Brain Research. Volume 61. Amsterdam: Elsevier Biomedical Press, 443-455.

Butler, S.R. and Glass, A. (1974). Asymmetries in the electroencephalogram associated with cerebral dominance. Electroencephalography and Clinical Neurophysiology, 36, 481-491.

Butler, S.R. and Glass, A. (1985). The validity of EEG alpha asymmetry as an index of the lateralisation of human cerebral function. In D. Papakostopoulos, S. Butler and I. Martin (Eds). Clinical and Experimental Neuropsychophysiology. London: Croom Helm, 370-394.

Butler, S.R., Glass, A. and Heffner, R. (1981). Asymmetries of the contingent negative variation (CNV) and its after positive wave (APW) related to differential hemispheric involvement in verbal and non-verbal tasks. Biological Psychology, 13, 157-171.

Butler, S.R., Glass, A. and Fisher, S. (1987). Effect of potential conversants (social facilitation) on resting EEG alpha asymmetry (in preparation).

Cohen, G. (1982). Theoretical interpretations of lateral asymmetries. In J.G. Beaumont (Ed). Divided Visual Field Studies of Cerebral Organisation. New York: Academic Press, 87-111.

Flor-Henry, P. (1969). Psychoses and temporal lobe epilepsy: a controlled investigation. Epilepsia, 19, 363-395.

Furst, C.J. (1976). EEG asymmetry and visuospatial performance. Nature, 260, 254-255.

Gazzaniga, M.S. (1970). Bisected Brain. New York: Appleton-Century-Crofts.

Glass, A. (1984). Round table on hemispheric specialization and lateral asymmetries in the EEG: Introduction. Biological Psychology, 19, 151-157.

Glass, A. and Butler, S.R. (1977). Alpha EEG asymmetry and speed of left hemisphere thinking. Neuroscience Letters, 4, 231-235.

Glass, A., Butler, S.R. and Carter, J.C. (1984). Hemispheric asymmetry of EEG alpha activation: effects of gender and familial handedness. Biological Psychology, 19, 169-187.

Gur, R.C. and Reivich, M. (1980). Cognitive task effects on hemispheric blood flow in humans. Brain and Language, 9, 78-93.

Harcum, E.R. and Dyer, D.W. (1962). Monocular and binocular reproduction of binary stimuli appearing right and left of fixation. American Journal of Psychology, 75, 56-65.

Hécaen, H. and Sauguet, J. (1971). Cerebral dominance in left handed subjects. Cortex, 7, 19-48.

Kimura, D. (1967). Functional asymmetry of the brain in dichotic listening. Cortex, 3, 163-178.

Lassen, N.A., Ingvar, D.H. and Skinhoj, E. (1978). Brain function and blood flow. Scientific American, (October), 50-59.

Laudan, L. (1984). Science and Values. London: University of California Press.

Levy, J. (1982). Handwriting posture and cerebral organization: how are they related? Psychological Bulletin, 91, 589-608.

Lishman, W.A. and McMeekan, E.R.L. (1977). Handedness in relation to direction and degree of cerebral dominance for language. Cortex, 13, 30-43.

Marr, D. (1982). Vision. A Computational Investigation into the Human Representation and Processing of Visual Information. San Francisco: W.H. Freeman & Co.

McKeever, W.F. and VanDeventer, A.D. (1977). Visual and auditory language processing asymmetries: influence of handedness, familial sinistrality and sex. Cortex, 13, 225-241.

Perelman, S.J. (1951). Crazy Like a Fox. Harmondsworth: Penguin Books.

Popper, K. (1972). The Logic of Scientific Discovery. London: Hutchinson. 3rd Edition.

Provins, K.A. and Jeeves, M.A. (1975). Hemisphere differences in response time to simple auditory stimuli. Neuropsychologia, 13, 207-211.

Raichle, M.E. (1985). Progress in brain imaging. Nature, 317, 574-576.

Reynolds, G.P. (1983). Increased concentrations and lateral asymmetry of amygdala dopamine in schizophrenia. Nature, 305, 527-529.

Rosenthal, R. and Bigelow, L.B. (1972). Quantitative brain measurement in chronic schizophrenia. British Journal of Psychiatry, 121, 259-264.

Satz, P., Achenbach, K., Pattishall, E. and Fennell, E. (1965). Order of report, ear asymmetry and handedness in dichotic listening. Cortex, 1, 377-396.

Satz, P., Achenbach, K. and Fennell, E. (1967). Correlations between assessed manual laterality and predicted speech laterality in a normal population. Neuropsychologia, 5, 295-310.

Segalowitz, S.J. and Bryden, M.P. (1983). Individual differences in hemispheric representation of language. In S.J. Segalowitz (Ed). Language Functions and Brain Organization. New York: Academic Press, 341-372.

Sperry, R.W. (1964). The great cerebral commissure. Scientific American, 210, 42-52.

Zangwill, O.L. (1960). Cerebral Dominance and its Relation to Psychological Function. Edingburgh: Oliver and Boyd.

Zangwill, O.L. (1982). On cerebral dominance. In R.N. Malatesha and L.C. Hartlage (Eds). Neuropsychology and Cognition- Volume I. The Hague: Martinus Nijhoff Publishers, 208-222.

Zurif, E.B. and Bryden, M.P. (1969). Familial handedness and left-right differences in auditory and visual perception. Neuropsychologia, 7, 179-187.

INDIVIDUAL DIFFERENCES IN HEMISPHERIC SPECIALIZATION: SOURCES AND MEASUREMENT

S.J.Segalowitz

Department of Psychology, Brock University
St. Catharines, Ontario
Canada L2S 3A1

INTRODUCTION

When we focus on individual differences in hemispheric specialization, we usually design our experiments to detect these differences as group effects. For example, we may include factors such as those listed in Table 1 in any particular study, and, of course, the list can be expanded at will. Although the factors must, for reasons of statistical analysis, be applied to groups of subjects, we often study them not because we are interested in groups of left handers, or schizophrenics, etc. *per se*, but rather because we want to make a statement about individuals. The individuals within the group are assumed (or hoped) to be alike on all other variables important for brain lateralization. For practical purposes, this is impossible since it would be impossible to form a group of subjects representing each intersection of all the factors listed. There are probably also sources of individual variation in lateralization tests that are beyond any divisions we have managed to make so far. With this perspective, each individual is seen to be a group unto himself to some extent. An intensive case study of the individual, although possible, and laudable (Dywan & Segalowitz, in press), would not satisfy our curiosity about the factors under examination (cf. Caramazza, 1986). We must remain with the traditional group paradigm, but we are left in the traditional, awkward position of having to accept the individual variation in lateralization not controlled for in our study as error variance. Thus, if we measured cerebral specialization in a fully crossed paradigm for the first three factors listed in Table 1, the others would add error variance. In this chapter, I outline some of these sources of uncontrolled variation and discuss a paradigm that allows us to measure it and separate such variation from the error variance measure.

SOURCES OF VARIANCE IN A LATERALITY TEST

There are at least five factors that contribute variance to lateralization scores. Some of these factors are relatively easy to incorporate into a research design and therefore to control. Others are less amenable to experimental control and yet are documented adequately to be a source of concern. The issue here is that whatever adds variance to the scores obtained on a test of lateralization is of concern, whether or not the particular factor relates to hemispheric specialization.

Table 1. A list of variables that researchers have suggested influence the asymmetry shown on various tests of hemisphere specialization.

HANDEDNESS
SEX
FAMILIAL SINISTRALITY
SCHIZOPHRENIA
DEPRESSION
ANXIETY
FATIGUE
AGE
SMOKING
INTROVERSION/EXTRAVERSION
.
.

1. Functional Asymmetries Common to the Group

The individual differences commonly associated with variation in brain lateralization are those due to handedness and sex. There is clear evidence that whereas right handers have speech representation in the left hemisphere to an overwhelming degree, left handers present a different distribution (Segalowitz & Bryden, 1983). For example, Rasmussen & Milner (1977) present data from Wada testing that illustrate the most commonly-held view: that whereas at least 96% of right handers are left-dominant for speech, only 70% of left handers are clearly so, with the remainder evenly split between right-dominant subjects and those bilaterally represented for speech. Similar group differences have been documented for behavioural tasks with normals, repeatedly (Zurif & Bryden, 1969; McKeever, VanDeventer & Suberi, 1973; Bryden, 1965). The issue of familial left-handedness is clear. Some claim that if a subject has left handedness in the family, then that subject is more likely to be bilaterally or right dominant for speech (Hannay & Malone, 1976; Varney & Benton, 1975). Others do not find this a robust effect (Bradshaw, Nettleton & Taylor, 1981; Bradshaw & Nettleton, 1979).

Sex differences in speech dominance have also been repeatedly reported (Lake & Bryden, 1976; Bradshaw, Gates & Nettleton, 1977; McKeever & Jackson, 1979). The common assertion is that males are more likely to be left dominant for speech, and that females, like left-handers, are more likely to show bilateral representation for language. The evidence is both clinical (McGlone, 1980; Inglis & Lawson, 1981) and experimental (Bryden, 1979). This picture is somewhat clouded by studies that have examined the factors of handedness, familial sinistrality and sex together. Piazza (1980), for example, found that whereas for men handedness and ear advantage on a verbal dichotic listening task interact (left handers being less asymmetric), the women showed no main effect or interaction. Women showed a significant handedness by ear interaction on the environmental-sounds dichotic task, while men showed no significant effects. A tachistoscopic face-recognition task produced a triple interaction among the factors of visual field, handedness and familial sinistrality, whereby only right handers with no familial sinistrality showed the expected left visual-field advantage. Thus, sex, modality, handedness and familial sinistrality interacted in this study. The situation may very well be more complicated. It may be necessary to include other factors in the equation such as the type of language task being examined, e.g. expressive versus receptive (Moore & Haynes, 1980; Orsini, Lewis & Satz, 1985), or possibly the within-hemisphere organization of language representation (Kimura, 1980). However, whatever the factors, as long as they can be catalogued and the subjects identified, they can be incorporated into the research design as a group factor.

2. Individual Differences in Neuropsychological Organization Beyond the Group Pattern

There is always residual variation in lateralization even within a supposedly homogenous group, e.g. of right-handed, non-schizophrenic, university age males without familial sinistrality. Could some of this remaining variation in lateralization be in some part due to other constitutional variables? There is good evidence that the temporal lobe differs in size between the hemispheres, (Geschwind and Levitsky, 1968) as indexed by the different directions of the curve of the Sylvian fissure (Rubens, 1977). As well, however, there is considerable individual variation in this asymmetry and therefore in the shape and size of the planum temporale, a key language area. Differences in dichotic listening scores could be related to such anatomical variations.

A number of workers have suggested that the asymmetries in size between the two hemispheres are an indicator of some gross functional differences in language skills, specifically in dyslexia (Hier, Lemay, Rosenberger & Perlo, 1978; Galaburda & Kemper, 1979). Indeed, they suggest that this may be a contributing factor to the syndrome of dyslexia. If specific language difficulties were related to this anatomical variation, we would expect that a group of subjects that otherwise appears homogeneous with respect to lateralization would on some measures (perhaps visual half-field reading tasks) show considerable variation. Similarly, Witelson (1985; this volume) has shown that hand preference differences correlate with size of the corpus callosum. There is every reason to expect that dichotic listening performance should be affected by such variation, especially since it is the anterior section of the corpus callosum that especially shows the effect.

There very well may be other constitutional factors that affect performance on lateralization tasks, factors that we may not be aware of yet. Ojemann and his colleagues (1983) have presented such evidence from their brain stimulation work in clinical patients. The specific posterior areas that when stimulated lead to language disruption vary considerably, although there is good consistency on the anterior area. Even within a bilingual individual, each language can be disrupted by stimulation to different areas.

3. Lateral Attentional Biases not related to Hemispheric Specialization

Individual subjects may have attentional biases that predispose them to respond more to stimuli on one side of their personal space. The source of this bias could be due to central or peripheral factors, including some that would be classified as dysfunctions. For example, there may be central dysfunctions that promote some form of hemispatial neglect that is not clinically serious but that can influence the result of any auditory or visual vigilance task. Also possible are peripheral dysfunctions that could be reflected in poorer visual acuity scores for one visual field or poorer hearing in one ear. Although normal subjects are usually screened for such factors, the procedure is usually relatively informal and often consists of self-reports. It is entirely possible that the subject is unaware of the slight perceptual asymmetry since it is only of interest in laboratory tests. One general instance of such lateral biases is related to hand preference. Bryden (1978) has pointed out that right handers have a greater right ear advantage on a dichotic listening task compared with left handers independent of the side of the speech dominance as assessed by Wada testing. We do not know whether or not this is a generalized directional bias that supersedes modality or is specific to listening tasks. The modality of testing may be critical and hearing, vision and motor asymmetries may not be consistent within individual subjects (Porac, Coren & Duncan, 1980).

4. Individual Differences in Hemispheric Activation

Just as the item above concerned some generalized perceptual bias because of some factor not related to hemispheric specialization specifically, it is also possible, though more controversial, that individuals may differ in asymmetric hemisphere activation. Asymmetries of activation are presumed to be present whenever the subject engages in some behaviour that requires the processes of one hemisphere more than the other (Galin & Ornstein, 1972; Ornstein, Johnstone, Herron & Swencionis, 1980; Moore, 1979; Moore & Haynes, 1980). It is also possible that because of structural asymmetries (including minor damage), individuals may differ in how relatively active each hemisphere is. This source of differences has not to my knowledge been documented, although one would suppose that frank asymmetric damage would affect the EEG output (cf. Heilman, 1979). It is clear, however, that within a group of subjects with left hemisphere speech, there is a wide variation of hemisphere activation (as reflected in asymmetric EEG alpha) across subjects performing a verbal task (Butler & Glass, 1976; this volume).

More controversial is the notion that some individuals habitually have asymmetric activation due to specific cognitive styles (Ornstein, 1972) or personality traits (Tyler & Tucker, 1982; Smokler & Shevrin, 1979). The suggestion is that while verbally oriented individuals and visuospatially oriented people are structurally similarly lateralized for speech as long as they have the same handedness characteristics, one group will have come to the testing situation with a heightened level of left hemisphere activation compared to the right, and the other group will have the opposite pattern. This difference may predispose them to utilize differing strategies on the lateralization task, even on tasks where we try to control the task requirements. Of course, there is no clear cause-and-effect relationship here: the choice of strategy may produce the asymmetric activation rather than any constitutional factor predisposing the subject towards one or the other. An example of this "hemisphericity" is illustrated in Levy, Heller, Banich & Burton (1983), who show that differences in VHF asymmetry can be attributed to strategy differences.

The strong hemisphericity hypothesis implies a long term effect: that subjects have a long-lasting activation asymmetry, e.g. that some people are chronically more verbal than others. It is also possible that any such predisposition is temporary and may even be confined to the testing situation. For example, some subjects may react to being in a psychology experiment by an increase in tension with, for some subjects, a concomitant increase in verbal mediation, while in other subjects there may be a decrease of the same proceses. The result would be a variation in lateralization scores that would look like an attentional bias that is not stimulus-specific. This effect would be hard to unconfound from the type of factor outlined in (3) above. For example, Levine, Banich and Koch-Weser (1984) found that the degree of asymmetry that subjects show in recognizing items that do not produce a VHF advantage among right handers (line drawings of chairs) generally correlates significantly with the degree of VHF advantage shown on lateralized tasks (e.g. recognising faces). This means that whatever predisposes a subject to recognize more information in one visual field on one task beyond the hemisphere-specific processing requirements generalizes to other tasks. They interpreted these results as indicating that subjects vary in their hemisphericity, i.e. asymmetry of activation. It could equally be the case that the stable asymmetry is due to lateral biases from other more peripheral sources including visual asymmetries.

5. Error Variance

Since there is a continual fluctuation in attention from internal and external factors during any vigilance task, and most lateralization tasks involve such demanding processes, there will always be some variation in the performance of any individual, even if the individual is tested repeatedly. This, of course, leads to a certain degree of variance unaccounted for - error variance - and adds to the unreliability of the tests (Segalowitz, in press).

In most lateralization studies, factors 2, 3, and 4 are considered to be nuisances, and the experimenters hope that a large sample size will reduce them, cancel them out and increase the relative strength of factor 1 to get the desired result. Although it is true that many of these factors will cancel themselves out given a large enough sample size, i.e. the asymmetry deviation due to them averaged over the entire group will be small, the variance they contribute is added to the error term. Not only does this inflate the Type II error, that is, the chances of not finding a significant effect when one is present, but masks what may be interesting facts about individual variation in hemispheric organization. The methods outlined here separate factor 1 from 2 and both from factors 3 and 4.

As an illustration of the difference between factors 1 and 2 on the one hand and factors 3 and 4 on the other, consider the following pair of subjects (Segalowitz & Orr, 1981; Segalowitz, 1983): Subject J showed right visual field advantages on both a verbal and on a spatial visual half-field task consistently over 6 test sessions on different days. Subject L showed a consistent LVF advantage. They both showed the expected trend for right handers: the verbal task producing a relatively higher RVF score, the spatial task a greater LVF score (see Figure 1). Thus, one could conclude that their pattern of hemispheric specialization is the same and conforms to the common one for right handers: LH dominance for verbal functions and RH dominance for visuospatial functions. This illustrates factors 1 and 2. The widely differing absolute scores, however, illustrate factors 3 and 4: whatever the pattern of hemispheric specialization, these subjects have VHF biases. The consistency of these VHF biases suggests we are not dealing

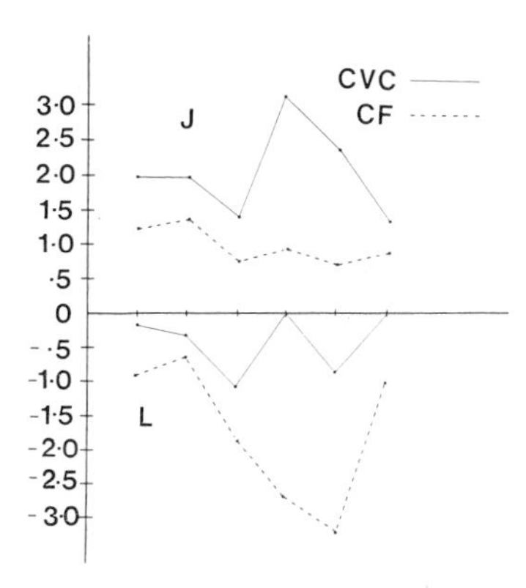

Fig. 1. VHF asymmetry scores for two subjects tested 6 times each over a 3 week period. The asymmetry score, lambda, is a ratio score, where positive values indicate a RVF advantage and negative scores a LVF advantage. Lambda is independent of total accuracy (see Bryden & Sprott, 1981). CVC stimuli are clockface times balanced for left/right asymmetries. The clockfaces have no digits on them. (Reproduced from Segalowitz & Orr, 1981).

Table 2. This analysis of variance model is for a design where Test Session and Condition are crossed. Only the variance components are included in order to simplify the table.

Effect	Mean Square Components							Error Term
Subjects	σ_e^2	$+\sigma_S^2$		$+\sigma_{ST}^2$				S x T
Conditions	σ_e^2		$+\sigma_C^2$		$+\sigma_{SC}^2$	$+\sigma_{CT}^2$	$+\sigma_{SCT}^2$	S x C or C x T (see Text)
S x C	σ_e^2				$+\sigma_{SC}^2$		$+\sigma_{SCT}^2$	S x C x T
Test Session	σ_e^2		$+\sigma_T^2$	$+\sigma_{ST}^2$				S x T
S x T	σ_e^2			$+\sigma_{ST}^2$				
C x T	σ_e^2					$+\sigma_{CT}^2$	$+\sigma_{SCT}^2$	S x C x T
S x C x T	σ_e^2						$+\sigma_{SCT}^2$	

with random error variance here. Also note, however, the critical use of the multiple measure paradigm: without the second task we would not be able to determine whether the consistent asymmetries were due to VHF bias factor (either 3 or 4) or to specialization of function (1 or 2). It is the pattern of the differences in VHF advantage between the two tasks that allows us to suggest similar specialization of function but differing hemisphericity or VHF bias.

This informal division of VHF results into cerebral specialization factors (1 & 2) and VHF bias factors (3 & 4) can be formalized in an experimental design. Consider the results of eight right handers being tested in the VHF paradigm illustrated with the 2 subjects J and L. They all received the verbal task and the spatial task on 6 occasions, spread over 3 weeks. By including more than one test session, we provide a basis for estimating the stability of the asymmetry scores. Coupling this with giving each subject more than one task (e.g. a verbal and a spatial task) allows us to estimate the degree to which the pattern of asymmetries for the two tasks varies over the subjects. This latter component is a measure of the degree of individual variation in hemisphere specialization. The analysis of variance model is outlined in Table 2. In it we see that rather than only being able to test the Subjects and the Conditions effects, as in the usual repeated-measures paradigm, we can now test 5 effects for significance. For simplicity, the dependent measure is an asymmetry score rather than raw scores for left and right sides.

(1) The mean square estimate from the Subjects factor can now be compared against the Subjects x Test Session (ST) effect.
(2) The Conditions factor is tested against either the Subjects x Condition (SC) interaction or the Condition x Test Session factor. Notice that the mean square estimate for this factor contains 5 components, while the comparison estimates each contain 3 factors. Any test of the Conditions effect involves an assumption that either the SC factor or the Conditions x Test Session (CT) factor is zero. Winer (1971) suggests that if either of these factors is not significant beyond the .25 level, then we may safely ignore it. Thus, if the CT factor is near zero with this test, then the appropriate error term for the Conditions factor is the SC interaction. Conversely, if the SC interaction is near zero, then the error term would be the CT interaction. If neither of these effects is near zero, then the Conditions factor must be addressed differently. In this case, we can average over test session and reduce the paradigm (for this factor only) to a standard repeated-measures design and the error term is the recalculated SC interaction.
(3) The SC interaction takes as its error term the Subjects x Condition x Test Session (SCT) interaction. This factor is usually not testable and represents the variation in pattern of lateralization across subjects.
(4) The Test Session factor examines any general shift in asymmetric response over the test session averaged across the various conditions.
(5) The CT interaction reflects differences in the change in asymmetry across the various tasks, e.g. whether or not one task changes in asymmetry score when other conditions do not.

When we apply these tests to the results of the study outlined above, we find the following results (Fig. 2):
(1) The Subjects effect accounted for 49% of the variance and was highly significant.
(2) The Conditions effect was also significant ($p < .025$), indicating that the verbal and spatial tasks produced differing lateralization as expected, and accounted for 11% of the variance. In this case the SC interaction was used as the error term since it was significant and the CT interaction was not.
(3) The SC interaction accounted for 8% of the variance in the scores and was significant ($p < .01$). This means that subjects significantly differed

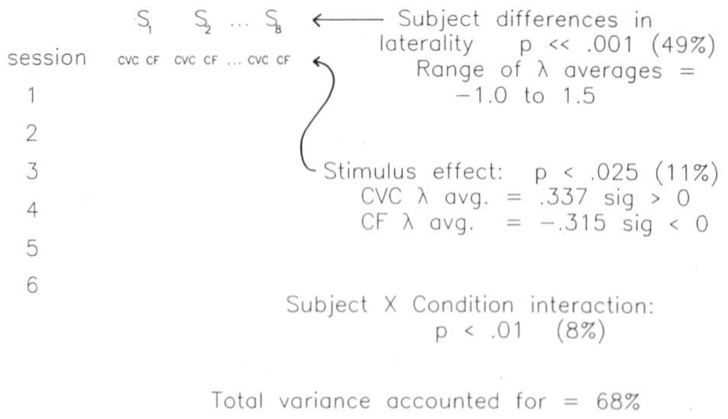

S1 S2 ... S8 ⟵ Subject differences in laterality p << .001 (49%)
Range of λ averages = −1.0 to 1.5

session CVC CF CVC CF ... CVC CF
1
2
3
4
5
6

Stimulus effect: p < .025 (11%)
CVC λ avg. = .337 sig > 0
CF λ avg. = −.315 sig < 0

Subject X Condition interaction: p < .01 (8%)

Total variance accounted for = 68%

Fig. 2. Results of a VHF study with 8 subjects tested 6 times each. See Figure 1 for description of stimuli and metric.

in the pattern of lateralization shown on the two tasks.
(4) The Test Session factor was not significant, i.e. there was not any significant shift in visual half-field advantage over both tasks.
(5) The CT interaction was not significant. A significant effect here would have been interesting and it may be useful to examine the interaction just to illustrate this factor. As can be seen in Figure 3, the verbal stimuli (CVCs) produced a consistent RVF advantage. The slope of the line indicating the CF asymmetry over time suggests that the task became more dependent on the right hemisphere as the subjects became familiar with it. If this effect had been significant, we could have claimed that the pattern of lateralization for the two tasks changed over time.

We can account for 68% of the variance in our study with significant effects. Much of this variance is due to individual differences in either cerebral organization (SC effect) or in VHF asymmetry bias (Subjects effect). When we test a model of lateralization, we usually consider the many subjects in the study to be simply replications of a prototype. For example, when we test a number of right-handed males, we do so because we consider this group to be homogenous with respect to cerebral organization. We now see that this assumption is not valid. With the paradigm outlined, however, we can measure the extent of this individual variation and take it into account, so that instead of accounting for only 11% of the variance with our Conditions effect, we could argue that we have accounted for 25% of the non-Subject variance with this effect (11/(100-49-8)).

A Single-Session Adaptation of the Paradigm

There are practical difficulties with the paradigm just outlined. First, it requires that subjects return for repeated testing over a number of test sessions. This is always problematic, as it becomes impossible in practical terms to equalize the inter-session period across subjects and this factor could conceivably influence the subject's performance on the tasks. Similarly, there is a practical problem of adding variance to scores because of different time-of-day for the test sessions. If it indeed is the case that there are bodily rhythms that influence hemisphere-specific activation levels (Klein & Armitage, 1979; Gordon, Frooman & Lavie, 1982), then when subjects return for further testing, we can expect increased variation in asymmetry scores simply due to this factor. Besides the possibility of this factor, we can also expect the subjects to be in a different mood and have a different level of energy. These factors may or may not influence performance levels (cf. Ogilvie & Segalowitz, 1974).

A second practical problem concerns the subject's behavior in the test situation. Lateralization measures involve cognitive processing at some

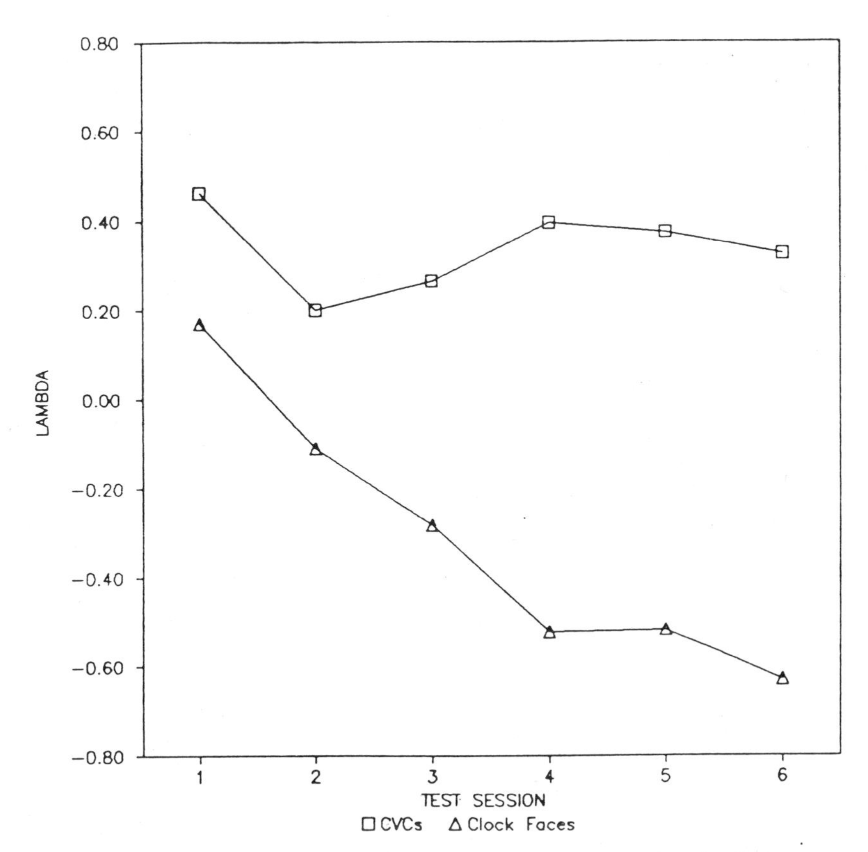

Fig. 3. Averages (right handers). The Test Session by Task interaction, not significant in this study but presented for illustrative purposes (see text).

level. The more complex the task, the more strategies the subjects may have available to them to use (Ross and Turkewitz, 1981). The problem of increased variance arises from this choice: As the subject becomes more familiar with the task and has the time to think about the experience, it may be that he or she changes the strategy applied during the test situation. To the extent that this occurs, the variance in performance increases because of the changes in hemisphere-specific strategies.

A solution to these problems is to gather all the data we need from the subject in a single session, as is traditional in experimental psychology. If we do this, however, and still incorporate the multiple-measure, repeated testing aspects of the paradigm outlined, we encounter some new issues. Consider the following example using EEG asymmetric alpha data as an illustration.

An EEG Example

Let us say we collect, in a single test session, EEG data from both sides of the head while the subject is engaged in a verbal task and again during a spatial task. We could take alpha asymmetry measures during, say, ten subsequent 4-second epochs during each condition. The repeated measurements allow for a measure of stability of the average asymmetry for each subject for each condition. However, the experimental design is not exactly the same as in the paradigm outlined earlier. In the earlier case, we had both conditions measured on a number of occasions, and so the conditions and test session factors were crossed. In this modification, the first condition is tested a number of times and then the second condition

Table 3. ANOVA results from an EEG alpha-asymmetry study, where the first Condition involved face perception and second Condition involved the reading and rehearsal of an abstract sentence (from Segalowitz, unpublished data).

Source	SS	d.f.	MS	error term	F	P
COND	0.061	1	0.061	C X S	9.92	< .006
SUBJ	1.098	18	0.061	Error	20.84	<< .0001
C X S	0.110	18	0.006	Error	2.09	< .006
Error	1.002	342	0.003			
Total	2.271					

follows. This means that the test sessions are nested within the Condition factor. One way of looking at this is that there is more linkage within each subject's condition than in the earlier paradigm where each condition received a data point before the next session. This makes the analysis of variance design a standard two-way ANOVA without crossed factors with the Condition factor as fixed and the Subjects factor as random.

To illustrate this with a set of EEG data, consider the alpha asymmetries produced when right-handed subjects in one condition read a long, abstract sentence, rehearse it, and then repeat it back, and in another condition watch an actress silently portray happy expressions (laughing, etc.). The results are given in Table 3. The two conditions produced different alpha asymmetries (right amplitude minus left divided by the sum of the two, recorded from and averaged by hemisphere over F3, F4, T3, T4, P3 and P4 sites): the sentence condition producing relatively less left hemisphere alpha compared with the face condition. This is as expected ($p < .006$) and yet accounts for only 2.9% of the variance in the total sums of squares. The Subjects effect was very significant ($p << .0001$), as one would expect, and accounted for 48% of the variance. This indicates that the subjects varied on overall alpha asymmetry, i.e. some show relatively more right hemisphere alpha and some relatively more left hemisphere alpha generally. These differences may be due to chronic hemisphericity, or due to some mechanical source such as differences in the asymmetry of the skull or the sensitivity of the electrode placement. Whatever the source, it is clear that subjects in our study varied considerably on their average alpha asymmetry.

The individual differences of EEG asymmetry in response to the verbal versus face condition is reflected in the Subject x Condition (SC) interaction, which is significant in this study ($p < .01$) and accounts for 4.8% of the variance in the asymmetry scores. Thus, the pattern of asymmetry across subjects varies significantly, adding a measure of individual variation beyond global subject differences.

Serial Correlations and the Independence of Data Points

One of the critical assumptions of any statistical test is that the data points are independent samples. That is, when we have sampled the phenomenon 100 times, we must be confident that we have not really sampled, say, 80 times and dressed up the sample to look like 100 data points by duplicating 20 of the points. This can happen unwittingly if the data are linked so that one sampling value predicts the next to some extent. To the extent that this prediction from one point to the next exists, we have a "serially correlated" data set, and instead of having as many independent samples as data points in our study, we may have only eighty percent as many. If this is the case, we must revise the critical value on the statistical distribution to conform with the true degrees of freedom we in fact have. The test for a serial correlation coefficient is simple: Calculate a Pearson correlation coefficient where the pairs consist of each data point and the subsequent one. In our EEG example, the first 4-second epoch is paired with the second, the second with the third, and so on. We can calculate in this way a serial correlation for each cell in the SC interaction. We do this because the nesting of these repeated samplings (of the same subjects) within the SC interaction suggests the possibility of non-independence. For example, what if there is a general increase of the asymmetry across the test session. Since all the values for one Condition precede the other condition, it may appear that there is a significant Condition effect (or an inflated SC interaction if the conditions were poorly counterbalanced). However, the gradual shift also would produce a significant correlation coefficient and we would know not to trust the degrees of freedom in the paradigm.

There are two possible ways of dealing with the data if the data do in fact prove to be dependent. First, one could run the study again allowing the subject to habituate more effectively before the data of interest are collected, or one could lengthen the period of each condition hoping that this would lower the serial correlations. A second possibility is to adjust the degrees of freedom to take into account the lack of independence between samplings. There is a complex formula for this (Cattell, 1963).

In the EEG example used to illustrate the paradigm, we found that of the 38 correlation coefficients calculated (19 subjects x 2 conditions), only one was significant at the .05 level. Since we would expect that by chance this would occur 1.9 times (5% of 38), we have no reason to suspect that our data points are anything but independent.

CONCLUSIONS

As the field of brain lateralization develops and we learn more about the functional asymmetries of hemispheres, the relative theoretical importance of individual variation will increase. It is now clear that handedness and sex do not account for all or nearly all of the variance found in lateralization tasks. Some of the remaining variance may be due to non-central factors, such as peripheral sensory asymmetries, or may be due to strategy biases that the subject brings to the testing situation. It is very difficult to disentangle these two issues. On the other hand, separating the variance due to the inter-subject variation in them is now possible.

The central issue of variation in lateralization within an epidemiologically homogenous group still remains. The paradigm outlined allow a direct examination of this factor.

REFERENCES

Bradshaw, J.L., Gates, A. & Nettleton, N.C. (1977). Bihemispheric involvement in lexical decisions: handedness and a possible sex difference. Neuropsychologia, 15, 277-286.

Bradshaw, J., Nettleton, N.C. & Taylor, M.J. (1981). Right hemisphere language and cognitive deficit in sinistrals. Neuropsychologia, 19, 113-131.

Bradshaw, J. & Taylor, M.J. (1979). A word-naming deficit in nonfamilial sinistrals Laterality effects of vocal responses to tachistoscopically

presented letter strings. Neuropsychologia, 13, 21-32.
Bryden, M.P. (1965). Tachistoscopic recognition, handedness and cerebral dominance. Neuropsychologia, 3, 1-8.
Bryden, M.P. (1978). Strategy effects in the assessment of hemispheric asymmetries. In G. Underwood (Ed.) Strategies of information processing. London: Academic Press.
Bryden, M.P. (1979). Evidence for sex-related differences in cerebral organization. In M. Wittig & A.C. Peterson (Eds.) Sex-related differences in cognitive functioning. New York: Academic Press.
Bryden, M.P. & Sprott, D.A. (1981). Statistical determination of degree of laterality. Neuropsychologia, 19, 571-581.
Butler, S.R. and Glass, A. (1976). EEG correlates of cerebral dominance. In A.H.Riesen and R.F. Thompson (Eds.) Advances in Psychobiology, vol.3 New York: John Wiley, 219-272.
Caramazza, A. (1986). On drawing inferences about the structure of normal cognitive system from the analysis of patterns of impaired performance: the case for single-patients studies. Brain and Cognition, 5, 41-66.
Cattell, R.B. (1963). The structuring of change by P-technique and incremental R-technique. In C.W. Harris (Ed.) Problems in measuring change. Madison: University of Winsconsin Press.
Dywan, J. & Segalowitz, S.J. (1986). The role of the case study in neuropsychological research. In J. Valsinjer (Ed.) The role of the individual subject in scientific psychology. New York: Plenum, in press.
Galaburda, A.M. & Kemper, T.L. (1979). Cytoarchitectonic abnormalities in developmental dyslexia: a case study. Annals of Neurology 6, 94-100.
Galin, D. & Ornstein, R. (1972). Lateral specialization of cognitive mode: an EEG study. Psychophysiology, 95, 412-419.
Geschwind, N. & Levitsky, W. (1968). Human brain: Left-right asymmetries in temporal speech region. Science, 161, 186-187.
Gordon, H.W., Frooman, B. & Lavie, P. (1982). Shift in cognitive asymmetries between wakings from REM and NREM sleep. Neuropsychologia, 20, 99-103.
Hannay, H.J. & Malone, D.R. (1976). Visual field effects and short term memory for verbal material. Neuropsychologia, 14, 203-209.
Heilman, K. (1979). Neglect and related disorders. In K. Heilman & E. Valenstein (Eds.) Clinical Neuropsychology. New York: Oxford.
Hier, D.B., Lemay, M., Rosenberger, P.B. & Perlo, V.P. (1978). Developmental dyslexia: Evidence for a subgroup with a reversal of cerebral asymmetry. Archives of Neurology 35, 90-92.
Inglis, J. & Lawson, J.S. (1981). Sex differences in the effects of unilateral brain damage on intelligence. Science, 212, 693-695.
Kimura, D. (1980). Sex differences in intrahemispheric organization of speech. Behavioural and Brain Sciences, 3, 240-241.
Klein, R. & Armitage, R. (1979). Rhythms in human performances: 1-1/2-hour oscillations in cognitive style. Science, 204. 1326-1328.
Lake, D. & Bryden, M.P. (1976). Handedness and sex differences in hemispheric asymmetry. Brain and Language, 3, 266-282.
Levine, S.C., Banich, M.T. & Koch-Weser, M. (1984). Variations in patterns of lateral asymmetry among dextrals. Brain and Cognition, 3, 317-334.
Levy, J., Heller, W., Banich, M.T. & Burton, L. (1983). Are variations among right-handers in perceptual asymmetries caused by characteristic arousal differences between hemispheres? Journal of Experimental Psychology: Human Perception and Performance, 9, 329-359.
McGlone, J. (1980). Sex differences in human brain asymmetry: a critical survey. The Behavioral and Brain Sciences, 3, 215-263.
McKeever, W.F. & Jackson, T.L. (1979). Cerebral dominance assessed by object- and color-naming latencies: sex and familial sinistrality effects. Brain and Language, 7, 175-190.
McKeever, W.F., VanDeventer, A.D. & Suberi, M. (1973). Avowed, assessed and familial handedness and differential processing of brief sequential and

nonsequential visual stimuli. Neuropsychologia, 11, 235-238.
Moore, W.H., Jr. (1979). Alpha hemispheric asymmetry of males and females on verbal and nonverbal tasks: some preliminary results. Cortex, 15, 321-326.
Moore, W.H. & Haynes, W.O. (1980). A study of alpha hemispheric asymmetries for verbal and nonverbal stimuli in males and females. Brain and Language, 9, 338-349.
Ogilvie, R. & Segalowitz, S.J. (1975). The effect of 20% sleep reduction upon mood, learning and performance measures. Sleep Research, 4, 242.
Ojemann, G.A. (1983). Brain organization for language from the perspective of electrical stimulation mapping. Behavioural and Brain Sciences, 6, 189-230.
Ornstein, R., Johnstone, J., Herron, J., and Swencionis, C. (1980). Differential right hemisphere engagement in visuospatial tasks. Neuropsychologia, 18, 49-64.
Orsini, D.L., Lewis, R.S. & Satz, P. (1985). Expressive vs. receptive language functions with regards to sex and handedness differences in cerebral lateralization. Unpublished manuscript.
Piazza, D.M. (1980). The influence of sex and handedness in the hemispheric specialization of verbal and nonverbal tasks. Neuropsychologia, 18, 163-176.
Porac, C., Coren, S. & Duncan, P. (1980). Life-span age trends in laterality. Journal of Gerontology, 35, 715-721.
Rasmussen, T. & Milner, B. (1977). The role of early left-brain injury in determining lateralization of cerebral speech functions. Annals of the New York Academy of Sciences, 299, 355-369.
Rubens, (1977). Anatomical asymmetries of human cerebral cortex. In S. Harnad, Doty. R., Goldstein, L., Jaynes, J & Krauthamer, G. Lateralization in the Nervous System. New York: Academic Press.
Ross, P. & Turkewitz, G. (1981). Individual differences in cerebral asymmetries for facial recognition. Cortex, 17, 199-214.
Segalowitz, S.J. (1983). Two sides of the brain. Englewood-Cliffs, NJ: Prentice-Hall.
Segalowitz, S.J. (1986). Validity and reliability of noninvasive measures of brain lateralization. In J. Obrzut & D. Hines (Eds.) Child Neuropsychology: Empirical Issues. New York: Academic Press, in press.
Segalowitz, S.J. & Bryden, M.P. (1983). Individual differences in hemispheric representation of language. In Segalowitz (ed.) Language Functions and Brain Organization. New York: Academic Press.
Segalowitz, S.J. & Orr, C. (1981). How to measure individual differences in brain lateralization: demonstration of a paradigm. International Neuropsychological Society, (February).
Smokler, I.A. & Shevrin, H. (1979). Cerebral lateralization and personality style. Archives of General Psychiatry, 36, 949-954.
Tyler, S.K. & Tucker, D.M. (1982). Anxiety and perceptual structure: individual differences in neuropsychological function. Journal of Abnormal Psychology, 9, 210-220.
Varney, N.R. & Benton, A.L. (1975). Tactile perception of direction in relation to handedness and familial handedness. Neuropsychologia, 13, 449-454.
Winer, B.J. (1971). Statistical Principles in Experimental Design. Second edition. New York: McGraw-Hill (Cattell, 1963).
Witelson, S.F. (1985). The brain connection: the corpus callosum is larger in left-handers. Science, 229, 665-668.
Zurif, E. & Bryden, M.P. (1969). Familial handedness and left-right differences in auditory and visual perception. Neuropsychologia, 7, 179-187.

CEREBRAL LATERALITY: RUBE GOLDBERG AT THE BAUHAUS?

John C. Marshall

Neuropsychology Unit
The Radcliffe Infirmary
Oxford, OX2 6HE, U.K.

INTRODUCTION

What could be simpler than a dichotomy? A whole, obviously. Or as Pierre Flourens put it, quoting Rene Descartes and attacking Franz-Joseph Gall:

> "I remark here ... that there is a great difference between the mind and the body, in that the body is, by its nature, always divisible, and the mind wholly indivisible. For, in fact, when I contemplate it - that is, when I contemplate my own self - and consider myself as a thing that thinks, I cannot discover in myself any parts, but that I clearly know that I am a thing absolutely one and complete" (Young, 1970, p.72).

O happy days! Today, of course, wholes are out. Even the most committed of dualists (e.g. Eccles, 1985) know that the brain is the organ of mind, and that the brain, like it or not, is _not_ a unitary organ, either anatomically, physiologically, or functionally. For Eccles, the immaterial will can act only _via_ the supplementary motor area, not _via_ the occipital lobes, for example.

The brain may be _unified_, in much the same sense that the (rest of the) body (or an automobile) is unified ...but ´unified´ does not mean ´having no parts´ (in Descartes´ sense). As one of Descartes´ followers, La Forge, elegantly phrased the matter: a machine is "a body composed of several organic parts which being united conspire to produce certain movements of which they would be incapable if separate" (Marshall, 1980, p.172). Thus, to pick an example at random, two separate hemispheres (with eyes closed) are incapable of making the movement "pen" when a pen is held in the left hand, an enterprise that occasions no difficulty when the neocommissures are intact (Bogen, 1985a). But long before such experiments were performed, the anatomy of the brain had convinced many physicians, Hippocrates included, that the brain was more akin to the kidneys or the lungs than to the heart:

> ".... the human brain, as in the case of all other animals, is double" (Chadwick and Mann, 1950, p.183).

Yet the analogy with the other paired organs of the body could not have been exact even for the ancients. For Lokhorst (1982) has recently drawn attention to a theory of hemispheric specialization that can be traced back to the Greek physician Soranus in the fourth-century B.C.E.:

> ".... there are two brains in the head, one which gives understanding, and another which provides sense perception. That is to say, the one which is lying on the right side is the one that perceives; with the left one, however, we understand" (Lokhorst, 1982, p. 34).

Needless to say, Soranus' theory has no parallel in accounts of the functioning of the kidneys. As far as I know, no-one has ever hypothesized that the left kidney had *different* functions from the right. The position taken by Soranus will, however, enable us to see why the further evidence for cerebral laterality, derived in the first place from study of the effects of unilateral cerebral lesions, does not unequivocally lead one to suppose that the mind-brain is *dual* (Benson and Zaidel, 1985). The relevant question is now: What's so special about two? And, pertinently enough, there are even two versions of what I shall call the duality hypothesis. In an exceptionally clear account of 'the dual brain' Bogen (1985b) distinguishes between hemispheric specialization and cerebral duality.

The first notion, hemispheric specialization, refers to the (putative) fact that the functional capacities of the left and the right brains are different (either quantitatively or qualitatively); the second, cerebral duality, implies that "each hemisphere can function to a significant extent independently of the other" (Bogen, 1985b, p.28), perhaps indeed to such an extent that we might be prepared to credit one (normal) person with two minds (Wigan, 1844). Phrased in this fashion one can immediately see that both hypotheses could, in point of logic, be true (or false), or, more interestingly, that either one could be true with the other one false. What seems to be critical is that the two hypotheses should not be conflated when their empirical adequacy is assessed.

HEMISPHERIC SPECIALIZATION

The number two undoubtedly has a special affinity for neuropsychologists. Throughout the nineteenth century, physicians and philosophers attempted to describe the functioning of the cerebral hemispheres by such polar contrasts as intelligence/emotion, reason/madness, male/female, and objective/subjective. (In each pair, the left hemisphere characterization is given first). Harrington (1985) provides a masterly survey of these dichotomies and the 'evidence' that was held to support them. Yet when one comes to the 'modern' work that supports our own central dogma of 'complementary hemispheric specialization' (Geschwind and Galaburda, 1984), one does not find Broca's children attempting to locate the faculties of reason and objectivity or madness and subjectivity (not even understanding and sensation *qua* Soranus) in one hemisphere or the other.

Rather, Broca (1965) argued that one aspect of the language faculty was (usually) localized in the third frontal convolution of the left hemisphere; that component was "the memory of the procedure that is employed to *articulate* language". And likewise for other (relatively) specific domains: language comprehension (Wernicke, 1874), and skilled praxis (Liepmann, 1908), for example. The association of deficits in these areas - expressive and receptive aphasia, ideomotor and ideational apraxia - with left hemisphere pathology contrasted with the *apparent* lack of any striking cognitive deficits subsequent to unilateral right hemisphere pathology led directly to the first modern conceptualization of hemispheric specialization: namely, that their relationship was that of master and slave. On this account, the left hemisphere was the source of all higher mental functions; when bilateral action or unilateral action of the left side of the body was required, it was the left hemisphere that determined the course of action and instructed the right hemisphere (as a purely executive organ) to act appropriately. Hemispheric specialization yes, cerebral duality no, for the right hemisphere as a truly passive, fully obedient slave clearly had ′no mind of its own′.

This picture, of course, changed dramatically upon the rediscovery and extension (Poetzl, 1928; McFie, Piercy, and Zangwill, 1950) of John Hughlings Jackson′s work. Jackson (1876) had reported that a patient with a large right temporal glioma (and smaller growths in the right hippocampal region) could no longer find her way in familiar surroundings and had great difficulty with the spatial component of dressing herself. These findings led Jackson to propose that posterior areas of the right hemisphere were "the seat of visual ideation". The right hemisphere now had a specialization of its own. Jackson had also observed that even in cases of very severe aphasia (consequent upon left hemisphere damage) some language ability was usually retained. In particular, overlearned, ′automatic′, and emotional speech (including obscenity and blasphemy) was spared. Accordingly, Jackson suggested that the left hemisphere was "leading" for language and the right "more automatic" whereas the relationship was reversed for visuo-spatial cognition. Thus was the notion of ′complementary hemispheric specialization′ born.

The question now become: In the light of subsequent research into the effects of unilateral and bilateral brain injury, does this <u>dichotomy</u> indeed serve to unify the respective functional specializations of the hemispheres?

Certainly, the generalization has some force. Most, if not all, ′core′ linguistic functions are left-lateralized in the vast majority of (right-handed) adults. The gross impairments of sentence structure, word finding, segmental phonology, and language comprehension seen after unilateral damage to the perisylvian region of the left-hemisphere are not (usually) found after comparable insult to the right. It would seem that the relevant generalization must refer to an ′abstract′ characterization of linguistic form (grammar) rather than to a modality-specific form of speech processing (i.e. auditory-vocal language). Thus the primary disorders of reading, writing and spelling are consequent upon left-hemisphere damage (Patterson, Marshall and Coltheart, 1985). Likewise, aphasic disorders of sign languages (i.e. languages that have no surface features in common with auditory-vocal languages) are found after left-hemisphere injury, despite the fact that sign-languages are executed in three-dimensional space and perceived visually (Damasio, Bellugi, Damasio, Poizner, and Van Gilder, 1986; Marshall, 1986). One could summarize these latter findings by saying that where there is conflict between characterizing the ′representational domain′ for which a hemisphere is specialized and the modality in which the domain is expressed, then the representational domain takes (biological) precedence.

With respect to the specialization of the right hemisphere, Jackson′s conjecture has received impressive confirmation <u>for some tasks</u>. There are a variety of visuo-spatial skills, including spatial orientation, learning and memory that seem to be preferentially impaired by unilateral right lesions. These include the ability to mentally rotate shapes (Ratcliff, 1979); the learning of routes through a visually-guided maze (Ratcliff and Newcombe, 1973); and memory for topographical locales and routes, when these cannot be verbally mediated (Whitty and Newcombe, 1973; Whiteley and Warrington, 1978; Hecaen, Tzortzis and Rondot, 1980; Landis, Cummings, Benson and Palmer, 1986).

There is also evidence that the fine discrimination of differences between visually-similar forms is maximally impaired by right hemisphere lesion (Warrington and James, 1967; Orgass, Poeck, Kerschensteiner and Hartje, 1972). Very substantial deficits in visuo-perceptual tasks that require gestalt integration, as in so-called 'closure' tests, are also found after right- but not left-sided injury (Kerschensteiner, Hartje, Orgass, and Poeck, 1972; Wasserstein, Zappulla, Rosen and Gerstman, 1984; Newcombe, Ratcliff and Damasio, 1986).

The problem, however, is that whilst the generalizations 'language is a left-hemisphere function' and 'visuo-spatial cognition is a right-hemisphere function' are good summaries they are far from perfect in encapsulating the entire range of lateralized deficits. A few examples will suffice to indicate the failures.

Some language skills are preferentially impaired after right-hemisphere damage. The expression and comprehension of prosody, both in the service of grammatical and affective functions (Weintraub, Mesulam and Kramer, 1981; Roberts, Kinsella and Wales, 1981; Ross, 1983) is perhaps the most striking example. This suggests a relationship between (some aspects of) language and music. Thus singing can be relatively well-preserved in cases of severe Broca's aphasia (Yamadori, Osumi, Masuhara and Okubo, 1977; see also Gordon and Bogen, 1974). There is also well-known evidence that the right temporal lobe is preferentially involved in the discrimination and recognition of timbre and melodic patterning (Milner, 1962; Zatorre, 1985). Musical abilities can be retained at a high level despite severe aphasia consequent upon left-hemisphere damage (Assal, 1973; Luria, Tsvetkova and Futer, 1965). By contrast, severe disorders of musical execution are found after unilateral right-hemisphere damage (Damasio and Damasio, 1977; McFarland and Fortin, 1982).

Many disorders of arithmetic knowledge and skill, disorders that are not secondary to either visuo-spatial impairment or aphasia, are preferentially associated with left posterior damage (Grafman, Passafiume, Faglioni and Boller, 1982; Warrington, 1982; see also McCloskey, Caramazza and Basili, 1985).

Some disorders of visuo-perceptual functioning, detecting the hidden figure, for example, in Gottschaldt's Embedded Figures Test have been found, contrary to expectation, only in left-hemisphere patients with aphasia (Orgass, Poeck, Kerschensteiner and Hartje, 1972). The capacity to generate and utilize visual imagery is usually thought of as a right-hemisphere skill. Yet recent reviews and studies provide little support for this position (Ehrlichman and Barrett, 1983). It rather seems to be the case that the generation of mental images at least is dependent upon the integrity of left posterior cortex (Farah, 1984).

Other visuo-perceptual and visuo-spatial disorders are seen only with bilateral damage. Examples include the visual agnosias (Davidoff and Wilson, 1985) and locomotor map-following deficit (Ratcliff and Newcombe, 1973). Frank clinical prosopagnosia is usually dependent upon bilateral lesions (Damasio, Damasio and Van Hoesen, 1982), although there is some recent evidence that the condition may be provoked by unilateral right posterior damage (Landis, Cummings, Christen, Bogen and Imhof, 1986; De Renzi, 1986).

Kurt Schwitters meets Le Corbusier

What price, then, hemispheric specialization? It is not in dispute that unilateral lesions of one hemisphere often provoke cognitive impairments that are qualitatively distinct from those provoked by lesion of the other hemisphere. But ... the notion of complementary hemispheric specializaton qua over-riding dichotomy surely implies that there is a unitary characterization of the left hemisphere's intrinsic competence such that all the superficially different manifestations of that competence fall under a deeper theoretical generalization. And likewise for the surface manifestations of right hemisphere competence. In short, the hypothesis is that the (anatomically) dual brain was designed (by Le Corbusier?) as an organized biological structure wherein cognitive functions could rationally co-habit: Similar functions should inhabit adjacent territory; mechanisms whose operations call upon the output of other mechanisms should likewise be neighbours. By contrast, radically different functions should be geographically quite separate, independent mechanisms likewise. And the great divide, the Danube separating the (quasi-) independent cites of Buda and Pest is the corpus callosum. Lashley's exposition of the problem is still classic:

> "...separate localization of functions is determined by the existence of diverse kinds of integrative mechanism which cannot function in the same nerve field without interference. (...) If temporal order is determined by space factors in the nervous system, the fields in which this type of organization is dominant cannot also serve other space systems. There is thus some reason to believe that the utilization of the spatial arrangement of excitations in the timing functions determines an additional group of isolated cerebral areas" (Lashley, 1937).

But what do we actually find? The left hemisphere seems to be preferentially specialized for: 'Core' language functions; skilled praxis; some aspects of arithmetic and calculation; some aspects of visual detection, face and object recognition, route-finding, spatial attention and visual imagery; and some aspects of autobiographical memory. The right seems preferentially specialized for 'core' components of topographical learning and memory; some aspects of spatial attention and computation, face and object recognition; prosodic aspects of speech perception and production; (many) aspects of music; and some aspects of autobiographical memory! What then is the generalization that would make one recognize the validity of hemispheric specialization over and above the unquestioned facts of focal specialization between and within hemispheres?

The 'design' looks as if it were put together by Kurt Schwitters. Or Rube Goldberg, as Gould (1977) suggests: "... the structures evolved ... are jerry-built out of available parts used by ancestors for other purposes". During the course of the last thirty years or so, many theoreticians dissatisfied with the language/visuo-spatial skills dichotomy have attempted to unify each hemisphere under a variety of more 'abstract' labels. These 'new' dichotomies (which have the flavour of their nineteenth century great-grandparents) include: Analytic/Holistic; Serial/Parallel; Focal/Diffuse; Temporal/Spatial; High (spatial) frequency/Low frequency. Yet there is as little evidence that these categories will suffice as there was for the generalization of Soranus (see Marshall, 1981). Their primary vice, as currently stated, is vagueness: their application to data-bases thus permits unconstrained, post-hoc, 'strategic' fudges that allow almost any pattern of results to fall under whatever label the theoretician chooses.

CONCLUSIONS

To repeat: I am not denying that there are (important) differences between the functions of the hemispheres. Neither am I denying that the human brain/mind manifests at least duality (Bogen, 1985b). As Bogen notes, the evidence from hemispherectomy and commissurotomy shows that one hemisphere can function with minimal dependence upon (cortical areas of) the other. My point is solely that the number two may well be too small. Franz-Joseph Gall postulated forty-eight (quasi-) independent cortical organs that serve human cognition; each organ in his theoretical account dealt with a different representational domain and drew upon its own resources of memory, attention, and volition (Marshall, 1984). And within the human population, Gall asserted, there are substantial individual differences in the functional efficiency and correlated (anatomical) size of these organs. I hold no brief for the claim that bigger means better, nor for the number forty-eight. I do, however, suspect that forty-eight is closer to the truth than is two.

Acknowledgement

I thank Drs. Freda Newcombe and Joseph Bogen for many discussions of these issues; neither friend should be held responsible for my continued failures of understanding.

REFERENCES

Assal, G. (1973). Aphasie de Wernicke sans amusie chez un pianiste. Revue Neurologique, 129 251-255.

Benson, D.F. and Zaidel, E., eds. (1985). The Dual Brain: Hemispheric Specialization in Humans. New York: The Guilford Press.

Bogen, J.E. (1985a): Split-brain syndromes. In J.A.M. Frederiks (ed.), Handbook of Clinical Neurology, Vol. 1 (45): Clinical Neuropsychology. Amsterdam: Elsevier.

Bogen, J.E. (1985b). The dual brain: some historical and methodological aspects. In D.F. Benson and E. Zaidel (eds.), The Dual Brain. New York: The Guilford Press.

Broca, P. (1865). Sur le siège de la faculté du langage articulé. Bulletin de la Société d'Anthropologie, 6 337-393.

Chadwick, J. and Mann, W.N. (1950). The Medical Works of Hippocrates. Oxford: Blackwell Scientific Publications.

Damasio, A., Bellugi, U., Damasio, H., Poizner, H. and Van Gilder, J. (1986). Sign language aphasia during left-hemisphere Amytal injection. Nature, 322, 363-365.

Damasio, A., Damasio, H., and Van Hoesen, G. (1982). Prosopagnosia: anatomic basis and behavioral mechanisms. Neurology, 32, 331-341.

Damasio, H. and Damasio, A. (1977). Musical faculty and cerebral dominance. In M. Critchley and P.A. Henderson (eds.), Music and the Brain. London: Heinemann Medical Books.

Davidoff, J. and Wilson, B. (1985). A case of visual agnosia showing a disorder of pre-semantic visual classification. Cortex, 21, 121-134.

De Renzi, E. (1986). Prosopagnosia in two patients with CT scan evidence of damage confined to the right hemisphere. Neuropsychologia, 24, 385-389.

Eccles, J.C. (1985). Mental summation: the timing of voluntary intentions by cortical activity. The Behavioral and Brain Sciences, 8, 542-543.

Ehrlichman, H. and Barrett, J. (1983). Right hemisphere specialization for mental imagery: a review of the evidence. Brain and Cognition, 2, 39-52.

Farah, M.J. (1984). The neurological basis of mental imagery: a componential analysis. Cognition, 18, 245-272.

Geschwind, N. and Galaburda, A.M., eds., (1984). Cerebral Dominance: The Biological Foundations. Cambridge, Mass.: Harvard University Press.

Gordon, H.W. and Bogen, J.E. (1974). Hemispheric lateralization of singing after intracarotid sodium amlyobarbitone. Journal of Neurology, Neurosurgery, and Psychiatry, 37, 727-738.

Gould, S.J. (1977). Ever since Darwin. New York: Norton.

Grafman, J., Passafiume, D., Faglioni, P., and Boller, F. (1982). Calculation disturbances in adults with focal hemispheric damage. Cortex, 18, 37-50.
Harrington, A. (1985). Nineteenth-century ideas on hemispheric differences and "duality of mind". The Behavioral and Brain Sciences, 8, 617-660.
Hecaen, H., Tzortzis, C. and Rondot, P. (1980). Loss of topographic memory with learning deficits. Cortex, 16, 525-542.
Jackson, J.H. (1876). Case of large cerebral tumour without optic neuritis and with left hemiplegia and imperception. Royal London Ophthalmological Hospital Reports, 8, 434-442.
Kerschensteiner, M., Hartje, W., Orgass, B., and Poeck, K. (1972). The recognition of simple and complex realistic figures in patients with unilateral brain lesion. Arch. Psychiat. Nervenkr., 216, 188-200.
Landis, T., Cummings, J.L., Benson, D.F. and Palmer, E.P. (1986). Loss of topographic familiarity: An environmental agnosia. Archives of Neurology, 43, 132-136.
Landis, T., Cummings, J.L., Christen, L., Bogen, J.E. and Imhof, H.-G. (1986). Are unilateral right posterior lesions sufficient to cause prosopagnosia? Clinical and radiological findings in six additional patients. Cortex, 22, 243-252.
Lashley, K.S. (1937). Functional determinants of cerebral localization. Archives of Neurology and Psychiatry, 38, 371-387.
Liepmann, H. (1908): Drei Aufsatze aus dem Apraxiegebiet. Berlin: Karger.
Lokhorst, G.-J. (1982). An ancient Greek theory of hemispheric specialization. Clio Medica, 17, 33-38.
Luria, A.R., Tsvetkova, L.S. and Futer, D.S. (1965). Aphasia in a composer. Journal of Neurological Sciences, 2, 288-292.
Marshall, J.C. (1980). On the biology of language acquisition. In D. Caplan (ed.), Biological Studies of Mental Processes. Cambridge, Mass: MIT Press.
Marshall, J.C. (1981). Hemispheric specialization: What, how and why. The Behavioral and Brain Sciences, 4, 72-73.
Marshall, J.C. (1984). Multiple perspectives on modularity. Cognition, 17, 209-242.
Marshall, J.C. (1986). Signs of language in the brain. Nature, 322, 307-308.
McCloskey, M., Caramazza, A. and Basili, A. (1985): Cognitive mechanisms in number processing and calculation: Evidence from dyscalculia. Brain and Cognition, 4, 171-196.
McFarland, H.R. and Fortin, D. (1982). Amusia due to right temporoparietal infarct. Archives of Neurology, 39, 725-727.
McFie, J., Piercy, M.F. and Zangwill, O.L. (1950). Visual-spatial agnosia associated with lesions of the right cerebral hemisphere. Brain, 73, 167-190.
Milner, B. (1962). Laterality effects in audition. In V.B. Mountcastle (ed.), Interhemispheric relations and cerebral dominance. Baltimore: Johns Hopkins Press.
Newcombe, F., Ratcliff, G. and Damasio, H. (1986). Dissociable visual and spatial impairments following right posterior cerebral lesions: clinical, neuropsychological and anatomical evidence. Neuropsychologia, 25, (in press).
Orgass, B., Poeck, K., Kerschensteiner, M. and Hartje, W. (1972). Visuo-cognitive performance in patients with unilateral hemispheric lesions. Zeitschrift fur Neurologie, 202, 177-195.
Patterson, K., Marshall, J.C. and Coltheart, M., eds. (1985). Surface Dyslexia. London: Erlbaum.
Poetzi, L. (1928). Die optisch-agnostischen Storungen. Leipzig. Denticke.
Ratcliff, G. (1979). Spatial thought, mental rotation and the right cerebral hemisphere. Neuropsychologia, 17, 49-54.
Ratcliff, G. and Newcombe, F. (1973). Spatial orientation in man: effects of left, right, and bilateral posterior cerebral lesions. Journal of Neurology, Neurosurgery and Psychiatry, 36, 448-454.

Roberts, C., Kinsella, G. and Wales, R. (1981). Disturbances in processing prosodic features of language following right hemisphere lesions. Supplement to the Bulletin of the Postgraduate Committee in Medicine, University of Sydney, December 1981, 172-183.

Ross, E.D. (1984). Right-hemisphere lesions in disorders of affective language. In A. Kertesz (ed.), Localization in Neuropsychology. New York: Academic Press.

Warrington, E.K. (1982). The fractionation of arithmetical skills: A single case study. Quarterly Journal of Experimental Psychology, 34A, 31-51.

Warrington, E.K. and James, M. (1967). An experimental investigation of facial recognition in patients with unilateral cerebral lesions. Cortex, 3, 317-326.

Wasserstein, J., Zapulla, R., Rosen, J. and Gerstman, L. (1984). Evidence for differentiation of right hemisphere visual-perceptual functions. Brain and Cognition, 3, 51-56.

Weintraub, S., Mesulam, M.-M. and Kramer, L. (1981). Disturbances in prosody: a right hemisphere contribution to language. Archives of Neurology, 38, 742-744.

Wernicke, C. (1874). Der aphasische Symptomenkomplex. Breslau: Cohn and Weigart.

Whiteley, A.M. and Warrington, E.K. (1978). Selective impairment of topographical memory: a single case study. Journal of Neurology, Neurosurgery and Psychiatry, 41, 575-578.

Whitty, C.W.M. and Newcombe, F. (1973). R.C. Oldfield's study of visual and topographic disturbances in a right occipito-parietal lesion of 30 years duration. Neuropsychologia, 11, 471-475.

Wigan, A.L. (1844). The Duality of the Mind. London: Longmans. Reprinted 1985. Malibu, California, J. Simon Press.

Yamadori, A., Osmi, Y., Masuhara, S. and Okubo, M. (1977). Preservation of singing in Broca's aphasia. Journal of Neurology, Neurosurgery and Psychiatry, 40, 221-224.

Young, R.M. (1970). Mind, Brain and Adaptation in the Nineteenth Century. Oxford: Clarendon Press.

Zatorre, R.J. (1985). Discrimination and recognition of tonal melodies after unilateral cerebral lesions. Neuropsychologia, 23, 31-41.

IMPLICATIONS OF THE RIGHT SHIFT THEORY OF HANDEDNESS FOR INDIVIDUAL DIFFERENCES IN HEMISPHERE SPECIALISATION

Marian Annett

Department of Applied Social Studies
Coventry (Lanchester) Polytechnic
Coventry, UK

Two main approaches to problems of individual differences in hemisphere specialisation are to be found in the literature. The first is avoidance: subjects are restricted to fully right-handed males, with no known left-handed relatives. It is assumed that such subjects are likely to be homogeneous for the typical pattern of cerebral specialisation. The second approach is to compare subjects for personal hand preference, or for the presence of left-handed relatives, usually taking care to treat the sexes separately, in the expectation that these variables will be associated with differing patterns of cerebral specialisation. The right shift (RS) theory of handedness (Annett, 1972) suggests that the homogeneity of subjects in the first approach, and the discriminating power of variables in the second approach, are overestimated. Some of the challenges of the RS theory were evident from its initial formulation, and others have been discovered in subsequent explorations of its implications. A brief review of the development of the theory was given by Annett (1981) and a full review by Annett (1985). This paper summarises implications of the theory for individual differences, giving first an overview, and then a selective review of evidence for the main assumptions.

Overview of implications of the right shift theory for individual differences

The RS theory grew from an analysis of hand preference and hand skill, but it led to a new way of thinking about hemisphere specialisation. The theory suggests that individual differences in brain specialisation depend first on chance and, secondly, on the presence or absence of a gene (rs+), and whether the gene is carried in single or double dose (rs--, rs+-, and rs++ genotypes). The gene is hypothesised to give to the left cerebral hemisphere in early life a relative advantage that promotes the development of speech control from that side. The human species bias to right hand preference is a by-product of the gene promoting left hemisphere speech.

In the rs--, there are no systematic biases to either side, but only chance biases for hemisphere lateralisation and also for handedness, and these chances are independent of each other. About 50% of the rs-- have right hemisphere speech and about 50% left hemisphere speech; some small unknown proportion may be classifiable as bilateral speakers. On objective measures of hand skill 50% of the rs-- are expected to be faster with the left hand and 50% faster with the right hand, but in many cases, the difference between the hands is so trivial that either hand could easily acquire the ability to perform skilled actions. Cultural pressures ensure that the majority of rs-- are right-handed. This leads to the implication that the majority of right brained speakers are right-handed.

In the rs+- and rs++, the presence of the gene is sufficient, in normal development, to ensure that speech will be controlled from the left side of the brain. Anything that slows or distorts the normal developmental pattern may hinder the expression of the gene, and lateral biases for hand and brain are expected to diminish, reverting toward chance levels. The relative advantage imparted to the left hemisphere by the gene is sufficient to increase the probability of right-handedness, but it does not determine dextrality. Some rs+- and rs++ genotypes are likely to develop left-handedness. Mixed handedness may develop in all genotypes. These relationships imply that there can be no strong associations between asymmetries of hand and brain; there can only be changes in relative probabilities. The chances of right brainedness for speech are certainly higher in left-handers than right-handers, but never higher than the theoretical maximum of 50%; the level of right brainedness predicted in left- and right-handers varies with criteria of hand preference, so there are no specific values that are true of all studies.

Sex and twinning influence the expression of the rs+ gene. The incidence of right-handedness is higher and the rate of speech acquisition faster in females than males and in the singleborn than in twins. These variables are also correlated with relative maturity at birth. This suggests that the gene works by modulating the relative growth of the left and right cerebral hemispheres in late fetal life. Left hemisphere specialisation for speech is likely to be stronger in early than late maturers, at least in the neonatal period, and possibly later (Netley and Rovet, 1983; Waber, 1976).

In addition to leading to a re-evaluation of the role of variables well established in the laterality literature, the RS theory leads to new fields of exploration in individual differences. It suggests that variability at the dextral side of the laterality continuum may be at least as important as at the sinistral side. There is considerable evidence that the rs+ gene has disadvantages as well as advantages, and that the genotype frequencies are stable in the population in relationships that can be described as a balanced polymorphism with heterozygote advantage. If the rs+- genotype is the most favoured, it is very probable that undergraduates in higher education are not representative of the general population for cerebral laterality. This would imply that studies based on undergraduates are sampling a restricted range of individual differences in hemisphere specialisation.

Why right shift ?

The right shift analysis depended on a coordination of three sets of data, hand preference in humans, hand and paw preferences in non-humans and the distribution of left minus right (L-R) hand skill. During the 1960s data were collected on hand preferences, by observation and by questionnaire, from several large samples of college students, schoolchildren and service recruits, who took part in class groups where there was little scope for volunteer effects. These samples were highly consistent when subjects were classified as pure left-, mixed- and pure right- handers. ('Pure' means no preference for the other hand for any of 12 items, and 'mixed' means a definite preference for the other hand for at least one item. Reports of 'either' hand preference were not counted as evidence of non-dextrality). Percentages in 7 large samples were about 4, 30, 66%, (ranges 2-5, 25-37, 58-71%) left-, mixed- and right-handers respectively (Annett, 1967).

Studies of hand and paw usage in other species suggested that if non-humans mammals, including primates, were classified on the criteria used above for humans, the corresponding percentages would be about 25, 50, 25%. The proportions for both humans and for non-humans posed a special puzzle because they fitted certain expectations of my first theory of the genetics of handedness (Annett, 1964) but other evidence showed that the first theory was wrong (Annett, 1967).

Measurements were made, also during the 1960s, of the skill of each hand on a peg moving task in schoolchildren and undergraduates. When L-R differences were plotted for large samples, it was evident that the distribution is continuous and takes a form approximating a normal curve. That there is a systematic relationship between degrees of hand preference and degrees of L-R skill was also clear for several years, (Annett, 1970a; 1970b; 1976).

The RS solution was discovered when I asked what areas, under the normal curve of L-R differences, would be required to represent the frequencies of L, M, and R handedness observed. In Figure 1, each L, M and R represents one per cent under the normal curve. The curve to the left gives the proportions expected for non-humans; the mean is at 0, or no difference between the hands in skill. Close to the mean on either side, L-R differences are small enough for the animal to develop mixed paw preferences. At some critical value or threshold (about .67 SD) along the continuum, the L-R difference will favour one hand or paw strongly enough for the animal to develop a consistent preference for that side. Hence, beyond the threshold to each side, all animals are represented as L or R.

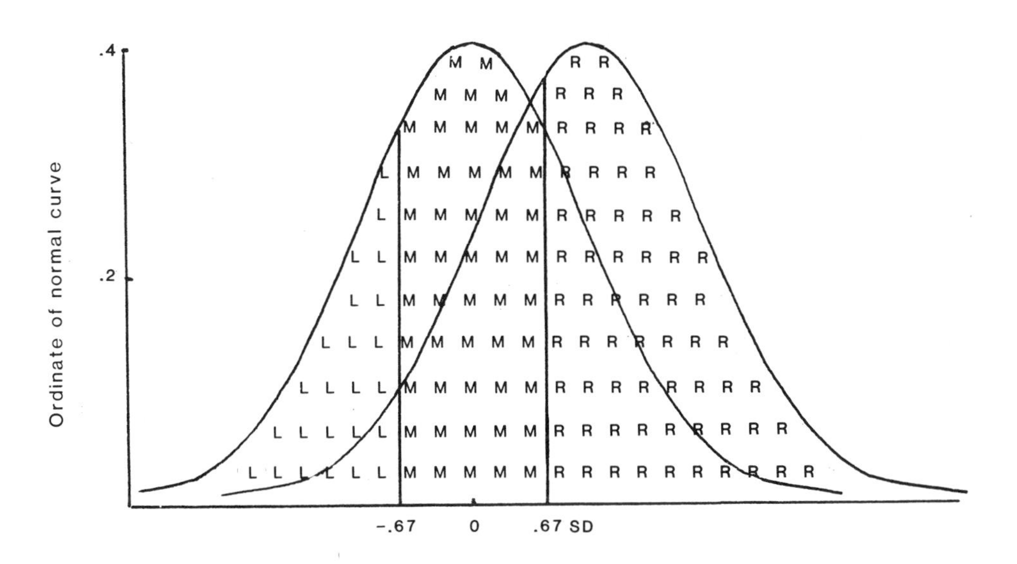

Fig.1 The proportions of left-(L), mixed- (M), and right- (R) handers expected if hand preference depends on differences in L-R skill. Each letter represents 1% under the normal curve. The thresholds for consistent L or R handedness are about 0.67 SD from 0. The curve with mean at 0 represents the distribution expected in non-human mammals; the curve with mean to the right of 0 represents the human distribution.

The key discovery on which the RS theory is based depends on what happens when the normal curve is moved slightly to the right, while all other relationships remain constant. The proportion of consistent left-handers falls, and also the proportion of mixed-handers, but in relationships that are mutually concordant for areas under the normal curve. The distribution for non-humans becomes the distribution for humans, given a shift of the curve to the right, and no other change. This implies that the thresholds required to represent the proportion of mixed-handed humans and of mixed-handed non-humans are identical. We can say that there is no substantial difference between the distributions of humans and non-humans except that the human curve is shifted such that the mean is to the right of 0, while the mean for non-humans is at 0.

This consistency has been demonstrated again in dyslexic children (Annett and Kilshaw, 1984). The proportions of pure left-handers and of mixed-handers were elevated in comparison with controls, but they were raised to the mutually consistent extents expected if a normal curve were shifted not quite so far to the right.

Figure 2 shows the observed L-R distributions for 617 males and 863 females. Both distributions are more peaked and negatively skewed than expected of a true normal curve. Both are consistent with expectations for the <u>sum</u> of 2 or 3 normal subdistributions, which might correspond to the genotypes hypothesised to make up the total population (Annétt and Kilshaw, 1983).

The female distribution is farther to the right than that of males, at almost all points to the right of 0. A stronger dextral bias in females than males has been found in all my samples, whether or not incidences of left hand preference were higher in males. It is especially worth noting that at the left side of the L - R continuum there are no sex differences. Whatever causes sex differences affects the right, but not the left, side of the distribution. This is just as expected if whatever causes the RS is expressed more effectively in females than males, but in the absence of RS

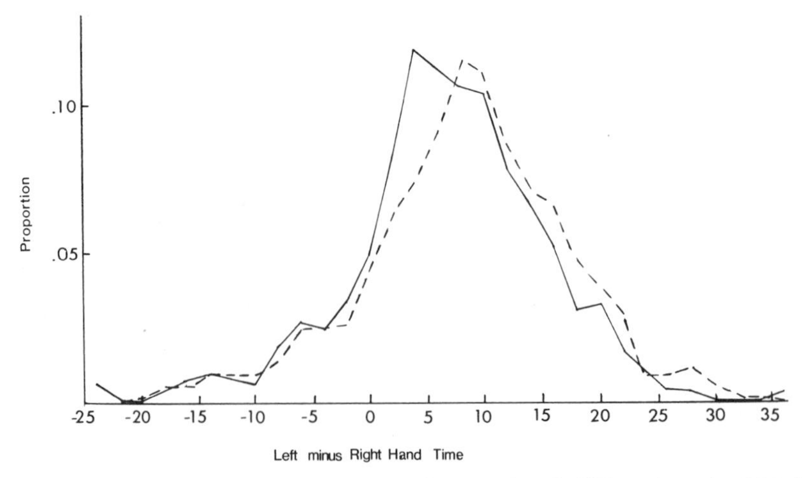

Fig.2 Observed L-R times on a peg moving task of 617 males (solid line) and 863 females (hatched line), showing the proportions of subjects at intervals of 2/10 s (from Annett and Kilshaw, 1983).

there are no sex differences for laterality. This interpretation is strengthened by the absence of sex differences in the children of two left-handed parents (Annett, 1983).

The RS analysis requires only one systematic influence on laterality, something which shifts the distribution to the right. There is nothing to suggest a factor that gives a systematic bias to the left, nor is there evidence for such a factor in the laterality literature. All claims to find atypical biases depend on very small numbers of cases where small biases to one side or the other could be due to chance. However, the laterality literature continues to search for the elusive essence of left-handedness. McManus (1985) claimed to discover a left biased subgroup in my samples, as represented in Figure 2. The claim applied to males only and seems to depend on the little dip in the curve at -1s. An 'eyeball' test is sufficient to suggest that this is trivial.

Since making this analysis in the early 1970s, I have been exploring its implications for puzzles about laterality. If there is only one systematic influence on human handedness, an influence that is not present in our closest primate cousins, the most probable source of influence is whatever biases the human left hemisphere to serve speech. The RS theory suggests that left hemisphere specialisation is the only specific factor involved. All the rest can depend on chance.

The right shift as due to a bias to left hemisphere speech

The longstanding puzzles about the relations between left and right hemisphere speech laterality and right- and left-handedness can be regarded as substantially solved, if the RS analysis is accepted (Annett, 1975, 1985). Independent support for the main assumptions has been given in data of Ratcliff, Dila, Taylor and Milner (1980). Patients identified by the Wada technique as left, right, or bilateral speakers were examined for the distribution of structural brain asymmetry, as shown in the angle of the posterior branches of the middle cerebral artery, visualised on carotid angiograms. In patients with left hemisphere speech the distribution of differences between the hemispheres was approximately normal and biased to smaller angles on the left side. In patients with bilateral or right sided speech, differences were also approximately normally distributed, but clearly centred at L = R; 45% were recorded as left-handed.

The idea that among right hemisphere speakers, the majority should be right-handed, not left-handed, follows from the very reasonable assumption that among people without strong biases to either side for skill, the majority will be persuaded by cultural pressures to use the right hand. The idea looks paradoxical in the context of the usual approach to atypical brainedness, but analyses of data for dysphasics in large consecutive series demonstrate that the expectation is fulfilled.

The widely held belief that left-handers are typically bilateral for cerebral speech representation is without secure foundation. In some of the major consecutive series of dysphasics compared for lesion laterality and handedness (Gloning and Quatember, 1966; Hecaen and Ajuriaguerra, 1964; Hecaen and Piercy, 1956) incidences of left-handedness for the total sample were comparable with expectation for the left threshold in Figure 1; for dysphasics incidences were closer to expectation for the right threshold. That is, the belief that right hemisphere speech is unlikely to be found in right-handers led to a shift of criterion of non-right handedness. The main evidence for bilateral brainedness in left-handers has arisen from an inflation of left-handers among dysphasics; there is no significant excess of dysphasics among the left-handed, as the bilaterality argument demanded.

The key question that needs to be asked, from the viewpoint of the RS theory, is not how many right brained speakers there are in left-handers, or how many in right-handers, but how many right brained speakers there are in the total population, irrespective of handedness. Using the combined data of 4 series Annett (1975) estimated that 9.27% of the population are wholly or partially right brained. Two checks on this estimate from other sources confirm that this is about correct (Annett, 1985). This implies that about 1 in 11 of the population have right or bilateral speech representation, and the majority of these individuals (perhaps 60-70%) are right-handed writers. If this is correct, the extent of individual variation of hemisphere specialisation has been greatly underestimated in the general population and in right-handers.

Why a genetic basis for the right shift?

The first support for the idea that the RS has a genetic basis came from the absence of RS in families in which both parents are left-handed (Annett, 1974). However, in families where parents reported significant personal birth stress, or where parents showed abnormally slow right hand peg moving times, (suggesting that the parental sinistrality could be pathological), the children were found to be strongly biased to the right-hand. The importance of these observations was such as to make it worth collecting a second sample, reported by Annett (1983). In the combined data of two samples, there were 20 children in families where parental left-handedness was judged possibly pathological. The L-R times of this subgroup showed a RS as strong as in general samples, in spite of being reared by LxL parents. In the remaining 95 children, there was a normal distribution of L-R times, with mean nonsignificantly to the right of 0, as expected if the rs+ gene were absent in the majority of children. A similar result was obtained if families were differentiated, not on grounds of parental pathology, but on the presence or absence of sinistral relatives of the parents. These results complement the findings of an adoption study (Carter-Saltzman, 1980) in showing that rearing by right- or left-handed parents is not a major determinant of hand preference.

The second support for the genetic hypothesis came from the discovery that the distribution for right- and left-handedness in families could be predicted, when parameters of the model were taken from data for dysphasia. The frequency of the rs-- genotype was inferred to be twice the incidence of right brainedness as deduced in the consecutive series, mentioned above.

$$2(9.27) = 18.54$$

The incidence of the rs-- gene is then the square root of this figure (0.43) and the incidence of the rs+ gene must be

$$1 - 0.43 = 0.57$$

The other main parameter of the model required for genetic calculations is the extent of RS; that is, how far is the right-shifted distribution to the right of the rs-- distribution? Estimates have been derived from data for dysphasics and also from distributions of L-R times in normal samples. The best estimates currently available are that for males, the mean of the rs+- distribution is at 1 standard deviation, and the mean of the rs++ distribution at 2 standard deviations to the right of the rs-- mean (which is 0 by definition). The shifts for females are expected to be slightly larger than those for males. For twins of both sexes, shifts must be smaller, to account for the slightly higher percentage of left handers among twins than singletons, and for the distribution of RR, RL and LL pairs. The same level of reduction is needed for monozygotic and for dizygotic pairs, showing that the lesser bias is a function of twinning itself, not zygosity.

Given these assumptions, the theory predicts the distributions of handedness observed in twins and in families, in all the major series of the literature, published before and after the initial discovery (Annett, 1978; 1979; 1985).

The details of genetic calculations depend on estimates of gene frequency, extent of shift, and on the incidences of left- or mixed-handedness recorded in any sample. Incidences vary as threshold criteria are more strict or more generous, and expectations for genotypes in families vary with changing thresholds. The distributions in Table 1 should be taken as an example that depends on the particular incidences reported by Ashton (1982) and a particular set of assumptions as to parameters of the model as discussed above. The example is given to illustrate the implications of familial sinistrality for individual differences in hemisphere specialisation. Table 1A shows the distribution of the three genotypes in the total population. Table 1B shows the distribution of genotypes in each of the 4 types of family classified for parental handedness, (distinguishing sex of parent who is left handed). The distribution is given for the whole population (summing to 1). Table 1C gives the same analysis, expressed as percentages within each family type.

Table I. Genotypes in Families

	Genotype proportions		
	rs + +	rs + -	rs - -
A. Total population	.3242	.4904	.1854
B. In families			
R x R	.3081	.4016	.1309
L x R	.0100	.0467	.0262
R x L	.0060	.0402	.0236
L x L	.0002	.0020	.0047
C. As percentage within families			
R x R	37	48	15
L x R	12	47	31
R x L	9	57	34
L x L	3	28	69

Calculations are based on assumptions of the RS model as described in the text, and on parental incidences reported by Ashton (1982) for left + ambidextrous fathers (8.97%) and mothers (7.64%). In A the genotype distribution is given as estimated for the total population. In B it is given for each family type (father X mother) over the total population, and in C as a percentage within each family type.

If the individual differences in hemisphere specialisation of interest are taken to be right and bilateral speech, then on the RS theory these occur as 50% of rs-- genotypes. Table 1B shows that the majority of all rs-- in the population (some 70%) occur in the families of RxR parents (in the same way that the great majority of left-handers in the population occur in this family type). Knowledge that one parent is left-handed raises the probability of atypical hemisphere specialisation in the children, but this is still expected in not more than about 1 in 6 cases. When both parents are left handed, atypical cerebral speech is expected in about 1 in 3 children.

If individual differences in hemisphere specialisation are also to be found in the rs++, as will be argued in the next section, it is worth noting in Table 1c that the presence to a left-handed parent is associated with a marked decrese in the frequency of this genotype. It may well be the case that effects associated with familial sinistrality, in the laterality literature, owe more to the fall in rs++ than to the rise in rs-- genotypes.

The rs+- genotype is the most frequent in the total population. It is also the most frequent in RxR, LxR and RxL families. In LxL families the rs+- genotype is present in more than a quarter of children (given the particular assumptions of these calculations). The main conclusion to be drawn from Table 1 is that knowing that someone has a left-handed parent does not make much difference to the probability that the individual will have right hemisphere speech. Similar considerations would apply to the presence of a left-handed sibling also.

A Balanced polymorphism for the rs+ gene

The most exciting idea to be prompted by the genotype frequencies (Table 1A) follows from the observation that the heterozygote, rs+-, is the most frequent and about as high as possible (maximum 50%). This observation suggests that the heterozygote is the most favourable genotype, and that the rs-- and the rs++ both have disadvantages and advantages that are balanced over the total population. Balanced polymorphisms are very numerous in human genetics, and the majority are not fully understood, in the sense that the costs and benefits of the genes involved are unknown. Since first deducing the genotype frequencies (Annett, 1978), I have looked for evidence as to the costs and benefits of the rs+ gene, and some progress has been made in understanding why the rs+ gene evolved but did not become universal in the population.

The rs+ gene probably evolved in early hominids who were already developing speech capacities. Speech is a human species universal; it develops in all but grossly abnormal individuals, whether the rs+ gene is present or absent. It is hypothesized that the function of the rs+ gene is to expedite speech development by making speech production and speech analysis occur on the same side of the brain. The infant corpus callosum is too immature to be an efficient channel for coordinating feedback from the mouth and the ear, if these two sorts of information are being analysed in different hemispheres. Some asymmetry in cerebral development was introduced by the gene to make one hemisphere, the left, more likely to control speech learning. The advantages of having speedy and clear speech, in young, who are otherwise helpless and dependent on the goodwill of adults, is sufficient to benefit those who carried the gene.

Although the gene mechanism is unknown, I believe that we can be fairly confident in the assumption that it works through cerebral maturation rates in late fetal life. Anything that affects maturation in that period affects the expression of the gene. The fact that females are a little more mature at birth than males is sufficient to account for their greater shift to the right. The fact that the growth of twins is slowed to allow two fetuses to be accommodated in the womb could explain why handedness in twin pairs can be fully predicted by the RS model, provided the extent of shift is reduced in twins compared with the singleborn. Twins are well known to risk delayed language growth in comparison with singletons (Mittler, 1971).

If the gene aids the development of speech, why is it not universal in the population? In nature there are few benefits without costs. An analysis of the actual peg moving times of left-handers, mixed-handers and consistent right-handers led to the surprising hypothesis that the cost of ensuring that speech control will develop in the left hemisphere is an _impairment of the right hemisphere_ (Annett, 1980; Kilshaw and Annett, 1983). In the rs+-, who have only one dose of the gene's effects, the right hemisphere impairment is relatively mild. In the rs++, there is a danger of significant handicap to right hemisphere function. Further research is planned to explore these possibilities.

What might this right hemisphere impairment by the rs+ gene entail? There is evidence, I believe, for three sorts of handicap. First, the left hand is weak, slow and not very useful for skilled tasks (Kilshaw and Annett, 1983). Second, there is impaired capacity for visuo-spatial and mathematical thinking (Annett and Kilshaw, 1982). The rs++ are not likely to be found among high level thinkers in any field requiring non-verbal intellectual skills. Evidence of poorer spatial ability in undergraduates who are strongly right-handed than in others has been found for both sexes (Burnett, Lane and Dratt, 1982). Third, when one side of the brain is underfunctioning, there is some finite loss of total intellectual capacity. The individual is likely to be less intelligent, not only in visuo-spatial directions, but in _verbal ones also_ (Annett and Kilshaw, 1984).

Right hemisphere impairment would give disadvantages in visuo-spatial skills and in activities where close cooperation between two sides of the brain, and two sides of the body is required. Mathematicians use linguistic symbols to represent visuo-spatial relationships. In many sports, either or both sides of the body must be able to react quickly. Surgeons must control both hands very well, and must also envisage three-dimensional relationships. Musical instruments may demand good control of both hands, playing separately and together. The highest levels of human performance in many skilled activities require freedom from the risks of the rs++ genotype. The slightly raised incidences of left- and mixed-handedness in mathematicians, sportspeople, surgeons and musicians is probably due to the _absence of the rs++ genotype_ (Annett and Kilshaw, 1982). A possible physical basis for the better coordination of the activities of the two sides of the brain in those not strongly biased to right-handedness has been found by Witelson (1985; and this volume).

It must be recognised that left-handedness, or a particular pattern of hemisphere specialisation, are not _causes_ of superior performance in any of the above activities. Rather, the rs+ gene _limits_ the level of possible achievement. This is why the gene has not become universal, and why some individuals must risk the problems of speech and language development associated with its absence. With regard to the favourite subjects of the laterality literature, students in higher education, what has this analysis to say? It seems probable that undergraduates selected for higher education are more likely to be of rs+- genotype than an unselected group of the general population. Hence, the generalisation of results obtained for undergraduates must be questioned.

With regard to sex differences, it was said above that the rs+ gene is expressed more effectively in females than males. This is contrary to the hypothesis of cerebral bilaterality in females. However, I believe that the difficulties encounted by psychologists trying to demonstrate laterality effects in females (usually undergraduates) may arise because females, even when heterozygote, are more likely to have an underfunctioning right hemisphere than males. That is, they try to solve all problems verbally, as suggested by Sherman (1978). The critical difference is not so much between a left-verbal and a right-nonverbal hemisphere, as between a better left and relatively poorer right hemisphere in females. These relationships do not occur only in females. They occur in rs++ males also, but such males are probably infrequent among right-handed undergraduates. The analysis of Inglis et al. (1982) is especially important in showing that females tend to rely on the left hemisphere for both verbal and nonverbal functions. This is not a function of sex as such, but of the stronger expression of the rs+ gene in females. The present analysis suggests that females who are rs--, or rs+- with relatively weak expression of the rs+ gene, might be as good in visuo-spatial and mathematical thinking as men. Their main problem could be that society does not expect them to be good, and opportunities for such girls to develop in these directions are limited.

CONCLUSIONS

This chapter has summarised some key ideas about individual differences in hemisphere specialisation, from the view point of the RS theory. The theory offers a new perspective on questions of human lateral asymmetry, provided the reader is prepared to accept the cognitive restructuring required. Since the RS theory suggests that the main variable involved is chance, and all systematic effects have to be detected against a random background, the number of subjects required for adequate tests of hypotheses must be substantially larger than have been used in the typical laterality experiment. The theory suggests that there is more intrinsic variability in the typical right-hander than is usually recognised, and that the effect of having sinistral tendencies in the subject or in relatives is smaller than might be hoped. The most exciting implications of the theory arise from the idea that human cerebral specialisation for speech is a human evolutionary adaptation which has costs as well as benefits. When relationships are found between hand preferences and factors associated with hemisphere specialisation (several chapters in this volume), the RS theory suggests that the most useful interpretations will prove to be in terms of the costs and benefits of the right shift gene.

REFERENCES

Annett, M. (1964). A model of the inheritance of handedness and cerebral dominance. _Nature_. _204_, 59-60.

Annett, M. (1967). The binomial distribution of right, mixed and left handedness. _Quarterly Journal of Experimental Psychology._, _29_, 327-333.

Annett, M. (1970b). The growth of manual preference and speed. _British Journal of Psychology_, _61_, 545-558.

Annett, M. (1972). The distribution of manual asymmetry. _British Journal of Psychology_, _63_, 343-358.

Annett, M. (1974). Handedness in the children of two left handed parents. _British Journal of Psychology_, _65_, 129-131.

Annett, M. (1975). Hand preference and the laterality of cerebral speech. _Cortex_, _11_, 305-328.

Annett, M. (1976). A coordination of hand preference and skill replicated. _British Journal of Psychology_, _67_, 587-592.

Annett, M. (1978). "A Single Gene Explanation of Right and Left Handedness and Brainedness". Lanchester Polytechnic, Coventry.

Annett, M. (1979). Family handedness in three generations predicted by the right shift theory. _Annals of Human Genetics_, _42_, 479-491.

Annett, M. (1981). The right shift theory of handedness and developmental language problems. Bulletin of the Orton Society, 31, 103-121.
Annett, M. (1983). Hand preference and skill in 115 children of two left handed parents. British Journal of Psychology, 74, 17-32.
Annett, M. (1985). "Left, Right, Hand and Brain: The Right Shift Theory", Lawrence Erlbaum, London.
Annett, M. and Kilshaw, D. (1982). Mathematical ability and lateral asymmetry. Cortex, 18, 547-568.
Annett, M. and Kilshaw, D. (1984). Lateral preference and skill in dyslexics: Implications of the right shift theory. Journal of Child Psychology and Psychiatry, 25, 357-377.
Burnett, S.A., Lane, D.M., and Dratt, L.M. (1982). Spatial ability and handedness. Intelligence, 6, 57-68.
Carter-Saltzman, L. (1980). Biological and sociocultural effects on handedness: Comparison between biological and adoptive families. Science, 209, 1263-1265.
Gloning, K. and Quatember, R. (1966). Statistical evidence of neuropsychological syndromes in left handed and ambidextrous patients. Cortex, 2, 484-488.
Hecaen, H. and Ajuriaguerra, J. (1964). "Left handedness: Manual Superiority and Cerebral Dominance". Grune and Stratton, New York.
Hecaen, H. and Piercy, M. (1956). Paroxysmal dysphasia and the problem of cerebral dominance. Journal of Neurology, Neurosurgery and Psychiatry 18, 194-201.
Inglis, J. Ruckman, M.S. Lawson, J.S. MacLean, A.W. and Monga, T.N. (1982). Sex difference in the cognitive effects of unilateral brain damage. Cortex, 18, 257-276.
Kilshaw, D. and Annett, M. (1983). Right and left hand skill I: Effects of age, sex, and hand preference, showing superior skill in left handers. British Journal of Psychology, 74, 253-268.
McManus, I.C. (1985). Right and left hand skill: Failure of the right shift model. British Journal of Psychology, 76, 1-16.
Netley, D. and Rovet, J. (1983). Relationships among brain organization, maturation rate and the development of verbal and nonverbal ability. In: "Language Functions and Brain Organization". S.J. Segalowitz, ed., Academic Press, New York. 245-266.
Ratcliff, G. Dila, C. Taylor, L. and Milner, B. (1980). The morphological asymmetry of the hemispheres and cerebral dominance for speech: A possible relationship. Brain and Language, 11, 87-98.
Sherman, J.A. (1978). "Sex-related Cognitive Differences: An Essay on Theory and Evidence". Charles C. Thomas, Springfield, Illinois.
Waber, D.P. (1976). Sex differences in cognition: A function of maturation rates. Science, 192, 572-574.
Witelson, S.F. (1985). The brain connection: The corpus callosum is larger in left-handers. Science, 229, 665-668.

IS CEREBRAL LATERALIZATION A GRADED OR A DISCRETE CHARACTERISTIC?

John C. Marshall

Neuropsychology Unit
The Radcliffe Infirmary
Oxford
OX2 6HE, U.K.

The human body is replete with paired organs, both externally (eyes and ears, for example), and internally (the kidneys). The anatomical similarity between the members of such pairs leads us to expect that they will have similar or even identical functions. This expectation is born out in fact, although some mechanisms of depth perception and sound localization demand that both eyes and both ears are respectively operative. At the level of _gross_ anatomy, the human brain likewise shows every appearance of being a _double_ organ, and it is thus hardly suprising that, until the time of Broca (1865), the two hemispheres were usually regarded as functional duplicates of each other. Broca's discovery that a left _unilateral_ lesion could severely impair speech production (and the many later reports of cognitive deficit subsequent upon either left or right unilateral damage) dealt the duplicate model a blow from which it has never recovered. It was initially replaced, however, by an equally simple dichotomous model. The notion of complementary specialization was often taken to imply that, for many higher functions, one hemisphere and one hemisphere alone possessed the relevant underlying computational capacities, the other hemisphere being totally inert within that domain of processing.

But many scholars now regard the strict dichotomy model as too extreme, not so much simple as simplistic. It is thus often argued that cerebral lateralization for many (all?) language and speech functions is a "graded characteristic, varying in scope and completeness from individual to individual" (Zangwill, 1960). More generally, Bradshaw and Nettleton (1981) have argued that for any of the 'Analytic Left, Holistic Right' kinds of partitioning, the notion of a true dichotomy is untenable. They claim that there is a "continuum of functions" such that the hemispheres should be regarded as differing in degree rather than kind. Although the 'graded' model of cerebral lateralization is currently the most acceptable version of brain specialization for most neuropsychologists, I find myself in the somewhat unfortunate position of having little (or no) idea of what the claim implies or even means (Marshall, 1973). The reason is that we have not succeeded in specifying the _metric_ over which degrees of lateralization should be computed.

Let us indulge ourselves then in a little numerology: Imagine that, in principle, both hemispheres of the human brain can support _all_ language functions with a greater or lesser degree of proficiency. We now define _maximal proficiency_ for a single hemisphere as 100%, _minimal proficiency_ for a single hemisphere as 0%. Without the imposition of any further constraints (i.e. assuming full independence between hemispheres), this would permit the relative efficiencies of total brains (i.e. left plus right hemispheres) to vary from 0% to 200% proficiency. Now assume a 'ceiling effect' at 100%, such that any capacity in excess of 100% is 'spare'. Is this the theoretical domain of 'brain-power' for language from which the actually-occurring population of human brains is a statistically-biased sample? Within the above model we could find one interpretation of the notion that the non-dominant hemisphere for language (i.e. the hemisphere for which $Language_1\% < Language_2\%$) is a 'back-up' device to be called into service consequent upon injury to the dominant hemisphere. Within this model, it follows that the greater the degree of bilateral representation (in an individual subject) the smaller the chance of any unilateral injury resulting in (permanent) aphasic disorder (Marshall, 1981a).

If, on the other hand, we assume that there is additionally a 'final common path' which is unique to the dominant hemisphere, then unilateral injury to that hemisphere will result in impairment irrespective of the degree of language proficiency or talent (0% to 100%) of the non-dominant hemisphere. Similarly, injury to the non-dominant hemisphere will result in no impairment, again irrespective of its degree of 'intrinsic' language-capacity.

Consider now an alternative model where, again, the proficiency of each hemisphere may vary between 0% and 100%, but subject to the constraint that their sum may not exceed 100%. This is to say that the hemispheres are no longer regarded as independent devices. If we allow the hemispheres to interact (communicate), then on this proposal injury to either hemisphere will produce some impairment in all cases save for the two extremes of left = 100%, right = 0%, and vice-versa. In other words, the model excludes any 'back-up' component; recovery from unilateral aphasia does not involve drawing upon 'uncommitted' tissue from the intact hemisphere. Is this the theoretical framework which underlies the claim that lateralization is a graded characteristic?

It might be objected at this point that the entire line of reasoning is specious in that it involves the assumption that language is a unitary or 'global' function. The objection can be countered by running the argument on any subfunction which contributes to the definition of language-abilities (Marshall, 1981b). Indeed, the argument can be run on any unit of analysis whatsoever (down at least to the level of an individual nerve cell, which we presume must have a discrete, not a graded, spatial location). Considerations of the aforementioned nature do, however, remind us of yet a further qualitatively distinct interpretation of lateralization; the notion, due in the first place to Hughlings Jackson, that some language functions may be more (and differently) lateralized than others. Such a concept would allow control of speech production, for instance, to be firmly left-lateralized whilst another function (e.g. perception of emotive aspects of prosody?) could be firmly right-lateralized. Likewise, comprehension of spoken language might, as Jackson suggested, have a relatively bilateral representation in the brain. If we insist upon capturing the notion 'degree of lateralization' by a single number (Marshall, Caplan and Holmes, 1975), should that number be the weighted product of a variety of differently lateralized subfunctions of language ability? Would such a number have any real meaning or value? Would a set of numbers be more appropriate?

One might now essay the more radical criticism that any kind of numerology along the above lines has no more scientific value than, say, the study of gematria*. Unfortunately, such a wholesale condemnation would seem to deny any possibility of ever interpreting the meaning of, say, dichotic listening or split visual-field scores; it would make it difficult to interpret quantitative estimates of degree of recovery from aphasia in terms of 'back-up' or relearning mechanisms in the non-dominant hemisphere. Likewise, it would seem to preclude serious study of, say, the differential effects of right- versus left-hemispherectomy. Indeed, the absolutely basic data of the field - the behavioral consequences of left- versus right-sided lesions - could become uninterpretable.

We thus appear to be committed to some form of numerology for the lateralization of psychological functions, whether we like it or no. The only concern is which numerology can be theoretically justified (Allen, 1983). One final point: The astute reader will have noticed that I have ignored (thus far) the, (possibly related) issue of degrees of anatomical asymmetry. I have done so for the following reason: Bigger may or may not turn out to be better, and it is only in the case that bigger is better that psychologists have any reason to get excited. Should it turn out that there is any necessary (or even interesting) correlation between size and proficiency, I hereby pledge my subscription for the erection of a statue to Franz-Joseph Gall in Harvard Square.

----

* A cabbalistic method of interpreting the Hebrew Scriptures by interchanging words whose letters have the same numerical value when added. (The Shorter O.E.D.) - Ed.

Acknowledgement

I am grateful to Dr. Patrick Hudson for many discussions on the issues raised here.

REFERENCES

Allen, M. (1983). Models of hemispheric specialization. Psychological Bulletin, 93, 73-104.

Bradshaw, J.L. and Nettleton, N.C. (1981). The nature of hemispheric specialization in man. The Behavioral and Brain Sciences, 4, 51-91.

Broca, P. (1985). Sur le faculté du langage articulé. Bulletin de la Societe d'Anthropologie de Paris, 6, 493-494.

Marshall, J.C. (1973). Some problems and paradoxes associated with recent accounts of hemispheric specialization. Neuropsychologia, 11, 463-470.

Marshall, J.C. (1981a). Lateral and focal organization in the human brain. In Y. Lebrun and O. Zangwill (Eds.) Lateralization of Language in the Child. Lisse: Swets and Zeitlinger.

Marshall, J.C. (1981b). Hemispheric specialization: What, how and why. The Behavioral and Brain Sciences, 4, 72-73.

Marshall, J.C., Caplan, D. and Holmes, J.M. (1975). The measure of laterality. Neuropsychologia, 13, 315-321.

Zangwill, O. (1960). Cerebral dominance and Its Relation to Psychological Function. Edinburgh: Oliver and Boyd.

INDIVIDUAL DIFFERENCES IN THE ANATOMY OF THE CORPUS CALLOSUM: SEX, HAND PREFERENCE, SCHIZOPHRENIA AND HEMISPHERE SPECIALIZATION

Sandra F. Witelson+ and Debra L. Kigar++

+Departments of Psychiatry, Psychology, and Neurosciences
McMaster University, Hamilton, Ontario, Canada
++Department of Psychiatry, McMaster University

ABSTRACT

Recent studies of the gross anatomy of the corpus callosum show that there is marked variation in its size and shape, but also considerable consistency in these variations across very diverse studies. One study to date has reported a larger callosum, particularly in the mid and anterior regions, in mixed and left handers compared to consistent right handers. Several reports have examined possible sex differences in callosal anatomy and have produced apparently inconsistent results. The evidence clearly does not support a larger posterior splenial region in absolute size in females. However, a minority of the studies suggest that the posterior region, proportional to the size of the total callosum, may be larger in females than in males. Further clarification is needed. The early studies of callosal anatomy in schizophrenia suggested a thicker callosum in schizophrenics. Subsequent studies do not support this finding and may be confounded by variables such as chronological age, body size, brain size, and type of control group. Any anatomical differences between schizophrenic and normal individuals may involve some interaction of callosal region, sex and hand preference. These results are discussed in relation to individual differences in hemisphere specialization and brain function.

INTRODUCTION

The corpus callosum is clearly an essential structure in the integration of the functioning of the two cerebral hemispheres (e.g., Leporé, Ptito & Jasper, 1986). Neurosurgical section of the callosum (by commissurotomy or callosotomy) results in the dramatic isolation phenomenon in which an individual behaves as if he were unaware of the incoming information and perceptions derived from one side of his sensory world, and manifests two separate unintegrated streams of consciousness (Sperry, 1974). The posterior segment of the callosum appears to be particularly important for the integration of sensory information and has been termed a sensory window between the hemispheres. The anterior regions of the callosum appear to be involved in the interhemispheric integration of higher level mental processes such as the interaction between perceptual and mnemonic rather than sensory information (Sidtis, Volpe, Holtzman, Wilson & Gazzaniga, 1981).

It has also been hypothesized that the corpus callosum plays a role in the manifestation, and possibly in the maintenance, of the functional specialization of the hemispheres (Witelson, 1985b). The development of hemisphere specialization is not included in this hypothesis as there is considerable evidence that hemisphere specialization--the differential capacities that the hemispheres have in mediating cognition--does not develop but is present from birth or very soon thereafter (e.g., Witelson, 1985b; 1987).

Within this framework, the study of the anatomy of the corpus callosum is relevant to the study of the functional integration and specialization of the hemispheres. A brief history of the early conceptualizations of the functions of the corpus callosum and an overview of the types of anatomical studies are presented elsewhere (Colonnier, 1986; Witelson, 1986). Any variation in the anatomy of the corpus callosum found to be correlated with aspects of functional brain organization would have theoretical significance for understanding the brain as the substrate of cognition. It could also serve as an anatomical marker in the clinical diagnosis of syndromes such as dyslexia in which atypical hemisphere specialization may be a factor (e.g., Hynd & Cohen, 1983) or schizophrenia in which interhemispheric integration may be relevant (e.g., Gur, Skolnick, Gur, Caroff, Rieger, Obrist, Younkin & Reivich, 1983), or in genetic studies of the heritability of hemisphere specialization.

This paper will review the available studies of the gross anatomy of the human corpus callosum. In some studies, particularly the older ones, statistical analyses were not done, but the raw data were often presented in the original reports. Using these data, the present authors carried out some simple statistical analyses relevant to the issues of this paper. The picture that emerges is one of marked variation in the anatomy of the corpus callosum, but also of considerable consistency in the variation, which was observed in different laboratories, in studies done for very different purposes, decades apart. In addition, even though the data are just beginning to be accumulated, the variation appears to be correlated to some extent with variables such as sex, hand preference, and schizophrenia. Clinical and experimental neuropsychological studies have found some correlation between individual differences in hemisphere specialization and hand preference and sex (e.g., Bryden, 1982; De Vries, De Bruin, Uylings & Corner, 1984). Thus the anatomical variation may be relevant to individual differences in hemisphere specialization and cognition.

In recent studies of callosal anatomy, the variables of handedness and sex have been investigated directly in relation to cerebral dominance. Some of the earliest work considered sex as a variable, but only in relation to possible group differences between the sexes or different races. The corpus callosum has also received considerable attention in the anatomical investigation of the brains of schizophrenics as a result of the recent interest in the neuroanatomical substrate of schizophrenia and the possibility of abnormal interhemispheric functioning (e.g., Gruzelier & Flor-Henry, 1979). Thus, it seemed worthwhile to review this literature in conjunction with the work on normal brains.

ANATOMY OF THE CORPUS CALLOSUM IN NORMAL ADULTS

The earliest anatomical studies of the corpus callosum appeared at the turn of the century. Spitzka (1902; 1904; 1907) and Bean (1906) were interested in the callosum in relation to race, heredity, and intellectual ability. Bean also looked at sex as a factor. Until the 1960's very little further work appeared--only sporadic reports on the gross anatomy of the callosum and the work of Tomasch (1954), which still remains some of the only available histological study of the human callosum. Then Rakic and

Yakovlev (1968) presented their detailed study of the gross morphometric changes of the callosum over development until maturity. In addition, Yakovlev and Lecours (1967) published their report of the myelogenetic course of fibre tracts, including the corpus callosum, over the life span, and documented the fact that myelination of the corpus callosum continues until at least age 10 years. It was not until the 1980's, in the wake of the extensive interest in neuropsychological research and its implications for the functions of the corpus callosum, that attention was redirected to the anatomy of the human corpus callosum. One of the first papers concerned a case of commissurotomy. It was noted that the inconsistency of the behavioral sequelae in apparently similar cases of commissurotomy could be, in part, related to the variability of the shape of the posterior rounded end of the callosum which is difficult to view during neurosurgery and, therefore, to section completely (Greenblatt, Saunders, Culver & Bogdanowicz, 1980). Lang and Ederer (1980), noting the need for an extensive study of the size and shape of the human corpus callosum, measured various aspects of the callosum in 100 postmortem brains. In the last few years several studies have started to look in more detail at the size and shape of the callosum and its subdivisions in relation to neuropsychological and neuropsychiatric variables, not only in postmortem brains but in vivo, by means of the new technology of magnetic resonance imaging (MRI) which produces computerized images of the living brain. The studies of neuropsychiatric patients have provided information not only about clinical populations, but also about control groups of normal individuals which can be used as further normative data.

The available studies deal with measurements of the maximal anteroposterior length of the callosum, its total area, the areas and widths of various subdivisions of the callosum along its mid-longitudinal or mid-sagittal axis, and widths of coronal sections of the callosum. All the anatomical features, subdivisions and dimensions measured are shown in Figures 1 and 2. Table 1 presents a summary of the main findings of the gross anatomy of the corpus callosum in normal human adults.

Total Corpus Callosum

Maximal anteroposterior length and total area have been measured in several studies. The results indicate a wide range of values, but much consistency between studies. All studies reported mean maximal lengths between 70-80 mm (see Table 1) with the exception of the study by Rakic and Yakovlev (1968) in which the value is about 15 percent less than in other studies.

This raises the issue of the importance of considering the method of tissue fixation when comparing values from different studies. All the measurements in the Rakic and Yakovlev (1968) study are smaller than those of other reports, due to the fact that their measurements were taken from celloidin-embedded sections rather than from formalin-fixed brains. Celloidin embedding results in considerable shrinkage (approximately 20 percent) whereas formalin-fixed tissue undergoes almost no shrinkage (van Buren & Burke, 1972). Formalin fixation was used in all the other postmortem studies reporting length. Tomasch (1954), who reported only area measurements, used paraffin-embedded sections which also involve considerable shrinkage and thus his values are also smaller than those reported in other studies.

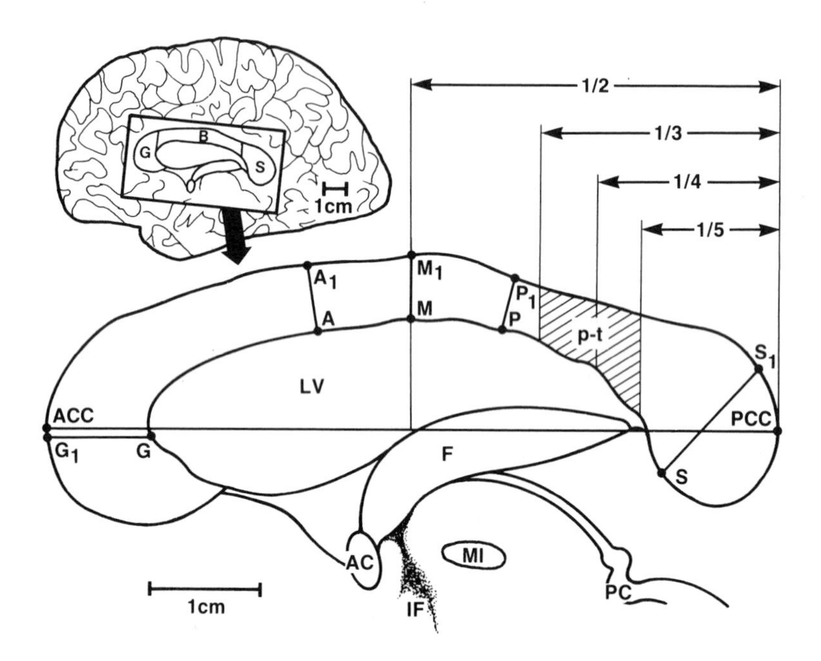

Figure 1. The corpus callosum of an adult human is shown in midsagittal view. The various measures and subdivisions referenced in Tables 1, 2, and 3 are indicated. Abbreviations: G, genu; B, body; S, splenium; ACC and PCC, anteriormost and posteriormost points of the corpus callosum, respectively, which form the line of maximal length used to obtain the callosal subdivisions (the anterior and posterior halves, and the posterior third, quartile, and fifth). The midposterior region (cross-hatched area), labelled the parietotemporal region (p-t), is defined as the posterior third minus the posterior fifth region; GG_1, maximal width of the genu; SS_1, maximal dorsoventral width of the splenium; AA_1, a width of the anterior part of the body of the callosum; MM_1, midbody width; PP_1, a width of the posterior part of the body; LV; lateral ventricle; F, fornix; IF, interventricular foramen; MI, massa intermedia; AC, anterior commissure; PC, posterior commissure.

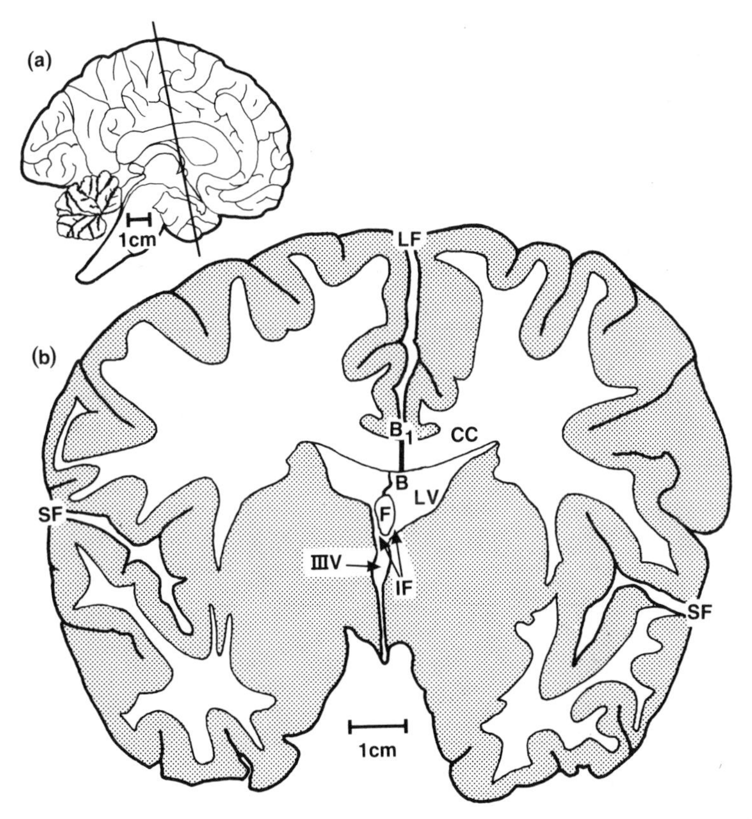

Figure 2. Coronal (frontal) section of a normal adult brain at the level of the interventricular foramen. (a) Axis of coronal section is shown. (b) Coronal section. Abbreviations: LF, dorsal aspect of the longitudinal (midsagittal) fissure; SF, Sylvian fissure; CC, corpus callosum; BB_1, midbody width of the callosum (this midbody width is taken in a plane perpendicular to those in Figure 1); LV, lateral ventricle; F, fornix; IF, interventricular foramen; IIIV, third ventricle.

Table 1. Results of Studies of the Anatomy of the Corpus Callosum in Normal Adults [a]

STUDY	MATERIAL	SAMPLE	SUBGROUPS	TOTAL CORPUS CALLOSUM(CC)		ANTERIOR HALF CC		POSTERIOR HALF CC		SPLENIUM		
				LENGTH	AREA	WIDTH	AREA	WIDTH	AREA	WIDTH[b]	AREA[c]	BULBOUS[d]
SPITZKA 1902	POST-MORTEM mid-sagittal drawings	N = 2 ESKIMO[e] $\overline{X}$ = 55 yr	1 Male(M)	80								M & F have thick splenia
			1 Female(F)	76[f]								
SPITZKA 1904	POST-MORTEM mid-sagittal drawings	N = 3 BROTHERS $\overline{X}$ = 24 yr	Male	72 in each case								
SPITZKA[g] 1907	POST-MORTEM mid-sagittal drawings	9 "SCHOLARS" $\overline{X}$ = 58 yr	Male	77.0	739							
		10 "ORDINARY" $\overline{X}$ = 33 yr		72.6*	563*							
BEAN[h] 1906	POST-MORTEM mid-sagittal drawings	67 CAUCASIAN(C) (includes 26 cases from Retzius,1900) $\overline{X}$ = 52 yr	53 Male		691		368[i]		301		173[j]	
			14 Female		634		314*		289		164	
		80 NEGRO(N) $\overline{X}$ = 44 yr	52 Male		635		321		316		186	
			28 Female		571*		286*		286*		173	
					*M;F: C>N		*Males: C>N				*Males: C<N	

STUDY	MATERIAL	SAMPLE	SUBGROUPS	TOTAL CORPUS CALLOSUM(CC)		ANTERIOR HALF CC		POSTERIOR HALF CC		SPLENIUM		
				LENGTH	AREA	WIDTH	AREA	WIDTH	AREA	WIDTH[b]	AREA[c]	BULBOUS[d]
BLINKOV & CHERNYSHEV 1936	POST-MORTEM mid-sagittal	N = 53 24-74 yr	k		753							
TOMASCH 1954	PARAFFIN-embedded sections mid-sagittal	N = 3 $\overline{X}$ = in 50s	Male		532[m]		277[n]		255[n]		149[p]	
RAKIC & YAKOVLEV 1968	CELLOIDIN-embedded sections mid-sagittal	N = 11 $\overline{X}$ = 59 yr		61.6	325.1	7.9[q] (genu)		3.5 (mid-body)		8.4		
GREENBLATT, SAUNDERS, CULVER & BOGDANOWICZ 1980	POST-MORTEM mid-sagittal	N = 10										Variable shape

Table 1 (Continued)

STUDY	MATERIAL	SAMPLE	SUBGROUPS	TOTAL CORPUS CALLOSUM(CC)		ANTERIOR HALF CC		POSTERIOR HALF CC		SPLENIUM		
				LENGTH	AREA	WIDTH	AREA	WIDTH	AREA	WIDTH[b]	AREA[c]	BULBOUS[d]
LANG & EDERER 1980	POST-MORTEM mid-sagittal	N = 100		73.7	622	11.7[q] (genu)		5.7 (mid-body)		12.0		
						6.4[r] (anterior body)		5.0[r] (posterior body)				
DE LACOSTE-UTAMSING, HOLLOWAY, KIRKPATRICK & ROSS 1981, Abstract	POST-MORTEM mid-sagittal	N = 28			Total group					Min/Max		
			15 Male		662.2					9/14		Females are more bulbous
			13 Female							14/18*	*F>M	
DE LACOSTE-UTAMSING & HOLLOWAY 1982	POST-MORTEM mid-sagittal	N = 14										
		$\bar{X}$ = 37 yr[s]	9 Male	No sex difference	704.3					11.4	186.1	Cylindrical
		$\bar{X}$ = 28 yr[s]	5 Female		708.3					16.4*	218.3	Bulbous
DEMETER, RINGO & DOTY 1985, Abstract	POST-MORTEM mid-sagittal	N = 31	19 Male							No sex difference	No sex difference	No sex difference
			12 Female									

STUDY	MATERIAL	SAMPLE	SUBGROUPS	TOTAL CORPUS CALLOSUM(CC)		ANTERIOR HALF CC		POSTERIOR HALF CC		SPLENIUM		
				LENGTH	AREA	WIDTH	AREA	WIDTH	AREA	WIDTH[b]	AREA[c]	BULBOUS[d]
WITELSON 1985a	POST-MORTEM mid-sagittal	N = 42 CAUCASIAN	Consistent Right Handers (CRH)									
		$\bar{X}$ = 48 yr	7 Male		672.1		367.4		304.6		180.7	
		$\bar{X}$ = 51 yr	20 Female		654.8		345.1		309.7		172.4	
			Mixed Handers (MH)									
		$\bar{X}$ = 49 yr	5 Male		800.6		423.2		377.4		202.4	
		$\bar{X}$ = 49 yr	10 Female		697.1		376.6		320.5		178.4	
					*CRH<MH		*CRH<MH		*CRH<MH			
WITELSON 1986	POST-MORTEM mid-sagittal	N = 42 CAUCASIAN						(Mid-body)				
		$\bar{X}$ = 50 yr	27 CRH (7M & 20F)	71.3	659.4		350.9	6.5	308.4		174.5	
		$\bar{X}$ = 49 yr	15 MH (5M & 10F)	74.3	731.6*		392.1*	6.6	339.5*		186.4	
										(Mid-posterior[t]) 62.3(CRH)		
										74.1*(MH)		

Table 1 (Continued)

STUDY	MATERIAL	SAMPLE	SUBGROUPS	TOTAL CORPUS CALLOSUM(CC)		ANTERIOR HALF CC		POSTERIOR HALF CC		SPLENIUM		BULBOUS[d]
				LENGTH	AREA	WIDTH	AREA	WIDTH	AREA	WIDTH[b]	AREA[c]	
					Min/Max		% of CC Area				% of CC Area	
CLARKE, KRAFTSIK, INNOCENTI & VAN DER LOOS 1986, Abstract	POST-MORTEM & MAGNETIC RESONANCE IMAGING (MRI)[u] combined mid-sagittal	N = 58	32 Male		480/930		No sex diff-erence			Females have greater splenial width relative to post-erior body width*	25.6	
			26 Female		490/760						27.9*	
											Absolute Area[v]	
											No sex differ-ence	
CONTROL GROUPS FROM STUDIES OF SCHIZOPHRENICS												
											% of CC Area	
NASRALLAH, ANDREASEN, OLSON, COFFMAN, COFFMAN, DUNN & EHRHARDT 1985, Abstract	MRI mid-sagittal	N = 41 NORMAL VOLUNTEERS	Right Handers(RH)									
			11 Male								33.4[p]	
			10 Female								36.3*	
			Left Handers(LH)									
			10M & 10F	(No values or comparisons given for LH groups)								

STUDY	MATERIAL	SAMPLE	SUBGROUPS	TOTAL CORPUS CALLOSUM(CC)		ANTERIOR HALF CC		POSTERIOR HALF CC		SPLENIUM		
				LENGTH	AREA	WIDTH	AREA	WIDTH	AREA	WIDTH[b]	AREA[c]	BULBOUS[d]
NASRALLAH, ANDREASEN, COFFMAN, OLSON, DUNN, EHRHARDT & CHAPMAN 1986	MRI mid-sagittal	N =41 NORMAL VOLUNTEERS $\bar{X}$ = 28 yr	RH			(Anterior body)		(Mid-body)	(Posterior body)			
			11 Male	80.0[w]	705	5.6		6.0	6.1			
			10 Female	77.6	605	4.6		5.1	5.2			
			LH									
			10 Male	79.0	660	5.6		5.2	4.5			
			10 Female	(No values or comparisons given)								

Table 1 (Continued)

Footnotes

* Indicates a statistically significant difference: either between the starred value and the one directly above it; or between the group comparisons as indicated. No asterisk indicates no statistical difference. Level of significance used is .05.

[a] All measurements are mean values in mm or mm^2. The anatomical locations are schematically represented in Figure 1.

[b] Maximal dorsoventral width of the splenium.

[c] Splenial area is defined as the posterior fifth region of the corpus callosum based on a linear subdivision as shown in Figure 1.

[d] Bulbous refers to the bulbousness of the posterior (splenium) part of the corpus callosum.

[e] Race is given only when reported.

[f] This value was measured and calculated for this report using information from Spitzka (1902), Figure 2, page 40.

[g] The statistical analyses were done by the present authors based on raw data presented in the original report in Tables A and B. Independent t-tests (two-tailed) were done for length, area and brain weight.

[h] The statistical analyses were done by the present authors based on raw data presented in the original report in Tables I, VI and VII. Cases 16 yr of age or less were excluded. Independent t-tests (two-tailed) were done for all callosal area measures presented and also for brain weight between sex and race subgroups. All significant results are indicated.

[i] N's for the anterior and posterior half areas are different than for the total corpus callosum: 38 Caucasian males, 8 Caucasian females, 54 Negro males, 25 Negro females.

[j] N's for the splenial area are different than for the other areas: 53 Caucasian males, 14 Caucasian females, 52 Negro males, 25 Negro females.

[k] Blank indicates no information available.

Table 1 (Continued)

[m] Tomasch calculated that these values were reduced by about 25% due to shrinkage during histological processing, which would make the true total area approximately 665 mm^2.

[n] The area of the anterior half was calculated for this report by adding the area of the genu plus the area of the anterior half of the body; the posterior half area was calculated by adding the splenial area plus the area of the posterior half of the body.

[p] Splenial area in this study was defined as the posterior quartile region based on a linear subdivision of the corpus callosum as shown in Figure 1.

[q] Maximal width of the genu.

[r] These values, calculated for this report using information from Lang and Ederer (1980), Figure 2, page 951, are mean values of several widths which were taken at different points either on the anterior or on the posterior body.

[s] Mean ages obtained from de Lacoste-Utamsing (June, 1983), personal communication.

[t] These are measurements of the mid-region of the posterior half of the corpus callosum (labelled the parietotemporal region) as shown in Figure 1.

[u] In contrast to postmortem material, midsagittal MRI scans may not be aligned with the true midsagittal plane. Scans also represent the maximal area over a 3-dimensional cut of the callosum.

[v] Personal communication; Clarke et al. (June, 1986).

[w] These numbers are based on a magnification of the MRI scans using a factor which gave approximately true values.

Since the values in the Rakic and Yakovlev paper are not true absolute measures they are not appropriate baseline measures, although in the past they have frequently been used as such, since for years the Rakic and Yakovlev report was the most recent extensive study. Their paper is unique in presenting information about the relative morphometric changes of the total callosum and its different subregions as it transforms in size and shape from gestation to maturity.

At this point it is useful to note the relatively small discrepancies between the mean maximal lengths obtained from formalin-fixed postmortem material, such as 73.7 mm based on 100 cases (Lang & Ederer, 1980) and 72.4 mm based on 42 cases (Witelson, 1986), and that of 78 mm obtained via MRI scans (Nasrallah, Andreasen, Coffman, Olson, Dunn, Ehrhardt & Chapman, 1986).

Midsagittal total corpus callosum area measurements vary across studies with means ranging from about 560 to 800 mm^2 (excluding the value of 325.1 mm^2 in the Rakic and Yakovlev study and the value of 532 mm^2 in the Tomasch report). Individual values range from approximately 400 to 1000 mm^2. Area measurements reflect the width and length of the callosum as well as the variation in the width and shape of the genu, body, and splenium. The greater variation in the area measures of the callosum suggests that area may be a better index of individual differences than measures of length.

Spitzka's (1902; 1904; 1907) reports were concerned with anatomical variation of the callosum in relation to heredity and intelligence. In one paper (1904), he studied the brains of three brothers and found the length of the callosum to be identical in each, in addition to similarities in other anatomical aspects of the brains. In his 1907 report, he compared a group of "scholars" and "ordinary" individuals. The conclusions Spitzka drew from these data were not based on statistical analyses. However, the inclusion of the raw data in Spitzka's papers allowed the present authors to do some analyses which tended to support some of his statements. Both callosal length and area (see Table 1) were found to be significantly larger in the scholar group ($t = 2.49$, $df = 17$, $p = .02$; $t = 3.46$, $df = 17$, $p = .003$, respectively). However, callosal size and brain weight are correlated. For example, as reported in Witelson (1985a), the correlation between callosal area and brain weight is $r = 0.51$ ($df = 40$, $p < .001$). Mean brain weights for the scholars and ordinary men were 1513 and 1443 grams, respectively, but analysis showed the difference to be nonsignificant ($t = 1.74$, $df = 17$, $p = .10$). Thus, the callosal differences between the scholars and ordinary men may not be completely accounted for by differences in brain weight. Callosal area proportional to brain weight was 49% for the scholar group and 39% for the ordinary group. Nor does chronological age and the associated decrease in brain weight appear to account for the difference in callosal anatomy as the mean age at death for the scholars was 58 years, and for the ordinary men, with a smaller callosum, 33 years.

Spitzka (1907) also looked at the length of the corpus callosum relative to the length of the hemispheres. Hemisphere length did not differ between the two groups (172 mm for each group). Thus the brains of the scholars not only had larger absolute callosa, but also larger callosa even when hemisphere length, brain weight and age were taken into account. It should be noted, however, that the group of ten ordinary men were prisoners and it is not clear what this represents in terms of the intellectual ability of the group. In addition, the cause of death was different for the groups. The scholars died from illnesses, the prisoners from electrocution. Whether this factor affects brain structure at postmortem is not clear. Since callosal size correlates with brain weight (Witelson, 1985a) and brain weight correlates with body size and height (Holloway, 1980), it is important to take into account the contribution to callosal size of these and other variables such as nutritional history.

Bean (1906) studied total callosal size as part of an extensive anatomical study of brains in different racial groups. Like Spitzka (1907), Bean drew many conclusions from his data based solely on comparisons of the relative value of mean scores without the precaution of statistical analyses. Moreover, he made strong statements regarding the correlation between anatomical and cognitive differences between the races in the absence of empirical evidence of either the existence of cognitive differences or of any correlation between anatomical and cognitive variables. Since Bean also included the raw data, some statistical analyses could be done by the present authors for those issues relevant to this chapter. Some of the anatomical differences proved to be statistically significant. The total callosal area was greater in the Caucasian than in the Negro group within each sex (males: $t = 2.63$, $df = 103$, $p = .01$; females: $t = 2.06$, $df = 40$, $p = .05$; see Table 1). However, comparison of absolute callosal size is questionable in this study since the mean brain weight of the groups differed significantly, at least for males. Using all the data available, the mean brain weight for the Caucasian and Negro male groups was 1304 and 1216 grams, respectively ($t = 3.34$, $df = 108$, $p = .001$); for the females, 1105 and 1068, respectively ($t = 0.98$, $df = 38$, $p = .33$). As indicated above, brain weight and callosal area are correlated. If callosal area is considered proportional to brain weight, the mean ratio scores are very similar in the two racial groups. For the male Caucasians and Negroes, the mean ratio scores are 53 and 52 percent, respectively; for the females, 57 and 54 percent. The callosal differences may be related to the brain weight differences, and the brain weight differences may be affected by any number of factors such as age at death, sex, height, body size, nutritional history, cause of death and brain removal and storage procedures.

Subdivisions of the Corpus Callosum

A few authors have reported measurements for subdivisions of the callosum, such as area of the anterior and posterior halves and width of different parts of the trunk or body of the callosum (see Figure 1). The two studies that have reported area measures for the anterior and posterior halves are quite consistent. The mean area for the anterior half reported for Caucasian males is 368 (Bean, 1906) and 367 mm^2 (Witelson, 1985a); and 314 and 345 mm^2 for females, respectively. The mean area for the posterior half for males is 301 and 305 mm^2, and for females 289 and 310 mm^2 in the two studies, respectively. In Bean's study, the area of the anterior half, like the results for total callosal area, was significantly greater in Caucasians than in Negroes, but this time only for males (males: $t = 3.56$, $df = 90$, $p = .0006$; females: $t = 1.16$, $df = 31$, $p = .26$.) (The issue of possible sex differences will be considered in a subsequent section.) No such difference between the races was obtained for the posterior half region (males: $t = 1.36$, $df = 90$, $p = .18$; females: $t = 17$, $df = 31$, $p = .87$; see Table 1). One simple approach to controlling for the possible contribution of different overall brain size to group differences in size of callosal subdivisions is to consider the relative size of one part to another, for example, the anterior half to the posterior half. Such ratio scores might reveal possible group differences. What, if any, functional significance such anatomical variation may have remains to be determined.

The values given in the three studies that measured the width of the callosum at the midpoint of the body or trunk (see Figure 1) were 5.7 mm for 100 postmortem brains (Lang & Ederer, 1980), 6.6 mm for 42 postmortem brains (Witelson, 1986), and 5.5 mm for 31 MRI scans of normal volunteers (Nasrallah et al., 1986). From data given in the reports of Lang and Ederer (1980) and Nasrallah et al. (1986), the maximal body width was 6.7 and 6.1 mm, respectively, and the minimal width was 4.5 mm in both studies.

Splenium

The posterior region of the callosum forms an expanded bulbous area relative to the body of the callosum and is referred to as the splenium. It has no clear anatomical landmarks to separate it from the body and, accordingly, it generally has been arbitrarily defined for measurement by geometrical definitions, such as the posterior fifth of the callosum, based on a linear division of the maximal anteroposterior length (see Figure 1). The splenium has received considerable attention, both with respect to function and anatomy (e.g., Leporé, Ptito & Jasper, 1986). The splenium has been demonstrated by various experimental techniques to house the interhemispheric fibres from the occipital prestriate cortex and from some temporal visual regions in rhesus monkey (e.g., Pandya, Karol & Heilbronn, 1971; Rockland & Pandya, 1986) and in humans (de Lacoste, Kirkpatrick & Ross, 1985).

The maximal width of the splenium (12-14 mm) appears to be very similar to that of the genu (the anterior knee-shaped region of the callosum) (Rakic & Yakovlev, 1968; Lang & Ederer, 1980; see Table 1). The splenium has been found to show considerable variation in size and shape. The variation in shape may explain some of the inconsistencies reported in the manifestation of deficits following commissurotomy (e.g., Myers, 1984). Greenblatt et al. (1980) reported a case in which interhemispheric transfer of visual and auditory sensory stimulation remained intact following splenial commissurotomy and this was attributed to inadvertent partial, rather than complete, sectioning of the splenium due to its variability in shape. In consideration of this hypothesis, Greenblatt et al. examined the shape of the splenium in ten normal brains and found that it varied from being barely bulbous and having almost no curve on the ventral border to being distinctly curved and bulbous. This anatomical variability has been described by others as well (de Lacoste-Utamsing & Holloway, 1982; Demeter, Ringo & Doty, 1985; Clarke, Kraftsik, Innocenti & van der Loos, 1986).

The absolute area of the posterior fifth region has been reported in a few studies (e.g., Bean, 1906; de Lacoste-Utamsing & Holloway, 1982; Witelson, 1985a; see Table 1) and tends to range from 160 to 220 mm^2. Several studies have looked at possible group differences. Since groups may differ in total callosal area, it is important to consider the posterior fifth region proportional to the total area. Such a ratio score helps to assess any internal callosal variation in shape without the confounding effects of different brain size, body size and height, chronological age, and fixation methods. Such analyses lead to surprisingly consistent results between most groups and studies. These results will be discussed in the following sections.

SEX DIFFERENCES AND CALLOSAL ANATOMY

Total Callosum and Sex

The size of the total callosum, measured by length and area, has been investigated for possible sex differences. Since the male brain is significantly larger than the female brain by about 10 to 15 percent (Holloway, 1980), one might expect that different regions, such as the callosum, might also be larger in males. Most researchers considering sex differences did not make explicit what difference, whether absolute or relative, might be expected.

The first study of the human corpus callosum to look at the sexes separately was likely that of Bean (1906), and this was done within large groups of Caucasian and Negro individuals. In 1982, another study looked at the sexes although this was done on small samples: 5 female and 9 male brains (de Lacoste-Utamsing & Holloway, 1982). Subsequently, several other studies have considered the possible effects of sex: Witelson (1985a) in a study on hand preference and the callosum; Nasrallah et al. (1986) in a study of schizophrenia; and a few studies specifically attempting to evaluate sex differences in callosal anatomy (Demeter, Ringo & Doty, 1985; Clarke et al., 1986). Several other studies are currently underway in different laboratories.

A few studies measured length. Although values for the male brain tend to be minimally larger, the sex difference is not significant. Nor has any sex difference been observed for total area. Again, values for the male brain are larger, but not significantly, neither in those studies that reported statistical analyses nor in those for which some statistical analyses were done in this chapter based on raw data included in the original reports.

Only in the small sample of de Lacoste-Utamsing and Holloway (1982) did the mean callosal area tend to be larger in females, but again the difference did not approach significance. However, de Lacoste-Utamsing and Holloway concluded that the total corpus callosum "was greater in females relative to brain weight" (p. 1432) on the basis that brain weight differed significantly as expected, but callosal area did not. No statistical analysis of any ratio score was done. Since brain weight and callosal area are not perfectly correlated, there is no reason to assume that the lack of a difference in callosal area in conjunction with the presence of a difference in brain weight indicates a significant difference in the relative size of the callosum.

In Witelson's (1985a) study of the callosum, which included 12 male and 30 female brains, sex differences were evaluated both for absolute area measures and for area in relation to cerebrum weight (weight of the hemispheres with the hindbrain removed) using an analysis of covariance. The absolute values of the males tended to be larger as in other studies, but not significantly so. However, even with the effect of cerebrum weight partialled out with an analysis of covariance, no significant difference in favor of females emerged. The results of Demeter et al. (1985) and Clarke et al. (1986) are still in short abstract form, but neither reported any sex difference in total callosal area.

In the above studies, the brains examined were either specified as Caucasian, or race was not indicated. Bean's study looked at sex differences within race. No sex difference in total callosal size was observed for Caucasians ($t = 1.73$, $df = 65$, $p = .09$); however, comparison of the male and female Negro groups indicated that Negro females had a significantly smaller total corpus callosum ($t = 2.75$, $df = 78$, $p = .008$). However, brain weight difference was particularly large between the Negro groups (1216 vs 1068 gm, $t = 5.06$, $df = 83$, $p < .00001$), and the callosal difference may be related to the difference in brain size. Further statistical analyses to determine the contribution of brain weight would be necessary.

Splenium and Sex

The absolute area of the posterior fifth region of the callosum has been investigated in five studies (Bean, 1906; Clarke et al., 1986; de

Lacoste-Utamsing & Holloway, 1982; Demeter et al., 1985; Witelson, 1985a). In no study was a significant sex difference found. In the Bean and Witelson studies, there was no statistical difference between the sexes in absolute area, and the male groups tended to have larger values (for Bean, Caucasians: $t = 0.85$, $df = 65$, $p = .40$; Negroes: $t = 1.61$, $df = 75$, $p = .11$; see Table 1). In the Clarke et al. and Demeter et al. studies, again no sex difference was observed in absolute size of the posterior fifth region, although mean scores were not reported in these abstracts (see Table 1). In the de Lacoste-Utamsing and Holloway (1982) study, the mean value for the female group tended to be larger. The difference was at the .08 level of probability and was reported as a significant finding. Maximal splenial width was also measured and found to be significantly greater in the female group. Demeter et al. also measured maximal splenial width but found no sex difference.

The results of the studies by de Lacoste-Utamsing and Holloway and by Witelson may superficially appear to be discrepant and difficult to reconcile. However, there is no true discrepancy. The reported sex difference by de Lacoste-Utamsing and Holloway was not a statistically significant one, and the results of Bean, Witelson, and Clarke et al. in larger samples, clearly indicated that females did not have a larger posterior fifth region compared to males. Although no significant sex difference was found in the size of the posterior fifth region in any of these studies, in only the de Lacoste-Utamsing and Holloway study did the female group tend to have a larger absolute splenial region. It is difficult to account for the unique direction of this variation. The female group of the de Lacoste-Utamsing and Holloway study is very small ($n = 5$) and one possibility is that for some unknown reason it is an atypical sample. Moreover, it is not clear how the 14 cases included in the de Lacoste-Utamsing and Holloway (1982) report were selected from the 28 cases of the earlier 1981 report (de Lacoste-Utamsing, Holloway, Kirkpatrick & Ross; see Table 1). In the 1982 report, the posterior fifth region tended to be larger in females, but so did the total corpus callosum. This is the only report in which the total corpus callosum tended to be larger in the female than the male group. Although the minimum and maximum values for splenial width in the female groups were the same in both the 1981 and 1982 de Lacoste-Utamsing et al. reports, the mean total callosal area of 662 mm^2 of the total group in the 1981 report is considerably less than that of either the male or female group (704 and 708mm^2, respectively) in the 1982 report, and the value of the earlier report is more comparable to those of other studies.

Since callosal size is correlated with brain size, although imperfectly, and brain size is different between the sexes, anatomical comparisons between the sexes become a complex issue. There are several statistical procedures that may help to remove the effect of variation in callosal size due to brain size: a proportional score of callosum to brain size or analysis of covariance, partialling out brain weight or possibly brain volume, raised to the two-thirds power, as an estimate of cortical surface. In Witelson's (1985a) report a second analysis of posterior fifth scores was done using an analysis of covariance for cerebrum weight. No sex difference was observed even with this analysis.

One further approach, which is also useful in helping to rule out differences due to different materials and methods in interstudy comparisons, is the use of a ratio score of the size of a callosal region to total callosum. As reported elsewhere (Witelson, 1985a, footnote 26), the ratio scores of the posterior fifth region to total callosal area were almost identical for all sex subgroups of the three studies except for the de Lacoste-Utamsing and Holloway (1982) female group. Table 2 presents a comparison of the mean ratio score for the posterior fifth region for all the studies which measured this region in the sexes separately.

Table 2. Absolute and relative size of the posterior fifth region of the callosum in males and females in different studies

STUDY	POSTERIOR FIFTH REGION (mm^2)					
	ABSOLUTE AREA				AREA PROPORTIONAL TO TOTAL CALLOSAL AREA	
	MALE		FEMALE		MALE	FEMALE
	N	$\bar{X}$	N	$\bar{X}$		
BEAN, 1906[a]						
Caucasian	53	173	14	164	.25	.26
Negro	52	186	25	173	.29	.30
DE LACOSTE-UTAMSING et al. 1981	15	--[b]	13	--	--	--
DE LACOSTE-UTAMSING & HOLLOWAY, 1982	9	186	5	218	.26	.31
WITELSON, 1985a						
Consistent-right-handers	7	181	20	172	.27	.26
Mixed handers	5	202	10	178	.25	.26
CLARKE et al. 1986	32	--	26	--	.26	.28

[a]Ratio scores are based on N's for which both total and posterior fifth areas were available.

[b]--No scores given.

The values in Table 2 reveal the consistently smaller absolute size of the posterior fifth region in the female group compared to the male group in all studies except the de Lacoste-Utamsing and Holloway (1982) report. In addition, the absolute values are quite consistent within each sex across studies, except for the female group of the de Lacoste-Utamsing and Holloway report.

The ratio scores also show remarkable consistency for all subgroups in all studies except for the de Lacoste-Utamsing and Holloway female group (score of .31) and Bean's Negro female group (score of .30). It should be noted that no statistical analyses were done to determine if these ratio scores are significantly higher than those for the male groups. However, the result of Bean's study may be relevant for some of the apparently discrepant sex difference findings in the different studies. It is noted that the de Lacoste-Utamsing and Holloway paper did not specify the racial origin of the groups, but did indicate that the brains were obtained from the Dallas Forensic Institute, Dallas, Texas. If callosal morphology should prove to be different in Caucasians and Negroes, one possible explanation of the discrepant results may be that the de Lacoste-Utamsing and Holloway sample of 5 female brains included a high proportion of Negro women, resulting in the proportionately larger splenial region.

On balance, the evidence does not support an absolutely larger splenial region in the female compared to the male brain. However, there may be some difference between the sexes in the morphology or relative proportions of different regions of the callosum. In Bean's study, within each racial group, females had a significantly smaller area compared to males for the anterior half of the callosum (Caucasians: $t = 2.14$, $df = 44$, $p = .04$; Negroes: $t = 2.50$, $df = 77$, $p = .02$; see Table 1). In contrast, there was no difference between the sexes for the absolute or relative posterior fifth region as indicated above, or for the posterior half in the Caucasian group ($t = 0.52$, $df = 44$, $p = .60$), but in the Negro group the posterior half was smaller in the females ($t = 2.36$, $df = 77$, $p = .02$; see Table 1). In Witelson's study, no sex difference was found for absolute area or for area corrected for brain size for the anterior half, posterior half, or for the splenial region. However, for the area of the posterior half, the interaction between the factors of sex and hand preference, with cerebrum weight as a covariate, almost reached statistical significance ($p = .08$). The analysis indicated that for the posterior half of the callosum, males had a proportionately larger region relative to brain weight than did females among mixed handers, but not among right handers. In contrast, Clarke et al. reported that the anterior half region proportional to the total callosum was not different between the sexes, although the splenial region proportional to the total callosum was. Nasrallah et al. in an MRI study of schizophrenics with normal volunteers as controls, reported that for the normal adults the area of the posterior quartile of the callosum relative to the total callosum was larger in females than males. Thus some evidence is accumulating to suggest that the posterior region in some respects may be proportionately larger in females.

HAND PREFERENCE AND CALLOSAL ANATOMY

Hand preference has been found to be correlated with direction and degree of hemisphere lateralization of function, both on the basis of clinical studies of brain-damaged patients and experimental neuropsychological studies. The correlation is not perfect. Like right handers, the majority of left handers appear to have language functions primarily mediated by the left hemisphere, although a larger proportion of left than right handers have language functions mediated primarily by the right hemisphere. Moreover, left handers, regardless of direction of hemisphere lateralization, appear to have less strong lateralization or greater bihemispheric representation of function than do right handers (Bryden, 1982). If the callosum plays some role in individual differences in brain organization, it might be expected that the greater bihemispheric representation in left handers might be associated with a larger corpus callosum. This could allow for greater interhemispheric communication, whether excitatory or inhibitory, at the physiological level.

In support of this hypothesis, Witelson (1985a) found that in a group of 42 brain specimens studied at postmortem, the area of the total callosum was 11 percent larger in a group of 15 individuals with mixed hand preference than in a group of 27 individuals with consistent right-hand preference. Handedness was classified on the basis of a 12-item hand preference test adapted from Annett (1967). Consistent right-handedness was defined as 100 percent right-hand preference; all others were classified as mixed (or non consistent-right-handed), regardless of writing hand. No individuals with consistent left-handedness, defined as 100 percent left-hand preference, were present in the sample. Such hand preference occurs in only about 4 percent of the population. (See Witelson, 1985a, for further detail.) The areas of the anterior and posterior halves were also larger in the mixed hand group. These differences were also obtained when sex and brain weight were controlled for.

Splenium and Hand Preference

The splenium (posterior fifth area) is unique in that, in contrast to the rest of the callosum, it did not differ in size between the two handedness groups. It may be noteworthy that the splenium is also unique in other respects. The more posterior regions of the callosum may have a different function than the anterior regions in interhemispheric transfer of information. They appear to transmit mainly sensory information, rather than more highly processed perceptual information (Sidtis et al., 1981). Based on experimental work with monkeys, it has been suggested that the presence of the splenium is associated with unilateral engrams, whereas the anterior commissure is associated with bilateral traces (Doty, Overman & Negrão, 1979). Consistent with these results are the findings of Macko and Mishkin (1985) based on metabolic mapping of the visual areas by the deoxyglucose method. They found functional differences in different parts of the commissural system and suggested that the callosal input to the prestriate visual cortex, which is known to course through the splenium, has suppressive rather than the facilitative electrophysiological influence that the anterior commissure appears to have.

In addition, the neural developmental course is different for the splenial region. After birth, the splenium undergoes the greatest relative growth in overall size compared to the rest of the callosum, tripling its width compared to its size at birth (Rakic & Yakovlev, 1968). It is also the first region to begin myelination (at about the fourth postnatal month), with callosal myelination spreading anteriorly from the splenium (Yakovlev & Lecours, 1967). Such anatomical differences may suggest some difference in the course of neural regressive events in the splenium compared to the rest of the callosum.

PARIETOTEMPORAL CALLOSAL REGION AND HEMISPHERE SPECIALIZATION

In a subsequent study (Witelson, 1986), further callosal subdivisions were examined. It was found that the posterior region of the body of the callosum, referred to as the parietotemporal region (see Figure 1), was markedly larger in the mixed than right-handed group (by 19 percent). This result is of particular interest for brain lateralization of function as this region has been found to house the fibres that cross from the parietotemporal regions of the two hemispheres, based on anatomical studies in monkeys (Caminiti & Sbriccoli, 1985; Cipolloni & Pandya, 1985; Pandya et al., 1971; Seltzer & Pandya, 1983) and in humans (de Lacoste et al., 1985). In humans, these cortical regions are crucial for language, praxis and visuospatial functions, functions which are typically lateralized and may be less lateralized in mixed and left handers. The larger parietotemporal region in the mixed handers may be related to greater bihemispheric representation of functions.

Hand Preference as a Dichotomy

Hand preference has been defined in different studies in different ways: for example, by the hand used specifically for writing, or the hand used for the majority of unimanual tasks. Some researchers have suggested that consistent or 100 percent right-hand preference may be different than even predominantly or moderate right-hand preference (e.g., Annett, 1972). In the callosal studies related to hand preference reviewed above, handedness was classified as a dichotomy as in Annett's (1972) model of hand preference: consistent right handers versus all others.

Callosal size was found to vary with the side of hand preference: consistent-right-preference versus non consistent-right-preference. Further analyses were done to determine whether the callosum varied with degree or magnitude of hand preference (scores could vary from +12 to -12). Since the 27 consistent right handers had almost identical scores, the correlations were calculated for only the 15 mixed handers which included members of each sex. The partial correlation (to rule out differing brain size between the sexes) for total callosal area and hand score, which reflects both degree and direction of hand preference, was $r = -0.03$ ($df = 12$, $p = .92$). The partial correlation for absolute hand score, which reflects only degree of hand preference, was $r = 0.26$ ($df = 12$, $p = .40$). Therefore, no evidence was obtained that callosal size is associated with degree of hand preference (see Witelson, 1985a; 1986). These anatomical findings support the biological validity of a simple functional dichotomy of hand preference: consistent right handedness versus mixed handers.

To ensure that the difference was not merely between consistent versus non consistent-hand-preference, MRI scans of the midsagittal view of the callosum were obtained for two normal male volunteers having consistent left-hand-preference, as no such individuals were available in the 42 cases available for study at postmortem. For both men, total callosal area measured from MRI scans appeared to be greater by about three standard deviations than the mean based on postmortem measures for the group of consistent right-handed males. Thus, consistent left handers do not appear to have as small a callosum as consistent right handers, but seem closer in size to mixed handers (Witelson, 1985a; 1986). These results further support a dichotomy between consistent right handers and all others in the classification of hand preference.

Only one other study has reported data on callosal size in right- and left-handed men. This is an MRI investigation of schizophrenics which examined normal volunteers as controls (Nasrallah et al., 1986; see Table 1). In this study the left handers tended to have a smaller total callosal area than the right handers. However, statistical analyses were not presented comparing the normal subgroups nor were the raw data given, so that no statistical analyses could be done by the present authors. No data were given to indicate whether the groups were comparable in chronological age. Total midsagittal cerebral area for each group could be calculated from the information given in their Tables 2 and 3; and the right handers showed a larger total midsagittal area which might account for any group differences in callosal size. Finally, the definition of right and left handed was not specified and right handed likely involved the typical liberal definition of right handedness, which would include some mixed handers.

Role of Experience

If hand preference varies with callosal anatomy, the question arises whether one is an antecedent factor of the other. The hypothesis could be raised that the experience of bimanual hand usage affects brain development such that a larger callosum results. Several findings argue against this suggestion. A further statistical analysis was done with respect to this issue (Witelson, 1986). The mixed handers were subdivided according to writing hand. The mixed handers who wrote with their right hand were very similar in manual preference to the consistent right handers in that both groups wrote with their right hand and the mixed group preferred their right hand for almost all tasks. The mixed handers who wrote with their left hand were much more bimanual. If the experience of bimanual practice has any effect on the callosum, it might be predicted that the left-writing mixed group, being more bimanual, would have a larger callosum than the right-writing mixed group. Analysis showed that the two mixed groups were almost identical in callosal size and both mixed groups differed from the consistent right handers. Such results suggest that callosal size is not related to the experience of bimanual hand usage.

From another field of research, recent findings in developmental neurobiology suggest that few, if any, additional callosal fibres cross the midline after birth. In fact, experimental evidence in different species suggests that within days after birth there is a period of intense loss of callosal fibres, and that this period of axonal elimination, part of the early regressive events in neural development, appears to end with the onset of callosal myelination and rapid synaptogenesis. An estimate of the period of axonal elimination in the callosum in humans is from age one to four postnatal months (Innocenti, 1986). This phenomenon of neural development suggests that the number of fibres in the callosum is set by early infancy. If the larger callosum of mixed and left handers is a reflection of more fibres and not merely thicker fibres or some other histological feature -- thicker myelin sheaths, for example -- then the anatomical difference is not likely the result of differential experience associated with different hand preference. The question then is not what biological factor results in the larger callosum of mixed and left handers, but what prevents them from not having a smaller callosum, comparable in size to right handers. As argued previously (Witelson, 1985a; 1986), there may be some mechanism, possibly genetically based, which results in less axonal elimination, a larger callosum, and the behavioral manifestation of non consistent-right-handedness.

These results, indicative of possible correlations between callosal anatomy and hand preference and sex, suggest that the individual differences in patterns of hemisphere specialization which have been found to be related to handedness and sex may have an anatomical substrate. Given that regions of the callosum show differential differences between the sexes and between hand groups, and given the recent neurobiological documentation of early neural development, these anatomical findings may have implications for the nature and origin of differences in hemisphere specialization. The finding that callosal size may vary in accordance with a dichotomous classification of hand preference suggests that such a model of hand preference may have biological validity and may be useful in other neuropsychological research. Such work illustrates the way in which neuroanatomical analysis may be a key to the elucidation of some psychological functions (Witelson, 1983).

Needless to say, the results to date are just the beginning and further research is needed. Histological analysis of the corpus callosum in different individuals in relation to psychological variables is essential. Further neuropsychological work is needed, for example, in evaluating the role of familial handedness in callosal size, and studying callosal size in relation to indices of hemisphere specialization, such as dichotic listening and electrophysiological measures, and in relation to level of performance on various cognitive tests. The issue of possible differences in fibre number or in fibre size may be of considerable functional importance. If the larger callosal size in some individuals is a reflection of more fibres, then this might result in more communication between the hemispheres and, accordingly, in better performance in some tasks requiring interhemispheric integration, such as bimanual motoric and sensory tasks, or in different cognitive profiles in some respect (see Witelson, 1986). If the increase in callosal size is a reflection of a thicker diameter of individual fibres or of a greater proportion of myelinated fibres, then such factors may result in more rapid transmission along the axons and faster interhemispheric conduction time as suggested by Green, Glass and O'Callaghan (1979).

SCHIZOPHRENIA AND CALLOSAL ANATOMY

In a series of studies done independently of those on normal adults, the callosum has been examined in the brains of schizophrenics. The initial impetus for this work was a postmortem anatomical investigation of the brains of schizophrenics by Rosenthal and Bigelow (1972), which indicated that the width of the middle of the callosum was thicker in a small group of schizophrenics than in a group of nonschizophrenic psychiatric patients. This anatomical finding, coupled with a growing body of neuropsychological findings suggesting atypical interhemispheric transfer in schizophrenics (e.g., Gruzelier & Flor-Henry, 1979), resulted in further studies. Other postmortem studies followed, involving various techniques and groups, and producing varying results. Some used postmortem coronal sections (Bigelow, Nasrallah & Rauscher, 1983; Brown, Colter, Corsellis, Crow, Frith, Jagoe, Johnstone & Marsh, 1986) and others the postmortem midsagittal surface (Machiyama, Watanabe & Machiyama, 1985; Nasrallah, McCalley-Whitters, Bigelow & Rauscher, 1983). Some of the recent studies have measured _in vivo_ midsagittal scans by means of magnetic resonance imaging (MRI) (Mathew, Partain, Prakash, Kulkarni, Logan & Wilson, 1985; Nasrallah, Andreasen, Olson, Coffman, Coffman, Dunn & Ehrhardt, 1985; Nasrallah et al., 1986). Most of this work has focussed on the width of the callosum measured at different points along its body. A few of the more recent studies also considered the variables of sex and hand preference. Table 3 presents a summary of the studies and their results.

Initially Rosenthal and Bigelow, and subsequently Bigelow et al. (1983) reported that the body of the callosum was wider in schizophrenic than in nonschizophrenic individuals (other psychiatric and neurological patients), particularly in the mid-region of the callosal body along its longitudinal axis. This difference in the callosum was the only one found in the ten measures examined in the Rosenthal and Bigelow study. In the subsequent study (Bigelow et al., 1983), only one subgroup of schizophrenics -- those having an early onset of the disease -- was found to have a wider callosum than the nonschizophrenic patients and this was in the anterior and mid-regions of the callosal body. Machiyama et al. also reported that a small group of schizophrenics had a significantly wider corpus callosum compared to a group of nonpsychiatric patients, this time only in the anterior body, although no measurements were given in this abstract.

In the Bigelow et al. report, early-onset versus late-onset schizophrenics were compared. As stated above, the early-onset group was reported to have a wider callosum in the midbody region compared to the late-onset group. However, the late-onset group had a significantly lower mean brain weight than the early-onset group and than either control group and this factor was not controlled for. In another study (Nasrallah et al., 1983), only the midbody was measured and no significant difference was found between schizophrenics of early versus late onset ($t = 1.97$, $df = 16$, $p = .07$; analysis by present authors). None of the other studies indicated this diagnostic variable in their groups of schizophrenics.

In contrast to the above studies, the remaining reports found that schizophrenics in general had either thinner callosal bodies or callosa of equal width compared to control groups. In the study by Nasrallah et al. each of three psychiatric subgroups, early- and late-onset schizophrenics

Table 3. Results of Studies of the Anatomy of the Corpus Callosum in Schizophrenic Individuals[a]

STUDY	MATERIAL	SAMPLE	SUBGROUPS	TOTAL CORPUS CALLOSUM(CC)		CALLOSUM BODY WIDTH			SPLENIUM
				LENGTH	AREA	ANTERIOR[b]	MID[c]	POSTERIOR[d]	AREA
ROSENTHAL & BIGELOW 1972	POSTMORTEM coronal sections	N = 10 SCHIZ. $\overline{X}$ = 56 yr	4 Male (M) & 6 Female (F)				6.1[e]		
		N = 10 NONSCHIZ. PSYCHIATRIC CONTROLS $\overline{X}$ = 50 yr	5M & 5F				5.2*		
BIGELOW, NASRALLAH & RAUSCHER 1983	POSTMORTEM coronal sections	N = 29 SCHIZ. $\overline{X}$ = 67 yr	21 early-onset			5.2[e]	5.1[e]	4.4[e]	
		$\overline{X}$ = 74 yr	8 late-onset			4.7	4.4*	4.2	
		N = 27 CONTROLS $\overline{X}$ = 73 yr	14 manic-depressive			4.4	4.4	4.0	
		$\overline{X}$ = 71 yr	13 neuro-logical			4.2	4.0	3.9	
						*early-onset> each control	*early-onset> each control		

Table 3 (Continued)

STUDY	MATERIAL	SAMPLE	SUBGROUPS	TOTAL CORPUS CALLOSUM(CC)		CALLOSUM BODY WIDTH			SPLENIUM
				LENGTH	AREA	ANTERIOR[b]	MID[c]	POSTERIOR[d]	AREA
NASRALLAH, McCALLEY-WHITTERS, BIGELOW & RAUSCHER 1983[f]	POSTMORTEM coronal sections	N = 18 SCHIZ.[g]							
		$\overline{X}$ = 66 yr	11 early-onset				5.1[h]		
		$\overline{X}$ = 73 yr	7 late-onset				4.3		
		N = 18 CONTROLS							
		$\overline{X}$ = 70 yr	7 manic-depressive				4.5		
		$\overline{X}$ = 64 yr	11 medical/surgical patients				9.2 *9.2 > all other groups		
NASRALLAH, ANDREASEN, OLSON, COFFMAN, DUNN & EHRHARDT 1984, Abstract	MAGNETIC RESONANCE IMAGING (MRI) mid-sagittal	N = 17 SCHIZ. 20-45 yr	Male Right Handers (RH) Left Handers (LH)		Smaller*	Thinner*	Thinner*	Thinner*	Smaller*[i]
				(No differences between total schiz. and NV groups)					
		N = 21 NORMAL VOLUNTEERS (NV)	Male RH LH		*LH: schiz. < NV				

Table 3 (Continued)

STUDY	MATERIAL	SAMPLE	SUBGROUPS	TOTAL CORPUS CALLOSUM(CC)		CALLOSUM BODY WIDTH			SPLENIUM
				LENGTH	AREA	ANTERIOR[b]	MID[c]	POSTERIOR[d]	AREA
NASRALLAH, ANDREASEN, OLSON, COFFMAN, COFFMAN, DUNN & EHRHARDT 1985, Abstract	MRI mid-sagittal	N = 38 SCHIZ.	RH						% of CC Area
			23 Male						34.4[i]
			10 Female						32.8
			LH						
			5 Male						
		N = 41 NORMAL VOLUNTEERS	RH						
			11 Male						33.4
			10 Female						36.3*
			LH 10M & 10F	(no values or comparisons given for any LH groups)					*Females: schiz. < NV
MATHEW, PARTAIN, PRAKASH, KULKARNI, LOGAN & WILSON 1985	MRI mid-sagittal	N = 18 SCHIZ. $\overline{X}$ = 38 yr	11M & 7F	33.8[j]	148				
		N = 18 NORMAL VOLUNTEERS $\overline{X}$ = 39 yr	10M & 8F	31.3*	141				

Table 3 (Continued)

STUDY	MATERIAL	SAMPLE	SUBGROUPS	TOTAL CORPUS CALLOSUM(CC)		CALLOSUM BODY WIDTH			SPLENIUM
				LENGTH	AREA	ANTERIOR[b]	MID[c]	POSTERIOR[d]	AREA
MACHIYAMA, WATANABE & MACHIYAMA 1985, Abstract	POSTMORTEM mid-sagittal	N = 5 SCHIZ. N = 7 NONPSYCHIATRIC PATIENTS				Thinner*	No difference	No difference	
						(Genu): no difference		(Splenium): no difference	
NASRALLAH, ANDREASEN, COFFMAN, OLSON, DUNN, EHRHARDT & CHAPMAN 1986	MRI mid-sagittal	N = 38 SCHIZ. $\overline{X}$ = 33 yr	Male						
			23RH	76	704	6.3	6.1	5.8	
			5LH	74	580*	4.9*	5.1*	5.5	
			Female						
			10RH	76.9	682	6.1	6.1	5.4	
		N = 41 NORMAL VOLUNTEERS $\overline{X}$ = 28 yr	Male						
			11RH	80	705	5.6	6.0	6.1	
			10LH	79	660	5.6	5.2	4.5	
			Female						
			10RH	77.6	605	4.6	5.1	5.2	
			10LH	(No values or comparisons given for this group)					
				*Male RH: schiz.<NV	*Male LH: schiz.<NV	*Female RH: schiz.>NV	*Female RH: schiz.>NV		

Table 3 (Continued)

STUDY	MATERIAL	SAMPLE	SUBGROUPS	TOTAL CORPUS CALLOSUM(CC)		CALLOSUM BODY WIDTH			SPLENIUM
				LENGTH	AREA	ANTERIOR[b]	MID[c]	POSTERIOR[d]	AREA
BROWN, COLTER, CORSELLIS, CROW, FRITH, JAGOE, JOHNSTONE & MARSH 1986	POSTMORTEM coronal sections	N = 41 SCHIZ. $\overline{X}$ = 67 yr	24 Male 17 Female				4.5[k]		
		N = 29 AFFECTIVE PSYCHOSIS $\overline{X}$ = 68 yr	12 Male 17 Female				4.7		
		N = 7 HUNTINGTON´S CHOREA $\overline{X}$ = 72 yr	6 Male 1 Female				3.9		
		N = 16 ALZHEIMER´S DISEASE $\overline{X}$ = 79 yr	9 Male 7 Female				4.5		
				(No values or comparisons given for any female groups)					

Table 3 (Continued)

Footnotes

* Indicates a statistically significant difference: either between the starred value and the one directly above it; or between the group comparisons as indicated. No asterisk indicates no statistical difference. Level of significance used is 0.05.

[a] All measurements are mean values in mm or mm^2. The anatomical locations are schematically represented in Figures 1 and 2.

[b] Anterior body width is the mean width measurement of several widths taken at different points on the anterior part of the corpus callosum body.

[c] The midbody width in these studies, in contrast to those given in Table 1, is usually not based on a single value, but is the average of several values. Different definitions of midbody width are indicated as necessary.

[d] Posterior body width is the mean width measurement of several widths taken at different points on the posterior part of the corpus callosum body.

[e] Value calculated from bilateral measures in coronal sections from two widths in each region of the body of the corpus callosum. In the Bigelow et al. (1983) study, the values are adjusted means for age at death and death autopsy interval.

[f] No statistics were reported in this paper comparing callosal width between different subgroups. The statistical analyses were done by the present authors based on the raw data reported in Table 2 of the original report. Independent t-tests (two-tailed) were done.

[g] The schizophrenic and manic-depressive subgroups were cases taken for further anatomical study from the Bigelow et al (1983) report. The medical/surgical control group was a new group.

[h] These mean values were calculated from the raw data given in Nasrallah et al (1983), Table 2. They are mean values of several anterior and posterior widths.

Table 3 (Continued)

[i] The splenium is defined as the posterior quartile region based on a linear subdivision of the corpus callosum as shown in Figure 1.

[j] Values are uncorrected measurements from MRI scans, whose magnification is approximately 0.5X true brain size.

[k] Values are means of widths taken at midline and at the midpoint of the roof of the lateral ventricle on either side of midline from a single coronal section at the level of the interventricular foramina. The means are adjusted scores for covariation of age at death and year of birth.

and a group of manic-depressives, was found to have a significantly thinner midbody width than the group of medical/surgical controls ($t = 5.9$, $df = 20$, $p < .0001$; $t = 6.0$, $df = 16$, $p < .0001$; $t = 7.0$, $df = 16$, $p < .0001$, respectively; analyses by present authors). However, the mean callosal width of 9.2 mm for the medical/surgical control group is inexplicably high compared to all other values for this measure in normal brains.

A subsequent study reported in abstract form was the first report to study a group of schizophrenics homogeneous for sex (Nasrallah et al., 1984). In this study male schizophrenics were found to be similar to normals. This work may be part of the subsequent report by Nasrallah et al. (1986) in which the total schizophrenic group was found to have a wider callosum than the normal group, but an analysis by gender indicated that this was true only for the females. Brown et al. studied a group of male schizophrenics compared to three other psychiatric and neurological male patient groups -- affective disorders, Huntington's chorea and Alzheimer's disease. The width of the callosal midbody of the schizophrenics did not differ from any of the other clinical groups. However, the brain weight of the Huntington's group was significantly less than the other groups. No normal control group was included in the report.

A possible sex difference in callosal anatomy among schizophrenics was suggested in the Nasrallah et al. (1986) report. The study included schizophrenic subgroups of each sex. The female schizophrenics were found to have a wider callosal body than normal females in the middle and anterior, but not posterior regions. However, these results may be confounded with brain size as the schizophrenic females had a larger mean cerebrum area, determined from the MRI scans, but this was not controlled for statistically. Moreover, the results may also be confounded with hand preference. The incidence of left handedness has been suggested to be relatively high in female schizophrenics (e.g., Hauser, Pollock, Finkelberg, McGrail, Voineskos & Seeman, 1985) and the definition of right handedness in this study was broad enough to include mixed handers who wrote with their right hand which may have resulted in the female schizophrenic group being a less homogeneous right-handed group.

In sum, these studies provide very inconsistent results concerning the width of the callosum in schizophrenic individuals compared to normals: in different studies the schizophrenics were found to have either larger, equal, or smaller callosa. Before considering the differences between and within these studies, the values for callosal width of the schizophrenics were compared to those available from the studies of normal individuals summarized in Table 1. The comparison revealed that the groups of schizophrenics had midbody widths of 6.1, 5.1, 4.4, 6.1, 5.1, 6.1 and 4.5 mm (see Table 3) which do not appear different from the midbody values for normal brains (see Table 1) such as 5.7 mm based on 100 cases (Lang & Ederer, 1980), 6.5 mm based on 42 cases (Witelson, 1986), and 6.0, 5.2 and 5.1 mm based on MRIs from normal volunteers (Nasrallah et al., 1986).

If there is any consistent finding in these studies, it may be that most of the control groups used in the studies of schizophrenia included other neuropsychiatric patients and these groups may have a thinner mid-callosal region than do schizophrenics or normals. Further scrutiny of these groups used as controls reveals that in almost all cases in which a group had a thin corpus callosum (taken arbitrarily as 5 mm or less based on the results for normal individuals from Table 1), the mean age of the group tended to be close to or greater than 70 years, somewhat older than the schizophrenic groups which had wider callosa, and considerably older than the normal groups listed in Table 1. Thus, chronological age, a variable which was not accounted for in most studies may be an important factor in the callosal differences reported between groups. The callosum may shrink with advancing age, and neither total brain size nor total callosum size were available in most of these studies to use as possible baseline values. One study of schizophrenics which did use younger patients and a group of normal controls well matched for age, did find callosal measures for both groups that were very comparable to those of other groups of normal individuals. In this study, a greater callosal width was found only for female schizophrenics compared to female controls (Nasrallah et al., 1986).

Only two studies of schizophrenics measured the total callosum. Mathew et al., in an MRI study, found no difference in callosal area between a group of schizophrenics, including both sexes, and a group of normal controls. Schizophrenics were found to have a greater callosal length, with no baseline difference in total brain midsagittal area. Unfortunately, callosal widths were not reported to complete the anatomical picture. Nasrallah et al. (1986) studied schizophrenic and normal individuals of comparable age to the subjects in the Mathew et al., study. They found no difference in area except for a small group of left-handed male schizophrenics who had a smaller callosum compared to matched controls, but this difference disappeared when total brain midsagittal area was taken into account. In contrast to Mathew et al., Nasrallah et al. found no difference in length between the total group of schizophrenics and normals, but did find the right-handed male schizophrenics to have a shorter callosum than matched controls.

Possible sex differences emerged. Female schizophrenics showed a greater callosal width in the anterior and mid-regions than normal females; no such difference was observed for the male schizophrenics (Nasrallah et al., 1986). In the posterior region -- specifically the posterior quartile area (see Figure 1) -- a smaller region was found in female schizophrenics compared to normals; not in males (Nasrallah et al., 1985). Right-handed male schizophrenics showed smaller values for total callosal length than matched normals; no such difference was observed for females (Nasrallah et al., 1986).

In sum, the possibility exists that the morphology of the callosum may be different in schizophrenics than in normals, but if so, it is clearly not as simple a difference as a thicker or thinner callosum. Much current research incorrectly assumes that the corpus callosum is enlarged in schizophrenics compared to normals (e.g., Schwartz, Winstead & Walker, 1984). If there is a difference between schizophrenics and normals, the difference may lie in the relative proportion of anterior to posterior regions, and this may differ between the sexes. Schizophrenic females may have thicker anterior, but proportionately smaller posterior callosal regions, compared to normal females. Male schizophrenics may be more equal to normal males. In other words, there may be a complex interaction involving region of the callosum, sex and possibly hand preference. Before any conclusions may be drawn about callosal anatomy in schizophrenia, further studies are needed which consider the many relevant variables unconfounded with each other: chronological age, body size, brain weight, sex, hand preference and possibly other disease-related variables such as age of onset of illness and symptomatology.

The recent studies implicating abnormalities in interhemispheric transfer of information in schizophrenia (e.g., Gruzelier & Flor-Henry, 1979), left hemisphere dysfunction or overactivation (e.g., Gur et al., 1983), and atypical activity in the frontal lobes (e.g., Gur et al., 1983; Weinberger, Berman & Zec, 1986), make both gross and microscopic investigations in different regions of the corpus callosum particularly interesting. Any atypical callosal anatomy documented in schizophrenia may be part of a neuroanatomical substrate of the disease. Such work is of particular interest for the etiology of schizophrenia in that given the current knowledge of neuroanatomical development, anatomical variation that may be found in schizophrenics may help point to the mechanisms and time course of the neurobiological aspects of the disorder.

ACKNOWLEDGEMENTS

Preparation of this paper was supported in part by U.S. NIH-NINCDS Contract NO1-NS-6-2344 and NINCDS Grant RO1-NS-18954 awarded to S.F.W.

The authors thank Toni Newman for library searches, Cheryl McCormick for helping with the statistical analyses of the raw data reported in other papers, Diane Clews for expert typing of the manuscript and tables, and Janice Swallow for invaluable editorial assistance.

REFERENCES

Annett, M. (1967). The binomial distribution of right, mixed and left handedness. Quarterly Journal of Experimental Psychology, 19, 327-333.

Annett, M. (1972). The distribution of manual asymmetry. British Journal of Psychology, 63, 343-358.

Bean, R.B. (1906). Some racial peculiarities of the Negro brain. American Journal of Anatomy, 5, 353-432.

Bigelow, L.B., Nasrallah, H.A., & Rauscher, F.P. (1983). Corpus callosum thickness in chronic schizophrenia. British Journal of Psychiatry, 142, 284-287.

Blinkov, S.M., & Chernyshev, A.S. (1936). Variations in the Human Corpus Callosum: Collection in Honor of P. I. Émdin. Rostov-on-Don.

Brown, R., Colter, N., Corsellis, N., Crow, T.J., Frith, C.D., Jagoe, R., Johnstone, E.C., & Marsh, L. (1968). Postmortem evidence of structural brain changes in schizophrenia. Archives of General Psychiatry, 43, 36-42.

Bryden, M.P. (1982). Laterality: Functional Asymmetry in the Intact Brain. Toronto: Academic Press.

Caminiti, R., & Sbriccoli, A. (1985). The callosal system of the superior parietal lobule in the monkey. Journal of Comparative Neurology, 237, 85-99.

Cipolloni, P.B., & Pandya, D.D. (1985). Topography and trajectories of commissural fibers of the superior temporal region in the rhesus monkey. Brain Research, 57, 381-389.

Clarke, S., Kraftsik, R., Innocenti, G.M., & van der Loos, H. (1986). Sexual dimorphism and development of the human corpus callosum. Neuroscience Letters, 26, S299, Abstract.

Colonnier, M. (1986). Notes on the early history of the corpus callosum with an introduction to the morphological papers published in this Festschrift. In F. Leporé, M. Ptito & H.H. Jasper (Eds.) Two Hemispheres - One Brain: Functions of the Corpus Callosum. Neurology and Neurobiology, 17, 37-45.

de Lacoste, M.C., Kirkpatrick, J.B., & Ross, E.D. (1985). Topography of the human corpus callosum. Journal of Neuropathology and Experimental Neurology, 44, 578-591.

de Lacoste-Utamsing, C., & Holloway, R.L. (1982). Sexual dimorphism in the human corpus callosum. Science, 216, 1431-1432.

de Lacoste-Utamsing, C., Holloway, R.L., Kirkpatrick, J.B., & Ross, E.D. (1981). Anatomical and quantitative aspects of the human corpus callosum. Society for Neuroscience, 7, Abstract No. 127.5.

Demeter, S., Ringo, J., & Doty, R.W. (1985). Sexual dimorphism in the human corpus callosum? Society for Neuroscience, 11, Abstract No. 254.12.

De Vries, G.J., De Bruin, J.D.C., Uylings, H.B.M., & Corner, M.A. (Eds.) (1984). Sex Differences in the Brain: The Relation between Structure

and Function. Progress in Brain Research, 61, 491-508.
Doty, R.W., Overman W.H., & Negrão, N. (1979). Role of forebrain commissures in hemispheric specialization and memory in macaques. In I.S. Russell, M.W. Van Hof, & G. Berlucchi (Eds.) Structure and Function of Cerebral Commissures. Baltimore: University Park Press, 333-342.
Green, P., Glass, A., & O'Callaghan, M.A.J. (1979). Some implications of abnormal hemisphere interaction in schizophrenia. In J. Gruzelier, & P. Flor-Henry (Eds.) Hemisphere Asymmetries of Function in Psychopathology. New York: Elsevier/North-Holland Biomedical Press, 431-448.
Greenblatt, S.H., Saunders, R.L., Culver, C.M., & Bogdanowicz, W. (1980). Normal interhemispheric visual transfer with incomplete section of the splenium. Archives of Neurology, 37, 567-571.
Gruzelier, J.H., & Flor-Henry, P. (Eds.) (1979). Hemisphere Asymmetry of Function in Psychopathology. New York: Elsevier/North-Holland Biomedical Press.
Gur, R.E., Skolnick, B.E., Gur, R.C., Caroff, S., Rieger, W., Obrist, W.D., Younkin, D., & Reivich, M. (1983). Brain function in psychiatric disorders. I. Regional cerebral blood flow in medicated schizophrenics. Archives of General Psychiatry, 40, 1250-1254.
Hauser, P., Pollock, B., Finkelberg, F., McGrail, S., Voineskos, G., & Seeman, M. (1985). On sinistrality and sex differences in schizophrenia. American Journal of Psychiatry, 142, 1228, Letter.
Holloway, R.L. (1980). Within-species brain-body weight variability: A reexamination of the Danish data and other primate species. American Journal of Physical Anthropology, 53, 109-121.
Hynd, G.W., & Cohen, M. (1983). Dyslexia: Neuropsychological Theory, Research, and Clinical Differentiation. Toronto: Academic Press.
Innocenti, G.M. (1986). The general organization of callosal connections. In E.G. Jones and A. Peters (Eds.) Cerebral Cortex, Volume 5. Sensory-motor areas and aspects of cortical connectivity. New York: Plenum Press.
Lang, J., & Ederer, M. (1980). Uber form und groBe des corpus callosum und das septum pellucidum. Gegenbaurs morph. Jahrb., Leipzig, 126, 949-958.
Leporé, F., Ptito, M., & Jasper, H.H. (Eds.). (1986). Two Hemispheres-One Brain: Functions of the Corpus Callosum. Neurology and Neurobiology, 17.
Machiyama, Y., Watanabe, Y., & Machiyama, R. (1985). Neuroanatomical abnormalities in the corpus callosum in schizophrenia. IVth World Congress of Biological Psychiatry, Abstract No. 424.7.
Macko, K.A., & Mishkin, M. (1985). Metabolic mapping of higher-order visual areas in the monkey. In L. Sokoloff (Ed.) Brain Imaging and Brain Function, New York: Raven Press, 73-86.
Mathew, R.J., Partain, C.L., Prakash, R., Kulkarni, M.V., Logan, T.P., & Wilson, W.H. (1985). A study of the septum pellucidum and corpus callosum in schizophrenia with MR imaging. Acta Psychiatrica Scandinavica, 72, 414-421.
Myers, J.J. (1984). Right hemisphere language: Science or fiction? American Psychologist, 39, 315-320.
Nasrallah, H.A., Andreasen, N.C., Coffman, J.A., Olson, S.C., Dunn, V.D., Ehrhardt, C., & Chapman, S.M. (1986). A controlled magnetic resonance imaging study of corpus callosum thickness in schizophrenia. Biological Psychiatry, 21, 272-282.
Nasrallah, H.A., Andreasen, N.C., Olson, S.C., Coffman, J.A., Coffman, C.E., Dunn, V.D., & Ehrhardt, J.C. (1985). Absence of sexual dimorphism of the corpus callosum in schizophrenia: A magnetic resonance imaging study. Society of Neuroscience, 11, Abstract No. 382.8.
Nasrallah, H.A., Andreasen, N.C., Olson, S.C., Coffman, J.A., Dunn, V.D., & Ehrhardt, J.C. (1984). A controlled magnetic resonance study of the corpus callosum in schizophrenia. American College of Neuropsychopharmacology, Abstract. p. 107.

Nasrallah, H.A., McCalley-Whitters, M., Bigelow, L.B., & Rauscher, F.P. (1983). A histological study of the corpus callosum in chronic schizophrenia. Psychiatry Research, 8, 251-260.

Pandya, D.N., Karol, E.A., & Heilbronn, D. (1971). The topographical distribution of interhemispheric projections in the corpus callosum of the rhesus monkey. Brain Research, 32, 31-43.

Rakic, P., & Yakovlev, P.I. (1986). Development of the corpus callosum and cavum septi in man. Journal of Comparative Neurology, 132, 45-72.

Retzius, G. (1900). Ueber das hirngewicht der schweden. Biologische Untersuchungen, 9, 51-68.

Rockland, K.S., & Pandya, D.N. (1986). Topography of occipital lobe commissural connections in the rhesus monkey. Brain Research, 365, 174-178.

Rosenthal, R., & Bigelow, L.B. (1972). Quantitative brain measurements in chronic schizophrenia. British Journal of Psychiatry, 121, 259-64.

Schwartz, B.D., Winstead, D.K., & Walker, W.G. (1984). A corpus callosal deficit in sequential analysis by schizophrenics. Biological Psychiatry, 19, 1667-1676.

Seltzer, B., & Pandya, D.N. (1983). The distribution of posterior parietal fibers in the corpus callosum of the rhesus monkey. Experimental Brain Research, 49, 147-150.

Sidtis, J.J., Volpe, B.T., Holtzman, J.E., Wilson, D.H., & Gazzaniga, M.S. (1981). Cognitive interaction after staged callosal section: Evidence for transfer of semantic activation. Science, 212, 344-346.

Sperry, R.W. (1974). Lateral specialization in the surgically separated hemispheres. In F.O. Schmitt, & F.G. Worden (Eds.) The Neurosciences: Third Study Program. Cambridge, Mass: MIT Press, 5-19.

Spitzka, E.A. (1902). Contributions to the encephalic anatomy of the races. American Journal of Anatomy, 2, 25-71.

Spitzka, E.A. (1904). Hereditary resemblances in the brains of three brothers. American Anthropologist, 6, 307-312.

Spitzka, E.A. (1907). A study of the brains of six eminent scientists and scholars belonging to the American Anthropometric Society, together with a description of the skull of Professor E.D. Cope. Transactions of the American Philosophical Society, 21, 175-308.

Tomasch, J. (1954). Size, distribution, and number of fibers in the human corpus callosum. Anatomical Record, 119, 119-135.

Van Buren, J.M., & Burke, R.C. (1972). Variations and Connections of the Human Thalamus. Vol. 1: The Nuclei and Cerebral Connections of the Human Thalamus. New York: Springer-Verlag.

Weinberger, D.R., Berman, K.F., & Zec, R.F. (1986). Physiologic dysfunction of dorsolateral prefrontal cortex in schizophrenia. I. Regional cerebral blood flow evidence. Archives of General Psychiatry, 43, 114-124.

Witelson, S.F. (1983). Bumps on the brain: Right-left anatomic asymmetry as a key to functional asymmetry. In S. Segalowitz (Ed.) Language Functions and Brain Organization. New York: Academic Press, 117-143.

Witelson, S.F. (1985a). The brain connection: The corpus callosum is larger in left handers. Science, 229, 665-668.

Witelson, S.F. (1985b). On hemisphere specialization and cerebral plasticity from birth: Mark II. In C. Best (Ed.) Hemispheric Function and Collaboration in the Child. New York: Academic Press, 33-85.

Witelson, S.F. (1986). Wires of the mind: Anatomical variation in the corpus callosum in relation to hemispheric specialization and integration In F. Leporé, M. Ptito, & H.H. Jasper (Eds.) Two Hemispheres - One Brain: Functions of the Corpus Callosum. Neurology and Neurobiology, 17, 117-137.

Witelson, S.F. (1987). Neurobiological aspects of language in children. Child Development, 58. 653-688.

Yakovlev, P.I., & Lecours, A-R. (1967). The myelogenetic cycles of regional maturation of the brain. In A. Minkowski (Ed.) Regional Development of the Brain in Early Life. London: Blackwell Scientific, 3-65.

HEMISPHERIC SPECIALIZATION AND REGIONAL CEREBRAL BLOOD FLOW

Ruben C. Gur and Raquel E. Gur *

Brain Behavior Laboratory, **
Departments of Psychiatry and Neurology
University of Pennsylvania

ABSTRACT

Many studies have demonstrated that the two hemispheres are specialized, in the majority of individuals, so that the left hemisphere regulates verbal-analytic cognitive abilities while the right hemisphere predominates in spatial-synthesis functions. The studies also suggested individual differences in the direction and degree of this hemispheric specialization. Thus, not all individuals have the same organization of cognitive processing in the two hemispheres. Our research program has examined these variations both in brain damaged patients and in normals. We have used the tachistoscopic and dichotic listening techniques, as well as paper-and-pencil tests, in the study of left-handed and right-handed males and females. These studies have helped identify reliable behavioural measures of hemispheric functioning. We also made initial steps toward applying the new techniques for measuring regional brain activity to the study of human cognitive functioning. Our studies with the 133-Xenon inhalation technique for measuring regional cerebral blood flow (rCBF) showed lateralized changes in rCBF for verbal and spatial tasks, and the effects were influenced by handedness and sex. We also performed a study of local cerebral glucose metabolism using Positron Emission Tomography (PET), which also found lateralized changes for the standardized cognitive tasks. A second line of research examined hemispheric specialization for emotional processing. Clinical case reports and tachistoscopic studies suggested right hemispheric superiority for emotional processing. Some evidence also suggested hemispheric asymmetry in emotional valence, with the left hemisphere showing a "bias" toward positive affect. These factors are yet to be examined with rCBF.

A. INTRODUCTION

Most of our knowledge on how behaviour is regulated by the brain in humans has come from observations on the effects of brain lesions and brain surgery and from studies of normal subjects using tachistoscopic and dichotic listening techniques (see Harnad et al., 1977, for reviews).

----

* Supported by NIMH Grant MH 30456, NIH Grant NS 19039, and The Spencer Foundation.

** Correspondence to: 205 Piersol Building/G1, Hospital of the University of Pennsylvania, Philadelphia, Pennsylvania, 19104

A congruent finding from this research is that in the majority of humans, the left and right cerebral hemispheres are specialized, respectively, for verbal-analytic and spatial-holistic functions. However, there are substantial individual differences in the direction and degree of hemispheric specialization, and these differences have been linked to variability in cognitive functioning (see Herron, 1978, for reviews).

Recent developments in isotopic clearance techniques enable the measurement of regional brain activity as it is reflected in regional cerebral blood flow and metabolism. These techniques make it possible to apply experimental manipulations of behaviour and examine their effect on regional brain activity (see Gur, 1983, for review). Potentially, this will vastly enhance the rate of acquisition of knowledge on brain behaviour relationships and will add experimental rigor and direct measurement of variables pertinent to functional brain mapping. During the past decade, our laboratory and a number of other laboratories here and abroad have demonstrated the sensitivity of isotopic clearance techniques to changes in regional brain activity induced by cognitive activity. Thus, rCBF was found to increase during cognitive activity, and to show greater increase to the left hemisphere for verbal tasks and to the right hemisphere for spatial tasks. Furthermore, variability in the direction and degree of these changes was associated with handedness and sex.

The missing link in this research is a systematic evaluation of individual differences in behaviour in relation to individual differences in the pattern of rCBF changes during cognitive activity. This is the objective of our ongoing research. In this chapter we will present briefly our procedures for obtaining measurements of rCBF and then summarize some of our findings to date in the area of hemispheric specialization for cognitive and emotional factors.

B. THE RESEARCH PARADIGM

1. The rCBF measurement

Upon arrival in the rCBF laboratory, subjects are first accomodated and acclimated to the laboratory environment. During this phase, subjects are shown the cerebrograph, placed on the bed, and administered a "dry" rCBF procedure. This is helpful in reducing anxiety and improving subjects' comfort. Effort is spent in creating a friendly atmosphere. This aspect of the procedure, by its nature, cannot be entirely standardized. The personnel of the rCBF laboratory have been trained in fostering open exchange of information with subjects while maintaining a professional attitude.

Following the accomodation period, subjects are explained the tasks and trained to use the response lever. The response lever is connected to a point light projector with which the subject can indicate his response to the stimuli. The rCBF is currently measured with a Novo Cerebrograph, a 32 detectors system which uses the 133-Xenon technique developed by Obrist and colleagues (Obrist, Thompson, Wang and others, 1975; Obrist and Wilkinson, 1979; Risberg et al., 1975). The technique provides a means of determining regional cerebral blood flow by the use of a trace amount of 133-Xenon in air (for inhalation) or in saline solution (for venous injection). A sample of venous blood is removed to obtain hemoglobin level. For the inhalation procedure, 5-7 mCi of 133-Xenon per liter is inhaled through a mouthpiece for one minute. The uptake and clearance of the isotope from the brain is monitored for 14 minutes by collimated NaI crystal detectors placed over the scalp. Cerebral blood flow is computed from the clearance rates as described by Obrist et al. (1975). All rCBF studies are evaluated for integrity of the blood flow measurements, which includes absence of gross artifact in the recorded curves, good count rates and curve fits, and adequate estimation of end-tidal CO_2. Less than 5 per cent of the flow curves are typically rejected.

Our standard procedure is to examine in detail two flow indices: IS (Obrist's initial slope - a noncompartmental index of grey matter flow) and CBF-15 (a noncompartmental index of mean flow of grey and white matter). IS is the initial slope (at time zero) of the mathematically equivalent instantaneous bolus injection. CBF-15 represents the mean flow of all tissues seen by the detector, including a small extracerebral (scalp) component. It is mathematically equivalent to the height-over-area method where the integration is carried out to 15 minutes. The latter time is preferred over infinity, since it reduces the contribution of slow extracerebral components (see Obrist and Wilkinson, 1979). Finally, a useful index is the relative size of the fast compartment, W1, which in normals predominately represents the percentage of the grey matter compartment (see Obrist et al., 1975; Obrist and Wilkinson, 1979). The CBF values are corrected for CO_2, based on correction factors obtained in normal control volunteers (3.0 percent/mmHg of change in PCO_2 for IS and 3.5 percent/mmHg of change in PCO_2 for CBF-15).

All CBF data and the computed blood flow data are stored by the computer on disks. The raw clearance curves can be displayed on the Televideo terminal. The final results are displayed in a table as well as on a topographical outline of the brain (See Figure 1).

Alternative methods are available for isotope delivery and monitoring of isotope removal from the lungs. One method of particular interest to studies involving psychological manipulations, particularly when it is desired that the subjects talk during the study, is the monitoring of isotope concentration from a lung curve. Lung monitoring enables evaluation of end-tidal expired air without having to use a mouthpiece or facemask (Jaggi and Obrist, 1983).

2. The standard cognitive activation procedures during rCBF measurement

Each task begins 5 min prior to isotope administration and continues

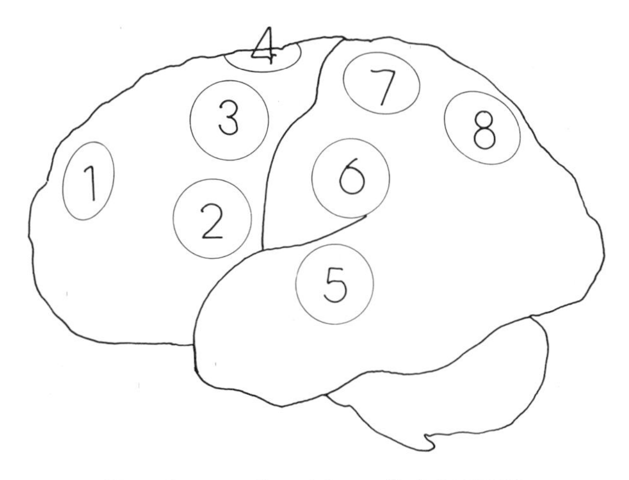

Fig. 1. Location of detectors.

for 20 min. There is a 10 min rest between conditions. This period permits return of background radiation to baseline. The tasks are:

a. Verbal. The verbal task consists of analogies adapted from the Miller Analogies Test, the Scholastic Aptitude Tests, and the Educational Testing Service's Kit of Factor Referenced Cognitive Tests. The subject is presented with an analogy (e.g. Bird relates to Eagle as Car relates to a. Engine; b. Wheel; c. Cadillac; d. Gasoline), and his task is to point to the letter corresponding to the correct answer. There are 5 practice trials before presentation of test trials. Subjects proceed at their own pace and the response activates presentation of the next trial.

b. Spatial. The spatial task consists of an adaptation of Benton's Line Orientation Test. The subject is presented with two lines and his task is to point to the digit adjacent to the lines on an array corresponding in orientation to the stimulus lines. The procedure is otherwise identical to Task a.

C. HEMISPHERIC SPECIALIZATION FOR COGNITIVE FUNCTIONS

The effects of cognitive tasks on regional brain activity have been measured by the 133-Xenon technique for measuring rCBF and by positron emission tomography (PET) for measuring glucose metabolism. The effects of handedness and sex on hemispheric activation has been specifically examined.

Gur and Reivich (1980) reported the effects of cognitive activity on rCBF in a sample of 36 right-handed undergraduate males. They found reliable effects of increased rCBF during cognitive activity. The verbal analogies task produced greater left hemispheric increase. The spatial task in that study was the Gestalt Completion Test. It produced no lateralized increase. However, the 17 subjects who showed greater right hemispheric increase performed better.

Gur et al. (1982) reported the effects of handedness and sex on hemispheric rCBF changes during cognitive activity. The sample consisted of 62 young, healthy undergraduate volunteers (15 right-handed males, 15 right-handed females, 15 left-handed males, 17 left-handed females). The main findings were: 1. The rCBF increased during cognitive activity compared to resting baseline. 2. The increase was higher to the left hemisphere for the verbal (analogies) task, and higher to the right hemisphere for the spatial (line-orientation) task. 3. The laterality effects were moderated by handedness; right-handers showed the effects while left-handers, as a group, did not. 4. Females had higher flows, both during rest and during cognitive activity. 5. Females showed more lateralized changes during the performance of the verbal and spatial tasks. 6. Females had a higher percent of fast-clearing tissue (wl) presumably grey-matter. 6. Left-handers have a higher wl than right-handers (Figures 2 and 3).

The second study (Gur, Gur, Rosen et al., 1983) reported the effects of the same verbal analogies and spatial line orientation tasks on local cerebral glucose metabolism measured with PET. The study demonstrated lateralized differences in areas corresponding to Wernicke's region for the verbal task, and in homotopic right hemispheric region for the spatial task. In addition, lateralized effects occurred in the area of the frontal eye fields, which control orientation. This finding was interpreted as supporting Trevarthen's (1972) hypothesized network linking cognitive activity to motor orientation (Figure 4).

D. HEMISPHERIC SPECIALIZATION FOR EMOTIONAL PROCESSING

This work has progressed from the assessment of lateral differences in

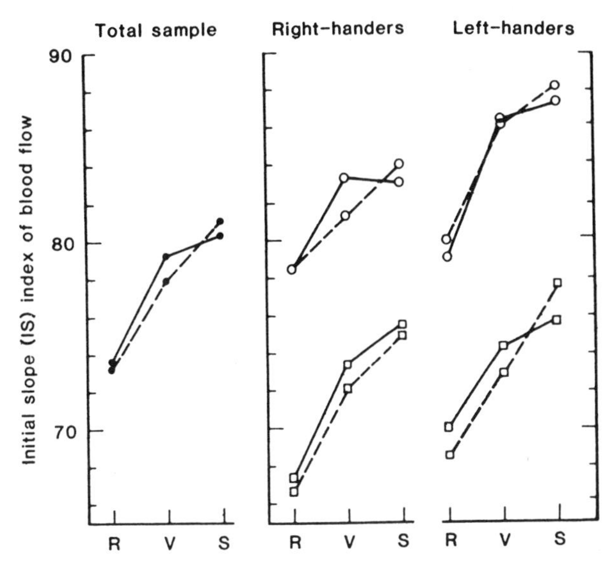

Fig. 2. Initial slope (IS) index of blood flow to the left (—) and right (----) hemispheres for the total sample and for right- and left-handed females (○) and right- and left-handed males (□) during resting (R), verbal (V), and spatial task performances (S).

the intensity of emotional expression (Sackeim, Gur and Saucy, 1978; Sackeim and Gur, 1978) to literature review of clinical case reports (Sackeim et al., 1982) and experimental studies involving the tachistoscopic technique (Natale, Gur and Gur, 1983). We are currently investigating the effects of emotional processing on rCBF.

Sackeim, Gur and Saucy (1978) obtained ratings of emotional intensity for faces expressing six emotions. The faces were presented in the original and in composites consisting of the left or the right side of the face. The left-left composites were judged more intense and hence it was concluded that emotions are expressed more intensely on the left side of the face. Sackeim and Gur (1978) examined the specificity of this effect and found that the significant difference in this direction was for the "negative" emotions of sadness, anger, disgust and fear, whereas for the "positive" emotions of happiness and surprise the trend was in the opposite direction. Sackeim et al. (1982) examined all published reports of:- 1. Outbursts of pathological crying or laughing following destructive lesions; 2. Laughing or crying occurring during epileptic seizures; 3. Mood changes following surgical removal of a cerebral hemisphere. The main findings were that crying outbursts occur with destructive lesions to the right hemisphere, while laughing outbursts occur following right hemispheric lesions. Correspondingly, left hemispheric activation during epileptic seizures was associated with ictal laughter, whereas four or the five reported cases of ictal crying had right hemispheric foci. Finally, right hemispherectomy resulted in postoperative increase in euphoric mood.

Natale, Gur and Gur (1983) performed three experiments using the tachistoscopic technique with emotional stimuli. Experiment 1 found a right visual field superiority for discriminating the emotional valence of faces expressing a range of emotions. This effect was moderated by handedness and writing posture; it was weaker for left-handers as a group and did not exist

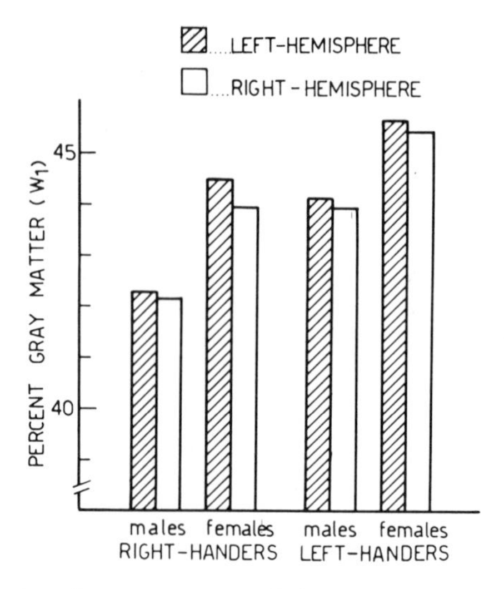

Fig. 3. Percentage of grey matter (W) in right- and left-handed males and females.

for left-handers who use the noninverted writing posture. Experiment 2 replicated the left visual field superiority effect for longer exposure durations and specifically for discrimination of happy from sad faces. In Experiment 3 chimeric faces composed of happy and sad expressions were presented. The subject was required to decide whether the mood expressed was positive or negative. Right visual field presentations produced a bias toward positive judgments, whereas left visual field presentations produced no bias (Figures 5 and 6). The paradigm of studying activation effects on rCBF has not yet been applied to investigation of emotion.

E. CONCLUSIONS

The introduction of neuroimaging techniques to the study of brain behavior relation provides new opportunities for a systematic research on individual differences in cerebral organization. The research paradigm presented here can yield data on the relationship between direct *in vivo* measures of regional brain activity and major behavioral factors related to cognition and affect. At present we are still in the very initial phase of this research. We have determined the sensitivity and reliability of the rCBF measures obtained with the 133-Xenon inhalation technique and found that they show consistent effects of activation with cognitive tasks. These initial studies are encouraging and open the way for expanding the range of factors to be examined. Such factors include emotion. The sensitivity of the technique to individual differences in brain organization such as are associated with sex and handedness suggests that the technique could be used in further examination of individual differences. We are currently studying "cognitive specialists", individuals with exceptional abilities in areas such as mathematics and art.

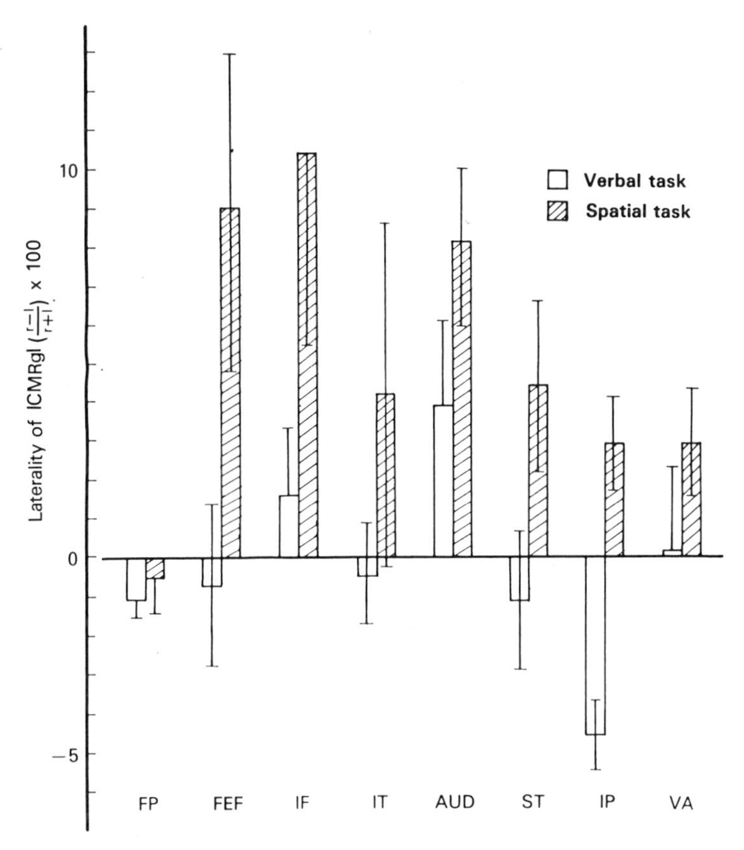

Fig. 4. Laterality scores of ICNRgl for the seven ROIs in the two groups of subjects (Verbal and Spatial). Error bars reflect standard error of the mean and are drawn in the direction allowing the necessary between-group comparisons.

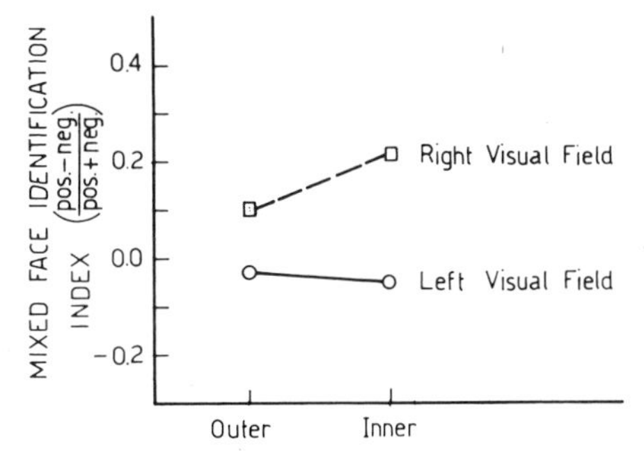

Fig. 5. Ratings of affect for the three groups of subjects and the six emotions presented in the left (L) and right (R) visual fields in Experiment 1.

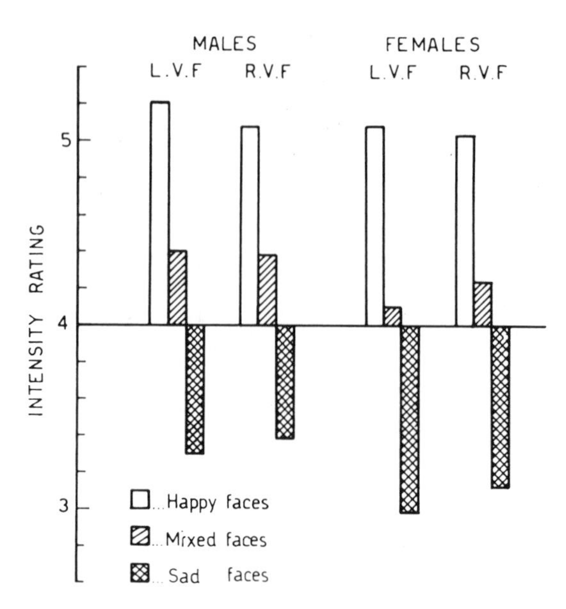

Fig. 6. Intensity ratings of chimeric emotional faces for males and females in the two visual fields (Experiment II).

REFERENCES

Gur, R.C. (1983). Measurement and imaging of regional brain function: Implications for neuropsychiatry. In P. Flor-Henry, and J. Gruzelier, Laterality and Psychopathology. Amsterdam: Elsevier.

Gur, R.C., Gur, R.E., Obrist, W.D., and others. (1982). Sex and handedness differences in cerebral blood flow during rest and cognitive activity. Science, 217, 659-661.

Gur, R.C., Gur, R.E., Rosen, A.D., and others. (1983). A cognitive motor network demonstrated by positive emission tomography. Neuropsychologia, 21, 601-606.

Gur, R.C. and Reivich, M. Cognitive task effects on hemispheric blood flow in humans. Brain and Language, 9, 78-93, 1980.

Harnad, S. and others. 1976. (Eds.) Lateralization in the nervous system. New York: Academic Press.

Herron, J. 1979. (Ed.) The Neuropsychology of Left-handedness. New York: Academic Press.

Jaggi, J.L. and Obrist, W.D. (1983). External monitoring of the lung as a substitute for end-tidal Xenon-133 sampling in noninvasive CBF studies. Journal of Cerebral Blood Flow and Metabolism, 3 (supplement).

Natale, M., Gur, R.E., and Gur, R.C. (1983). Hemispheric asymmetries in processing emotional expressions. Neuropsychologia, 21, 555-565.

Obrist, W.D., Thompson, H.K., Wang, H.S., and others. (1975). Regional cerebral blood flow estimated by 133 Xenon inhalation. Stroke, 6, 245-256.

Obrist, W.D. and Wilkinson, W.E. (1979). The non-invasive Xe-133 method: evaluation of CBF indices. In A. Bes and G. Geraud (Eds.) Cerebral Circulation. Amsterdam: Elsevier.

Risberg, J., Ali, Z., Wilson, E.M., et al. (1975) Regional cerebral blood flow by 133 Xenon inhalation - preliminary evaluation of an initial

slope index in patients with unstable flow compartments. Stroke, 6, 142-148.
Rubin, D.A. and Rubin, R.T. (1980). Differences in asymmetry of facial expression between left and right handed children. Neuropsychologia, 18, 373-377.
Sackeim, H.A., Greenberg, M.S., Weiman, A.L., and others. (1982). Hemispheric asymmetry in the expression of positive and negative emotions: neurological evidence. Archives of Neurology, 39, 210-218.
Sackeim, H.A., Gur, R.C., and Saucy, M.C. (1978). Emotions are expressed more intensely on the left side of the face. Science, 202, 434-436.
Sackeim, H.A., and Gur, R.C. (1978). Lateral asymmetry in intensity of emotional expression. Neuropsychologia, 16, 473-481.
Trevarthen, C. (1972). Brain bisymmetry and the role of the corpus callosum in behavior and conscious experience. In J. Cernacvec and F. Podivinsky (Eds.) Cerebral Hemispheric Relations. Bratislava: The Publishing House of the Slovak Academy of Sciences.

INDIVIDUAL DIFFERENCES IN THE ASYMMETRY OF ALPHA ACTIVATION

Stuart Butler and Alan Glass

Department of Anatomy
The Medical School
Birmingham
U.K. B15 2TJ

Asymmetry in the distribution of the alpha rhythm first received attention during the early seventies (Morgan, McDonald and MacDonald, 1971; Galin and Ornstein, 1972a & b; Glass and Butler, 1973; Butler and Glass, 1974). The effect was attributed to the activation of lateralised mechanisms serving language and 'visuospatial' functions and was therefore perceived as a non-invasive method of investigating hemispheric specialisation. It has recently become clear that alpha asymmetry varies systematically as a function of gender and handedness. In this chapter we shall consider the significance of this variation in terms of individual differences in the structural and dynamic aspects of hemispheric specialisation.

STRUCTURAL AND DYNAMIC ASPECTS OF HEMISPHERIC SPECIALISATION

Evidence for some form of variation in hemispheric specialisation is provided by several sources:

(a) Gender differences exist in functions whose neural mechanisms are normally lateralised. Certain verbal skills are acquired earlier and to a higher level by females (Herzberg and Lepkin, 1954; Gates, 1961; Hutt, 1972; Bryden, 1979), some aspects of hand preference are more strongly expressed in females (Hicks and Kinsbourne, 1976); Annett, 1982;1985) whereas spatial abilities are generally higher in males (Harris, 1978; Kimura and Harshman, 1984).

(b) Dichotic listening and split-field tachistoscopy reveal significant left field superiority for the perception of non-verbal information and superiority of the right field for the detection of verbal material. They clearly reflect the advantage of direct access to lateralised verbal and spatial mechanisms. The effects are subject to significant variation as a function of handedness and gender (McKeever and Van Deventer, 1977; Annett, 1982; Fairweather, 1982).

(c) The incidence of aphasia after unilateral lesions is more common in males than in females (for example: Landsell, 1962; McGlone, 1980; Inglis, Ruckman, Lawson, MacLean and Monga, 1981; Basso, Capitani and Marascini, 1982; Sundet, 1986) and more common in left handers than in right handers (Warrington and Pratt, 1973).

Observations such as these have been interpreted to mean that the pattern on which cognitive mechanisms are lateralised is subject to individual variation, that lateralisation is more complete in males than in females, and that reversal of the classical pattern of hemispheric dominance is more common among left handers than right handers. These imply that group differences can be accounted for by variation at the structural level, that is to say, in the anatomical distribution of the neural mechanisms responsible for cognitive processes. An early hypothesis of this type was put forward by Levy (1969). This holds that the superiority of females in verbal but not spatial skills is due to the recruitment of cortex of the right hemisphere for linguistic processes (i.e. language is bilateral). The involvement of the right side in language occurs at the expense of its commitment to spatial skills.

There is no doubt that the anatomical location of mechanisms serving verbal and spatial skills varies within the population. There are several, independent sources of evidence. Firstly, the system is remarkably resistant to unilateral brain damage in the early years of life (Basser, 1962; Rasmussen, 1964; Rasmussen and Milner, 1977). Extensive left hemisphere damage may be sustained without clinically detectable language abnormality in adulthood. The inference is that areas of cortex of the undamaged right hemisphere take on the function. Secondly, amytal studies have shown language to be dependent on the right hemisphere in a significant number of patients especially females and left handers (Wada and Rasmussen, 1960; Strauss and Wada, 1983). Similarly, aphasia is present in a significant proportion of patients presenting with disorders of the right hemisphere (Bryden, Hecaen and De Agostini, 1983; Basso, Capitani, Laracona and Zanobio, 1985). Finally, there is evidence for sex differences in the intrahemispheric organisation of the speech areas. Kimura (1983b) finds a dissociation between the aphasic disorders of males and females with lesions confined to anterior or posterior cortical territories.

The existence of such cases has encouraged the view that variation in the lateralisation of language and spatial functions may be fairly common in the population as a whole. However, recent evidence suggests the contrary. When patients with early brain damage and slowly progressive disorders such as tumours are excluded there is very little evidence for right hemisphere language in the population. Kimura (1983a) reports the effects of unilateral lesions in a large group of patients, the majority of whom had late-onset brain damage (strokes). Off 244 patients with left sided lesions, 100 were aphasic. Lesions believed to be confined to the right hemisphere were accompanied by aphasia in only two cases out of 179. Among the 40 left handers, five were aphasic and all these had left sided lesions. Similarly in a study by Strauss and Wada (1983) the proportion of patients with right sided language was largely, if not entirely,due to those with brain damage in childhood.

Further investigation of verbal and spatial abilities following stroke and head injury is required to confirm the picture emerging from these recent studies. If their findings are supported, then variation in the structural aspects of hemispheric specialisation is unlikely to account for group differences in verbal and spatial skills, hand preference, and the recovery of cognitive functions after brain damage because variation on the standard pattern is much less common than was formerly believed.

As long ago as 1972, Bogen and his colleagues introduced another dimension to this question with the concept of 'hemisphericity' (Bogen, DeZure, Tenhouten and Marsh, 1972). This referred to the tendency of different individuals to make use of alternative cognitive strategies in processing the same material. The activation of left or right hemisphere mechanisms was no longer seen to be determined solely by external events, by the nature of the stimuli or the type of output required, but by the individual's preference for particular cognitive strategies. Many types of problem may be solved either by symbolic logic (verbal or mathematical) or alternatively by imagery of spatio-temporal relationships. A trade-off will exist between the appropriateness of the strategy adopted and the individual's proficiency with it. Moreover, the role of one hemisphere in any given task will depend of the individual's preference for verbal or spatial modes of thinking, and the facility with which he or she interposes and integrates these complementary modes. The form that such variation in cognitive dynamics might take has been analysed by Cohen (1982).

The significance of 'hemisphericity' in the present context is that there may be systematic differences in cognitive style as a function of gender, hand preference and even ethnic background (Bogen et al., 1972; Bryden, 1978; Wolff, Hurwitz, Imamura and Lee, 1983). In other words the group differences with which we are concerned may exist in the dynamics of mental activity rather than in the neural hardware.

GROUP DIFFERENCES IN ALPHA ASYMMETRY

Evidence for a connection between EEG asymmetry and hemispheric specialisation is provided by the observation that, averaged over a sufficient number of subjects, the alpha rhythm suppresses to a greater extent over the left hemisphere than the right during the performance of verbal and mathematical tasks (Butler and Glass, 1974; Butler and Glass, 1976; Donchin, Kutas and McCarthy, 1977, for review of the earlier findings). More recently it has been confirmed that alpha asymmetry is associated with cognitive processes and is not due to any asymmetry in the load on sensory or motor systems (Butler and Glass, 1985).

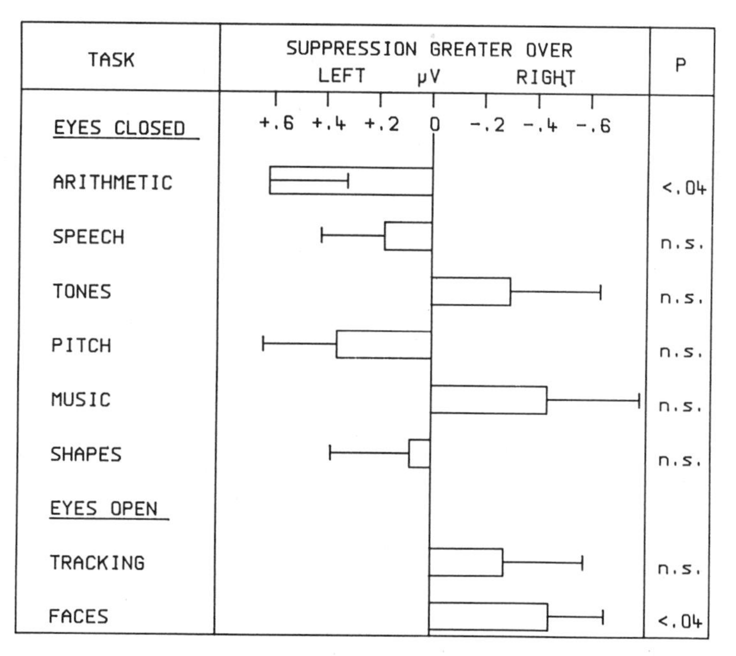

Fig.1. Asymmetry in the suppression of alpha rhythm in a group of 10 right handed subjects on tasks designed to engage the cognitive mechanisms of left or right hemispheres. See text for details. Recordings from 01,02 with respect to mastoids in average common reference.

It is unlikely that alpha asymmetry provides a direct measure of structural lateralisation in the sense that the intracarotid amytal test identifies the hemisphere responsible for language (Branch, Milner and Rasmussen, 1964; Rasmussen and Milner, 1977). Primarily, the amplitude of the alpha rhythm is related to levels of activation and it cannot be assumed that the distribution of cortical activity is determined simply by the type of information or material which a subject has been asked to process. As we have seen, cognitive strategies may be adopted which are inappropriate to the type of information to be processed. Indeed, many investigators have been unable to detect alpha asymmetry during the performance of tasks which it was thought would engage only one hemisphere (*inter alia*: Provins and Cunliffe, 1972*; Beaumont, Mayes and Rugg, 1978; Gevins, Zeitlin, Doyle, Schaffer, Callaway and Yeager, 1979; Rugg and Dickens, 1982; John Shaw, personal communication).

In particular, tasks which might be expected to rely on the cognitive specialisation of the right hemisphere seem to be accompanied by small and less consistent EEG asymmetries than those which accompany mental arithmetic. For example, Figure 1 shows asymmetry in the suppression of occipital alpha rhythm averaged over ten right handed subjects for a number of different tasks. This work was carried out in our laboratory some years ago by Peter Nava (Nava, Butler and Glass, 1975). The tasks included serial subtraction and speech comprehension, intended to engage the left hemisphere and a variety of visuospatial, stereognostic and musical tasks designed to engage the right hemisphere, including pitch and tone discrimination, listening to music, face recognition, a visual tracking task and somaesthetic shape recognition. The EEG was recorded from O1 and O2 referred to average common reference electrodes on the mastoids. The r.m.s. amplitude in the alpha band was computed from the amplitude spectrum for each channel for each task and while subjects relaxed with eyes open and closed. Asymmetry in alpha suppression was calculated using the following formula which measures the asymmetry during the task against a resting baseline:

$$(\text{Left} - \text{Right})_{\text{Rest}} - (\text{Left} - \text{Right})_{\text{Task}}$$

A positive value thus signifies that the alpha rhythm declined more over the left hemisphere than the right during the task. In Figure 1 the greater suppression of alpha rhythm over the left hemisphere during mental arithmetic with eyes closed is significant ($P<0.04$). Of the non-verbal tasks only one, face recognition with eyes open is consistently accompanied by greater suppression over the right hemisphere ($P<0.04$). Recordings were also made from parietal and central regions. The only 'right hemisphere' task which included statistically significant asymmetry over these areas was visual tracking, and only in parietal regions. Similar findings for the weakness of right hemisphere effects have been reported by others. Failure to control the natural human tendency to adopt verbal strategies and to interpose verbal trains of thought even while required to tackle visuospatial problems could well be sufficient to account for such failure to observe right hemisphere suppression.

Further evidence that alpha asymmetry is not a simple index of structural lateralisation is provided by the pattern of its variation among

----

* However, these authors presented evidence that total EEG power from symmetrical posterior parietal electrodes was suppressed more over the left cerebral hemisphere than the right during silent reading in right handed not left handed subjects. All their EEG asymmetries (including alpha asymmetries) were rejected from further statistical analysis because of their apparent intertrial unreliability.

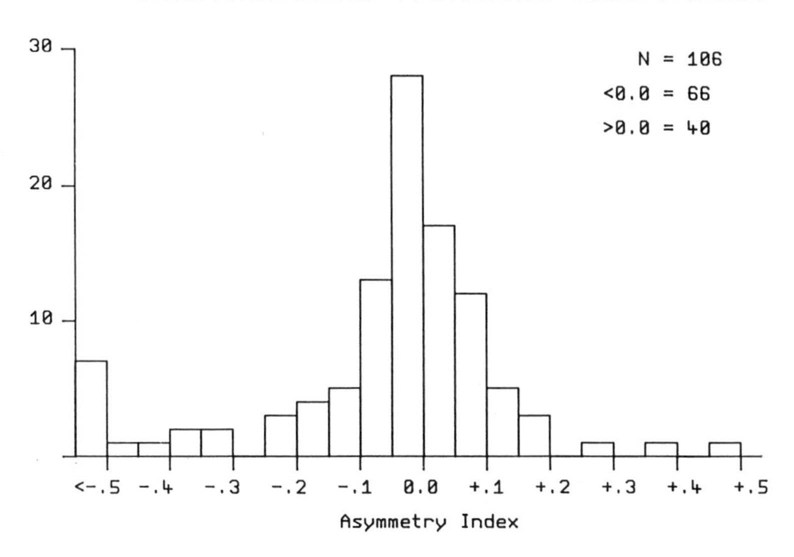

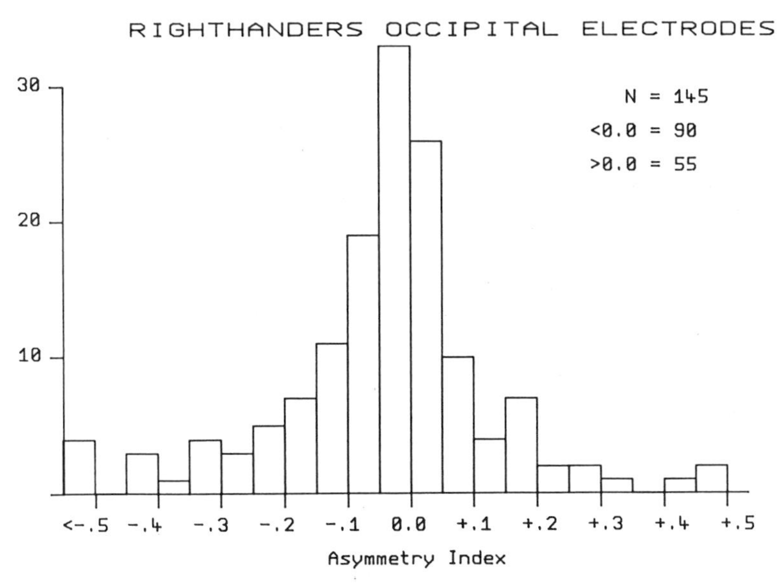

Fig. 2. Asymmetry in the suppression of alpha rhythm during the performance of mental arithmetic by right handed subjects. Recordings from electrodes 01,02,P3 and P4 with respect to mastoids in common reference.

individuals. We have studied this most extensively in subjects performing mental arithmetic. Mental arithmetic appears to be dependent upon mechanisms resident in the left hemisphere (Grewel, 1952; 1969; Levin, 1979). It is not clear whether this is because the task is verbally mediated or because it represents an additional mode of symbolic operation for which the left hemisphere is specialised. In any event, we find that this task is more consistently associated with left hemisphere suppression than overtly verbal tasks such as a speech comprehension test (Figure 1). Perhaps this is because speech more potently evokes parallel imagery in the right hemisphere.

Fig. 2 enables us to examine the variation in alpha suppression during mental arithmetic tasks in a large group of right handed subjects. The histograms plot a measure derived by the following expression:

$$\text{Asymmetry Index} = \left(\frac{\text{Right} - \text{Left}}{\text{Right} + \text{Left}}\right)_{\text{Rest}} - \left(\frac{\text{Right} - \text{Left}}{\text{Right} + \text{Left}}\right)_{\text{Task}}$$

The numerators represent the difference in alpha power over opposite hemispheres measured at rest and during the task. The denominators serve to normalise for overall change in alpha power between the rest condition and the task. As before, the use of a resting baseline controls for asymmetries introduced by factors such as local variation in skull thickness since the latter may have large effects on EEG power (Fisher, Butler and Glass, 1986). The calculation of the index is such that negative values signify greater suppression of the alpha over the left hemisphere during the task than at rest.

Fig. 2 shows that suppression is greater over the left hemisphere in the majority of these right handed subjects over both occipital and parietal regions during mental arithmetic. However the features of particular interest here are that the values are apparently normally distributed (even in spite of the fact that the values of the index must vary between limits of +1 and -1), and that suppression is greater over the right hemisphere in a third of the subjects. As we have seen, this is not what would be expected on the basis of the clinical findings in patients without early brain damage. There is no indication from those studies of any graded effect in the lateralisation of symbolic processes. Even in studies which do not exclude patients with early onset lesions, there is no indication that the right hemisphere plays the major role in symbolic processing in such a high proportion of individuals (Bryden, Hecaen and Agostini, 1983). Moreover, on our curve a large number of subjects have asymmetry indices close to zero indicating almost equal suppression of alpha activity over left and right sides, yet the clinical lesion studies show that a bilateral representation of language is the least common variant of hemispheric specialisation.

A similar mismatch between the distribution of alpha asymmetry and lateral dominance is to be found in left handed males. Estimates of the right dominance for language in this group varies between 18% (Bryden, Hecaen and DeAgostini, 1983) and virtually nil (Kimura, 1983a), and depending on whether patients with long standing brain damage are excluded. We should therefore expect a large majority of normal left handed subjects to be left cerebral dominant for symbolic processes. In Figure 3, the alpha asymmetry indices for this group are almost equally divided between left and right hemisphere suppression during mental arithmetic.

Given that left hemisphere suppression during overtly verbal tasks is even weaker than that observed during mental arithmetic, the crucial role of the left hemisphere in these symbolic processes is clearly not reflected in the alpha asymmetry. If, as we have suggested, alpha asymmetry reflects the dynamics of cortical activation rather than simply an underlying structural asymmetry, it reveals group differences in cognitive style and not hemispheric specialisation. Indeed the difference between the alpha asymmetry distributions for left and right handed subjects accords well with laterality effects in divided field techniques (Annett, 1982, for review) which must similarly be subject to the effects of cognitive strategy.

In view of the several lines of evidence for sex differences in laterality, we may expect to see similar group differences in EEG activation. In fact, the evidence for sex differences in alpha asymmetry is not consistent. The studies by Tucker (1976), Davidson, Schwartz,

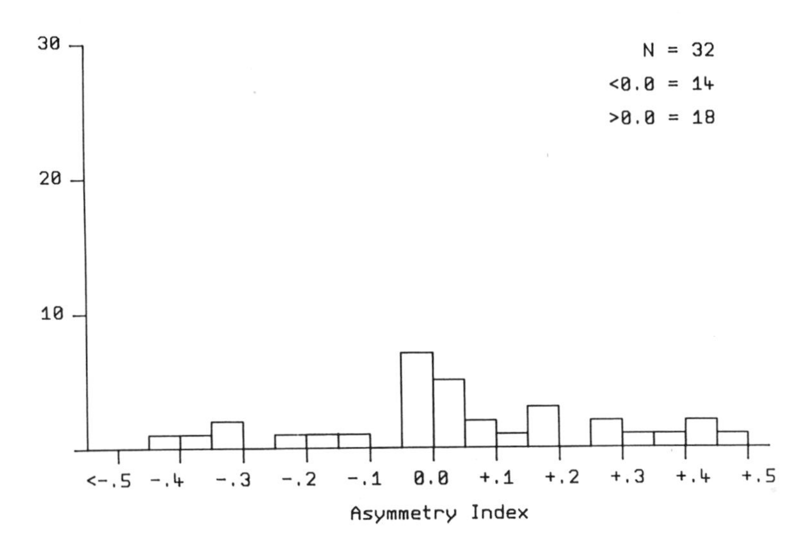

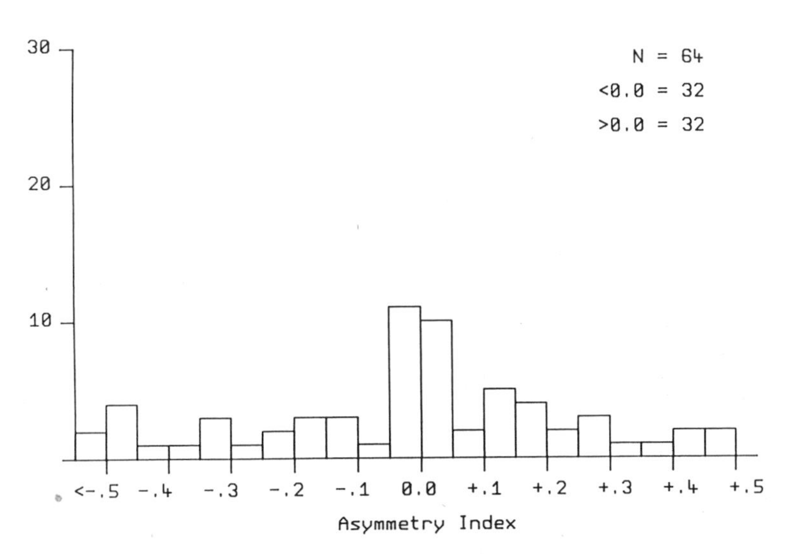

Fig. 3. Asymmetry in the suppression of alpha rhythm in left handed male subjects. Other details in Figure 2.

Pugash and Bromfield (1976), Ehrlichman and Wiener (1980), Rebert and Mahoney (1978), Haynes (1980), Haynes and Moore, (1981a and b), Ornstein, Johnstone, Herron and Swencionis (1980) and Galin, Ornstein, Herron and Johnstone (1982) all failed to find any simple relationship between gender and alpha asymmetry in any task. Others using somewhat different procedures or different methods for quantifying the asymmetry have found that the suppression of alpha activity over the left hemisphere is greater in males (Glass, Butler and Allen, 1975; Ray, Morrell, Frediani and Tucker, 1976; Trotman and Hammond, 1979; Ray, Newcombe, Semon and Cole, 1981, see Ray, this volume). A collation of our own recent studies also reveals no simple difference in the distribution of alpha asymmetry during mental arithmetic between males and females. The data in Figure 2 refer to a

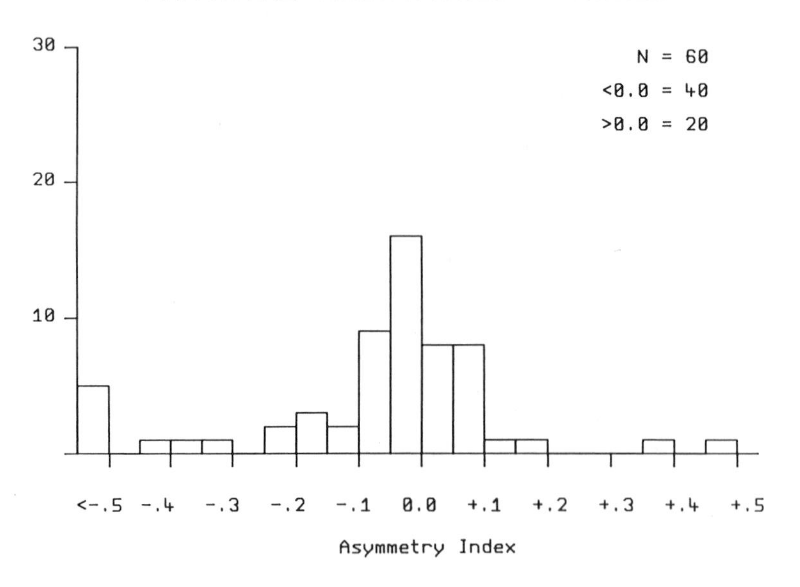

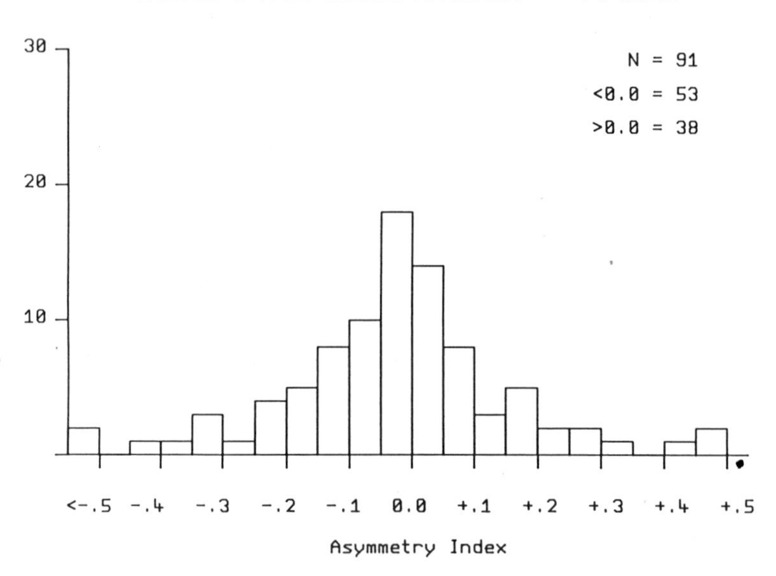

Fig. 4. Alpha asymmetry in right handed males, a subset of the data in Fig. 2.

mixed but predominantly male group of subjects. Although the number of females is small (and records of gender unfortunately do not survive for some of the early participants) there is nothing in the separate distributions for males and females (Figures 4 and 5) to suggest that the groups differ in alpha asymmetry during mental arithmetic. Assessed by parietal electrodes, 67% of the males showed greater suppression over the left hemisphere during mental arithmetic and, occipitally, 58% of the males showed greater left alpha suppression. The incidences in the females were almost exactly similar. Parietally, 56% of the females suppressed more over the left hemisphere whereas occipitally 67% showed greater left hemisphere suppression.

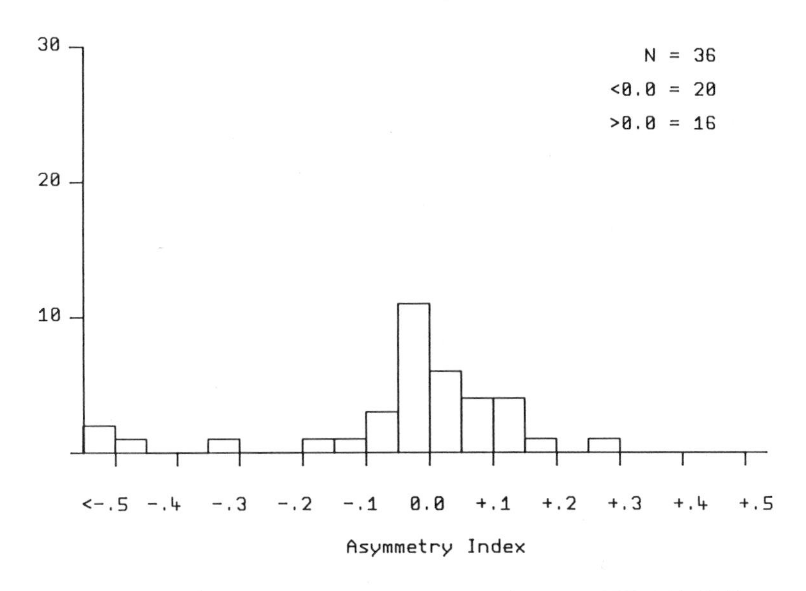

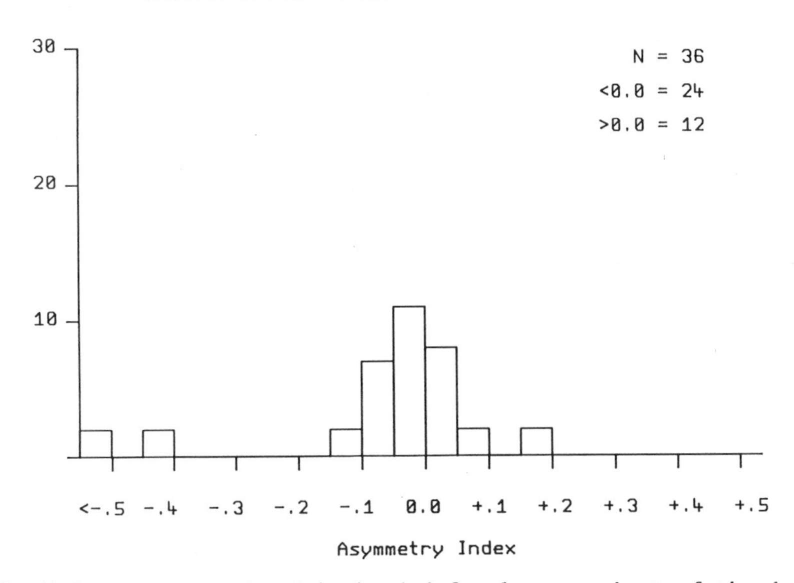

Fig. 5. Alpha asymmetry in right handed females, a subset of the data in Figure 2.

However, a somewhat surprising finding emerges if we look at alpha asymmetry during additional tasks. We studied alpha asymmetry in 24 males and 24 females during the performance of two tasks: mental arithmetic and face recognition, carried out with the eyes open (Glass, Butler and Carter, 1984). Figure 6 shows the asymmetry of alpha suppression over parietal regions during these tasks. This figure uses the same measure of asymmetry as in Figure 1. It can be seen that both males and females show left hemisphere suppression on the maths task and, as in Figure 4 and 5, there is little difference between the means for the two groups. But the females also show left hemisphere suppression on the face recognition task whereas this is accompanied by right hemisphere suppression in the males.

At first this appears to confirm the conclusions of many investigators, namely that lateral asymmetry is more marked in males than in females (McGlone, 1977; 1978; 1980).

Further examination of the data leads to a somewhat different conclusion. The two groups of 24 subjects were each divided into two subgroups; one comprised subjects with no left handed parents or siblings (by convention referred to as Familial Sinistral Negative, FS-), in the other subjects had at least one left hander among their close relatives (FS+). The patterns of alpha suppression in these subgroups have been analysed in detail by Glass et al. (1984) and here we shall only summarise the main finding. Figure 7 shows the direction of alpha asymmetry in each subgroup on each task. Where the hemispheres are shaded black and white, there was a significant difference in the suppression of parietal alpha, greater over the side shaded black ($p<.033$, or less). Where the hemispheres are shown stippled, there was no significant difference in suppression over the two sides.

A complex pattern of individual differences emerges. First, it appears that the tendency toward left hemisphere activation is increased if the task is mathematical, if the subject has no left handed relatives (FS-) and if she is female. Right hemisphere activation is shown only by the group which is male, has left handed relatives (FS+) and then only on the face recognition task. A similar interaction between gender and handedness (overt not familial, however) has been reported by Galin et al. (1982) though simple effects of gender were not obtained in that study. While it is true that our males show a greater change in asymmetry from the maths task to the faces task than the females; no group shows a reversal of hemispheric activation as would be predicted by models of lateralisation which assume complementarity of hemispheric specialisation. This should not be taken as evidence, one way or the other, on the question of complementarity (Bryden et al., 1983). It could equally well be the result

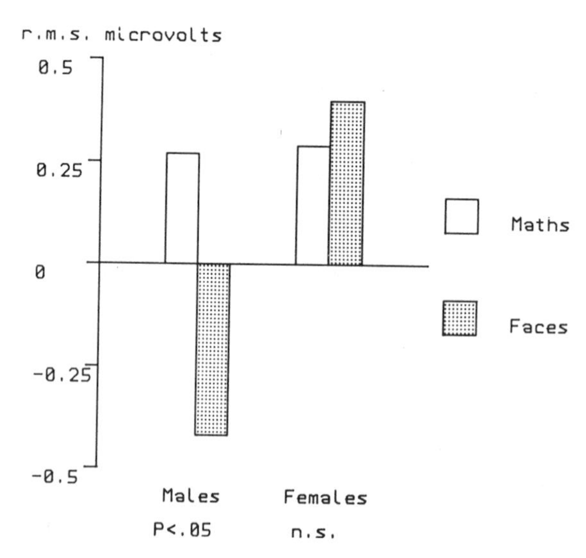

Fig. 6. Asymmetry of suppression of alpha activity in parietal regions (P3, P4, referred to mastoids in common reference) in 24 males and 24 females during mental arithmetic and a face recognition task. Negative values signify that the suppression was greater over the right hemisphere during the task.

of the intrusion of verbal strategies into all forms of human cognitive activity to an extent which varies with gender and other factors.

THE DYNAMICS OF HEMISPHERIC ACTIVATION

We have seen that individual and group differences in alpha asymmetry do not correspond to what is known of individual differences in the structural features of lateralisation. We have seen that they would be more readily accounted for by supposing that they reflect the dynamics of cortical activation, i.e. differences in cognitive style. Such an interpretation would make more sense in terms of physiological first principles, in so far as we understand the significance of alpha rhythm at this level. We shall now consider evidence that alpha asymmetry is indeed sensitive to cognitive dynamics.

Effect of Cognitive Set

Early searches for the effects of cognitive style on alpha asymmetry were inconclusive (Robbins and McAdam, 1974, Doktor and Bloom, 1977). More recently it has been shown that task related asymmetry is revealed most powerfully if carry-over effects from the previous task are controlled for (Grabow, Aronson, Greene and Offord, 1979) suggesting that alpha asymmetry is sensitive to factors such as cognitive set.

We have confirmation of this in a retrospective analysis of data obtained in our laboratory (Stern, Glass & Butler, 1981). Figure 8 shows the order of tasks undertaken by 16 subjects in an experiment in which the EEG was recorded from left and right occipital regions (O_1, O_2). The order in which the first three tasks (drawing with the eyes closed, writing a letter with the eyes closed, and watching a cartoon film) were presented was balanced across subjects. Mental arithmetic was included as a final task for all subjects, chiefly as a "litmus" to check that the system replicated earlier findings. Figure 9 shows that the alpha asymmetry averaged over all subjects (histograms labelled A) reached significant proportions only for the drawing task. However, the magnitude of these effects was influenced by some form of carry-over from preceding tasks. Writing was immediately preceded by drawing (intended to engage the right hemisphere) for eight subjects; similarly maths was preceded by drawing for four subjects, and drawing was preceded by writing (intended to engage the left hemisphere) in eight subjects. If we exclude from the analysis those subjects doing a 'left hemisphere' task after drawing (histograms labelled B) and those doing the right hemisphere task immediately after writing (the

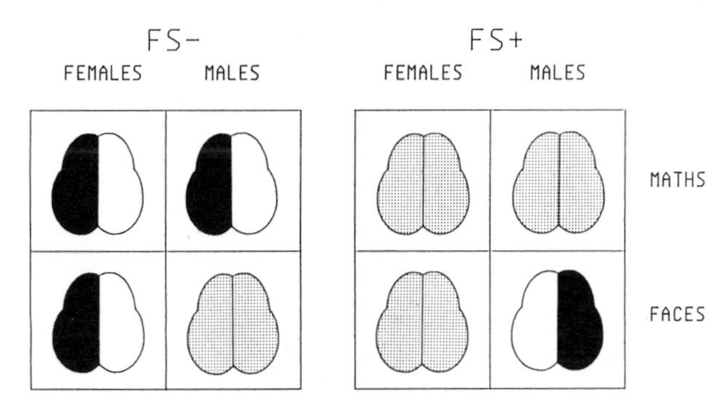

Fig. 7. The pattern of parietal alpha asymmetry in 48 subjects as a function of task, gender and familial handedness. See text for explanation of shading.

N	TASK ORDER			
4	CARTOONS	WRITING	DRAWING	MATHS
4	CARTOONS	DRAWING	WRITING	MATHS
4	DRAWING	WRITING	CARTOONS	MATHS
4	WRITING	DRAWING	CARTOONS	MATHS

Fig. 8. Task order which revealed carry-over effects in alpha activation see text for details and Figure 9.

histogram labelled C) then the mean alpha asymmetries all increase and the probability of the effects being due to chance decreases in spite of the reduced values of n (number of subjects).

Several mechanisms may be responsible for the sensitivity of alpha asymmetry to such carry-over. The use of the cognitive strategies of one hemisphere may increase the probability of their use in a subsequent task. Alternatively, subjects may simply interpose thoughts about the previous task during the early performance of a new one. Both amount to a preservation of mental set, and both would reduce alpha asymmetry.

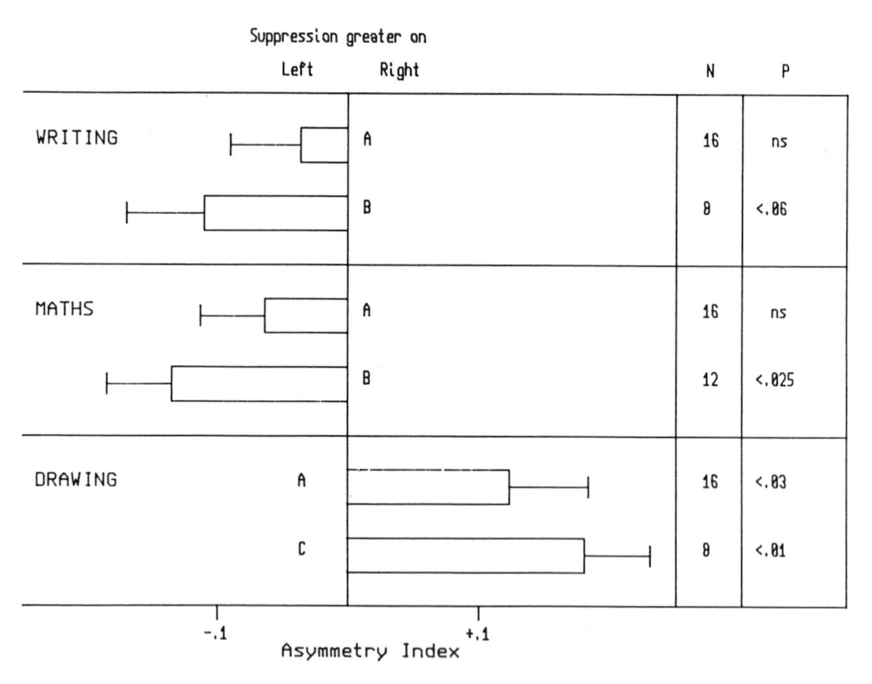

Fig. 9. Effect of set on alpha asymmetry. Alpha asymmetry is reduced when a task is preceded by one which calls for the engagement of a different hemisphere. The tasks are those listed in Fig. 8. Histograms labelled A represent alpha asymmetry averaged over all subjects; those labelled B or C show the effects of removing this serial order effect.

Effect of a potential conversant

Further evidence for the effects of mental set comes from an unexpected direction. In an earlier review of task specific alpha asymmetries (Butler and Glass, 1976), we suggested that the failure of some investigators to replicate the effects might be due to the conditions of social interaction under which the recordings were made. We suggested that the common factor in failures to replicate was that the subject was in the same room as the experimenter, or was in some other way primed for verbal communication. The set to process verbal information that this would engender would reveal itself as unreactive alpha power over the left hemisphere, irrespective of the given task.

We have recently investigated this prediction directly (Butler, Glass and Fisher, 1986) in an experiment in which the EEG was recorded from left and right parietal and occipital regions in 18 subjects under two conditions. In one, they sat in a sound isolated chamber having been asked simply to relax with their eyes open, after being assured that they would not be disturbed for several minutes. In the second condition, they sat in the same chamber, also with their eyes open with instructions to relax, but in this case the door of the chamber had been left open after the experimenter entered on the pretext of checking a troublesome electrode. In this condition the experimenter remained in the view of the subject and would have been heard had he spoken. The two conditions were randomised across subjects. Figure 10 shows the mean power in the alpha band (in arbitrary units) in each channel for the two conditions. The alpha rhythm is present with approximately equal power over left and right hemispheres when the subject is isolated, but is suppressed over the left hemisphere in both parietal and occipital regions while the door is open. Figure 11 shows the change in asymmetry between conditions and the high levels of statistical significance attached to this effect. Set clearly influences EEG asymmetry whether it is induced by carry over from previous activity or by preparation to respond with a certain strategy. The effect has important implications for the interpretation of individual differences in alpha asymmetry.

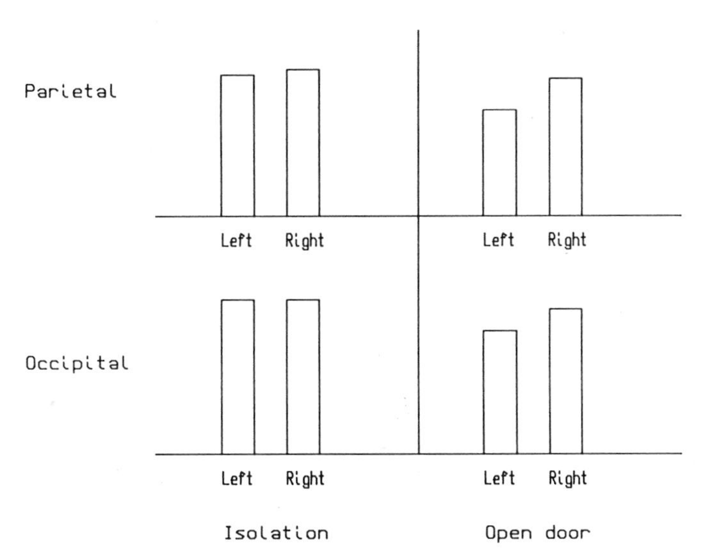

Fig. 10. Effect of social facilitation on EEG asymmetry (Eyes open, relaxed). Alpha power over left and right hemispheres (O1,O2,P3,P4, with respect of mastoids in common reference) while a subject relaxes, isolated in a sound proof chamber or in view of the experimenter.

Firstly, it shows that alpha asymmetry is not simply material specific. This alone is sufficient to account for the finding that the relationship between EEG asymmetry and structural lateralisation is a statistical, stochastic one; the effect cannot be used to determine the localisation of particular functions with confidence in a given individual. Secondly, it may provide an explanation for the differences we have described in alpha asymmetry as a function of gender, handedness and familial handedness. We have seen that the data do not correspond with individual differences in structural lateralisation believed to exist in these groups. Instead the effects may be due to group differences in cognitive style. In other words the inclination (or set) to use particular cognitive strategies may differ between males and females, and between left and right handers. Because the alpha rhythm reflects levels of cortical activation, asymmetry in its distribution reflects group differences in the usage of available cognitive mechanisms (the strategies of left and right hemispheres). These are effectively superimposed upon asymmetries of a purely structural nature.

CONCLUSION

In reviewing this data we have tried to show that the behaviour of the alpha rhythm during mental activity reflects the dynamics of cortical processing. Its distribution reveals a statistically significant asymmetry which reflects bias in the cognitive sequence due to the underlying structural asymmetry. Our data do not enable us, as yet, to say anything about the fine structure of that cognitive flow. In particular we are unable to address the issues raised by Cohen (1982) in her analysis of the forms of interaction between structural asymmetry and cognitive processing because our experiments have had more limited objectives, and because the technology has not existed. With the advent of compressed spectral arrays we should now be able to follow the rapid time course of patterns of cortical activation and so characterise individual differences in cognitive style.

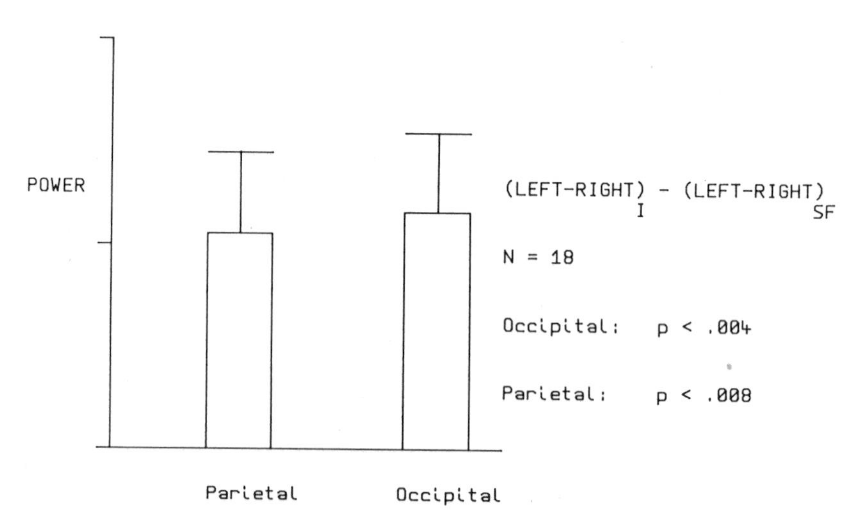

Fig. 11. EEG asymmetry: isolation versus social facilitation. The data of Figure 10 expressed as the differences in asymmetry, showing the effect of social facilitation in bringing about suppression of the alpha rhythm over the left hemisphere. (I = Isolated; SF = Socially Facilitated)

REFERENCES

Annett, M. (1982). Handedness. In J.G. Beaumont (Ed.) Divided Visual Field Studies of Cerebral Organisation London: Academic Press, 195-215.

Annett, M. (1985). Left, Right, Hand and Brain: The Right Shift Theory. London: Lawrence Erlbaum.

Basser, L.S. (1962). Hemiplegia of early onset and the faculty of speech with special reference to the effects of hemispherectomy. Brain, 85, 427-460.

Basso,. A., Capitani, E. and Marascini, S. (1982). Sex differences in recovery from aphasia. Cortex, 18, 469-475.

Basso, A., Capitani, E., Laracona, M. and Zanobio, M.E. (1985). Crossed aphasia: or more syndromes? Cortex, 21, 25-45.

Beaumont, J.G., Mayes, A.R. and Rugg, M.D. (1978). Asymmetry in EEG alpha coherence and power. Effects of task and sex. Electroencephalography and Clinical Neurophysiology, 45, 393-401.

Bogen, J.E., DeZure, R., Tenhouten, W.D. and Marsh S.F. (1972). The other side of the brain IV: The A/P ratio. Bulletin of the Los Angeles Neurological Society, 37, 49-61.

Branch, C., Milner, B. and Rasmussen, T. (1964). Intracarotid sodium amytal for the lateralisation of cerebral speech dominance. Neurology, 25, 907-910.

Bryden, M.P. (1978). Strategy effects in the assessment of hemipheric asymmetry. In G. Underwood, (Ed.) Strategies of Information Processing. New York: Academic Press: 117-149.

Bryden, M.P. (1979). Evidence for sex related differences in cerebral organisation. In M. Wittig and A.C. Peterson, (Eds.) Sex-related Differences in Cognitive Functioning. New York: Academic Press, 121-139.

Bryden, M.P., Hecaen, H. and De Agostini, M. (1983). Patterns of cerebral organisation. Brain and Language 20, 249-262.

Butler, S.R. and Glass, A. (1974). Asymmetries in the electroencephalogram associated with cerebral dominance. Electroencephalography and Clinical Neurophysiology, 36, 481-491.

Butler, S.R. and Glass, A. (1976). EEG correlates of cerebral dominance. In A.H. Riesen and R.F. Thompson (Eds.) Advances in Psychobiology Vol.3. New York: John Wiley, 219-272.

Butler, S.R. and Glass, A. (1985) The validity of EEG alpha asymmetry as an index of the lateralisation of human cerebral function. In D. Papakostopoulos, S. Butler and I. Martin. (Eds.) Clinical and Experimental Neuropsychology. London: Croom Helm, 370-395.

Butler, S.R., Glass, A. and Fisher, S. (1986) Effect of potential conversants (social facilitation) on resting EEG alpha asymmetry. In preparation.

Cohen, G. (1982). Theoretical interpretations of lateral asymmetries. In J.G. Beaumont (Ed). Divided Visual Field Studies of Cerebral Organisation. London: Academic Press, 87-111.

Davidson, R.J., Schwartz, G.E., Pugash, E. and Bromfield, E (1976). Sex differences in patterns of EEG asymmetry. Biological Psychology 4, 11-138.

Doktor, R. and Bloom, D.M. (1977). Selective lateralization of cognitive style related to occupation as determined by EEG asymmetry. Psychophysiology, 14, 385-387.

Donchin, E., Kutas, M. and McCarthy, G. (1977). Electrocortical indices of hemispheric utilization. In S. Harnad, R.W. Doty, L. Goldstein, J.Jaynes and G. Krauthamer (Eds.) Lateralization in the Nervous System. New York: Academic Press, 339-384.

Doyle, J.C., Ornstein, R.E. and Galin, D. (1974). Lateral specialisation of cognitive mode: EEG frequency analysis. Psychophysiology, 11 567-578.

Ehrlichman, H. and Wiener, M.S. (1980). EEG asymmetry during covert mental activity. Psychophysiology, 17, 228-235.

Fairweather, H. (1982). Sex differences: little reason for females to play midfield. In J.G. Beaumont (Ed). Divided Visual Field Studies of Cerebral Organisation. London: Academic Press, 147-194.

Fisher, S., Butler, S.R. and Glass, A. (1986). Asymmetries of the human skull in relation to EEG alpha asymmetries. Electroencephalography and Clinical Neurophysiology, In Press.

Galin, D. and Ornstein, R.E. (1972a). Lateral specialisation of cognitive mode: an EEG Study. Psychophysiology, 9, 412-418.

Galin, D. and Ornstein, R.E. (1972b). Lateral specialisation of cognitive mode: II. An EEG study. Psychophysiology, 9, 412-418.

Galin, D., Ornstein, R.E., Herron, J. and Johnstone, J. (1982). Sex and handedness differences in EEG measures of hemispheric specialization. Brain and Language, 16, 19-55.

Gates, A.I. (1961). Sex differences in reading ability. Elementary School Journal, 61, 431-434.

Gevins, A.S., Zeitlin, G.M., Doyle, J.C., Yingling, C.D., Schaffer, R.E., Callaway, E. and Yeager, C.L. (1979). Electroencephalogram correlates of higher cortical functions. Science, 203, 665-668.

Glass, A. and Butler, S.R. (1973). Asymmetries in suppression of alpha rhythm possibly related to cerebral dominance. Electroencephalography and Clinical Neurophysiology, 34, 729.

Glass, A., Butler, S.R. and Allen, D. (1975). Sex differences in the functional specialisation of the cerebral hemispheres. Proceedings of the Tenth International Congress of Anatomists, Tokyo. Science Council of Japan, 204.

Glass, A., Butler, S.R. and Carter, J.C. (1984). Hemispheric asymmetry of EEG alpha activation: effects of gender and familial handedness. Biological Psychology, 19, 169-187.

Grabow, J.D., Aronson, A.E., Greene, K.L. and Offord, K.P. (1979). A comparison of EEG activity in the left and right cerebral hemispheres by power spectrum-analysis during language and non-language tasks. Electroencephalography and Clinical Neurophysiology, 47, 460-472.

Grewel, F. (1952). Acalculia. Brain, 75, 397-407.

Grewel, F. (1969). The acalculias. In P.J. Vinken and G.W. Bruyn, (Eds.) Handbook of Clinical Neurology, 4. Amsterdam: North Holland: 181-194.

Harris, L.J. (1978) Sex differences in spatial ability: Possible environments, genetic and neurological factors. In M. Kinsbourne, (Ed.) Asymmetrical Function of the Brain. Cambridge: Cambridge University Press, 405-522.

Haynes, W.O. (1980). Task effect and EEG alpha asymmetry: an analysis of linguistic processing in two modes. Cortex, 16, 95-102.

Haynes, W.O. and Moore, W.H., Jr. (1981a). Sentence imagery and recall: An electroencephalographic evaluation of hemispheric processing in males and females. Cortex, 17, 49-62.

Haynes, W.O. and Moore, W.H., Jr. (1981b). Recognition and recall: An electroencephalographic investigation of hemispheric alpha asymmetries for males and females on perceptual and retrieval tasks. Perceptual and Motor Skills, 53, 283-290.

Herzberg, F. and Lepkin, M.A. (1954). A study of sex differences on the Primary Mental Abilities Test. Educational Psychology Measures, 14, 687-689.

Hicks, R.E. and Kinsbourne, M. (1976). Human handedness: A partial cross-fostering study. Science, 192, 908-910.

Hutt, C. (1972). Sex differences in human development. Human Development, 15, 153-170.

Inglis, J., Ruckman, M., Lawson, S.J., MacLean, A.W. and Monga, T.N. (1982). Sex differences in the cognitive effects of unilateral brain damage. Cortex 18, 257-276.

Kimura, D. (1983a). Speech representation in an unbiased sample of left handers. Human Neurobiology, 2, 147-154.
Kimura, D. (1983b). Sex differences in cerebral organisation for speech and praxic functions. Canadian Journal of Psychology, 37, 19-35.
Kimura, D. and Harshman, R.A. (1984). Sex differences in brain organisation for verbal and non-verbal functions. In G.J. De Vries, et al. (Eds.) Progress in Brain Research, 61, Amsterdam: Elsevier, 423-441.
Landsell, H. (1962). Laterality of verbal intelligence in the brain. Science, 135, 922-923.
Levin, H.S. (1979) The acalculias. In K.M. Heilman and E. Valenstein (Eds.) Clinical Neuropsychology. New York: Oxford University Press, 128-140.
Levy, J. (1969). Possible basis for the evolution of lateral specialisation of the human brain. Nature, 224, 614-615.
McGlone, J. (1977). Sex differences in organisation of verbal functions in patients with unilateral brain lesions. Brain, 100, 775-793.
McGlone, J. (1978). Sex differences in functional brain asymmetry. Cortex, 14, 122-128.
McGlone, J. (1980). Sex differences in human brain organisation: A critical survey. The Behavioural and Brain Sciences, 3, 215-227.
McKeever, W.F. and Van Deventer, A.D. (1977). Visual and auditory language processing asymmetries: Influence of handedness, familial sinistrality and sex. Cortex, 13, 225-241.
Morgan, A.H., McDonald, P.J. and MacDonald, H. (1971). Differences in bilateral alpha activity as a function of experimental task, with a note on lateral eye movements and hypnotisability. Neuropsychologia, 9, 459-469.
Nava, P.N., Butler, S.R. and Glass, A. (1975). Asymmetries of the alpha rhythm associated with functions of the right hemisphere. Electroencephalography and Clinical Neurophysiology, 43, 582.
Ornstein, R.E., Johnstone, J., Herron, J. and Swencionis, C. (1980). Differential right hemisphere engagement in visuospatial tasks. Neuropsychologia, 18, 49-64.
Provins and Cunliffe, P. (1982). The relationship between EEG activity and handedness. Cortex, 8, 136-146.
Rasmussen, T. (1964). Discussion on the current status of cerebral dominance. Research Publications of the Association for Research in Nervous and Mental Disorders, 42, 113-15.
Rasmussen, T. and Milner, B. (1977). The role of early left-brain injury in determining lateralisation of cerebral speech function. Annals of the New York Academy of Sciences, 229, 355-369.
Ray, W.J., Morrell, M., Frediani, A. and Tucker, D.M. (1976). Sex differences and lateral specialisation of hemispheric functioning. Neuropsychologia, 14, 391-394.
Ray, W.J., Newcombe, N., Semon, J. and Cole, P.M. (1981). Spatial abilities, sex differences and EEG functioning. Neuropsychologia, 19, 719-722.
Rebert, C. and Mahoney, R.A. (1978). Functional cerebral asymmetry and performance III: Reaction time as a function of task, hand, sex and EEG asymmetry. Psychophysiology, 15, 9-16.
Robbins, K.I. and McAdam, D.W. (1974). Inter-hemispheric alpha asymmetry and imagery mode. Brain & Language, 1, 189-193.
Rugg, M.D. and Dickens, A.M.J. (1982). Dissociation of alpha and theta activity as a function of verbal and visuospatial tasks. Electroencephalography and Clinical Neurophysiology, 53, 201-207.
Stern, A., Glass, A. and Butler, S.R. (1981). Sensory, motor and cognitive influences of task induced lateral asymmetries in the distribution of the alpha rhythm. Electroencephalography and Clinical Neurophysiology, 52, 103p.
Strauss, E. and Wada, J. (1983). Lateral preferences and cerebral speech

dominance. Cortex, 19, 165-177.

Sundet, K. (1986). Sex differences in cognitive impairment following unilateral brain damage. Journal of Clinical and Experimental Neuropsychology, 8, 51-61.

Trotman, S.C.A. and Hammond, G.R. (1979). Sex differences in task-dependent EEG asymmetries. Psychophysiology, 16, 429-431.

Tucker, D.M. (1976). Sex differences in hemispheric specialisation for synthetic visuospatial functions. Neuropsychologia, 14, 447-454.

Wada, J. and Rasmussen, T. (1960). Intracarotid injection of sodium amytal for the lateralisation of cerebral speech dominance: experimental and clinical observations. Journal of Neurosurgery, 17, 266-282.

Warrington, E.K. and Pratt, R.T. (1973). Language laterality in left handers assessed by unilateral ECT. Neuropsychologia, 11, 423-428.

Wolff, P.H., Hurwitz, I., Imamura, S. and Lee, K.W. (1983). Sex differences in speed of automatized naming. Neuropsychologia, 21, 283-288.

Zurif, E.B. and Bryden, M.P. (1969). Familial handedness and left-right differences in auditory and visual perception. Neuropsychologia, 7, 179-188.

AGE AND SEX RELATED EEG CONFIGURATIONS IN NORMAL SUBJECTS

P. Flor-Henry,* Z.J. Koles+ and J.R. Reddon*

* Alberta Hospital
Edmonton, Box 307
Edmonton,
Alberta, T6G 2G3
Canada

\+ University of Alberta
Department of Applied Sciences in Medicine,
Edmonton,
T5J 2J7
Canada

INTRODUCTION

In an earlier study of the EEG characteristics of normal subjects with a 4-electrode montage (Flor-Henry and Koles, 1982) we reported that women had more EEG power than men, increased EEG coherence and a pattern of relatively greater left hemispheric activation, men showing relatively greater right hemispheric activation. These effects were independent of handedness. In the present investigation, in order to examine further the influence of age and sex on the EEG, we analysed the EEG of a new sample of normal subjects with an 8-electrode montage during various cognitive conditions. The samples consist entirely of dextrals. A future study will evaluate the influence of handedness on the EEG.

The general characteristics of the population are shown in Table 1:

Table 1.

	N	Age	S.D.	Range
Dextral Males	57	32.9	10.7	18-58
Dextral Females	56	32.3	9.6	18-59

(Mean age differences between males and females was not statistically significant, $p > .05$).

METHODS

The EEG was recorded from 57 normal male subjects and 56 normal female subjects at locations P_4, P_3, T_4, T_3, F_8, F_7, T_6, and T_5 referred to Cz (International 10-20 system) for two minutes during the following mental conditions: at rest with eyes open, at rest with eyes closed, a vocabulary exercise, a word fluency exercise and a block design exercise. Two of the three exercises described were subtests of the Wechsler Adult Intelligence Scale (WAIS-R). The recordings from each scalp location were converted to digital form directly during the sessions using 12 bit conversions at the rate of 120 per second. A ninth channel recorded with the EEG was used to indicate operator-flagged artifacts. A hand-held button was used to create

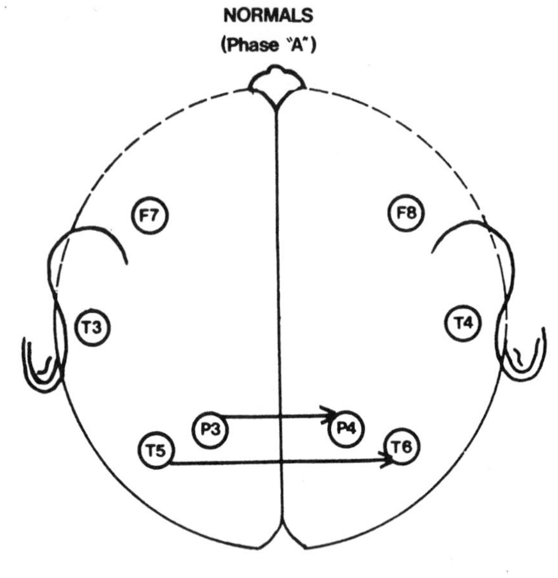

Fig. 1

a reject-on signal and this was recorded to indicate that the accompanying EEG should be excluded from the analysis because of excessive movement or eye-blink artifacts. Digitized recordings were stored on a disk and transferred later to a VAX 11/750 computer for subsequent analysis. The analysis consisted of dividing the recordings from each location into epochs of 128 consecutive samples, tapering these with a Hanning data window to

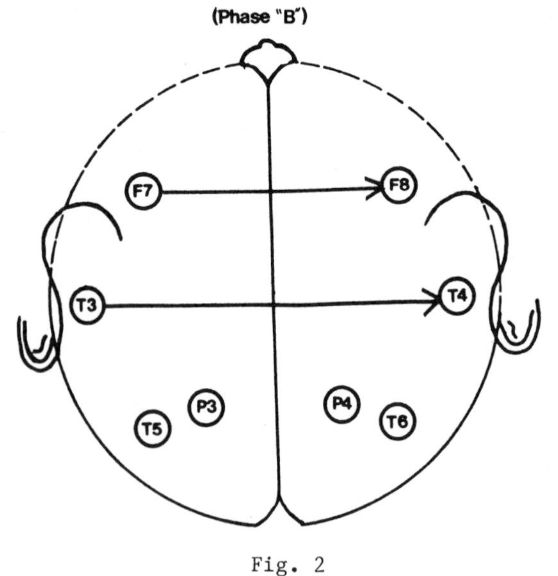

Fig. 2

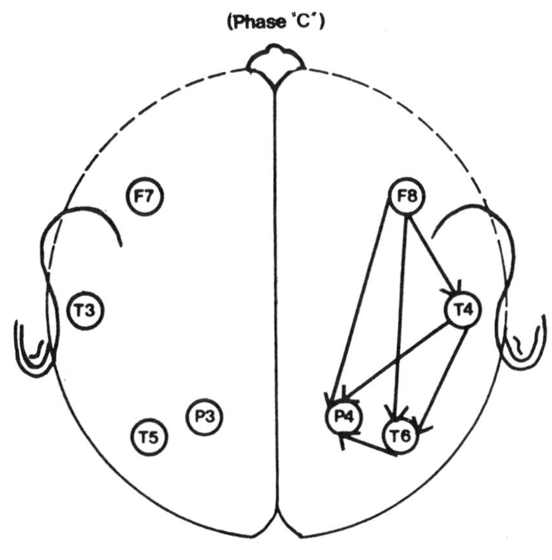

Fig. 3

restrict spectral leakage and Fourier transformation. The coefficients of each transformed epoch were grouped into the frequency bands of interest namely, coefficients 3 and 4 for 1 to 3 cycles/sec, 6, 7 and 8 for 4 to 7 cycles/sec, 10 to 14 for 8 to 13 cycles/sec and 23 to 43 for 20 to 40 cycles/sec and the remaining coefficients discarded. For each of the retained coefficients an 8 x 8 spectral matrix was formed using the relation

$$S_{xy} = XY^*$$

where X is a particular Fourier coefficient from scalp location X and Y is the same numbered coefficient from location Y. The asterisk indicates that the complex conjugate of Y should be used (both X and Y are complex numbers consisting of in-phase and in-quadrature components). In general, the elements on the diagonal of the spectral matrix are the variances of the potentials obtained from each scalp location in the frequency band related to the particular Fourier coefficient. Off-diagonal elements are the covariances between the scalp locations in the same frequency band. Unlike the diagonal elements, the off-diagonal elements of the spectral matrix are complex numbers.

The spectral matrices constructed as described for the retained Fourier coefficients were combined to form a single matrix for each of the four frequency bands of interest. This was done by averaging the respective elements in the spectral matrices of all the Fourier coefficients in the band. Specifically, for the 1 to 3 cycle/sec band the spectral matrices of coefficients 3 and 4 were added element by element and the result divided by 2. For the 4 to 7 cycle/sec band spectral matrices of coefficients 6, 7 and 8 were added and divided by 3. This process was repeated for the 8 to 13 and the 20 to 40 cycle/sec bands as well, with the division being by 5 and 21, respectively. In summary, then, coincidental 128 sample epochs from the 8 scalp locations were Fourier transformed and reduced to a single 8 x 8 spectral matrix for each of 4 frequency bands. The diagonal elements of a particular matrix are measures of the power (variance) in the electrical activity recorded from the brain during the epoch at each of the 8 scalp locations in one frequency band. The off-diagonal elements are measures of the cross-power (covariance) in the electrical activity between locations in the same frequency band.

An entire 2 minute recording of the EEG was processed as described in 128 sample epochs. Processing progressed from the beginning to the end of the recording using overlapped epochs shifted from one another by 8 samples (120 samples overlapped). Epochs containing samples which occurred with the reject signal on were discarded. The 4 spectral matrices resulting from all of the accepted epochs in a recording were then averaged together (element by element) to produce 4 composite matrices each describing the covariance structure of the recording in a particular frequency band. In addition, the diagonal elements from the spectral matrices of each epoch were individually stored to enable an analysis of temporal variations in right-left power symmetry through a recording. Overlapping of epochs as described enabled a measurement of the power in each of the 4 frequency bands every 8 sample points or 120/8 = 15 times each second.

The composite matrices were used to obtain estimates of the coherence and phase between the scalp locations as described before (Koles and Flor-Henry, 1981). Using the formula given by Welch (1967) with a Hanning data window and an 8 point epoch shift, estimates of coherence and phase obtained in this way contained over 400 degrees of freedom and therefore were essentially unbiased (Benignus, 1969). Rejection of half of a 2 minute recording due to the presence of artifacts would leave the estimates with 200 degrees of freedom still more than adequate to all but eliminate bias. Irrespective of this, no recording was utilized which contained artifacts lasting more than 20 seconds.

RESULTS

In order to derive a measurement model for the EEG montage and the resulting measures, a principal components decomposition was performed in the sample of normal controls consisting of 57 males and 56 females. The analysis was performed separately in each of the frequency bands on variables that had been separately standardized to zero mean and unit variance. A 13-dimensional components solution replicated reasonably well across the four bands in the varimax orientation and manifested good simple structure. Consequently the same model was chosen to represent all four bands.

The detailed structure of the fundamental EEG parameters emerging from the components analysis were:

1. POWER a) frontal F_7 F_8 — Anterior

   b) parietal P_3 P_4]

   posterior temporal T_5 T_6] — Posterior

   mid-temporal T_3 T_4]

2. COHERENCE General — Intra]

   Inter] Hemispheric

3. PHASE:

   a) ($T_5 \blacktriangleright T_6$); ($P_3 \blacktriangleright P_4$)

b) ($F_7 \blacktriangleright F_8$); ($T_3 \blacktriangleright T_4$)

c) ($T_4 \blacktriangleright P_4$); ($F_8 \blacktriangleright T_4$); ($F_8 \blacktriangleright P_4$); ($T_6 \blacktriangleright P_4$); ($F_8 \blacktriangleright T_6$); ($T_4 \blacktriangleright T_6$)

d) ($F_7 \blacktriangleright P_3$); ($F_7 \blacktriangleright T_3$); ($T_3 \blacktriangleright P_3$); ($T_5 \blacktriangleright P_3$); ($F_7 \blacktriangleright T_5$); ($T_3 \blacktriangleright T_5$)

e) ($F_7 \blacktriangleright P_4$); ($F_7 \blacktriangleright T_4$); ($T_3 \blacktriangleright P_4$); ($T_5 \blacktriangleright P_4$); ($F_7 \blacktriangleright T_6$); ($T_3 \blacktriangleright T_6$)

d) ($F_8 \blacktriangleright P_3$); ($T_4 \blacktriangleright P_3$); ($F_8 \blacktriangleright T_3$); ($T_6 \blacktriangleright P_3$); ($F_8 \blacktriangleright T_5$); ($T_4 \blacktriangleright T_5$)

4. a) log. right/left FRONTAL POWER RATIO

   b) log. right/left TEMPORAL-PARIETAL POWER RATIO

5. OSCILLATIONS* (9)

   a) Temporal-parietal

   b) Frontal

Figures 1 - 6 illustrate the patterns of phase relationships. It is noteworthy that the right and left intrahemispheric, anterior ► posterior and the interhemispheric phase leads are symmetrical, right and left whereas the homologous left ► right interhemispheric phase relation is unidirectional.

Age Effects

While the measures used in the present study are based on a unipolar reference system, these measures are compared with an average reference system (in Appendix I). With the exception of the phase measures, there was a good deal of correspondence between the two reference systems. Phase, however, is relatively meaningless in an average reference system.

Table 2 shows the overall age x factor (EEG) correlations for males and females combined and across tasks in the four frequency bands. More detailed results by sex and task separately are given in Appendix II. The detailed results show task and sex interactions in the evolution of the EEG with age. In general, with increasing age, power diminishes in the lower but increases in the fastest frequencies. Similarly there is a progressive decrease in coherence with increasing age in the lower frequency bands: delta, theta and alpha. Further with age the left ► right, homologous interhemispheric phase lead is reduced in the theta and alpha. Further with age the left ► right, homologous interhemispheric phase lead is reduced in the theta range; a trend in the same direction is also seen in the delta band. Interestingly the left hemisphere is again implicated in the anterior ► posterior phase reduction (alpha range). As subjects get older the log. of the right/left frontal power ratio becomes smaller in all the frequency bands, except for delta. Finally, with age there is a slowing of the right/left hemispheric energy oscillations in the delta, alpha and beta bands.

Repeated measures analysis of variance was used to evaluate effects due to sex, task, and task by sex interactions. The significant results are summarized in Table 3. Cell means for each measure by task and sex are given in Appendix III. Task effects in 1-3 Hz were significant for all EEG measures but left ► right posterior phase. In the 4 - 7 Hz band the task

---

* a new EEG measure estimating the frequency of right/left hemispheric energy shifts through time, calculated 15 times per second for every 2 minute epochs (see Flor-Henry et al., 1984).

Table 2. Age x Factor Correlation

EEG FACTORS	(1 - 3Hz)	(4 7Hz)	(8 - 13 Hz)	(20 - 40Hz)
1. Power				
a) Frontal	-0.21*	-0.29*	-0.18*	0.12*
b) Temporal-Parietal	-0.11*	-0.30*	-0.19*	0.06
2. Coherence	-0.12*	-0.32*	-0.17*	0.07*
3. Phase				
a) Left►Right Posterior	-0.06*	-0.14*	-0.01	-0.02
b) Left►Right Anterior	0.00	-0.09*	0.01	0.09*
c) Right Intra-hemispheric	0.06	-0.03	-0.04	-0.04
d) Left Intra-hemispheric	0.00	0.00	-0.07*	0.01
e) Left►Right Inter-hemispheric	0.02	-0.06	-0.03	-0.04
f) Right►Left Inter-hemispheric	0.01	0.11*	-0.04	-0.05
4. Log. Right/Left Power Ratio				
a) Frontal	-0.05	-0.10*	-0.07*	-0.15*
b) Temporal-Parietal	0.07*	0.02	-0.03	0.02
5. Oscillations				
a) Temporal-Parietal	-0.04	0.01	-0.12*	-0.20*
b) Frontal	-0.07*	0.15*	0.03	-0.15*

Note: * denotes $p < .05$.

effects for all EEG measures were significant. In 8;13 Hz only left►right anterior phase and frontal log. of right/left power ratio did not demonstrate significant task effects. In the 20 - 40 Hz band left ► right anterior phase, left intrahemispheric phase and frontal log. of right/left power ratio were not associated with task effect. Main effects for sex appeared for frontal power in all frequency bands except 20 - 40 Hz. In 4 - 7 Hz there was also a main effect for sex for the left► right posterior phase measure. In 20 - 40 Hz there was also a main effect for sex for posterior power. There was a task x sex interaction in 1 - 3 Hz and 4 - 7 Hz for frontal power. In 4 - 7 Hz there was also a task x sex interaction for the following phase measures: right intrahemispheric, left intrahemispheric and right interhemispheric. In 20 - 40 Hz there was also a task x sex interaction for left interhemispheric phase measure.

A univariate follow-up analysis of EEG measures by sex within each task, in the four frequency bands, was then carried out. (See Tables 4 - 8). This showed that, in the eyes open condition, women had more EEG power than men in the frontal region (delta, theta). Women had a systematically less pronounced left► right interhemispheric phase lead in the theta band, when compared to men. Further, women exhibited faster frontal oscillations (alpha). Similar findings occurred in the other conditions. In the eyes-closed situation women showed increased coherence (alpha), reduced left ► right interhemispheric phase (beta) and a reduction in anterior► posterior phase relationships (delta) in the left hemisphere. In the vocabulary task, frontal power was increased in all the frequency bands and posterior power was increased in the beta range. Coherence was increased (beta) and right► left interhemispheric and left anterior posterior phase was less pronounced (theta). During Oral Word Fluency, frontal power was increased (delta, alpha) and posterior power was increased in beta. Left anterior ► posterior phase were reduced (theta) but the left ► right

Table 3. Summary of Significant Analysis of Variance Results

Measure	1 - 3 Hz			4 - 7 Hz			8 - 13 Hz			20 - 40 Hz		
	S	T	TxS	S	T	TxS	S	T	TxS	S	T	TxS
Power												
a) Frontal	x	x	x	x	x	x	x	x			x	
b) Temporal-Parietal		x			x			x		x	x	
Coherence		x			x			x			x	
Phase												
a) Left ➤ Right posterior				x	x			x			x	
b) Left ➤ Right Anterior		x			x							
c) Right Intra-hemispheric		x			x	x		x			x	
d) Left Intra-hemispheric		x			x	x		x				
e) Left Inter-hemispheric		x			x			x			x	x
f) Right Inter-hemispheric		x			x	x		x			x	
Log Right/left power ratio												
a) Frontal		x			x						x	
b) Temporal-Parietal		x			x			x				
Oscillations												
a) Temporal-Parietal		x			x			x			x	
b) Frontal		x			x			x			x	

Note: S = Sex; T = Task; TxS = Task x Sex Interaction

x denotes results significant at $p < .05$

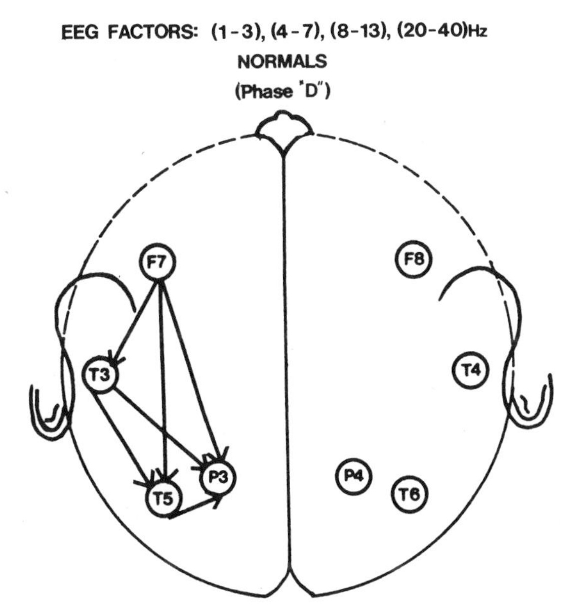

Fig. 4

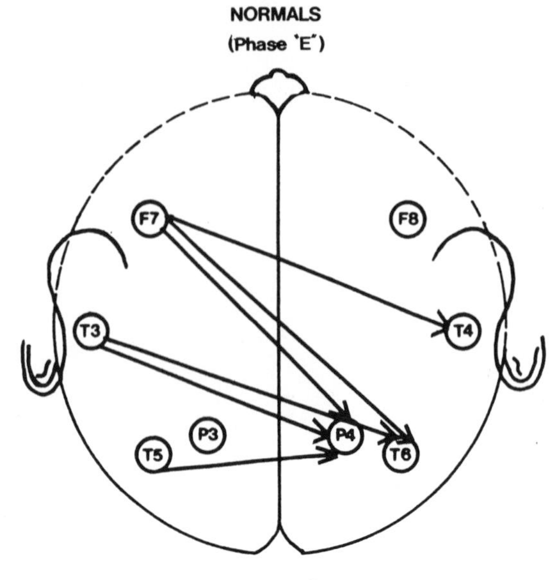

Fig. 5

posterior interhemispheric phase was reduced (delta, alpha) while the right intrahemispheric phase was reduced (theta). Lastly in the spatial processing task (Block-Design), frontal power was increased (delta, theta and alpha) as was posterior power (alpha). The left ► right homologous interhemispheric phase was reduced (theta and alpha), as was the right intrahemispheric phase (alpha) and the posterior oscillations were increased in delta.

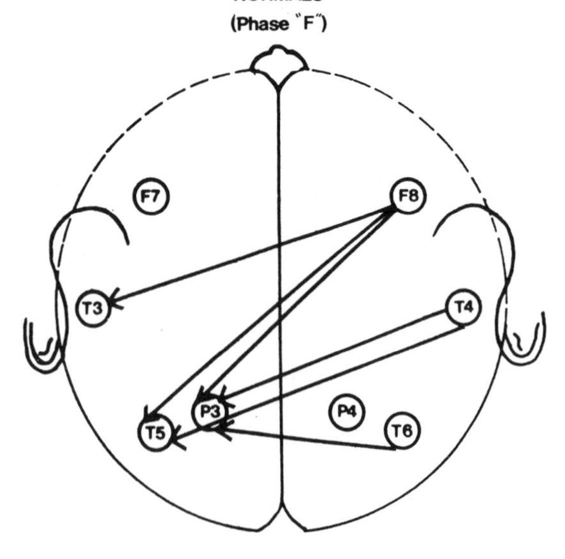

Fig. 6

Table 4. Eyes Open Condition

Males VS. Females

EEG FACTOR	HZ= 1-3	4-7	8-13	20-40	Male	Female
1. Power						
a) Frontal	p<0.05	p<0.05				▲
b) Temporal-parietal						
2. Coherence						
3. Phase						
a) Left►Right Posterior Homologous		p<0.05				▼
b) Left►Right Anterior Homologous						
c) Left►Right Interhemispheric		p<0.05				▼
5. Oscillations						
b) Frontal			p<0.05			▲

Table 5. Eyes Closed Condition. Males vs. Females

EEG FACTOR	Hz= 1-3	4-7	8-13	20-40	Male	Female
2. Coherence			p<0.05			▲
3. Phase						
a) Left ► Right Posterior Homologous				p<0.05		▼
d) Left Intrahemispheric	p<0.05					▼

Table 6. Vocabulary Condition. Males vs. Females.

EEG FACTOR	Hz= 1-3	4-7	8-13	20-40	Male	Female
1. Power						
a) Frontal	p<0.05	p<0.05	p<0.05	p<0.05		▲
b) Temporal-parietal				p<0.05		▲
2. Coherence				p<0.05		▲
3. Phase						
d) Left Intra-hemipheric		p<0.05				▼
f) Right ► Left Interhemispheric		p<0.05				▼

Table 7. Oral Word Fluency Condition Males vs. Females

EEG FACTOR	Hz=	1-3	4-7	8-13	20-40	Male	Female
1. Power							
a) Frontal		$p<0.05$		$p<0.05$			▲
b) Temporal-parietal					$p<0.05$		▲
3. Phase							
c) Right Intra-hemispheric			$p<0.05$				▼
d) Left Intra-hemispheric			$p<0.05$				▼
e) Left►Right Interhemispheric					$p<0.05$		▼

Table 8. Block Design Condition Males vs. Females

EEG FACTOR	Hz=	1-3	4-7	8-13	20-40	Male	Female
1. Power							
a) Frontal		$p<0.05$	$p<0.05$	$p<0.05$			▲
b) Temporal-parietal				$p<0.05$			▲
3. Phase							
a) Left ►Right Posterior Homologous			$p<0.05$	$p<0.05$			▼
c) Right Intra-hemispheric				$p<0.05$			▼
5. Oscillations							
a) Temporal-parietal		$p<0.05$		$p<0.09$			▲

DISCUSSION

The age dependent changes in EEG organization found in this investigation are that, with increasing age, there is a progressive reduction in slow activity and an increase in the fast frequencies in the 20-40 Hz bands. This reduction in slow and increase in fast activity is paralleled by a decrease in EEG coherence in the delta, theta and alpha bands. Moreover with CNS maturation the phase lead of the left hemisphere, left►right and left anterior►posterior becomes significantly less pronounced while the log. of the right/left frontal power ratio becomes smaller. This might suggest an increasing preponderance of right hemispheric processes as individuals become older. In this context it is interesting that Duffy et al. (1984) in the study of age related differences in the EEG of 63 healthy, dextral males ranging in age from 30 to 80 years found, as in our material, a decrease in delta, theta (and to a lesser extent in alpha) and in increase in beta activity as a function of age. Furthermore, the age related EEG features were overwhelmingly right hemispheric (n=33), a few were bilateral (n=9) and only 3 were predominantly left-sided. These authors conclude that "the neurophysiological data suggest that the ageing process affects the right hemisphere more than the left" and wonder if this might not reflect compensatory changes for less efficient left hemisphere functions. The majority of studies on the influence of age and sex on the EEG are, in general, consistent with the principal findings reported here. Greenblatt et al. (1944) studied the problems related to the age factor in the interpretation of EEG abnormalities in neuropsychiatric patients. There were 1593 acute neuropsychiatric patients and 240 normal controls. The ages ranged from 25 to 55+. Although at a given age very different values held for the various neuropsychiatric disorders and normals and specific abnormalities occurred differentially in the various psychiatric disorders, there was underlying all the groups, including the normals, an "age abnormality" curve characterized by a sharp decline in slow activity (below 8-12Hz) to the age of 35 and an increase in fast activity (over 8-12Hz) reaching its maximum between the ages of 45 and 55. Matsuura et al. (1984) review the major subsequent studies on the age development of the EEG. Gibbs and Knott (1949) find that delta and theta activity shows a general decline with age and that beta increases with age. Corbin and Bickford (1955) concluded that, in the evolution of the EEG in childhood, delta, theta and alpha bands showed independent behavior. Matousek and Petersen (1973) reported that the age development of various spectral parameters was linear in childhood and logarithmic in older subjects. Ahn et al. (1980) in his neurometric analysis of 2 large series in Sweden (n = 561) and United States (n = 750) of normal subjects between the ages of 1 and 21 notes that relative power in the delta and theta diminishes, while in the alpha and beta bands it increases with age. In the examination of EEG power density in frequency classes of 1 Hz width (frequency 1 to 15Hz) Colon et al. (1979) report a general decrease in power with increasing age in 47 children age 8, 9 and 10 years. Matsuura et al. (1984) investigated a population of 1416 healthy subjects between the ages of 6 and 39 and found that the age dependent effects were more important than the sex related ones. With a technique of quantification by a computerized wave form recognition method they found that the average alpha amplitude decreased with age, reaching a minimum at the age of 21, the same being true for theta amplitudes. Average delta amplitudes decreased with age until the age of 17. With respect to sex differences they reported that after puberty occipital percentage alpha time was higher in males but that the average alpha amplitude, in that region, was higher in females after the ages of 18 - 21. After the age of 22, the percentage time of theta was higher in females although no sex differences in theta amplitude were observed. Percentage beta time was higher in females than in males at all ages and after the age of 21 the average beta amplitude was also higher in the females. It should be noted that these Japanese workers found a significantly higher alpha amplitude in females than in males in the occipital region, but no difference in alpha amplitude in the other two cerebral sites in the left hemisphere which they monitored (F_p1 and C3), a result which cannot readily be attributed to the differential skull thickness. These findings are very similar to our own, except for the fact that, in several instances, in the Japanese material the age dependent changes cease after the age of 21. This is probably due to the composition of their population which, compared to ours, is heavily skewed towards a much younger age: 88.2% are below the age of 25 and only 11.8% over that age. Duffy et al. (1984) in their "Brain Electrical Activity Mapping" (BEAM) study of age related EEG effects, discussed earlier, found that the largest change occurred between the ages of 30-40 and 50-59 years. Beaumont et al. (1978) studied the effects of task and sex on alpha coherence and found that females showed higher coherence than males overall, independent of task. In their small sample (n=16) the males showed more power asymmetry than the females, once more a result not attributable to skull dimensions. In both sexes the spatial task was associated with an increase in right intrahemispheric coherence, but there was no corresponding coherence change in the left hemisphere during the verbal task. However Tucker et al. (1985) found in the intensive study of 2 subjects tested weekly in a controlled laboratory, and examined for several months, that the major change from resting state was an increase in left anterior coherence during word fluency tasks. In this study, the analysis of variance showed the presence of main effects for task and significant EEG measures x task interactions in all four frequency bands. The EEG correlates of male-female differences in verbally based and spatially based cognitive tasks have been reported elsewhere (Koles and Flor-Henry, 1986, a,b). Briefly, taking bipolar derivations of the EEG and using measurements of coherence to evaluate synchrony between brain regions it was found that comparing resting state (eyes open) with verbal and spatial states, that the spatial state was the most distinct. A stepwise discriminant function analysis showed that a left posterior and right frontal component most strongly distinguished between the verbal and spatial conditions and that while change in pattern from resting state was stronger in men than in women for the spatial state and stronger for women than men for the verbal state, the only significant difference was in the spatial condition.

The analysis of variance across all bands (EEG measures x task x sex) showed a significant main effect for task (theta, alpha and beta) and significant task x measures interactions in all bands. However, analysing the bands separately (males vs. females by task on the 13 EEG measures) significant main effects for sex were found in 10 EEG measures. All the EEG measures x sex interactions were significantly different in men and women in each of the four bands, although here the only significant main effect for sex was in the beta range. Significant main effects for task were found in all bands. It can be concluded that, although task related EEG effects are stronger than sex related ones, there are nevertheless very systematic changes in EEG organization that are gender dependent. These relate principally to differences in power; right and left intrahemispheric anterior ➤ posterior phase lead, and left ➤ right interhemispheric phase relationships. Also EEG coherence was greater in women than in men. Without exception, in women the phase relationships were less pronounced and in particular, the lead of the left hemisphere over the right was reduced. If we ignore a few discordant findings in the beta frequencies in which, despite careful effort, there is always the ever present problem of possible myogenic infiltration of the EEG signal, the directionality of these differences in EEG organization by gender are remarkably constant in all the four bands and across the five cognitive states. The overall increase in general coherence and the lesser degree of antero-posterior and left ➤ right interhemispheric phase gradients suggest, in women a more synchronized and less lateralized pattern of neuroelectric cerebral organization. This is in keeping, of course, with the numerous cognitive studies which have shown,. statistically, that both verbal and spatial cognition engages more bilateral brain systems in women than in men, who on the other hand show greater lateralization for both verbal and spatial processing. The recent demonstration by de Lacoste-Utamsing and Holloway, (1982) that women have more abundant fibres in the posterior part of the corpus callosum than men provides a possible neuroanatomical substrate for the greater bilateralization in female cognitive organization. It is noteworthy that sinistrals compared to dextrals tend also to be more bilateral in their cognitive processing and Witelson (1985) has observed a significantly larger corpus callosum in sinistrals than in dextrals. The present study, as did our previous investigation of male/female EEG differences, finds greater EEG power in women. This was also found by Perris et al. (1981). A problem of interpretation lies in the fact that the skull thickness of men is greater than that of women and therefore this would attenuate the amplitude of the EEG in men. However the brain mass in men is also greater, which might compensate for the differential skull thickness. Furthermore the increase in power is found in some conditions, in particular frequency bands - and not in others - would suggest a neural, rather than an anatomico-physical causal mechanism. In any event, the increased coherence cannot be attributed to differential skull effects and, in a general sense power and coherence are positively correlated. It seems probable that the increase in power in women is related to the fact that women have a thicker cortical gray mantle, and increased rate of cerebral circulation than men (Gur et al., 1982).

CONCLUSIONS

Factorial techniques demonstrate that the EEG of normal subjects consists of 5 fundamental EEG parameters: power, coherence, phase factors: anteroposterior and interhemispheric, log. of right/left power ratios and oscillations. Analysed in terms of these parameters, the EEG organization in men and women is different, women exhibiting more EEG power, a greater degree of EEG coherence and a lesser degree of antero-posterior (bilateral) and left ➤ right interhemispheric phase relationships. This configuration is consistent with known neuroanatomical and cognitive differences in the two genders. Ageing is accompanied by a reduction in the slow, and an increase in the fast frequencies of the EEG power spectrum together with a reduction in the degree with which the left hemisphere leads the right and greater relative right frontal activation. It would appear therefore that, in a certain sense, the ageing brain approximates the female pattern of cerebral organization, in both sexes with increased right hemispheric preponderance.

REFERENCES

Ahn, H., Prichep, L., John, E.R., Trepetin, M., Brown, M.D., and Kaye, H. (1980). Developmental equations for the electroencephalogram, Science, 210, 1255 - 1258.

Beaumont, J.G., Mayes, A.R. and Rugg, M.D. (1978). Asymmetry in EEG alpha coherence and power: effects of task and sex. Electroencephalography and Clinical Neurophysiology, 45, 393-401.

Benignus, V.A. (1969). Estimation of the coherence spectrum and its confidence intervals using the Fast Fourier Transform. IEEE Transactions Audio and Electroacoustics, AU-17, 145.

Corbin, H.P.F. and Bickford, R.G. (1955). Studies of the electroencephalogram of normal children: comparison of visual and automatic frequency analysis. Electroencephalography and Clinical Neurophysiology, 40, 113-131.

Colon, E.J., deWeerd, J.P.C., Notermans, S.L.H. and de Graaf, R. (1979). EEG spectra in children aged 8, 9 and 10 years. Reference values. Journal of Neurology, 221, 263-268.

Duffy, F.H., Albert, M.S., McAnulty, G. and Garvey, A.J. (1984). Age-related differences in brain electrical activity of healthy subjects. Annals of Neurology, 16, 430-438.

De Lacoste-Utamsing, C. & Holloway, R.L. (1982). Sexual dimorphism in the human corpus callosum. Science, 216, 1431-1432.

Flor-Henry, P., Koles, Z.J. and Reddon, J. (1985). EEG studies of sex differences, cognitive and age effects in normals (age range 18 - 60 years). Electroencephalography and Clinical Neurophysiology. 61, S160.

Flor-Henry, P., Koles, Z.J. and Sussman, P.S. (1984). Further observations on right/left hemispheric energy oscillations in the endogenous psychoses. Advances in Biological Psychiatry, 15, 1-11.

Flor-Henry, P. and Koles, Z.J. (1982). EEG characteristics in normal subjects: a comparison of men and women and of dextrals and sinistrals. Research Communications in Psychology, Psychiatry, and Behavior, 7, 21-38.

Gibbs, F.A. and Knott, J.R. (1949). Growth of the electrical activity of the cortex. Electroencephalography and Clinical Neurophysiology, 1, 223-229.

Greenblatt, M., Healey, M.M., & Jones, G.A. (1944). Age and electroencephalographic abnormality in neuropsychiatric patients. American Journal of Psychiatry, 101, 82-90.

Gur, R.C., Gur, R.E., Obrist, W.D. Hungerbuhler, J.P., Younkin, D., Rosen, A.D., Skolnick, B.E. and Reivich, M. (1982). Sex and handedness differences in cerebral blood flow during rest and cognitive

activity. Science, 217, 659-661.
Koles, Z.J. and Flor-Henry, P. (1981). Mental activity and the EEG: task and workload related effects. Medical and Biological Engineering and Computing, 19, 185-194.
Koles, Z.J. and Flor-Henry, P. (1986a). EEG correlates of male-female differences in verbally and spatially-based cognitive tasks. Abstract for the 12th C.M.B.E.C./1st Pan. Pacific Symposium, Vancouver, Canada.
Koles, Z.J. and Flor-Henry, P. (1986b). The effect of brain function on coherence patterns in the bipolar EEG. Unpublished paper.
Lehmann, D. (1981). Spatial analysis of evoked and spontaneous EEG potential fields. In N. Yamaguchi and K. Fujisawa (Eds.) Recent Advances in EEG and EMG Data Processing. Amsterdam: Elsevier North Holland Biomedical Press, 117-132.
Matousek, M. & Peterson, I. (1973). Automatic evaluation of EEG background activity by means of age-dependent EEG quotients. Electroencephalography and Clinical Neurophysiology, 35, 603-612.
Matsuura, M., Yamamoto, K., Fukuzawa, H., Okubo, Y., Uesugi, H., Moriiwa, M., Kojima, T. & Shimazono, Y. (1985). Age development and sex differences of various EEG elements in healthy children and adults - quantification by a computerized wave form recognition method. Electroencephalography and Clinical Neurophysiology, 60, 394-406.
Perris, C., von Knorring, L., Cumberpatch, J. and Marciano, F. (1981). Further studies of depressed patients by means of computerized EEG. Advances in Biological Psychiatry, 6, 41-49.
Tucker, D.M., Dawson, S.L., Roth, D.L. & Penland, J.G. (1985). Regional changes in EEG power and coherence during cognition: intensive study of two individuals. Behavioral Neuroscience, 99, 564-577.
Witelson, S.F. (1985). The brain connection: the corpus callosum is larger in left-handers. Science, 229, 665-668.
Welch, P.D. (1967). The use of the Fast Fourier Transform for the estimation of power spectra: a method based on time averaging over short, modified periodograms. IEEE Transactions Audio and Electroacoustics, 15, 70-73.

APPENDIX I

Dietrich Lehmann (1981) has persuasively argued that amplitude and phase information in any given recording channel is ambiguous information for depending on the chosen reference the value will differ. Average reference is however "reference-free" since in such a system the EEG potential fields are not determined by any particular montage. The following two tables indicate the correspondence between the montage used in the study and reconstructed average reference. It is seen that although the global predictability of one system with respect to the other is modest, 15 - 20%, the correlations for the EEG frequencies between the two systems are generally robust, except for the phase.

See over/...

Pearson product moment correlations between measures derived from unipolar and average references

Measure	1 - 3Hz	4 - 7Hz	8 - 13Hz	20 - 40Hz
1. Power				
a) Frontal	.801	.693	.642	.635
b) Temporal-Parietal	.360	.358	.744	.829
2. Coherence	.366	.234	.665	.647
3. Phase				
a) Left→Right Posterior	.056	.304	.264	.077
b) Left→Right Anterior	.226	.234	.122	.076
c) Right Intrahemispheric	.054	.221	.274	.168
d) Left Intrahemispheric	.094	.175	.211	.183
e) Left→Right Interhemispheric	.038	.138	-.038	-.061
f) Right→Left Intrahemispheric	.097	.142	.039	.046
4. Log Right/Left Power Ratio				
a) Frontal	.367	.426	.341	.315
b) Temporal-Parietal	.323	.332	.327	.414
5. Oscillations				
a) Temporal-Parietal	.515	.244	.625	.564
b) Frontal	.356	.290	.515	.515

Canonical Redundancy Between Unipolar (UR) and Average (AR) Reference

Canonical Correlations

	1 - 3 Hz	4 - 7 Hz	8 - 13 Hz	20 - 40 Hz
1	.83	.77	.78	.87
2	.58	.65	.72	.73
3	.53	.54	.52	.58
4	.40	.46	.51	.52
5	.35	.38	.43	.42
6	.30	.32	.41	.34
7	.25 $p < .05$	.26	.36	.31
8		.25	.28	.28 $p < .05$
9		.19 $p < .05$	.26	
10			.19 $p < .05$	

Canonical Redundancy

Percent	UR	AR	UR	AR	UR	AR	UR	AR
%	15.2	20.9	14.9	18.4	24.5	25.8	22.9	25.1

i.e. From Unipolar 20% of Average Reference Predictable

i.e. From Average 15% of Unipolar Reference Predictable.

Appendix II.

Males Age x Factor Correlations

Eyes Open

Factor	Frequency Bands (Hz) 1 - 3	4 - 7	8 - 13	20 - 40
1. Power				
a) Frontal	-0.25*	-0.37*	0.04	0.14
b) Temporal-Parietal	-0.26*	-0.32*	-0.05	-0.08
2. Coherence	-0.20	-0.31*	0.00	-0.01
3. Phase				
a) Left→Right Posterior	-0.05	0.07	0.03	-0.14*
b) Left→Right Anterior	-0.02	0.18	0.15	0.03
c) Right Intrahemispheric	0.22*	0.16	-0.16	-0.07
d) Left Intrahemispheric	-0.05	0.11	-0.18	-0.03
e) Left Interhemispheric	-0.12	0.25*	-0.05	0.01
f) Right Interhemispheric	0.20	0.21	-0.19	0.11
4. Average log Right/left Power Ratio				
a) Frontal	-0.26*	-0.27*	-0.20	-0.22*
b) Temporal-Parietal	0.16	0.03	0.03	-0.13
5. Oscillations				
a) Temporal-Parietal	0.13	-0.09	-0.22*	0.01
b) Frontal	0.16	0.16	-0.01	-0.06

Males Age x Factor Correlations

Eyes Closed

Factor	Frequency Bands (Hz) 1 - 3	4 - 7	8 - 13	20 - 40
1. Power				
a) Frontal	-0.31*	-0.27*	0.05	0.13
b) Temporal-Parietal	-0.24*	-0.27*	-0.18	-0.06
2. Coherence	-0.31*	-0.28*	-0.13	0.01
3. Phase				
a) Left→Right Posterior	-0.07	-0.27*	-0.04	-0.05
b Left→Right Anterior	0.08	-0.38*	-0.16	0.22*
c) Right Intrahemispheric	0.01	0.19	0.05	-0.21
d) Left Intrahemispheric	-0.09	0.06	-0.04	-0.28*
e) Left Interhemispheric	-0.18	-0.14	0.03	-0.03
f) Right Interhemispheric	-0.13	0.34	-0.01	-0.19
4. Average log Right/left Power Ratio				
a) Frontal	-0.27*	-0.32*	0.09	-0.18
b) Temporal-Parietal	0.09	-0.15	-0.08	-0.01
5. Oscillations				
a) Temporal-Parietal	0.04	0.26*	-0.01	-0.16
b) Frontal	0.08	0.13	-0.24*	-0.26*

Males Age x Factor Correlations

Vocabulary

Factor	Frequency Bands (Hz) 1 - 3	4 - 7	8 - 13	20 - 40
1. Power				
a) Frontal	-0.29*	-0.32*	-0.14	0.03
b) Temporal-Parietal	-0.09	-0.28*	-0.13	-0.11
2. Coherence	-0.10	-0.26*	-0.08	-0.03
3. Phase				
a) Left ➤Right Posterior	0.06	0.05	0.03	-0.10
b) Left ➤Right Anterior	0.03	0.05	-0.01	0.22*
c) Right Intrahemispheric	-0.01	0.08	-0.13	-0.12
d) Left Intrahemispheric	0.00	0.03	-0.14	0.05
e) Left Interhemispheric	0.13	0.01	-0.14	0.03
f) Right Interhemispheric	-0.05	0.05	0.06	0.00
4. Average log Right/Left Power Ratio				
a) Frontal	-0.13	-0.19	-0.08	-0.18
b) Temporal-Parietal	0.10	0.13	-0.14	-0.04
5. Oscillations				
a) Temporal-Parietal	0.02	0.17	-0.02	-0.08
b) Frontal	-0.03	0.03	-0.14	-0.18

Males Age x Factor Correlations

Oral Word Fluency

Factor	Frequency Bands (Hz) 1 - 3	4 - 7	8 - 13	20 - 40
1. Power				
a) Frontal	-0.025*	-0.32*	-0.13	0.21
b) Temporal-Parietal	-0.11	-0.32*	0.20	-0.17
2. Coherence	-0.12	-0.32*	-0.13	0.04
2. Phase				
a) Left ➤Right Posterior	-0.11	-0.02	0.11	0.00
b) Left ➤Right Anterior	-0.02	-0.15	-0.10	0.18
c) Right Intrahemispheric	0.17	0.14	-0.16	-0.03
d) Left Intrahemispheric	-0.03	-0.03	-0.22*	0.04
e) Left Interhemispheric	0.25*	0.03	-0.07	-0.02
f) Right Interhemispheric	-0.01	0.35*	-0.05	0.02
4. Average log Right/Left Power Ratio				
a) Frontal	-0.24*	-0.23	-0.03	-0.12
b) Temporal-Parietal	0.12	0.08	-0.12	0.06
5. Oscillations				
a) Temporal-Parietal	-0.11	0.20	-0.04	-0.07
b) Frontal	-0.16	0.03	-0.20	-0.23*

Males Age x Factor Correlations

Block Design

Factor	Frequency Bands (Hz) 1 - 3	4 - 7	8 - 13	20 - 40
1. Power				
a) Frontal	-0.34*	-0.41*	-0.34*	0.12
b) Temporal-Parietal	-0.10	-0.21	-0.22*	0.14
2. Coherence	-0.10	-0.19	-0.18	-0.02
3. Phase				
a) Left→Right Posterior	0.20	-0.03	0.09	0.00
b) Left→Right Anterior	0.00	-0.32*	-0.03	0.01
c) Right Intrahemispheric	-0.07	-0.31*	-0.14	0.28*
d) Left Intrahemispheric	0.08	-0.09	-0.03	-0.09
e) Left Interhemispheric	-0.03	-0.31*	-0.10	-0.08
f) Right Interhemispheric	-0.22*	-0.08	-0.17	-0.18
4. Average log Right/Left Power Ratio				
a) Frontal	-0.01	-0.16	-0.22*	-0.28*
b) Temporal-Parietal	0.07	0.03	0.01	0.03
5. Oscillations				
a) Temporal-Parietal	-0.08	-0.07	-0.02	-0.21
b) Frontal	-0.07	-0.10	-0.02	-0.23*

Females Age x Factor Correlations

Eyes Open

Factor	Frequency Bands (Hz) 1 - 3	4 - 7	8 - 13	20 - 40
1. Power				
a) Frontal	-0.23*	-0.40*	-0.35*	-0.20
b) Temporal-Parietal	-0.31*	-0.45*	-0.36*	0.05
2. Coherence	-0.19	-0.44*	-0.40*	-0.08
3. Phase				
a) Left→Right Posterior	-0.16	-0.38*	0.04	0.08
b) Left→Right Anterior	-0.09	-0.24*	0.03	0.14
c) Right Intrahemispheric	0.03	-0.22*	0.09	0.05
d) Left Intrahemispheric	-0.23*	-0.35*	0.06	-0.02
e) Left Interhemispheric	-0.06	-0.54*	0.12	0.04
f) Right Interhemispheric	0.20	0.08	-0.05	-0.16
4. Average log Right/Left Power Ratio				
a) Frontal	-0.08	-0.13	-0.19	-0.19
b) Temporal-Parietal	-0.11	-0.04	-0.20	0.15
5. Oscillations				
a) Temporal-Parietal	-0.37*	0.16	0.22*	-0.42*
b) Frontal	-0.07	-0.07	0.06	0.00

Females Age x Factor Correlations

Eyes Closed

Factor	Frequency Bands (Hz) 1 - 3	4 - 7	8 - 13	20 - 40
1. Power				
a) Frontal	-0.42*	-0.54*	-0.29*	-0.06
b) Temporal	-0.55*	-0.52*	-0.31*	-0.02
2. Coherence	-0.54*	-0.54*	-0.26*	-0.03
3. Phase				
a) Left→Right Posterior	-0.40*	-0.07	0.13	-0.13
b) Left→Right Anterior	0.00	-0.10	0.15	-0.03
c) Right Intrahemispheric	-0.05	0.09	0.18	-0.11
d) Left Intrahemispheric	0.02	-0.05	0.14	-0.01
e) Left Interhemispheric	-0.10	0.02	0.29	-0.06
f) Right Interhemispheric	-0.03	0.29*	0.00	-0.06
4. Average log Right/Left Power Ratio				
a) Frontal	0.04	-0.05	-0.05	-0.15
b) Temporal-Parietal	-0.11	-0.11	0.03	-0.10
5. Oscillations				
a) Temporal-Parietal	-0.11	-0.13	0.13	-0.30*
b) Frontal	-0.01	0.19	0.03	-0.15

Females Age x Factor Correlations

Vocabulary

Factor	Frequency Bands (Hz) 1 - 3	4 - 7	8 - 13	20 - 40
1. Power				
a) Frontal	-0.19	-0.25*	-0.26*	0.16
b) Temporal-Parietal	-0.18	-0.37*	-0.34*	0.04
2. Coherence	-0.20	-0.37*	-0.31*	0.08
3. Phase				
a) Left→Right Posterior	-0.16	-0.30*	-0.13	0.27*
b) Left→Right Anterior	-0.19	-0.04	0.06	-0.10
c) Right Intrahemispheric	0.22*	-0.15	-0.01	-0.23*
d) Left Intrahemispheric	0.05	0.07	-0.16	0.08
e) Left Interhemispheric	-0.07	-0.07	-0.13	-0.11
f) Right Interhemispheric	0.25*	-0.10	-0.13	0.00
4. Average log Right/left Power Ratio				
a) Frontal	0.13	0.06	0.06	-0.03
b) Temporal-Parietal	0.26*	0.22*	0.14	0.22*
5. Oscillations				
a) Temporal-Parietal	-0.18	0.19	0.18	-0.14
b) Frontal	-0.14	-0.08	-0.07	-0.26*

Females Age x Factor Correlations

Oral Word Fluency

Factor	Frequency Bands (Hz) 1 - 3	4 - 7	8 - 13	20 - 40
1. Power				
a) Frontal	-0.23*	-0.31*	-0.27*	0.24*
b) Temporal-Parietal	-0.09	-0.37*	-0.33*	0.12
2. Coherence	-0.12	-0.38*	-0.31*	0.28*
3. Phase				
a) Left→Right Posterior	-0.22	-0.30*	-0.23*	-0.06
b) Left→Right Anterior	-0.07	-0.10	-0.15	-0.08
c) Right Intrahemispheric	0.04	-0.20	-0.10	-0.27*
d) Left Intrahemispheric	0.05	0.00	-0.16	-0.04
e) Left Interhemispheric	-0.05	-0.01	-0.20	-0.09
f) Right Interhemispheric	0.11	0.16	0.13	-0.09
4. Average log Right/left Power Ratio				
a) Frontal	0.14	0.07	-0.03	0.04
b) Temporal-Parietal	0.17	0.07	-0.03	0.12
5. Oscillations				
a) Temporal-Parietal	-0.36*	0.18	0.15	-0.18
b) Frontal	-0.18	-0.04	-0.07	-0.38*

Females Age x Factor Correlations

Block Design

Factor	Frequency Bands (Hz) 1 - 3	4 - 7	8 - 13	20 - 40
1. Power				
a) Frontal	-0.40*	-0.44*	-0.25*	0.26*
b) Temporal-Parietal	-0.07	-0.16	0.04	0.30*
2. Coherence	-0.32*	-0.37*	-0.22*	0.26*
3. Phase				
a) Left→Right Posterior	0.01	-0.19	-0.17	-0.19
b) Left→Right Anterior	0.19	0.09	0.12	0.25*
c) Right Intrahemispheric	-0.09	-0.17	-0.25*	0.02
d) Left Intrahemispheric	0.05	0.17	0.11	0.13
e) Left Interhemispheric	0.16	0.10	-0.02	-0.05
f) Right Interhemispheric	-0.06	0.01	-0.02	0.04
4. Average log Right/left Power Ratio				
a) Frontal	0.17	0.09	0.03	-0.12
b) Temporal-Parietal	-0.01	-0.01	0.05	0.04
5. Oscillations				
a) Temporal-Parietal	0.07	0.15	-0.29*	-0.26*
b) Frontal	-0.08	-0.12	-0.45*	-0.34*

Appendix III.

CELL MEANS AND STANDARD DEVIATIONS (S.D.) BY TASK (1 - 3 Hz)

For Normal Males (n=57) and Normal Females (n=56 on =3 EEG Measures)

EEG Measure and Task	Normal Males (n=57) Cell Mean	Normal Males (n=57) Standard Deviation	Normal Females (n=56) Cell Mean	Normal Females (n=56) Standard Deviation
1. Power				
a) Frontal				
Eyes Open	-0.577	0.439	-0.357	0.531
Eyes Closed	-0.782	0.174	-0.782	0.153
Vocabulary	0.074	0.688	0.720	1.243
Oral Word Fluency	0.035	0.783	0.437	1.118
Block Design	0.233	0.928	0.911	1.210
b) Temporal-Parietal				
Eyes Open	-0.406	0.273	-0.355	0.221
Eyes Closed	-0.368	0.623	-0.460	0.198
Vocabulary	0.249	1.075	0.300	0.767
Oral Word Fluency	0.302	1.348	0.297	0.876
Block Design	0.170	1.968	0.224	0.786
2. Coherence				
Eyes Open	-0.390	0.310	-0.311	0.294
Eyes Closed	-0.455	0.332	-0.465	0.183
Vocabulary	0.116	0.667	0.340	0.835
Oral Word Fluency	0.204	1.138	0.260	0.795
Block Design	0.325	2.344	0.341	0.772
3. Phase				
a) Left→Right Posterior				
Eyes Open	-0.030	0.454	0.073	0.279
Eyes Closed	-0.014	0.442	-0.046	0.223
Vocabulary	-0.049	0.966	0.002	0.469
Oral Word Fluency	-0.072	1.187	0.278	2.286
Block Design	-0.080	0.604	-0.198	1.468
b) Left→Right Anterior				
Eyes Open	0.081	0.668	0.064	0.945
Eyes Closed	0.250	0.697	0.222	0.439
Vocabulary	0.053	0.822	0.020	1.129
Oral Word Fluency	0.086	0.639	0.121	0.769
Block Design	-0.418	1.531	-0.578	1.578
c) Right Intrahemispheric				
Eyes Open	0.173	1.133	0.002	0.697
Eyes Closed	0.385	0.512	0.339	0.404
Vocabulary	-0.360	1.146	-0.362	1.127
Oral Word Fluency	-0.127	1.323	-0.450	1.019
Block Design	0.238	0.532	0.061	1.317
d) Left Intrahemispheric				
Eyes Open	-0.161	1.022	-0.069	0.695
Eyes Closed	0.417	0.401	0.263	0.367
Vocabulary	-0.059	1.334	-0.267	1.331
Oral Word Fluency	-0.105	1.289	-0.518	1.144
Block Design	0.150	0.628	0.204	1.039
e) Left Interhemispheric				
Eyes Open	0.011	0.722	0.110	0.776
Eyes Closed	0.524	0.534	0.410	0.446
Vocabulary	-0.077	0.964	-0.230	1.058
Oral Word Fluency	0.056	0.913	-0.116	1.004
Block Design	-0.445	1.154	-0.512	1.570

CELL MEANS AND STANDARD DEVIATIONS (S.D.) BY TASK (1 - 3 Hz)

For Normal Males (n=57) and Normal Females (n=56) on 13 EEG Measures

	Normal Males (n=57)		Normal Females (n=56)	
EEG Measure and Task	Cell Mean	Standard Deviation	Cell Mean	Standard Deviation
f) Right Interhemispheric				
Eyes Open	0.093	0.841	0.060	0.913
Eyes Closed	0.217	0.723	0.301	0.449
Vocabulary	-0.224	0.970	-0.091	1.110
Oral Word Fluency	-0.100	0.934	-0.254	0.680
Block Design	0.109	1.227	-0.018	1.752
4. Average log Right/left Power Ratio				
a) Frontal				
Eyes Open	-0.119	1.041	-0.150	1.095
Eyes Closed	0.195	0.588	0.213	0.625
Vocabulary	-0.103	1.203	-0.144	0.938
Oral Word Fluency	-0.205	1.152	-0.417	1.198
Block Design	0.336	1.110	0.205	0.875
b) Temporal-Parietal				
Eyes Open	-0.180	0.691	-0.039	0.726
Eyes Closed	-0.177	0.876	-0.119	0.804
Vocabulary	0.148	0.888	-0.131	1.021
Oral Word Fluency	0.020	0.953	-0.063	1.035
Block Design	0.281	1.357	0.197	1.512
5. Oscillations				
a) Temporal-Parietal				
Eyes Open	0.279	0.986	0.164	0.793
Eyes Closed	0.867	1.129	0.997	1.212
Vocabulary	-0.263	0.926	-0.357	0.705
Oral Word Fluency	-0.238	0.829	-0.231	0.725
Block Design	-0.631	0.701	-0.397	0.485
b) Frontal				
Eyes Open	-0.020	1.054	-0.002	1.026
Eyes Closed	0.765	1.090	0.712	0.921
Vocabulary	-0.263	0.866	-0.198	0.971
Oral Word Fluency	-0.319	0.895	-0.546	0.917
Block Design	-0.101	0.841	0.010	0.799

CELL MEANS AND STANDARD DEVIATIONS (S.D.) BY TASK (4 - 7 Hz)

For Normal Males (n=57) and Normal Females (n=56) on 13 EEG Measures

EEG Measure and Task	Normal Males (n=57) Cell Mean	Normal Males (n=57) Standard Deviation	Normal Females (n=56) Cell Mean	Normal Females (n=56) Standard Deviation
1. Power				
a) Frontal				
Eyes Open	-0.600	0.453	-0.370	0.511
Eyes Closed	-0.652	0.591	-0.666	0.349
Vocabulary	0.116	1.211	0.563	1.024
Oral Word Fluency	0.040	1.131	0.356	0.910
Block Design	0.262	0.842	0.869	0.986
b) Temporal-Parietal				
Eyes Open	-0.444	0.557	-0.240	0.783
Eyes Closed	-0.064	1.285	-0.079	0.887
Vocabulary	0.073	1.313	0.613	0.850
Oral Word Fluency	0.056	1.155	0.201	0.846
Block Design	0.002	1.178	0.342	1.000
2. Coherence				
Eyes Open	-0.382	0.563	-0.125	0.919
Eyes Closed	-0.088	1.136	-0.002	0.882
Vocabulary	0.003	1.335	0.126	0.893
Oral Word Fluency	0.001	1.066	0.147	0.835
Block Design	0.125	1.494	0.203	0.817
3. Phase				
a) Left→Right Posterior				
Eyes Open	0.199	0.792	-0.164	0.775
Eyes Closed	-0.068	1.420	-0.147	0.987
Vocabulary	0.192	0.768	0.073	0.725
Oral Word Fluency	0.245	0.960	-0.049	1.429
Block Design	0.015	0.605	-0.412	1.294
b) Left→Right Anterior				
Eyes Open	0.076	0.595	0.005	0.542
Eyes Closed	-0.003	0.738	0.054	0.175
Vocabulary	-0.006	0.960	0.248	0.761
Oral Word Fluency	0.156	0.744	0.258	0.622
Block Design	-0.338	1.700	-0.403	1.809
c) Right Intrahemispheric				
Eyes Open	0.101	0.859	0.130	0.628
Eyes Closed	0.056	0.699	0.249	0.624
Vocabulary	0.100	1.217	-0.048	0.925
Oral Word Fluency	0.229	1.323	-0.344	1.325
Block Design	-0.299	0.721	-0.305	1.149
d) Left Intrahemispheric				
Eyes Open	0.045	0.829	-0.072	0.810
Eyes Closed	-0.046	0.764	0.092	0.516
Vocabulary	0.441	1.161	-0.046	0.928
Oral Word Fluency	0.205	1.400	-0.352	1.278
Block Design	-0.146	0.751	-0.192	1.200
e) Left Interhemispheric				
Eyes Open	0.264	0.662	-0.042	0.679
Eyes Closed	0.037	0.721	0.117	0.409
Vocabulary	0.428	0.958	0.254	0.913
Oral Word Fluency	0.302	0.969	0.055	1.124
Block Design	-0.680	1.477	-0.811	1.073

CELL MEANS AND STANDARD DEVIATIONS (S.D.) BY TASK (4 - 7 Hz)

For Normal Males (n=57) and Normal Females (n=56) on 13 EEG Measures

	Normal Females (n=57)		Normal Females (n=56)	
EEG Measure and Task	Cell Mean	Standard Deviation	Cell Mean	Standard Deviation
f) Right Interhemispheric				
Eyes Open	0.136	0.647	0.172	0.597
Eyes Closed	0.187	1.084	0.195	0.459
Vocabulary	0.242	1.068	-0.187	0.936
Oral Word Fluency	0.149	1.004	-0.196	0.954
Block Design	-0.471	1.191	-0.223	1.613
4. Average log Right/left Power Ratio				
a) Frontal				
Eyes Open	-0.058	0.969	-0.186	1.092
Eyes Closed	0.255	0.431	0.171	0.475
Vocabulary	-0.125	1.185	-0.150	1.000
Oral Word Fluency	-0.269	1.067	-0.366	1.434
Block Design	0.217	1.067	0.243	0.881
b) Temporal-Parietal				
Eyes Open	-0.205	0.782	-0.226	0.781
Eyes Closed	-0.112	1.034	-0.057	0.968
Vocabulary	0.118	0.859	-0.081	0.997
Oral Word Fluency	-0.011	0.936	0.068	1.029
Block Design	0.227	1.207	0.199	1.389
5. Oscillations				
a) Temporal-Parietal				
Eyes Open	-0.332	0.897	-0.401	0.925
Eyes Closed	-0.003	1.344	-0.162	1.111
Vocabulary	0.231	1.058	0.312	0.741
Oral Word Fluency	0.434	1.094	0.273	0.896
Block Design	-0.302	0.892	-0.213	0.605
b) Frontal				
Eyes Open	-0.573	0.820	-0.491	1.109
Eyes Closed	0.310	1.270	0.215	0.912
Vocabulary	-0.070	1.076	0.153	0.788
Oral Word Fluency	0.153	1.108	-0.012	0.853
Block Design	0.117	0.778	0.071	0.814

CELL MEANS AND STANDARD DEVIATIONS (S.D.) BY TASK (8 - 13 Hz)

For Normal Males (n=57) and Normal Females (n=56) on 13 EEG Measures

EEG Measures and Task	Normal Males (n=57) Cell Mean	Normal Males (n=57) Standard Deviation	Normal Females (n=56) Cell Mean	Normal Females (n=56) Standard Deviation
1. Power				
a) Frontal				
Eyes Open	-0.456	0.755	-0.240	0.960
Eyes Closed	0.008	1.090	0.390	1.409
Vocabulary	-0.259	0.687	0.310	1.249
Oral Word Fluency	-0.258	0.716	0.161	1.209
Block Design	-0.012	0.629	0.461	0.799
b) Temporal-Parietal				
Eyes Open	-0.281	0.580	-0.158	0.816
Eyes Closed	0.511	1.314	1.071	1.981
Vocabulary	-0.258	0.545	-0.042	0.723
Oral Word Fluency	-0.225	0.511	-0.032	0.718
Block Design	-0.360	0.389	-0.186	0.468
2. Coherence				
Eyes Open	-0.243	0.565	-0.121	0.781
Eyes Closed	0.470	1.213	1.064	1.914
Vocabulary	-0.254	0.610	-0.043	0.898
Oral Word Fluency	-0.214	0.525	0.001	0.967
Block Design	-0.304	0.564	-0.299	0.335
3. Phase				
a) Left ➤ Right Posterior				
Eyes Open	0.232	1.097	-0.085	1.019
Eyes Closed	-0.185	1.039	-0.039	1.127
Vocabulary	0.229	1.034	0.086	0.896
Oral Word Fluency	0.231	0.906	0.048	1.005
Block Design	-0.047	0.655	-0.478	1.312
b) Left ➤ Right Anterior				
Eyes Open	0.174	0.512	0.105	0.471
Eyes Closed	-0.063	0.435	-0.070	0.358
Vocabulary	0.026	0.843	0.021	0.904
Oral Word Fluency	0.027	0.641	-0.000	0.721
Block Design	-0.160	2.113	-0.085	1.626
c) Right Intrahemispheric				
Eyes Open	0.171	0.895	0.150	0.873
Eyes Closed	-0.264	1.245	-0.471	1.235
Vocabulary	0.108	0.960	0.106	1.273
Oral Word Fluency	0.055	0.984	0.081	1.274
Block Design	-0.122	0.340	0.068	0.638
d) Left Intrahemispheric				
Eyes Open	0.130	0.814	0.023	0.773
Eyes Closed	-0.312	1.224	-0.254	1.280
Vocabulary	0.147	0.953	0.190	1.286
Oral Word Fluency	0.047	1.010	0.249	1.256
Block Design	-0.211	0.452	-0.134	0.729
e) Left Interhemispheric				
Eyes Open	0.400	0.887	0.253	0.892
Eyes Closed	-0.320	0.853	-0.092	1.057
Vocabulary	0.147	1.200	0.051	0.963
Oral Word Fluency	0.033	0.983	0.180	0.923
Block Design	-0.336	1.052	-0.441	1.103

CELL MEANS AND STANDARD DEVIATIONS (S.D.) BY TASK (8 - 13 Hz)

For Normal Males (n=57) and Normal Females (n=56) on 13 EEG Measures

	Normal Males (n=57)		Normal Females (n=56)	
EEG Measure and Task	Cell Mean	Standard Deviation	Cell Mean	Standard Deviation
3. Phase				
f) Right Interhemispheric				
Eyes Open	-0.055	0.950	0.211	0.786
Eyes Closed	-0.075	1.148	-0.107	1.082
Vocabulary	-0.013	1.122	0.177	1.066
Oral Word Fluency	0.039	1.079	0.126	1.008
Block Design	-0.275	1.026	-0.190	0.937
4. Average log Right/Left Power Ratio				
a) Frontal				
Eyes Open	0.168	0.767	-0.085	0.986
Eyes Closed	0.183	0.717	-0.015	0.739
Vocabulary	-0.025	1.099	0.067	0.851
Oral Word Fluency	-0.087	0.894	-0.206	1.094
Block Design	-0.242	1.520	-0.013	1.026
b) Temporal-Parietal				
Eyes Open	0.050	0.744	-0.128	0.817
Eyes Closed	-0.008	1.159	0.168	1.249
Vocabulary	0.174	0.929	0.089	1.004
Oral Word Fluency	0.143	1.093	0.186	1.048
Block Design	-0.228	0.891	-0.344	0.985
5. Oscillations				
a) Temporal-Parietal				
Eyes Open	-0.280	0.939	-0.267	0.852
Eyes Closed	-0.946	0.749	-1.143	0.826
Vocabulary	0.474	0.896	0.436	0.853
Oral Word Fluency	0.424	0.932	0.351	0.830
Block Design	0.429	0.612	0.305	0.487
b) Frontal				
Eyes Open	-0.477	0.887	-0.422	0.754
Eyes Closed	-0.375	1.281	-0.722	1.040
Vocabulary	0.594	0.787	0.678	0.817
Oral Word Fluency	0.353	0.791	0.529	0.808
Block Design	-0.190	0.679	-0.087	0.831

CELL MEANS AND STANDARD DEVIATIONS (S.D.) BY TASK (20 - 40 Hz)

For Normal Males (n=57) and Normal Females (n=56) on 13 EEG Measures

	Normal Males (n=57)		Normal Females (n=56)	
EEG Measure and Task	Cell Mean	Standard Deviation	Cell Mean	Standard Deviation
1. Power				
a) Frontal				
Eyes Open	-0.366	0.488	-0.357	0.275
Eyes Closed	-0.416	0.281	-0.379	0.262
Vocabulary	-0.011	0.717	0.487	1.344
Oral Word Fluency	0.005	0.660	0.195	1.032
Block Design	0.226	1.084	0.693	1.986
b) Temporal-Parietal				
Eyes Open	-0.408	0.212	-0.341	0.180
Eyes Closed	-0.339	0.359	-0.307	0.214
Vocabulary	-0.158	0.421	0.230	0.744
Oral Word Fluency	-0.107	0.429	0.204	0.837
Block Design	0.285	1.031	0.929	2.367
2. Coherence				
Eyes Open	-0.283	0.482	-0.238	0.335
Eyes Closed	-0.157	0.673	-0.151	0.440
Vocabulary	-0.132	0.614	0.214	0.994
Oral Word Fluency	-0.129	0.519	0.080	0.757
Block Design	0.197	1.241	0.656	2.299
3. Phase				
a) Left➤Right Posterior				
Eyes Open	0.164	0.427	0.108	0.369
Eyes Closed	0.072	0.132	-0.074	0.414
Vocabulary	-0.023	0.836	0.318	1.366
Oral Word Fluency	0.255	0.558	0.057	1.066
Block Design	-0.177	1.572	-0.619	1.697
b) Left➤Right Anterior				
Eyes Open	0.066	0.244	-0.077	0.553
Eyes Closed	0.024	0.462	-0.027	0.321
Vocabulary	-0.034	1.142	-0.071	0.912
Oral Word Fluency	-0.055	1.350	0.143	0.898
Block Design	0.080	1.775	-0.108	1.420
c) Right Intrahemispheric				
Eyes Open	-0.026	0.566	-0.022	0.308
Eyes Closed	-0.021	0.202	-0.145	0.501
Vocabulary	-0.136	1.071	0.050	1.140
Oral Word Fluency	-0.210	1.110	0.011	1.112
Block Design	0.261	1.254	0.367	1.687
d) Left Intrahemispheric				
Eyes Open	-0.062	0.458	-0.193	0.572
Eyes Closed	-0.019	0.340	-0.165	0.610
Vocabulary	0.070	0.908	0.160	1.062
Oral Word Fluency	0.018	0.592	0.143	1.137
Block Design	0.042	1.146	0.276	1.748
e) Left Interhemispheric				
Eyes Open	-0.053	0.263	-0.166	0.442
Eyes Closed	-0.147	0.264	-0.250	0.309
Vocabulary	0.101	1.013	-0.111	1.029
Oral Word Fluency	-0.400	1.133	0.102	1.067
Block Design	0.478	1.618	0.364	1.518

CELL MEANS AND STANDARD DEVIATIONS (S.D.) BY TASK (20 - 40 Hz)

For Normal Males (n=57) and Normal Females (n=56) on 13 EEG Measures

EEG Measure and Task	Normal Males (n=57) Cell Mean	Normal Males (n=57) Standard Deviation	Normal Females (n=56) Cell Mean	Normal Females (n=56) Standard Deviation
3. Phase				
f) Right Interhemispheric				
Eyes Open	-0.084	0.232	-0.165	0.504
Eyes Closed	-0.190	0.334	-0.136	0.433
Vocabulary	-0.151	0.875	-0.212	0.713
Oral Word Fluency	0.004	0.820	-0.034	0.967
Block Design	0.402	1.739	0.620	1.566
4. Average log Right/Left Power Ratio				
a) Frontal				
Eyes Open	0.223	1.360	0.097	0.784
Eyes Closed	-0.133	0.963	-0.167	0.766
Vocabulary	-0.084	1.159	0.105	0.808
Oral Word Fluency	0.058	1.109	0.042	1.093
Block Design	-0.097	1.203	0.200	0.589
b) Temporal-Parietal				
Eyes Open	-0.005	0.769	-0.063	0.901
Eyes Closed	-0.084	0.704	0.011	0.774
Vocabulary	-0.063	1.062	-0.062	1.187
Oral Word Fluency	0.043	1.105	0.143	1.044
Block Design	0.043	1.202	0.283	1.323
5. Oscillations				
a) Temporal-Parietal				
Eyes Open	0.328	0.811	0.194	0.717
Eyes Closed	0.745	0.905	0.611	0.881
Vocabulary	-0.051	0.930	-0.262	1.100
Oral Word Fluency	0.071	0.921	-0.211	1.046
Block Design	-0.732	0.829	-0.769	0.775
b) Frontal				
Eyes Open	-0.330	1.005	0.044	1.024
Eyes Closed	0.244	1.138	0.283	0.865
Vocabulary	-0.164	1.037	0.082	1.087
Oral Word Fluency	-0.072	0.659	0.187	1.015
Block Design	-0.252	1.010	-0.012	1.056

ATTENTIONAL FACTORS AND INDIVIDUAL DIFFERENCES REFLECTED IN THE EEG

William J. Ray

Penn State University
University Park
Pa. 16802

INTRODUCTION

For the last few years the work in my laboratory has been directed toward understanding how individual differences are reflected in psychophysiological measures of cognitive and emotional processing. This program has brought together two somewhat different approaches to research. On the one hand we have sought to find the relationship between such psychophysiological processes as EEG and the processing of cognitive and emotional material (e.g., Ray & Cole, 1985). In this traditional approach to experimental research, the relationship between the dependent and independent variable is emphasized and individual differences minimized or ignored. On the other hand we have also used the opposite research approach which emphasizes individual differences. With this approach we seek to find situations which allow us to demonstrate performance differences between individuals or groups of individuals. These differences have been organized in terms of sex, gender, anxiety, introversion/extraversion, or spatial ability and are also present in behavioral measures and the EEG (cf., Ray, Newcombe, Semon and Cole, 1981; Ray & Geselowitz, 1985; Berfield, Ray & Newcombe, 1986). In this chapter I will discuss research from each of these parallel lines of inquiry. After a brief overview of the EEG hemispheric activation literature, two recent studies from my laboratory examining the role of basic attentional mechanisms will be presented. This will be followed by presentation of work in progress showing how EEG measures may aid in an understanding of individual differences. The emphasis throughout the presentation will be on the limits of our present understanding of EEG as well as potentially important new directions for research. The research presented in this chapter benefited from the careful thought and hard work of two graduate students. Work from the first tract was performed in collaboration with Harry Cole and from the second with Lola Geselowitz.

The long term overall goal of combining these two different research approaches is to define at what point or points in cognitive and emotional processing do individual or personality differences play a role. It is assumed that by examining the entire experimental session including periods before information is presented as well as periods in between formal tasks, it will be possible to obtain a clearer picture of how personality and other individual differences play a role in the manner in which a given individual approaches these tasks, in particular, and life in general.

PREPARATORY	PERCEPTUAL	PROCESSING	PRODUCTION
General State	What is seen	Type of logic	Response style

Fig. 1. Necessary aspects to consider when examining individual differences in information processing.

There exists a number of separate processes in which personality differences may play a part. For organizational purposes I will use Figure 1 to represent some of the main potential areas in which personality and individual differences may play a role which in turn suggests specific research questions to answer.

First, one may ask if different personality groups approach a task differently. This question asks if there is a particular set of preparation which a particular individual, "high anxious", for example, brings to the experimental situation as opposed to another individual, such as a "low anxious" person. This could also include the general state of the organism as well as the meaning of the event for the person. Second, one may ask if different individuals or personality groups perceive the situation or tasks differently. This raises the possibility that personality differences may be explained in terms of perceptual or cognitive structure differences. That is to say, some individuals may focus on the emotional characteristics of a situation whereas others may pay more attention to particular details. It is also the case that "experts" in a particular area (e.g., a professional) would have a more differentiated perception than a non-expert or novice. On another level, it is at this point that a metacontrol system may bias the organism for a particular type of processing ranging from orientating responses or defensive responses to hemispheric activation. Third, one may ask if personality differences are the resultant of different ways of processing the information or task. In studying spatial abilities for example, there is evidence to suggest that high spatial ability individuals use a more synthetic approach whereas low spatial ability individuals use a more analytic manner of process. It is at this point that particular types of logic such as analytic or synthetic processing would be available for study. Fourth, one may ask if different personality groups produce or report their processing of cognitive, emotional, and motor tasks differently. That is to say, it may be that there exists individual differences in the ability to report responses (i.e., response styles) which are separate from perceptual or processing differences. What remains uncertain at this time is whether individual or personality differences may be accounted for solely in terms of preparation, perception, processing or production or some combination of these or whether individual or personality differences pervade all the activity that a person engages in. Although still in the early stages, it is our goal to apply this type of approach to electrocortical measures of cognitive and emotional processing with both normal and psychopathological populations.

Historical Views of EEG Activity

The initial reports by both Berger (reprinted 1969) and Adrian and Matthews (1934) make for interesting reading. Berger in his original reports suggested that EEG alpha activity (8-12 Hz) would offer insights into the neural mechanisms of attention. This was related to his overall desire to find physical measurements which represented the processing of the mind. In his second report he stated:

"..I am nevertheless attracted by the view that the alpha waves probably represent concomitant phenomena of those nervous processes which have been termed psychophysical i.e., of those material cortical processes which under certain circumstances can also be associated with phenomena of consciousness. Among the psychophysical processes one does not only include those which in fact are associated with conscious phenomena, but also those which perform the so-called unconscious cortical activity. Frequently only the results of the latter enter consciousness". (p.91)

Later Berger suggested that alpha activity is to be found in all areas of the cortex during periods in which one is passively allowing thoughts and images to arise spontaneously. When attention is actively focused, as in the solving of a problem, alpha is arrested. This was seen to represent a mechanism in which psychophysiological processes were localized, coupled with a general inhibition in surrounding areas. The diffuse inhibitory process is seen in the EEG as a blocking of alpha rhythm.

Adrian and Matthews agree with Berger and report that attention given to a problem (e.g., mental arithmetic) even with the eyes closed will abolish alpha "but the rhythm returns as soon as his attention wanders (p.370)." Adrian and Matthews further note that "powerful muscular contractions do not interfere with the rhythm although the subject exerts his full strength (p.371)." In the conclusion to their paper, Adrian and Matthews present alpha activity as an idling of the brain. In particular they stated, "it was true that, in our view, the rhythm shows the negative rather than the positive side of cerebral activity, it shows what happens in an area of the cortex which has nothing to do, and it disappears as soon as the areas resumes its normal work (p.383)." Given these differences in perspective, it is surprising that more attention has not been paid to the implications of these underlying meta-assumptions for EEG research. It is equally interesting to note the similarity of the Adrian and Matthews assumption that EEG alpha reflects "an area of the cortex which has nothing to do" and the modern assumptions of the EEG hemispheric lateralization researchers that alpha activity in the one hemisphere suggests processing in the opposite hemisphere.

EEG Measures of Hemispheric Lateralization

One basic conceptualization of the role of the central nervous system has been derived from the work of Roger Sperry and his colleagues with split brain patients (see Sperry, 1982 for an overview). This research has led to the conclusion that the left hemisphere is more involved in the processing of verbal/analytic material whereas the right hemisphere is more involved in the processing of visuospatial/synthetic material. Although initially based on research with epileptic patients, other research using a variety of techniques including dichotic listening, tachistoscopic presentation, lateral eye movements, blood flow measures, evoked potentials as well as EEG has given support to the original formulation (see Springer & Deutsch 1981; Bryden, 1982; and Corballis, 1983; for general reviews of these areas).

In one of the first studies using EEG to reference hemispheric processing, Morgan, McDonald & Macdonald (1971) reported proportionally less alpha activity over the right occipital area (as compared to the left) during spatial tasks (imagining scenes) as compared to analytic ones (mental arithmetic and word construction). Galin and Ornstein (1972) using an undifferentiated EEG frequency band width ranging between 1 and 35 Hz reported EEG hemispheric differences in the parietal and temporal areas between spatial right hemispheric tasks (solving Kohs blocks and Minnesota

Form Board) and verbal left hemispheric tasks (writing a letter and mentally composing a letter). A later study (Doyle, Ornstein, & Galin, 1974) reported that the task dependent asymmetries were strongest in the alpha band and numerous researchers have limited their research to this band. In general researchers have reported EEG results consistent with the hemispheric lateralization hypothesis (see Donchin, Kutas, & McCarthy, 1977; Davidson & Ehrlichman, 1980; Yingling, 1980 for a presentation of these studies).

Like other laboratories, in our early work we found differences in wide band EEG between everyday tasks such as listening to music (right hemispheric) and mental arithmetic (left hemispheric). However, we found the expected asymmetry only for males and not for females (Ray, Morell, Frediani, & Tucker, 1976). While other studies have also reported sex differences consistent with our results (e.g., Trotman & Hammond, 1979; Wogan, Kaplan, Moore and Epro, 1979), others (e.g., Davidson, Schwartz, Pugash and Bromfield, 1976) have suggested an opposite relationship with laterality and sex differences. In spite of much theoretical debate concerning the broad question of sex differences and neurological organization (cf. McGlone, 1980; see also Butler, 1984), the sex differences question especially in relation to EEG work remains an open one, although recent work by Glass, Butler and Carter (1984) suggest that familial handedness may be one mediating variable.

One of the interesting individual differences measures we have followed-up is the relationship between sex and spatial ability. It is well known that males score higher on spatial ability tests than females and there has been some speculation that this is related to neurological organization and thus hemispheric differentiation (cf. McGlone, 1980). Based on some initial work by Furst (1976) with EEG and Gur and Reivich (1980) with blood flow, we (Ray, Newcombe, Semon and Cole, 1981) sought to determine if differential EEG activity in high and low spatial ability males and females was associated with successful solving of spatial problems. The answer to this question was, "yes", but only for males. In addition, high and low spatial ability males showed equal but opposite correlations with performance. EEG ratios (right-left / right + left) with performances were negative (-.77 during baseline and -.53 during tasks) for high spatial ability males but positive (.77 and .56 respectively) for low spatial ability males suggesting that different areas of the brain are associated with successful performance in each group. This suggested to us that high and low spatial ability males adopted opposite approaches to solving spatial tasks, one a more right hemisphere (spatial/synthetic) and the other a more left hemisphere (verbal/analytic). In addition, the higher baseline correlations suggest the EEG may be a useful measure for examining the set or preparation that an individual brings to a task. It would also be interesting to determine if these baseline differences are related to the slow wave potentials seen as a subject prepares for a task presentation (cf. Birbaumer, Elbert, Rockstroh, Lutzenbeger and Schwartz, 1981).

However, there are a number of methodological issues ranging from the site of the reference electrode to the mode of analysis to the meaning of the EEG itself that remain open issues at this time. Donchin, Kutas, & McCarthy (1977) in their review of electrocortical measures of hemispheric lateralization suggested that there existed a number of logical and methodological issues which made interpretation of the existing studies difficult. Gevins, Zeitlin, Doyle, Yingling, Schaffer, Callaway & Yeager (1979) suggested that previously discovered lateralized differences in electrocortical measures did not relate to cognitive processing but rather reflected inconsistencies in stimulus properties, limb movements during tasks, and/or performances factors such as subject's ability or engagement in a given task. This sort of methodological criticism suggested to us that simple EEG ratio measures of alpha activity along a single verbal/spatial continuum might not be presenting the entire story, and, partly motivated the research discussed in the chapter. The positive side of the picture is that whatever EEG is measuring, it is stable over time both in terms of task related EEG asymmetries over sessions (Amochaev & Salamy, 1979; Ehrlichman & Wiener, 1979) and in terms of a stable intraindividual trait (Gasser, Bacher & Steinberg, 1985). Thus, there remains much research and conceptualization to be done to answer such important questions as to what exactly EEG activity is reflecting in the studies of hemispheric activation and to delineate the manner in which individual differences play a role.

HEMISPHERIC LATERALIZATION AND EMOTIONALITY

Not only have cognitive processes been associated with differential hemispheric involvement, but also emotionality. Evidence has accumulated from a variety of sources to support the idea that lateralized central nervous system activity and emotionality are functionally related (see Tucker, 1981 for a review). In addition, recent evidence suggests that a rostral/caudal dimension must also be considered. For example, Robinson, Kubos, Starr, Rao and Price (1984) have reported differential anterior/posterior as well as hemispheric involvement with mood disorders. In an examination of stroke patient groups, Robinson et al. found that patients with left frontal lesions reported significantly more severe depression whereas less depression was noted in patients with lesions in the posterior left hemisphere. A different pattern was seen in the right hemisphere with depression being more pronounced in patients with lesions in the right posterior areas while patients with anterior lesions were more cheerful.

Research with neurologically intact subjects has also suggested differential hemispheric activity in terms of emotionality. Safer and Leventhal (1977) employed a dichotic listening paradigm and reported greater right hemispheric involvement with subjective/emotional information. Sackheim, Gur & Saucy (1978) had subjects judge facial expression and interpreted their results to suggest right hemisphere involvement in the recognition of emotions. Recording lateral eye movement, Schwartz, Davidson, & Maer (1975) suggested that greater right hemisphere involvement was associated with the recall of facts about emotional experiences or the differentiation of emotional words.

Electrocortical research has also suggested differential hemispheric activation related to emotionality with both psychiatric (Flor-Henry, 1979) and normal populations (Harman & Ray, 1977; Davidson, Schwartz, Saron, Bennett and Goldman, 1979; Ehrlichman and Weiner, 1980; Tucker, Stenslie, Roth & Shearer, 1981) using a variety of emotionally valenced tasks. The particular emotional valence associated with each hemisphere has varied. For example, in one of the first studies to examine EEG correlates of emotional functioning, we (Harman and Ray, 1977) reported differential hemispheric changes over time with the recall of positive and negative past events. Left hemispheric temporal power (3-30 Hz) increased during recall of positive affect and decreased during recall of negative affect. Using similar tasks and recording sites, Ehrlichman and Wiener (1980) also reported greater right hemispheric activation during positive emotional tasks.

Other research such as that requiring subjects to evaluate emotional stimuli has also shown hemispheric differences. Davidson et al. (1978) required subjects while watching a television show to subjectively rate, using a pressure sensitive device, how well they liked/disliked various parts of the program. Bilateral frontal and parietal EEG activity in the alpha range was compared during liked and disliked segments. The results indicated relatively greater right frontal alpha activity during liked segments and relatively greater left frontal alpha activity during disliked segments. No significant differences were reported for parietal activity. Tucker et al. (1981) used a mood induction procedure and then presented subjects with cognitive tasks. These results were interpreted as similar to Davidson et al.'s, (1979) since negative mood was associated with greater left frontal EEG alpha activity.

Methodological and Theoretical Questions in Relation to Emotionality

Since the EEG studies examining emotion, mood, and preference have used a variety of tasks and procedures involving different frequency bands and areas of the cortex, it remains unclear as to whether or not the different EEG results may be attributed to variation in the EEG measures being recorded (e.g. different frequency bands and/or data reduction), the manner in which different areas of the cortex (e.g. frontal, parietal and temporal) are involved in the processing of emotional material, or the task requirements themselves (e.g. the validity of the emotions produced). Part of the problem is also one of definition since emotionality and arousal have been linked for years in the experimental literature. In fact, emotionality as defined in many of the experimental studies dating back to the 1930s meant nothing more than activity on the part of an organism. Today, there is often little distinction between constructs such as emotionality, feelings and moods.

Theoretically, the most consistent view of EEG functioning in relation to emotionality has been in terms of arousal. The traditional arousal model has assumed a unitary continuum ranging from sleep to high activity which were thought to be indexed by any one of a number of psychophysiological measures (cf., Duffy, 1962). With the discovery of the role of the reticular activating system in desynchronization of EEG (cf., Moruzzi & Magoun, 1949), arousal seemed to have a firm theoretical basis on which to stand and a physiological mechanism by which to explain emotionality. EEG offered a convenient measure reflecting the activity of the reticular activating system. In EEG research with awake subjects the common assumption was that alpha and beta activity reflected arousal. A separate assumption was that alpha activity had an inverse relationship with beta and with mental processes. However, there are reasons to examine these assumptions and to suggest that this simple view of EEG activity is limited from both a theoretical and and empirical viewpoint. In particular the arousal assumption places a much too important role on EEG alpha and assumes less specificity in the cortex than is usually assumed to be the case. Indeed, although cognitive psychology has moved beyond a simple arousal model, this transition has not been clearly reflected in EEG research.

DIFFERENTIATION OF AROUSAL

General theoretical presentations have been made suggesting the replacement of the unitary arousal model with a more differentiated one distinguishing motor, perceptual, and cognitive involvement (Kahneman, 1973). Posner (1975) suggests that the notion of arousal is related to a general drive theory which is more compatible with stimulus response psychology than to information processing formulations. Unlike the general drive/arousal formulation, the information processing perspective suggests that it is possible for a person to be in an optimal state for one type of activity such as the processing of external signals and not for another such as recall from long term memory. Some tasks such as sensory intake and motor outflow may even be antagonistic (Routtenberg, 1971). With the transition to an information processing perspective, Posner replaces the concept of arousal with the general concept of attention. Attention is further described in terms of alertness, selection of information, and degree of effort. Posner uses the term alertness to refer to the reception of external signals. Posner suggests that this is manifested physiologically by a decrease in EEG alpha activity and cardiac deceleration as well as motor inhibition. However, Posner is uncertain how to describe the attentional mechanisms related to internal processes as there exists little research directed toward this topic. This of course offers an important area for future research especially since many clinical processes (hypochondriasis, for example) are related to either an over or undersensitivity to internal processes and their control.

Pribram and McGuinness (1975); McGuinness and Pribram (1980) have also discussed the shortcoming of the unitary arousal conceptualization and stressed the importance of differentiating arousal into at least three components. The first component is that of a phasic response which they refer to as arousal, and in many ways is similar to Sokolov's orientating response (Sokolov, 1965) or Broadbent's notion of pre-attention (Broadbent, 1971). The second is a more tonic component referred to as activation. In other formulations this is called attention (e.g. Broadbent). Activation is a component which involves the voluntary direction of attention once the organism has been physically aroused. The third component is an effort component utilized to maintain the relationship between arousal and activation. Pribram (1981) suggests that each of these aspects involves different brain processes.

Although the various models which differentiate arousal differ in particular aspects, Tucker and Williamson (1984) have recently suggested an underlying similarity of processes represented in these models related to hemispheric functioning. Although somewhat complex, their model does suggest a rationale for the differentiation of emotional/feeling processes from attentional factors, a distinction difficult to make within a unitary conceptualization of arousal.

Direction of Attention

Following Darrow (1929) and Lacey (1967) a distinction can be made between two types of attentional tasks. The first type of task Lacey referred to as "intake". These are tasks involving passive observation of environmental stimuli. The second type of task Lacey called "rejection". Rejection tasks are those such as mental arithmetic in which a person must "reject" external input which might interfere with performing the task. Different cardiovascular responses are associated with intake (heart rate deceleration) and rejection tasks (heart rate acceleration) (Lacey, Kagan, Lacey & Moss, 1963). At the physiological level, the Laceys (Lacey & Lacey, 1978) have suggested a link between autonomic arousal as represented by cardiovascular functioning and cortical arousal as represented by electrocortical measures involving baroreceptor activity in the carotid sinus although this position is not without criticism (e.g., Carroll & Anastasiades, 1978; Green, 1980). The complexity of the cardiovascular-cortex interaction is further demonstrated by the recently discovered mediational link related to hormones produced in the heart itself (e.g., the atrial natriuretic factor (Cantin and Genest, 1986). Whatever the exact mechanism there does exist a two-way street between the heart and the brain. For example, Walker and Sandman (1979) have investigated changes in average evoked potentials (AEP) and heart rate and report that heart rate changes are differentially reflected in the two hemispheres with lower heart rate being associated with greater amplitude right hemispheric AEP's and higher heart rate associated with larger amplitude left hemispheric AEP's. It remains an open question whether these same type of relationships might be seen also with EEG activity.

Empirical Studies of Cognitive and Emotional Processing

In order to better understand the role of attentional processes within the hemispheric lateralization concept, we designed two studies. In the first study performed (Ray & Cole, 1985), we examined cognitive processes and in the second study (Cole & Ray, 1985) emotional ones. In the first experiment, eighteen right-handed subjects (9 males, 9 females) of college age were given 2 trials of 8 cognitive tasks on each of three days. The second and third days differed from the first only in order of presentation of the tasks and the specific items given. Presentations were counterbalanced within sex. The tasks were of the type used in previous hemispheric lateralization studies shown to reflect left and right hemispheric processing and not requiring overt motor responses. The verbal/analytic (left hemisphere) tasks and the spatial/synthetic (right hemisphere) tasks were further divided into "intake" tasks requiring the use of external environmental information and "reject" tasks which did not. Thus, the hemispheric dimension was crossed with the intake/rejection attentional dimension in a 2 x 2 design. The particular tasks were as follows:

1. INTAKE TASKS (External Attention)
    A. Verbal-Analytic (Left Hemisphere)
        1. Task 1 -- Count verbs in a passage.
        2. Task 2 -- Find the error in a maths problem.
    B. Spatial-Synthetic (Right Hemisphere)
        1. Task 3 -- Paper-folding--choose the correct three dimension representation of a geometric design presented as a blueprint.
        2. Task 4 -- Mooney Faces--the subject determines the location of a face in a high contrast presentation of irregular shapes which initially appear only as contours of light and dark (Mooney, 1957). Right hemisphere damage inhibits performance on this task (Lansdell, 1967).

II REJECT TASKS (Internal Attention)
    A. Verbal Analytic
        1. Task 5 -- Mental Arithmetic.
        2. Task 6 -- Create sentences which begin with a certain letter.
    B. Spatial Synthetic
        1. Task 7 -- Mental Rotation of a geometric object.
        2. Task 8 -- Visualization.

All intake tasks were presented on a screen in front of the subject with the tasks being matched for visual angle and relative brightness. During the reject tasks subjects were instructed to keep their eyes open and to look at the screen in front of them. In the first experiment, EEG was recorded for 25 secs. per trial from F3, F4, P3 and P4 referenced to linked ears. The EEG was Fourier analysed and estimates of spectral power computed for 4Hz frequency bands from .5-28 Hz. These data were analysed with analysis of

Table 1. Mean Power Estimates for Intake and Rejection Tasks for Frequencies with Significant Attention by Hemisphere Interaction

	INTAKE TASKS			REJECTION TASKS		
	LEFT		RIGHT	LEFT		RIGHT
Frequency		Site			Site	
Experiment 1		Parietal			Parietal	
8-12 Hz**	540.5		649	1244.5		1791.5
12-16 Hz**	196		256	234		327
16-20 Hz**	97		126	118		165.5
Experiment 2		Parietal			Parietal	
8-15 Hz**	272.2		319.6	721.2		892.2
		Temporal			Temporal	
8-15 Hz**	127		188.6	227.1		353.8

Note: Power divided by 100000
** $P<.01$

variance (ANOVA) using sex as a between-subjects variable and day (3 days), task (analytic/synthetic), attentional demand (intake/reject) and side (right/left hemisphere) as within-subject variables. Separate analyses were performed for each frequency band by site (frontal/parietal). Ratio scores (left hemisphere EEG - right hemisphere EEG / left + right) were also computed in each experiment.

In the second study, emotionally valenced material was presented to 40 right handed males of college age. They completed eight separate tasks (two trials each) on each of two days. We used a 2 x 2 design in which intake/rejection demands were "crossed" with positively and negatively valenced tasks. The rejection tasks were based on those used in previous research (Harman & Ray, 1977; Ehrlichman & Wiener, 1980). These included remembering a happy and sad event from one's past as well as imagining future pleasant and unpleasant events. The intake tasks consisted of the presentation of slides considered to evoke positive and negative affect (Hare, Wood, Britian and Sholman, 1970; Ekman & Friesen, 1975; Safer, 1981). These slides included landscapes, happy faces, sad faces and accident scenes. During all tasks subjects were instructed to keep their eyes open and to focus on the screen in front of them. The duration of each trial was 30 seconds. In this second study, EEG activity from F3, F4, T3, T4, P3 and P4 (referenced to linked ears) was Fourier analysed and power estimates for a low band including theta (2-7 Hz), a mid-range band including alpha (8-15 Hz) and a higher band composed of beta (16-24 Hz) computed. These bands were chosen for computational ease since the first experiment suggested that little information would be lost by combining the bands in this way. These data from the second experiment were analysed with a day, by attentional focus, by emotional valence, by hemisphere, analysis of variance and by a three mode factor analytic technique (PARAFAC) (see Harshman & Berenbaum (1981) for a detailed description of this technique). Briefly, PARAFAC described the different relative involvement of EEG in relation to three separate modes: (1) left and right frontal, temporal and parietal sites; (2) the eight tasks on each of two days; and (3) the 40 subjects. The subject mode allowed us to determine if any given subject contributed differently from the others and thus represented an outlier. Mean heart rate during the tasks was also computed.

Table 2. Mean Power in Parietal Areas on Verbal and Spatial Tasks by Frequencies with Significant Task by Hemisphere Interaction

	SPATIAL TASKS		VERBAL TASKS	
	Left	Right	Left	Right
FREQUENCY BETA				
16-20 Hz*	108.5	152	106.5	139.5
20-24 Hz**	104	127	108	124
24-28 Hz*	27.5	35.5	27	32.5

Note: Power divided by 100000
* P< .05
**P< .01

From these two experiments we can report that attentional, cognitive, and emotional factors are differentially represented in terms of EEG frequency and site. In both studies, we found that the intake/rejection dimension was reflected in parietal areas for the middle frequencies including alpha. This finding is supported by a significant intake/reject interaction with side (right/left hemisphere) in both experiments. The ratio data also showed the same significant pattern. Consistently in both studies there were higher levels of alpha during the rejection as compared to the intake tasks in both hemispheres as well as an interaction showing differentially more right hemispheric alpha during rejection tasks. Specifically, this interaction was found in the first experiment for parietal mid-frequencies (8-12 Hz, $F(1,16)=11.657$, $p< .004$; 12-16 Hz , $F(1,16)=11.894$; $p< .003$; 16-20 Hz , $F(1,16)=11.026$, $p< .004$) and in the second experiment for the mid-frequencies (8-15 Hz) for both the parietal ($F(1,39)=10.891$; $p< .002$) and temporal ($F(1,39)=10.068$; $p< .003$) areas (see Table 1). This interaction was not found in the frontal areas in either experiment.

Whereas the attentional demands of the experiments were reflected in the middle frequencies including alpha, the task demands (both cognitive and emotional) were reflected in beta. In the first experiment this finding was supported by a significant task (verbal/analytic vs. spatial/synthetic) by side (right/left hemisphere) interaction in the three upper beta bands (16-20 Hz , $F(1,16)=5.762$, $p< .029$; 20-24 Hz , $F(1,16)=8.968$, $p< .009$; 24-28 Hz , $F(1,16)=6.335$, $p< .023$) for the parietal areas (see Table 2). This same pattern was reflected in the ratio and was also statistically significant. These data show more beta activity in the right parietal area during spatial as opposed to verbal tasks. In the second experiment there was a significant main effect for emotional valence in the temporal ($F(1,39)=7.91$; $p< .008$) and parietal ($F(1,39)=6.328$; $p< .016$) areas with there being more beta during positive as compared to negative tasks. Differential hemispheric emotional involvement is clearly described by the PARAFAC analysis with a factor being formed between the positively valenced tasks and beta in the right temporal area (i.e., there was more beta in the right temporal during positive than negative emotional tasks). These data are presented in Table 3.

Consistent with previous cardiovascular research (see Molen, Somsen & Oriebeke, 1984 for a review), there was a significant difference using analysis of covariance ($F(1,39)=39.14$; $p< .0001$) in heart rate between the intake and rejection tasks in the second experiment with the intake tasks showing a lower heart rate ($\bar{x}$ = 72.17) than rejection tasks ($\bar{x}$ = 73.79). Intriguing findings were significant correlations between alpha EEG ratio (R-L / R+L) activity in the temporal area and mean heart rate for each of

Table 3. Parafac Analysis
Factors Involving Tasks Under Study

MEASURE	ALPHA	BETA
Site		
F 3	.60	.03
F 4	.59	.16
T 3	.40	.60
T 4	.68	2.36
P 3	1.47	.09
P 4	1.59	.20
TASK		
Day 1		
REJECT		
Positive (past event)	.97	1.76
Positive (future)	.99	2.02
Negative (past)	1.11	.60
Negative (future)	1.04	.46
INTAKE		
Positive (landscape)	.35	.92
Positive (face)	.45	1.24
Negative (face)	.43	.20
Negative (accident)	.34	.70
Day 2		
REJECT		
Positive (past)	1.61	1.38
Positive (future)	1.67	1.06
Negative (past)	1.45	.58
Negative (future)	1.60	.17
INTAKE		
Positive (landscape)	.44	.71
Positive (face)	.59	1.14
Negative (face)	.60	.01
Negative (accident)	.45	.38

Note: These are the first factors extracted from two separate PARAFACs, one with alpha and one with beta. Higher numbers suggest more relative involvement with that particular factor.

the four tasks (reject r=.484**; reject negative r=.394*; intake positive r=.446**; intake negative r=.462**; *=P>.05, **=P>.01).

Since the intake/rejection difference appeared in both heart rate and alpha activity in the present study, this raises the question of a possible relationship between cardiovascular activity and hemispheric alpha. One possibility lies in the report of Walker & Walker (1983) in which they reported a relationship between EEG activity and carotid pressure. The results of their study showed that readings from the carotid artery were time locked to the rhythmic oscillations of the EEG especially in the alpha band. This suggests that carotid activity may play a role in synchronizing alpha. It might also be speculated that it is through this mechanism that cardiovascular activity has an inhibitory effect on cortical processing. The particular candidate for such a mechanism would be the baroreceptor system since research with animals has shown baroreceptor stimulation produces cortical and behavioural inhibition (Bonvallet, Dell & Hiebel, 1954; Bonvallet & Allen, 1963; Dworkin, Filewich, Miller, Craigmyle & Pickering, 1979). The Dworkin et al. study leads one to the conclusion that such disorders as hypertension may even function as a means of reducing the experience of external stimulation in times of stress. With humans, Lacey and others have speculated that a link exists between cardiovascular changes in response to intake and rejection tasks and the baroreceptor system. Thus, it is possible that increased alpha during rejection tasks in this, present, study was associated with such a mechanism whose purpose was to reduce unneeded external stimulation and allow for more efficient internal processing (e.g. recall of past events).

In terms of emotionality the PARAFAC analysis suggested an unique involvement of the right temporal areas in positive emotionally valenced tasks. Examining the report of beta activity in Table 3 it is possible to note that all positive tasks (in comparison to negative ones) had higher loadings and that the brain area with similar loadings was the right temporal lobe. This was true both across type of task (i.e., intake and reject) and across days. This finding is particularly interesting since this study held the processing demands of the positive and negatively valenced tasks constant. In other words, recalling a positive or negative event from one's past should require similar cognitive processes and should not require overt or differential motor responses. Likewise attending to a face displaying positive or negative emotional expression should involve similar task requirements. Thus EEG activity in the present study may be seen to reflect more than just motor task requirements as has been suggested previously (Gevins et al., 1979). We also found similar results along the emotional dimension from both the intake and rejection tasks in the present study, which suggests both the perception of emotion in the intake tasks and the emotional valence memory tasks in the rejection condition are reflecting the same underlying process.

The contribution of the right temporal area to those processes involving positive emotionality are intriguing, seen in the light of research with temporal lobe epilepsy. Patients with right temporal lobe epilepsy have described themselves as more elated than patients with left temporal lobe epilepsy who described themselves as more angry, paranoid, and dependent (Bear & Fedio, 1977). Thus in both the Bear and Fedio study and in this one there is a suggestion that the subjective experience of negative emotion of mood involves the left temporal areas whereas the subjective experience of positive affect involves the right hemisphere. However, the situation may be more complicated for in the Bear & Fedio study the appearance of emotional behaviour produced a different relationship. The same right temporal lobe patients who described themselves as more elated were described by raters as being more emotional and displaying periods of sadness. The left temporal lobe group on the other hand described themselves as more negative yet were described by others as more ideational and contemplative. Thus, there exists the possibility that subjective emotion as experienced by the person and expressive emotion as observed by another person may lead to differing and opposite conclusions concerning differential hemispheric involvement in emotionality. Since the observation of emotional response was not considered in the present research, an interesting possibility in a future study would be to allow for this condition.

Perhaps, there exist separate aspects of emotional processing differently reflected in the brain. The first aspect would be a subjective dimension in the emotional "flavor" of the task (e.g., remembering happy or sad events) which is reflected in the temporal area. The second aspect might be an evaluation factor (e.g., rate the pleasantness of a TV programme) reflected in the frontal areas. There may also exist the possibility of a third factor which would reflect an expressive dimension of emotion which would serve to communicate with others and which must on some level involve the motor and language areas of the brain. To date, research addressing this third area of emotional expression has used electromyographic rather than electrocortical measurements. However, future research combining both EMG and EEG approaches should prove fruitful.

Implications for Hemispheric Lateralization Research

In research using both emotional and cognitive tasks, we demonstrated that EEG alpha activity reflected attentional factors although it is not possible at this time to suggest which aspects of attention (i.e., perceptual, motor preparation, etc.,) are represented by alpha activity. It is also possible to suggest that the emotional valence of a task is seen in more posterior beta activity. The exact mechanism involved again waits to be described. However, in light of the results of these studies it becomes important that we clarify the traditionally held view of EEG alpha as reflecting arousal as well as attempting better conceptualiztion of our constructs of arousal, attention, and emotionality.

The EEG alpha differences in response to attentional demands and the lack of any alpha differences in response to the cognitive or emotional tasks further raises the possibility that previous hemispheric lateralization research may have unknowingly confounded the external (intake) and internal (rejection) attentional dimension with right and left hemispheric processing demands. For example, a verbal/analytic (i.e., left hemispheric) task that asks subjects to create sentences that begin with a certain letter and a spatial/synthetic (i.e., right hemispheric) task in which subjects were to solve a spatial problem presented on a table in front of them would confound the task dimension (verbal/spatial) with the attentional (intake/reject) demands of the task. The same possibility exists for emotional situations and it is important that especially with children the positive and negative tasks have a similar attentional appeal. Our experiments point out the importance of controlling for these attentional factors in the future.

In terms of previous criticism of EEG hemispheric lateralization research (e.g., Gevins et al., 1979), we can lend support to the possibility of potential problems in earlier studies which used alpha as a measure of cognitive or emotional processing. Whereas Gevins pointed out differential motor requirements in previously employed tasks, we point out differential attentional demands as a potential confound. In the two experiments reported in this paper we controlled for both attentional factors and for motor requirements, with the resultant being no differences in alpha activity in terms of the cognitive and emotional processes. However, EEG alpha activity is important in its ability to reflect attentional processes. In addition, even with the motor and attentional controls we were able to report beta differences reflecting both cognitive and emotional dimensions suggesting that EEG beta may be a potentially useful measure. Since EEG beta activity has not received extension definitive research, an exact interpretation of these data await further interpretation.

INDIVIDUAL DIFFERENCES

One of the intriguing aspects of EEG research has been the possibility of understanding how personality or individual differences might be manifested. Since both arousal and attentional mechanisms have been used as a basis for personality theories, it is not surprising that EEG would be a measure of choice for support of these theories. In my laboratory we have been asking if there are particular individuals who would be expected to differ in the manner in which they would direct their attention and how the direction of attention might be used in a defensive or psychopathological manner. This of course is not a new question. Wundt, for example, suggested that some individuals turn inward in response to the outside world whereas others move toward the stimulus. Jung used a similar idea when he suggested that some individuals are directed toward internal functioning such as dreams, ideas, and so forth whereas other individuals are more concerned with the external functioning of the world. The former he referred to as introverts and the latter extraverts (Jung, 1971, reprint).

Pavlov characterised the individual differences issue in terms of the strength of the nervous system and borrows from the terminology of the temperaments. Pavlov states, "We had good reason to distinguish four different types of cerebral hemispheres: two extreme ones, the excitatory and the inhibitory; and two central, balanced ones, the calm and the lively". (Teplov, 1964 p.21) In terms of the temperaments Pavlov described the four types as follows: the choleric individual is passionate and quickly irritated; the melancholic responds to events as inhibitory agents; the phlegmatic is self-contained and quiet whereas the sanguine individual is energetic and productive but only when the work is interesting. In terms of the extreme types, the choleric individual displays a predominance of excitation whereas the melancholic displays inhibition. It was also suggested that excitation types were easier to condition than inhibitory types.

In his 1967 book, Biological Basis of Personality, H.J. Eysenck developed an individual difference perspective based on excitability of the nervous system utilizing the constructs of introversion and extraversion (Eysenck, 1967). His system centered around the concept of arousal which was seen to be mediated by the brain stem reticular system. The reticular system, in turn, was seen to have an reciprocal relationship with the cortex which could be either inhibitory or excitatory. For Eysenck, extraverts were born with a lower level of arousal within the reticular system which in turn resulted in their seeking stimulation to increase arousal. Introverts on the other hand had a higher level of arousal and sought to reduce stimulation. This formulation has led to a number of EEG studies asking the question who is more aroused introverts or extraverts? Gale (1981) has extensively reviewed these studies and the methodological problems they contain. For example, few, if any, of these studies utilized multiple channels of EEG, making any type of hemispheric comparison impossible. Likewise there was little attempt to manipulate the arousal level of the tasks presented. However, Gale does point to some testable implications of Eysenck's theory if the arousal level were to be manipulated. For example, Gale (1981) suggests that, in low arousing conditions, extraverts will seek ways to stimulate themselves whereas in high arousing conditions introverts will withdraw attention in order to induce a state of calm.

Based on our own work in relation to EEG and attention, it seemed reasonable to record EEG while specific groups of subjects were either actively engaged in a task or waiting between trials (Ray & Geselowitz, 1985). The between-trial sessions would show any differences in the subject's more normal mode of processing. It would also offer the beginnings of studies designed to determine whether certain personality groups focused more internally or externally. It seemed to us that the construct of introversion and extraversion might offer one logical place to begin such a search since there existed a large amount of EEG work connected with this distinction. We had also been interested in the construct of locus of control as related to physiological functioning. From our preliminary work we can report no EEG differences between locus of control groups (i.e., internal/external) either during tasks or during baseline periods. However, the results with extraversion/introversion display a more interesting picture which will be presented in this section.

In this study, we gave the subjects (6 male extraverts & 6 male introverts) four visual tasks which allowed for various levels of involvement. The tasks were all projected on a screen in front of the subjects. In the first task the subjects were shown a slide of a single color. At the end of each presentation the subject was asked to rate how much he liked that color on a 1 to 9 scale. The second task was a traditional paper folding problem in which the subject was shown a three-dimensional figure presented in the form of a blueprint and asked to pick out which of four alternatives represented the constructed figure. The third task was a Rorschach-like task in which subjects were shown inkblots and asked to say what they saw or what they looked like to them. The fourth task was a series of photographs chosen to include an emotional component. In between each series of tasks, as well as at the beginning and end of the experiment, a baseline was taken.

During the baselines and tasks, EEG activity was recorded from bilateral frontal (F3 & F4) and parietal (P3 & P4) sites referenced to linked ears. The duration of each trial was 25 secs divided into 5 sec epochs. An online Fast Fourier program (FFT) was used to construct a frequency spectrum consisting of the following frequencies: 1.5-4 Hz, 4-8 Hz, 8-12 Hz, 12-17 Hz, and 17-30 Hz.

In terms of the baseline data, performing a group (extravert/introvert), by site (frontal/parietal), by trial (5 baselines), by hemisphere (left/right), analysis of variance for each frequency band displayed significant group main effects in the bands below alpha. Performing a similar ANOVA for the task data showed significant main effects for group in all bands except beta (17-30 Hz). In each case the extraverts show more power in the bands.

However, our interests in performing this study was not which group had more power but to determine if there was a pattern to the EEG of the subjects in each group. To answer this question we simply plotted the means. Since alpha is the most commonly used frequency band in studies of introversion/extraversion, we began there. For simplicity, we limited our discussion in terms of site and frequency. It is interesting to note the left frontal area baseline and tasks (Fig. 2). In this case the introverts and extraverts showed very similar baseline means. However, the task means for left frontal alpha dropped significantly for the introverts. What is more interesting is the pattern. It is clear from both these figures that for introverts, alpha drops during task presentation relative to baseline whereas this is less the case for extraverts. This same pattern although not as clear is present in other sites with alpha. In the right parietal area, for example, the extraverts also show a change between baseline and tasks although not as great as the introverts. We can also note that the left frontal beta activity follows a pattern similiar to that of left frontal alpha suggesting that the idea of an inverse relationship between alpha and beta is clearly not supported in our data as well as in other data (e.g., Daniel, 1965).

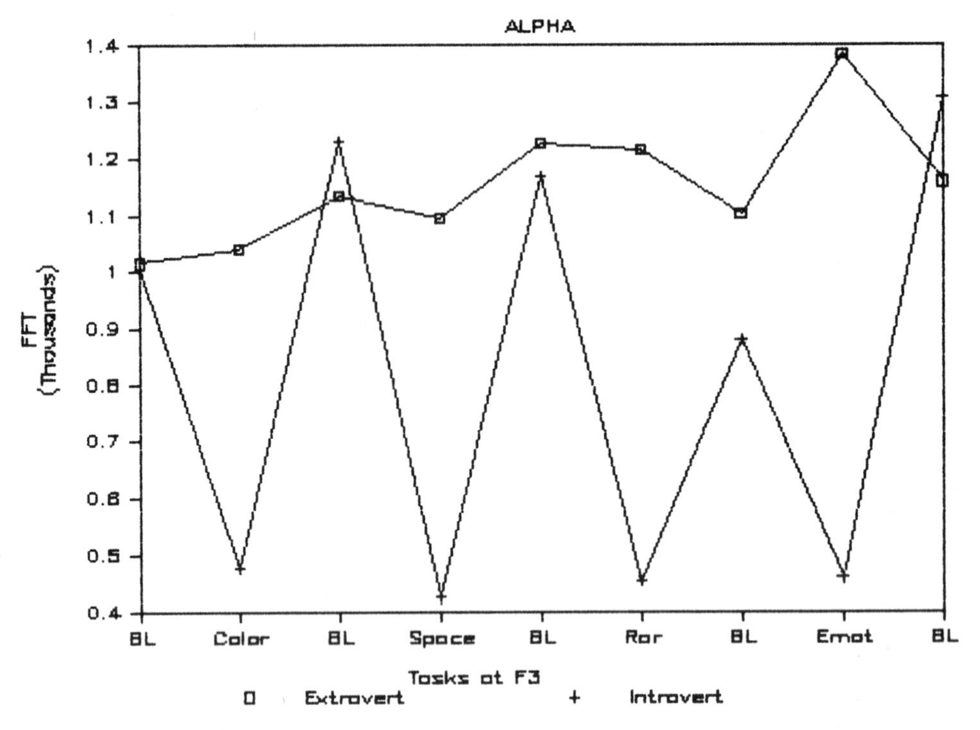

Fig. 2. Extravert and Introvert Left Frontal (F_3)

It is important to understand that we see these data as preliminary and at present are replicating the results. However, the patterns that are present in these data are suggestive, not only for understanding extraversion and introversion but also for direction in designing future studies in the area. In particular, the data suggests the importance of including both situations in which subjects are allowed to do nothing (i.e., display their personality) such as baseline periods as well as tasks varying in degree of required involvement. From our own results, it is clear that the empirical presentation is more complicated than generally assumed. However, to understand this complexity, researchers need to ask questions more sophisticated than - who are more aroused, introverts or extraverts? They need to ask a complex series of questions on how each group uses his or her mental processing in performing various tasks as well doing nothing (i.e., baseline). As Gale has previously suggested (1983), it is also important to vary the difficulty and involvement of the subjects with the task. It may also be useful to create situations which overload the subjects as one might find with a difficult video game. Technically, new studies must look at multiple sites, multiple frequencies, multiple measures (both physiological and behavioral performance ones) as well as different techniques of analysis (FFT, coherence, etc.,). We would suggest also that since individual differences are the topic of interest that future research analyze the data in these terms. One possibility would be to examine the data in two ways: first to analyze group data as one would normally do in the experimental studies, and second, to examine the individuals in the group as individuals and ask as with a case study approach how the individual patterns coalesce to form the group pattern.

CONCLUSION

In this chapter I have sought to overview EEG research which my students and I have been concerned with over the past few years. In the process I sought to point out some of the problems and promises of the area. EEG is a seductive area in that it holds out the potential for understanding mechanisms of cognitive and emotional processes not available in traditional psychological studies. The question before us now is how to ask empirical questions which are consistent with our theoretical understanding of cognitive and emotional processes. From our perspective, attention appears to be an important dimension which should be included in future studies. EEG also offers the potential for mapping how different types of individuals approach a specific task although it might be too simple an approach to expect a direct one-to-one mapping of physiological responses to behavioral measures. This particular study that was concerned with introverts and extraverts suggests a more "ecological" approach that seeks to undertake more than just task differences between groups but looks instead for patterns of EEG processing. We will have to await further research to determine if this is, indeed, a fruitful approach to individual differences.

REFERENCES

Adrian, E.D. & Matthews, B.H.C. (1934). Berger rhythm: Potential changes from the occipital lobes of man. Brain, 57, 355-385.

Amochaev, A., & Salamy, A. (1979). Stability of EEG laterality effects. Psychophysiology, 16, 242-246.

Bear, D.M., & Fedio, P. (1977). Quantitative analysis of interictal behavior in temporal lobe epilepsy. Archives of Neurology, 34, 454-467.

Berfield, K., Ray, W.J., & Newcombe, N. (1986). Sex role and spatial ability: an EEG study. Neuropsychologia, in press.

Berger, H. (1929). Über des Elektrenkephalogramm des Menschen. Translated and reprinted in Pierre Gloor, Hans Berger on the electroencephalogram of man. Electroencephalography and Clinical Neurophysiology, supplement 28. (1969) Amsterdam: Elsevier Publishing.

Birbaumer, N., Elbert, T., Rockstroh, B., Lutzenberger, W., & Schwartz, J. (1981). EEG and slow cortical potentials in anticipation of mental tasks with different hemispheric involvement. Biological Psychology, 13, 251-160.

Bonvallet, M., & Allen, M.B. (1963). Prolonged spontaneous and evoked reticular activation following discrete bulbar lesions. Electroencephalography and Clinical Neurophysiology, 15, 969-988.

Bonvallet, M., Dell, P., & Hiebel, G. (1954). Tonus sympathique et activite electrique corticale. Electroencephalography and Clinical Neurophysiology, 6, 119-144.

Broadbent, D.E. (1971). Decision and Stress, Academic Press, London.

Bryden, M.P. (1982). Laterality, Academic Press, New York.

Butler, S.R. (1984). Sex differences in human cerebral function. In G.D. De Vries et al., (Eds.) Progress in Brain Research, 61, Amsterdam: Elsevier Biomedical Press, 443-455.

Cantin, M., & Genest, J. (1986). The heart as an endocrine gland. Scientific American, 254, 76-81.

Carroll, D., & Anastasiades, P. (1978). The behavioral significance of heart rate: The Laceys' hypothesis. Biological Psychology, 7, 249-275.

Cole, H.C., & Ray, W.J. (1986). Emotional Valencies and EEG activity: Influence of attentional demands. International Journal of Psychophysiology, in press.

Corballis, M.C. (1983). Human Laterality, New York: Academic Press.

Daniel, R.S. (1965). Electroencephalographic pattern quantification and the arousal continuum. Psychophysiology, 2, 146-160.

Davidson, R.J., & Ehrlichman, H. (1980). Lateralized cognitive processes and the electroencephalogram. Science, 207, 1005-1006.

Davidson, R.J., Schwartz, G.E., Pugash, E., & Bromfield, E. (1976). Sex differences in patterns of EEG asymmetry. Biological Psychology, 4, 119-138.

Davidson, R., Schwartz, G., Saron, C., Bennett, J., & Goldman, D. (1979). Frontal versus parietal EEG asymmetry during positive and negative affect. Psychophysiology, 16, 202-203.
Darrow, C.W. (1929). Differences in the physiological reactions to sensory and ideational stimuli. Psychological Bulletin, 26, 185-201.
Donchin, E., Kutas, M., & McCarthy, G. (1977) Electrocortical indices of hemispheric utilization. In S. Harnad et al., (Eds.). Lateralization in the Nervous System. New York: Academic Press, 339-384.
Doyle J., Ornstein, R., & Galin, D. (1974). Lateral specialization of cognitive mode: II EEG frequency analysis. Psychophysiology, 11, 567-578.
Duffy, E. (1957). The psychological significance of the concept of "arousal" or "activation". Psychological Review, 64, 265-275.
Duffy, E. (1962). Activation and Behavior. New York: Wiley.
Dworkin, B.R., Filewich, R.J., Miller, N.E., Craigmyle, N., & Pickering, T.G. (1979). Baroreceptor activation reduces reactivity to noxious stimulation: Implications for hypertension. Science, 205, 1299-1301.
Ehrlichman, H., & Wiener, M. (1979). Consistency of task related EEG asymmetries. Psychophysiology, 16, 247-252.
Ehrlichman, H., & Wiener, M. (1980). EEG asymmetry during covert mental activity. Psychophysiology, 17, 228-235.
Ekman, P. & Friesen, W. (1975). Unmasking the Face. Englewood Cliffs: N.J.: Prentice-Hall.
Eysenck, H.J. (1967). The Biological Bases of Personality. Springfield, 111: C.C. Thomas.
Flor-Henry, P. (1979). On certain aspects of the localization of the cerebral systems regulating and determining emotion. Biological Psychiatry, 14, 677-698.
Furst, C. (1976). EEG alpha asymmetry and visuospatial performance. Nature, 260, 254-255.
Gale, A. (1981). EEG studies of extraversion-introversion: what's the next step? In R. Lynn (ed.), Dimensions of Personality: Essays in Honour of H.J. Eysenck. Oxford: Pergamon Press.
Galin, D., & Ornstein, R. (1972). Lateral specialization of cognitive mode: An EEG study. Psychophysiology, 9, 412-418.
Gasser, T., Bacher, P., & Steinberg, H. (1985). Test-retest reliability of spectral parameters of the EEG. Electroencephalography and Clinical Neurophysiology, 60, 312-319.
Gevins, A.S. (1983). Brain potential (BP) evidence for lateralization of higher cognitive functions. In J.B. Hellige (Ed.), Cerebral hemisphere asymmetry: Method, theory and application. New York: Praeger.
Gevins, A., Zeitlin, G., Doyle, J., Yingling, C.D., Schaffer, R.E., Callaway, E., & Yeager, C.L. (1979). Electroencephalogram correlates of higher cortical functions. Science, 203, 655-668.
Glass, A., Butler, S.R., & Carter, J.C. (1984). Hemispheric asymmetry of EEG alpha activation: Effects of gender and familial handedness. Biological Psychology 19, 169-187.
Gray, J.A. (1964). Pavlov's Typology. New York: MacMillan.
Green, J. (1980). A review of the Lacey's physiological hypothesis of heart rate change. Bioogical Psychology, 11, 63-68.
Gur, R.C., & Reivich, M. (1980). Cognitive effects on hemispheric blood flow in humans: evidence for individual differences in hemispheric activation. Brain and Language, 9, 78-92.
Hare, R., Wood, K., Britian, S., & Shadman, J. (1970). Autonomic responses to affective visual stimulation. Psychophysiology, 7, 408-417.
Harman, D. & Ray, W.J. (1977). Hemispheric activity during affective verbal stimuli: An EEG study. Neuropsychologia, 15, 457-460.
Harshman, R.A., & Berenbaum, S.A. (1981). Basic concepts underlying the PARAFAC-CANDECOMP three way factor analysis model and its application to longitudinal data. In D.H. Eichorn, P.H. Mussen, J.A. Clausen,

N. Haan, & M.P. Honzik (Eds.), Present and past in middle life. New York: Academic Press, 435-459.
Kahneman, D. (1973). Attention and Effort. New York: Prentice Hall.
Lacey, J. (1959). Psychophysiological approaches to the evaluation of psychotherapeutic process and outcome. In E.A. Rubinstein & M.B. Parloff (Eds.). Research in Psychotherapy. Washington, D.C.: American Psychological Association, 160-208.
Lacey, J. (1967). Somatic response patterning and stress: Some revision of activation theory. In M.H. Appley & R. Trumbull (Eds.). Psychological Stress: Issues in Research, New York: Appleton-Century-Crofts.
Lacey, J., Kagan, J., Lacey, B., & Moss, H. (1963). The visceral level: Situational determinants and behavioral correlates of autonomic response patterns. In P.H. Knapp (Ed.), Expression of the emotions in man. New York: International Universities Press, 161-196.
Lacey, B.C., & Lacey, J.I. (1978). Two-way communication between the heart and the brain. American Psychologist, 33, 99-113.
McGlone, J. (1980). Sex differences in human brain asymmetry: A critical survey. The Behavioral and Brain Sciences, 3, 215-263.
McGuinness, D., & Pribram, K. (1980). The neuropsychology of attention: Emotional and motivational controls. In M.C. Wittrock (Ed.), The Brain and Psychology. New York: Academic Press, 95-139.
Molen, M.W. van der, Somsen, R.J.M., & Oriebeke, J.F. (1986). The rhythm of the heart beat in information processing. In P.K. Ackles, J.R. Jennings, and M.G.H. Coles (Eds.). Advances in Psychophysiology. Greenwich, Conn: JAI Press.
Mooney, C.M. (1957). Age in the development of closure ability in children. Canadian Journal of Psychology, 11, 219-226.
Morgan, A., McDonald, P., & Macdonald, H. (1971). Differences in bilateral alpha activity as a function of experimental task, with a note on lateral eye movements and hypnotizability. Neuropsychologia, 9, 459-469.
Moruzzi, G., & Magoun, H.W. (1949). Brain stem reticular formation and activation of the EEG. Electroencephalography and Clinical Neurophysiology, 1, 455-473.
Posner, M. (1975). Psychology of attention. In M. Gazzaniga (Ed.) Handbook of Psychobiology. New York: Academic Press, 441-480.
Pribram, K. (1981). Emotions. In S.K. Filskov & T.J. Boll (Eds.). Handbook of Clinical Neuropsychology, New York: Wiley Interscience. 102-134.
Pribram, K.H., & McGuinness, D. (1975). Arousal, activation, and effort in the control of attention. Psychological Review, 82, 116-149.
Ray, W.J., & Cole, H.C. (1985). EEG alpha activity reflects attentional demands, and beta activity reflects emotional and cognitive processes. Science, 228, 750-752.
Ray, W.J., Morell, M., Frediani, A.W., & Tucker, D. (1976). Sex differences and lateral specialization of hemispheric functioning. Neuropsychologia, 14, 391-394.
Ray, W.J., & Geselowitz, L. (1985). Patterns of EEG activity and individual differences in response to baseline and spatial tasks. Electroencephalography and Clinical Neurophysiology, 61, 5152.
Ray, W.J., Newcombe, N., Semon, J., & Cole, P. (1981). Spatial abilities, sex differences, and EEG functioning. Neuropsychologia, 19, 719-722.
Robinson, R., Kubos, K., Starr, L., Rao, J., Price, T. (1984). Mood disorders in stroke patients. Brain, 107, 81-93.
Routtenberg, A. (1971). Stimulus processing and response execution: A neurobehavioral theory. Physiology and Behavior, 6, 589-596.
Sackeim, H., Gur, R.C., & Saucy, M. (1978). Emotions are expressed more intensely on the left side of the face. Science, 202, 434-436.
Sackeim, H., Greenberg, M., Weiman, A., Gur, R.C., Hungerbuhler, J., & Geschwind, N. (1982). Hemispheric asymmetry in the expression of positive and negative emotions. Archives of Neurology, 39, 210-218.

Safer, M. (1981). Sex and hemisphere differences in access to codes for processing emotional expressions and faces. Journal of Experimental Psychology: General, 110, 86-100.

Safer, M., and Levental, H. (1977). Ear differences in evaluating emotional tone of voice and of emotional content. Journal of Experimental Psychology: Human Perception and Performance, 3, 75-82.

Schwartz, G., Davidson, R., & Maer, F., (1975). Right hemisphere lateralization for emotion in the human brain: Interactions with cognition. Science, 190, 286-288.

Sokolov, E.N. (1965). The orientating reflex, its structure and mechanism. In L.G. Veronin, A.N. Leotrev, A.R. Luria, E.N. Sokolov, and O.S. Vinogradova (Eds.). Orientating Reflex and Exploratory Behavior. Washington, D.C.: American Institute of Biological Sciences. 141-151.

Sperry, R. (1982). Some effects of disconnecting the cerebral hemispheres. Science, 217, 1223-1226.

Springer, S.P., & Deutsch, G. (1981). Left Brain, Right Brain, San Francisco: W.H. Freeman and Co.

Teplov, B.M. (1964). Problems in the study of general types of higher nervious activity in man and animals. In J.A. Gray (Ed.) Pavlov's Typology. New York: MacMillan.

Trotman, S., & Hammond, G. (1979). Sex differences in task dependent EEG asymmetries. Psychophysiology, 16, 429-431.

Wogan, M., Kaplan, C.D., Moore, S.F., & Epro, R. (1979). Sex differences and task effects in lateralization of EEG alpha. International Journal of Neuroscience, 8, 219-223.

Tucker, D.M. (1981). Lateral brain function emotion, and conceptualization. Psychological Bulletin, 89, 19-46.

Tucker, D.M., Stenslie, C., Roth, R., & Shearer, S. (1981). Right frontal lobe activation and right hemisphere performance decrement during a depressed mood. Archives of General Psychiatry, 38, 169-174.

Tucker, D.M., & Williamson, P.A. (1984). Asymetric neural control systems in human self-regulation. Psychological Review, 91, 185-215.

Walker, B.B., & Sandman, C.A. (1979). Human visual evoked responses are related to heart rate. Journal of comparative and physiological psychology, 19, 520-527.

Walker, B.B., & Walker, J.M. (1983). Phase relations between carotid pressure and ongoing electrocortical activity. International Journal of Psychophysiology, 1, 65-73.

Wertheim, A.H. (1981). Occipital alpha activity as a measure of retinal involvement in oculomotor control. Pyschophysiology, 18, 432-439.

Yingling, C.D. (1980). Cognition, action, and mechanisms of EEG asymmetry. In G. Pfurtscheller et al., (Eds.) Rhythmic EEG Activities and Cortical Functioning. Amsterdam: Elsevier/North-Holland Biomedical Press. 79-90.

LONGITUDINAL AUDITORY EVOKED RESPONSES AND THE DEVELOPMENT OF LANGUAGE

Dennis L. Molfese and Victoria J. Molfese

Department of Psychology and School of Medicine
Southern Illinois University
Carbondale, IL 62901

INTRODUCTION

Over the past 50 years, researchers have been interested in problems involving specific brain-language relationships. Issues pertaining to the interrelationship of hemispheric specialization, cerebral lateralization and language development have been explored by researchers utilizing a variety of methodologies. One methodology that has produced promising results involves the use of auditory evoked response (AER) techniques. AERs refers to the portion of the on-going electrical activity of the brain which is time locked to the onset of some external auditory event. Using AER techniques, researchers have been able to report evidence of specialization and cerebral lateralization in human infants (Barnet, de Sotillo, and Campos, 1974; Crowell, Jones, Kapuniai and Nakagawa, 1973; Molfese, 1972; Molfese, Freeman and Palermo, 1975; Molfese and Molfese, 1979a, 1979b, 1980; Schucard, Schucard, Cummins and Campos, 1981). In recent years, multivariate analyses of data obtained using AER techniques have yielded information concerning the development of hemisphere differences. These procedures have also enabled researchers to isolate and identify specific components of the AER across hemispheres which appear to reflect responses to specific language-relevant acoustic and phonetic cues.

This chapter has two parts. Section One contains a review of evoked potential techniques as applied in brain-language investigations. In Section Two the relationship between hemispheric responses to speech cues recorded via auditory evoked response techniques early in development and later language performance is described.

I. EVOKED POTENTIAL TECHNIQUE

Evoked potential (EP) techniques attempt to establish strict temporal relationships between the onset of some stimulus event and the onset of changes in the various portions of the following EEG pattern. Because of the small size of these electrical patterns (5 to 15 microvolts) relative to other electrical noise sources, researchers must usually repeat the stimuli a number of times in order to evoke further replications of the brain potentials. The process of averaging, including the adding together of evoked potentials, is used to extract the EP from the background noise of non-replicable or non-redundant information in the brain activity. The final quotient (the averaged evoked potential) is then expected to reflect reliable brain activity elicited by some specific stimulus event. Once the averaged evoked potentials are obtained, a variety of different analyses can be utilized in order to determine whether changes in stimulus features might produce corresponding changes in various proportions of the brain response. In general, analyses focus on certain peaks in the waveforms that occur at certain time intervals or latencies following the onset of a stimulus event (Picton & Stuss, 1980; Vaughan, 1969). Such analyses may be based on procedures as simple as direct time or amplitude measures between two peaks or involve a complex factor analysis to reduce the complex waveform to a smaller set of simpler components.

There is a great deal of evidence that the averaged evoked potential can reflect changes in the neural activity of the brain during sensory processing (Regan, 1972, pages 31-116) and cognitive processing (Donchin, Ritter and McCallum, 1978, pages 349-411). In the first case, EP components associated with sensory processing appear to be relatively "impervious to variations of the psychological state of the observer" (Hillyard & Woods, 1979, p.346). These components (which for the most part occur prior to 100 msec following stimulus onset) seem to be very stable from one individual to the next and are not usually altered by subject state. Changes in some portion of these waveforms, whether in terms of the amplitude or latency of waveform peaks or the absence of certain usually present waves, signal some problem with a receptor, pathway, or brain area represented by that component (Rockstroh, Elbert, Birbaumer & Lutzenberger, 1982, p.3). These components have been referred to as "exogenous" components. On the other hand, EP waveforms which are associated with cognitive or perceptual processes of the brain are usually referred to as "endogenous components" (Hillyard & Woods, 1979, p.346). In general, the portion of the waveform which occurs after 100 msec will reflect endogenous activity. Here, the various characteristics of the EP waveforms, although triggered by some stimulus, are affected by the cognitive-perceptual processes involved in processing the stimulus. While the exogenous components are relatively stable across different subject states and individuals, the endogenous components can show great inter-subject variability and may change across different tasks and subject states. For the research to be outlined below it would appear that the categorical-like effects elicit exogenous type of activity, given the consistent pattern noted across ages and tasks.

In general, two different approaches have been used to study evoked potentials: (1) "defining the neuronal substrates of EPs <u>and</u> their relationship to behavioral events" and (2) "...a purely empirical relation of EPs and behaviour without recourse to neuronal mechanisms" (Purpura, 1978, p.83). While there are a number of excellent chapters documenting the first approach described above which attempt to identify discrete elements within the brain that are responsible for generating the various components of the EP (Goff, Allison & Vaughan, 1978; Klee & Rall, 1977), the exact nature of these brain mechanisms remains in doubt and under discussion. However, at a grosser level, topographic studies of human EPs do indicate that major portions of the EPs originate in the primary and secondary cortical areas that are specific to the modality involved in the detection of the stimuli presented (Vaughan, 1969; Simson, Vaughan & Ritter, 1977a). In this way, evoked potentials associated with auditory and visual presentations appear to be generated in the secondary cortex of the auditory and visual systems, respectively, as well as in the parietal association cortex (Simson et al., 1977a, 1977b). At a grosser level, EP activity recorded over the left side of the head originates for the most part in the left hemisphere while EP activity recorded from over the right hemisphere originates in the right hemisphere of the brain.

Advantages of the Evoked Potential Technique

Evoked potential procedures offer a number of methodological advantages. First, with the use of the EP techniques, there is opportunity to test children of different ages using one method that can be applied in a consistent fashion across the different ages. Routinely, investigators in the past have shown concern for possible "ceiling" and "floor" effects in which one set of problems may appear very difficult for young children (and consequently lead to a large number of errors) while these same problems may be very simple for older children (who would then not generate many errors). If analyses rely heavily on comparisons of error data, results from these two groups of subjects could not be considered comparable since the level of difficulty varies between groups. For example, one criticism of the dichotic listening procedure when used with subjects ranging in age from childhood to adulthood has been that it is difficult to control for stimulus and response difficulty across ages (Bryden, 1982). The EP techniques, however, can be applied in the same manner to all subject populations, regardless of age. Second, these procedures can be used with a variety of populations. Since they do not require an overt response, the investigator is not limited by the usual dependency of age related measures such as reaction time or language. Moreover, the procedures can be used as readily with brain damaged populations (whose motor or verbal skills have been impaired) as with normal intact populations. Third, with suitable controls the investigator can collect the electrophysiological data while recording other measures such as reaction time (which correlates well with components of the evoked potential such as the P300), choice responses and errors. The addition of this measure provides supplementary information which can be used to evaluate the state and strategies of the organism. Fourth, for those interested in hemisphere related effects, this procedure allows investigators not only to note possible differences in responding _between_ hemispheres but also differences _within_ hemispheres as well.

Concerns for Adequate Controls

There are a number of design factors that must be considered when utilizing the evoked potential procedures if the results from such studies are to be validly interpreted. While these concerns may cover a number of issues, three areas deserve special attention as they relate to the study of hemisphere differences: task sensitivity, localization, and stimulus/subject controls. In terms of task sensitivity, it is clear that the evoked potential is indeed sensitive to differences in task demands. The waveshape of the evoked potential changes dramatically when the frequency of some stimulus events are varied as in the case of the P300 component. Likewise, both waveshape and amplitude of the evoked potential appear to change as a function of the task demands (see Neville, 1980). This also appears to be true for the magnitude of the hemisphere differences reported between waveforms recorded from different hemispheres. As task demands increase, laterality effects appear to increase. In the interpretation of results, then, care must be taken to evaluate the possible contributions of task related factors.

There are severe limits on the use of the scalp recorded evoked potential as a means of localizing the source of some process within the brain. Although it appears that electrical activity recorded from locations over the left hemisphere does originate from within that hemisphere (similarly, the right hemisphere serves as a source for electrical activity which can be detected with electrodes placed over the right side of the head), scientists are unable at this time to localize the source of these currents much beyond this superficial level. Consequently, discussions as to the nature of the cortical structures involved in producing the EP and their role in the task involved in eliciting these responses must be considered with great caution.

Finally, it has been known for some time that stimulus and subject factors can also influence the waveform of the evoked potential. For example, there are a number of reports that the shape of the evoked potential will change as a function of an individual's hand preferences (Hillyard & Woods, 1979; Molfese, 1978b; Molfese, Linnville, Wetzel & Leicht, 1985). A variety of stimulus factors can also cause marked changes in the evoked potential. Factors such as stimulus rise and decay time, duration and loudness can all differentially affect different segments of the evoked potential (see Renau & Hnatiow, 1975, for a review of stimulus effects known to affect the auditory evoked potential).

2. EVOKED POTENTIAL RESPONSES AND LANGUAGE DEVELOPMENT

Our laboratory in the past 10 years has used evoked potential techniques to study the importance of lateralization and brain organization for language development. In addition to investigating changes in the developmental patterns of lateralization across the life span, these studies have attempted to isolate and identify the electrophysiological correlates of various speech perception cues across a number of developmental periods. One perceptual cue studied is PLACE OF ARTICULATION (PLACE). The PLACE cue is important for discriminating between consonants such as /b/ and /g/, consonant sounds that are produced in different portions or places of the vocal tract. The consonant /b/ is referred to as a "front" consonant because it is produced at the very front of the vocal tract with two lips. The consonant /g/, on the other hand, is produced in the back of the vocal tract and is labelled a "back" consonant. When the following vowel sounds are the same, the second formant transition (the rapid change in frequency which occurs for the second concentrated band of acoustic energy) as depicted in a sound spectrograph signals the place of articulation for the consonant. In the case of the syllable, /ba/, the second formant transition would be rising from its initial frequency to that of the steady state vowel formant. This formant transition falls, however, for the initial /g/ of the syllable /ga/. In general, our studies note that patterns of electrical activity recorded from different areas of the scalp change as a function of Place of Articulation cues. Furthermore, these patterns of discrimination do not appear to change to any great extent from early in infancy (Molfese & Molfese, 1979b, 1980) into adulthood (Molfese, 1978, 1980b; Molfese & Schmidt, 1983; Molfese, 1983).

Studies with adults note that portions of the AER which occur 300 msec after stimulus onset and recorded from electrodes placed over the left hemisphere discriminate between consonants such as /b/ and /g/. That is, the amplitude of this portion of the waveform as measured from the positive peak occurring at 300 msec to the following negative peak at 400 msec is reliably larger for the AERs elicited by the /b/ initial syllables than for the /g/ initial syllables (Molfese, 1978a, 1980b, 1983; Molfese & Schmidt, 1983). In studies with adults in which electrodes are placed at multiple sites over the two hemispheres (Molfese, 1980b, 1983, 1984; Molfese & Schmidt, 1983), an earlier occurring portion of the AER located at approximately 100 msec following stimulus onset varies reliably when evoked by a /b/ versus a /g/ initial syllable. This component behaves in a similar fashion to the one occurring at 300 msec with the exception that it occurs simultaneously over both hemispheres. In this way the AERs elicited by speech syllables reliably produce changes in two portions of the waveform in response to the /b/ and /g/ syllables. One area of waveform discriminates between PLACE differences at one point in time only over the left hemisphere electrode sites while a second portion of the AER changes systematically over both hemispheres in response to PLACE changes.

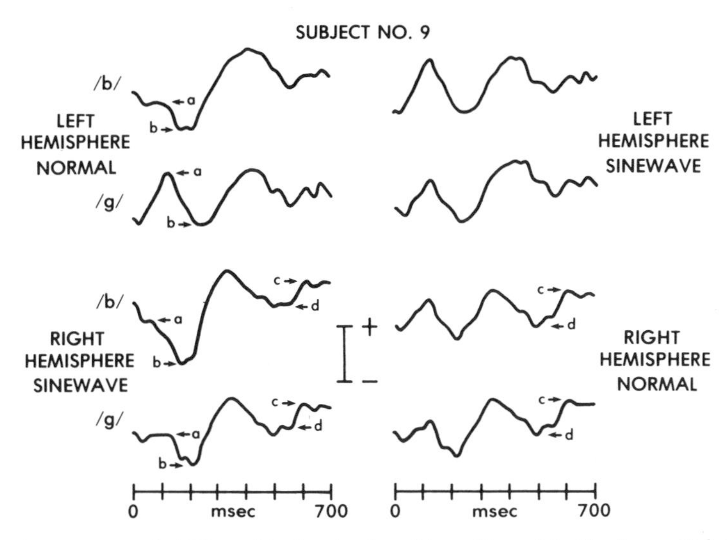

Fig. 1. The averaged AERs elicited shortly after birth from Subject #9, a member of the HIGH MCVP group, in response to the speech and nonspeech syllables beginning with the consonants /b/ and /g/. AER deviation is 704 msec with positive deflection up. The calibration marker is 10 μV.

Building on this work with adults, several recent papers have reported comparable findings with newborn infants. Molfese and Molfese (1979b) in a report based on tests with 16 newborn infants, reported that the initial large negative deflection or peak of the newborns' AERs which occurred approximately 230 msec following stimulus onset (N230) was larger in size when evoked by /b/ initial syllables than by /g/ initial syllables. This response difference occurred only over the left hemisphere electrode site. As was the case with adults, however, a second portion of the AERs detected at electrodes placed over both hemispheres were also noted to discriminate between the /b/ and /g/ initial syllables. A later study with preterm infants (Molfese & Molfese, 1980) reported comparable lateralized and bilateral responses that discriminated PLACE cues. This pattern of bilateral and lateralized discrimination of the PLACE cue as reflected in different portions of the AERs occurs, then, in both young infants as well as adults.

Auditory Evoked Responses as Predictors of Later Language Development

One major issue recently raised concerns the implications of these lateralized and bilateral patterns of brain responses to speech sounds for later language development (Molfese, 1983; Corballis, 1983). While data have continued to accumulate in the past decade which support the general view that lateralization of function is present in young infants, the question concerning the relevance of such early patterns of lateralization to later language acquisition has remained unanswered. Do these patterns of responses have any implications for later language development or do they reflect some basic patterns of auditory processing in the brain that have little relation to language development?

This interest in the possibility of biological precursors of language can be traced to a number of research findings published over the last two decades. Lenneberg (1967) argued that a biological substrate exists that subserves language abilities. Evidence for such a view, he noted, could be seen at a number of levels in humans. For example, even at the gross morphological level of the vocal tract, humans (unlike other primates) are structured in a certain way to produce a wide variety of speech sounds (Lieberman, 1977). The pinna itself is structured to favor the perception of sound frequencies that characterize the majority of important speech cues. At the neurological level, Lenneberg argued that language acquisition was linked to brain organization. For Lenneberg, lateralizaton of brain functions was a biological sign of language ability (Lenneberg, 1967, page 67). In this view, the presence of early lateralized processes for language could influence later language outcomes (Basser, 1962). Children who demonstrated language skills early in development were thought to have language already lateralized to one hemisphere. If the infant should then suffer cerebral damage that affected the language hemisphere, the infant's ability to recover language functions was thought to be more limited than a child whose language system had not yet lateralized. Although investigators have challenged some of Lenneberg's specific hypotheses on lateralization and language development (Molfese, 1972; Molfese, Freeman & Palermo, 1975), his general view that there are specific biological underpinnings for language that may facilitate language development continues to be supported (Dennis & Whitaker, 1975; Molfese & Molfese, 1979a, 1979b, 1980; Segalowitz & Gruber, 1977; Segalowitz, 1983). Given Lenneberg's (1967) notion that lateralization is a "biological sign" of language, could such early patterns of lateralized and bilateral discrimination of speech sounds predict later language outcomes? This was the major aim of a longitudinal study by Molfese and Molfese (1985a, 1985b). This project attempted to establish the predictive validity of demographic variables, behaviour scales and auditory evoked potentials for identifying developmental deviations in language abilities. The specific issue addressed concerned whether general hemisphere effects *per se* or hemisphere differences that interacted with specific stimulus characteristics would discriminate between children who later differed in language skills.

In this longitudinal study, sixteen infants were tested at birth. For each subject the following information was obtained: sex, birthweight, length, gestational age; the ages, income level, education and occupation of both parents; scores on the Obstetric Complications Scale (Littman & Parmelee, 1978); scores on the Brazelton Neonatal Assessment Scale (using scores on each of four *a priori* dimensions [Als, Tronick, Lester and Brazelton, 1977] and on the overall profile based on ratings for the 26 items of scale); mental subscale scores on the Bayley Scales of Infant Developmental (Bayley, 1969); and scores on two language tests administered at 36 months of age (the Peabody Picture Vocabulary Test, Dunn, 1965 and the McCarthy Scales of Children's Abilities, McCarthy, 1972). Auditory evoked responses were recorded at each testing period using recording electrodes placed on the scalp over left and right temporal areas. Synthetic speech syllables varying in PLACE OF ARTICULATION cues were presented to these babies and children since these materials had been shown to generate both bilateral and lateralized stimulus related effects as well as the more general hemisphere non-stimulus related effects (Molfese & Molfese, 1979a). Eight related stimulus items were added in order to test the generalizability of the findings for consonants across different vowel sounds.

Predicting language performance at 3 years of age from AERs obtained at birth

Analyses of the electrophysiological data led to the identification of the electrophysiological response correlates of specific stimulus features that appeared to predict later language performance from brain responses recorded early in development and from behavioural responses. Several portions of the brain response systematically changed across hemispheres and differentiated between specific speech and non-speech sounds.

One component of the AER that occurred between 88 and 240 msec reliably discriminated between the High and Low Language Performance groups at certain electrode (hemisphere) locations and under certain stimulus conditions. Only the AER waveforms recorded from over the left hemisphere of the High group systematically varied as a function of consonant sounds. A right hemisphere effect was noted for this same group of subjects for nonspeech stimuli.

The individual averages for a High MCVP Infant (S#9) are depicted in Figure 1. Here, the left hemisphere responses to the different stimuli are presented at the top of the figure while the right hemisphere responses are presented at the bottom. The region of the AER between 88 and 240 msec changed systematically as a function of stimulus features. The peak amplitudes as measured between points "a" and "b" were smaller for the /b/ initial syllables than for the /g/ syllables when the syllables were composed of normally structured speech formants. This pattern was reversed in the set of waveforms recorded from the right hemisphere in response to the sinewave formant stimuli (depicted in the lower left of this figure). No such amplitude variations are apparent in the two sets of waveforms presented on the right side of the figure. No differences were noted for the low group for these electrode sites or stimulus dimensions.

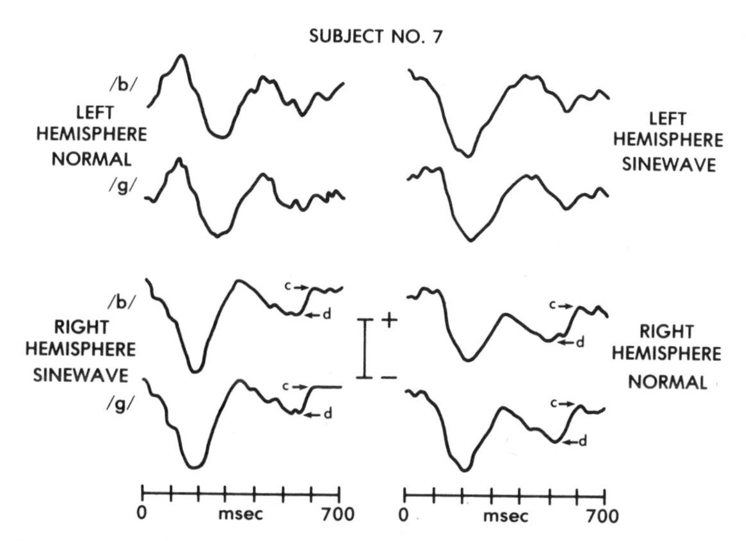

Fig. 2. The averaged AERs elicited shortly after birth from Subject #7, a member of the LOW MCVP group, in response to the speech and nonspeech syllables beginning with the consonants /b/ and /g/. AER deviation is 704 msec with positive deflection up. The calibration marker is 10 μV.

The individual averages for Infant #7, a member of the Low MCVP group, are depicted in Figure 2. No comparable /b/ versus /g/ differences can be noted for this infant across the left hemisphere normal formant and right hemisphere sinewave format conditions.

One other portion of the brain response that changed across hemispheres approximately 450 msec after the sound began failed to discriminate between the two language groups. No other group related effects were noted. Thus, it appears that at birth there were hemisphere responses differentially sensitive to specific stimulus characteristics which discriminated between children at three years of age who appeared to have different levels of language skills.

A second portion of the AER (as represented by Factor 7) with a late peak latency of 664 msec also discriminated between the High and Low groups. This component was recorded by electrodes placed over both sides of the head and consequently reflected bilateral rather than only lateralized activity that discriminated between the two consonant sounds. This component, however, did not behave in the same manner as that first described in that it was not affected by the speech-like quality of the stimuli. Furthermore, the consonant discriminations were dependent on the following vowel sounds. Thus, it appears that lateralized responses alone were not the sole discriminators between children who later differed in language abilities. These data suggest that the AERs are reflecting brain responsiveness to specific language-relevant speech cues rather than readiness of the brain to respond to any general stimulus.

The findings of this last study of a left hemisphere response that discriminates between the speech consonants /b/ and /g/ replicates an effect reported earlier by Molfese and Molfese (1979a). Both studies note that these left hemisphere lateralized responses are reflected in amplitude changes for the first large negative deflection in the AER. Furthermore, the bilateral effect that was noted in the present study (peak latency = 664 msec) behaved in a similar fashion in terms of latency and response characteristics to the bilateral response (peak latency = 630 msec) reported earlier in the Molfese and Molfese paper. Importantly, these replications hold across changes in the stimulus set, thereby indicating the generalizability of the findings across different vowel sounds.

Hemisphere effects were also noted that did not interact with stimulus or subject variables. The presence of such Hemisphere effects, both interacting with stimulus characteristics as well as independent of such variables, suggests that hemisphere effects should be treated as a multidimensional concept when applied to infants (Molfese & Molfese, 1985a and b; Moscovitch, 1977).

One interpretation for the results of this study is that early discrimination abilities may relate to later language development. The High MCVP infants in the present study not only discriminated between consonants alone and consonants in different vowel environments, they also discriminated between variants of the speech and nonspeech stimuli. This pattern of responding could suggest that the High MCVP infants were advantaged in the process of language development because their nervous systems were either more sensitive to or could make finer discriminations between a variety of auditory events, that share some commonality with speech perception events. Perhaps the earlier an infant can detect and discriminate between patterns of sounds in their language environment, the better able that infant will be to utilize such information as the extensive process of language acquisition begins.

An alternative interpretation of these data might suggest that the children who scored higher on the language test at three years of age were generally brighter, more attentive, and more responsive infants at birth. Such infants might respond in a more consistent manner to the speech and nonspeech stimuli at birth, thereby presenting a pattern of more clearly defined auditory discriminations. Because of these factors, these infants could have been consistently more testable at every age. While this is a viable explanation for these data we feel that it is not supported by the results of the behavioural tests administered to these infants at birth. As noted earlier, the Brazelton Neonatal Assessment Scale had been administered to these infants at birth and the Bayley Scales of Infant Development at 6- and 12-months of age. t-tests comparing the total scale scores on the Brazelton show no differences between groups at birth nor were there differences in developmental quotients on the Bayley for the High and Low groups at 6 or 12 months of age. The lack of differences on these developmental tests at birth and during the first year of life suggests that both groups of infants were equally responsive and achieved the same level of performance in at least these types of testing situations. It is, of course, possible that differences in testability might have been missed because the tests used were not sensitive to the relevant dimensions.

Behavioural Measures as Predictors of Language Development

In other analyses, the contributions of behavioural measures to predicting language performance at age three years have been determined. As has been found in prior research by other investigators, the correlations between the perinatal variables and the infant and child variables were low. In this case, few of the correlations reached significance. Significant correlations were found between Brazelton Neonatal Assessment Scale scores and 6-month (.41) and 12-month (.53) Bayley mental subscale scores. Obstetric Complication Scale scores were significantly correlated with 6-month Bayley mental subscale scores (.66) but not at other ages. The Bayley mental subscale scores from age 18-months on showed stronger correlations with the 3 year language measures. The 18-month, and 24-month Bayley mental subscale scores significantly correlate (.71 and .58, respectively) with the 36-month Peabody score. The 18- and 24-month Bayley scores significantly correlate (.63 and .73, respectively) with the 36-month McCarthy Scale scores. The Peabody scores and the McCarthy Scale scores significantly correlate at .64.

Demographic characteristics were not significant correlates of infant development and language performance scores. This may be due to the relatively homogenous nature of the families and infants involved in the study and the small number of characteristics measured. The families were middle class with average incomes of $20,000 to $25,000, and both parents had at least completed high school.

Regression models were constructed to test hypotheses concerning the usefulness of perinatal, demographic and infant development tests to predict language performance at 36 months. The models showed that McCarthy scores can be predicted from individual and combinations of the following variables: birthweight, length, gestational age, labor length, 18 and 24 month Bayley scores and Peabody scores (best full model is birthweight, length, gestational age, 18 month Bayley). Peabody scores can be predicted from individual and combinations of the following variables: labor length, Bayley at 18 and 24 months, McCarthy scales (best models are Bayley-18 and McCarthy, and Bayley-24 and McCarthy). When only birth scores (i.e., BNAS, birthweight, gestational age, length, obstetrical events and complications) were used to predict language scores at three years of age, the regression models were not significant.

Regression models were also constructed to test hypotheses concerning the usefulness of all the independent variables (i.e., perinatal, demographic, infant measures and AER factor scores) to predict the language performance scores. The only regression model which was significant involved predicting McCarthy Scale scores from Brazelton scores, Obstetric Complication Scale scores and AER factor scores. This regression model accounts for 3% more of the variance using perinatal measures and AER factor scores than was accounted for by using AER factor scores alone. Perinatal measures alone were not significant predictors. Clearly, the addition of non-brain related measures did little to improve predictions of language performance based on brain wave data alone.

Implications

From these data, it seems clear that brain activity measure at birth can be used with a high degree of accuracy to predict performance on two language tests at age three years. Children who later developed into better performers on language tests produced brain responses to speech consonant sounds at birth that were distinctly different from the responses produced by children who did not perform as well. The predictive accuracy achieved using the AER responses exceeds by far the predictive abilities of the other perinatal and behavioural measures used in this study and in other studies (Molfese & Molfese, 1985b).

It is our belief that hemispheric lateralization when coupled with the discriminating abilities for certain types of sounds would provide the best predictors of later language development. While Lenneberg has suggested that lateralization *per se* would provide a general indicator of later language development, our point in light of the data we have obtained is that lateralization *plus* some specific discriminatory abilities are necessary for proper language development. The High language performance group did not simply differ from the Low group in their ability to discriminate human from nonhuman sounds. Rather, the former group showed a discrimination in the left hemisphere for language sounds (specifically, place of articulation contrasts for stop consonants) and a right hemisphere discrimination for nonlanguage sounds (rapid frequency transitions with formant bandwidths not characteristic of speech) that were not shared by the Low language group. The presence of certain early discrimination abilities, then, are seen as facilitating the process of language acquisition. Infants without those early abilities or who are slower in acquiring those abilities would face more difficulty in acquiring language.

The data reported in the present study do not provide a sufficient base on which to ground claims concerning a single cause for the variety of language deficits which can occur throughout development. We are not claiming that an infant's ability to discriminate between two stop consonants is a sufficient basis for identifying the potential for later language disorders. The present study, however, does report data which show a definite relationship between brain responses recorded shortly after birth and relatively small differences in language performance three years later between children who apparently followed a normal pattern of development and who could clearly communicate with those around them. We have no data at this time as to whether these procedures might discriminate between children with language skills in the normal range and children who have clear language deficits. A replication of this study with a much larger and diverse population is now under way to address such issues.

REFERENCES

Als, H., Tronick, E., Lester, B. and Brazelton, T. (1977). The Brazelton Neonatal Behavioural Assessment Scale (BNAS). Journal of Abnormal Child Psychology, 5, 215-231.

Barnett, A., de Sotillo, M. & Campos, M. (1974). EEG sensory evoked potentials in early infancy malnutrition. Paper presented at the meeting of the Society for Neurosciences, St. Louis, Missouri, November.

Basser, L. (1962). Hemiplegia of early onset and the faculty of speech with special reference to the effects of hemispherectomy. Brain, 85 427-460.

Bayley, N. (1969). Bayley Scales of Infant Development; Birth to Two Years. New York: Psychological Corps.

Bryden, M. (1982). Laterality: Functional asymmetry in the intact brain. New York: Academic Press.

Corballis, M. (1983). Human Laterality. New York: Academic Press.

Crowell, D.H., Jones, R.H., Kapuniai, L.E. & Nakagawa, J.K. (1973). Unilateral cortical activity in newborn humans: An early index of cerebral dominance? Science, 180, 205-208.

Dennis, M. & Whitaker, H. (1975). Hemispheric equipotentiality and lanuage acquisition. In S. Segalowitz and F. Gruber (Eds.) Language development and neurological theory. New York: Academic Press.

Donchin, E., Ritter, W. & McCallum, W.C. (1978). Cognitive psychophysiology: The endogenous components of the ERP. In E. Callaway, P. Teuting and S. H. Koslow, (Ed.), Event related potentials in man. New York: Academic Press, pages 349-411.

Dunn, L. (1965). Peabody Picture Vocabulary Test. Circle Pines, MN: American Guidance Service.

Goff, E., Allison, T. & Vaughan, H. (1978). The functional neuroanatomy of event related potentials. In E. Callaway, E. Teuting and S. Koslow (Eds.), Event-related brain potentials in man. New York: Academic Press, pp. 1-79.

Hillyard, S.A. & Woods, D.L. (1979). Electrophysiological analysis of human brain function. In M. S. Gazzaniga (Ed.) Handbook of Behavioral Neurobiology, Vol. 2. New York: Plenum Press, pages 345-378.

Klee, M. & Rall, W. (1977). Computed potentials of cortically arranged populations of neurons. Journal of Neurophysiology, 40, 647-666.

Lenneberg, E. (1967). Biological foundations of language. New York: Wiley.

Lieberman, A. (1977). On the origins of language. New York: MacMillan.

Littman, B. & Parmelee, A. (1978). Medical correlates of infant development. Pediatrics, 61, 470-474.

McCarthy, D. (1972) Manual for the McCarthy Scales of Children's Abilities. New York: Psychological Corporation.

Molfese, D.L. (1972). Cerebral asymmetry in infants, children and adults: Auditory evoked responses to speech and music stimuli. (Doctoral dissertation, Pennsylvania State University, 1972). Dissertation Abstracts International, 33. (University Microfilms No. 72-48, 394).

Molfese, D.L. (1978a). Neuroelectrical correlates of categorical speech perception in adults. Brain and Language, 5 25-35.

Molfese, D.L. (1978b). Left and right hemisphere involvement in speech perception: Electrophysiological correlates. Perception and Psychophysics, 28, 237-243.

Molfese, D.L. (1980a). Hemispheric specialization for temporal information: Implications for the perception of voicing cues during speech perception. Brain and Language, 285-299.

Molfese, D.L. (1980b). The phoneme and the engram: Electrophysiological evidence for the acoustic invariant in stop consonants. Brain and Language, 9, 372-376.

Molfese, D.L. (1983). Event related potentials and language processes. In Gaillard, A. and W. Ritter (Eds.) Tutorials in ERPs. Holland: Elsevier.

Molfese, D.L. (1984). Left hemisphere sensitivity to consonant sounds not displayed by the right hemisphere: Electrophysiological correlates. Brain and Language, 22 109-127.

Molfese, D.L., Freeman, R.B. & Palermo, D.S. (1975). The ontogeny of brain lateralization for speech and nonspeech stimuli. Brain and Language, 2, 356-368.

Molfese, D.L., Linnville, S.E., Wetzel, F. & Leicht, D. (1985). Electrophysiological correlates of handedness and speech perception contrasts. Neuropsychologia, 23, 77-86.

Molfese, D.L. & Molfese, V.J. (1979a). VOT distinctions in infants: Learned or innate? In H.A. Whitaker & H. Whitaker (Eds.) Studies in Neurolinguistics (Vol. 4). New York: Academic Press.

Molfese, D.L. & Molfese, V.J. (1979b). Hemisphere and stimulus differences as reflected in the cortical responses of newborn infants to speech stimuli. Developmental Psychology, 15, 505-511.

Molfese, D.L. & Molfese, V.J. (1980). Cortical responses of preterm infants to phonetic and nonphonetic speech stimuli. Developmental Psychology, 16 574-581.

Molfese, D.L. & Molfese, V.J. (1985a). Electrophysiological indices of auditory discrimination in newborn infants: The bases for predicting later language development? Infant behaviour and Development, 8, 197-211.

Molfese, V. J. & Molfese, D.L. (1985b). Predicting a child's preschool language performance from perinatal variables. In R. Dillon (Ed.) Individual Differences in Cognition, Volume 2. New York: Academic Press.

Molfese, D.L. & Schmidt, A. (1983). An auditory evoked potential study of consonant perception. Brain and Language, 18, 57-70.

Moscovitch, L. (1977). The development of lateralization of language functions and its relation to cognitive and linguistic development: A review and some theoretical speculations, pages 194-212. In S. Segalowitz and F. Gruber (Eds.) Language Development and Neurological Theory. New York: Academic Press.

Neville, H. (1980). Event-related potentials in neuropsychological studies of language. Brain and Language, 11, 300-318.

Picton, T. W. & Stuss, D.T. (1980). The component structure of the human event-related potential. In H.H. Kornhuber and L. Deecke, (Eds.) Motivation, Motor and Sensory Processes of the Brain. Electrical Potentials Behavior and Clinical Use. Amsterdam, Elsevier, 17-42.

Purpura, D. (1978). Commentary. In E. Callaway, P. Tueting & S. Koslow (Eds.) Event-related brain potentials in man. New York: Academic Press, pp. 81-87.

Regan, D. (1972). Evoked potentials in psychology, sensory physiological, and clinical medicine. New York: John Wiley & Sons.

Reneau, J.P. & Hnatiow, G.Z. (1975). Evoked Response Audiometry: A Topical and Historical Review. Baltimore: University Park Press.

Rockstroh, B., Elbert, T., Birbaumer, N. & Lutzenberger, W. (1982). Slow brain potentials and behaviour. Baltimore: Urban & Schwarzenberg, Inc.

Segalowitz, S. (1983) Language functions and brain organization. New York: Academic Press.

Segalowitz, S. & Gruber, F. (1977). Language development and neurological theory. New York: Academic Press.

Shucard, J.L., Shucard, D.W., Cummings, K.R. & Campos, J.J. (1981). Auditory evoked potentials and sex related differences in brain development. Brain and Language, 13 91-102.

Simson, R., Vaughan, H.G. Jr., & Ritter, W. (1977a). The scalp topography of potentials in auditory and visual discrimination tasks. Electroencephalography and Clinical Neurophysiology, 42 528-535.

Simson, R., Vaughan, H.G. Jr., & Ritter, W. (1977b). The scalp topography of potentials in auditory and visual go/no go tasks. Electroencephaplography and Clinical Neurophysiology, 43, 864-875.

Trehub, S. & Rabinovitch, S. (1972). Auditory-linguistic sensitivity in early infancy. Developmental Psychology, 6, 74-77.

Vaughan, H.G. Jr. (1969). The analysis of brain activity to scalp recordings of event-related potentials. In E. Donchin and D.B. Lindsley (Eds.) Averaged evoked potentials: Methods, results, evaluations. Washington, D.C.: NASA, pages 45-94.

DIFFERENCES BETWEEN ANHEDONIC AND CONTROL SUBJECTS IN BRAIN HEMISPHERIC SPECIALIZATION AS REVEALED BY BRAIN POTENTIALS*+

Brigitte Rockstroh and Werner Lutzenberger

Arbeitsbereich Klinische und Physiologische Psychologie
Gartenstr. 29, D-4700 Tubingen, F.R.Germany

+ Research was supported by the Deutsche Forschungsgemeinschaft

INTRODUCTION

The purpose of the present paper is to discuss evidence for hemisphere specific information processing in anhedonic subjects as revealed by event related slow brain potentials. Anhedonia refers to a profound deficit in pleasure experience (see below) and is considered a risk factor for the development of a psychiatric disorder. The investigation of humans supposed to be at risk for a psychopathological development may offer insights into (deviant) human information processing. It may also uncover determining factors or processes underlying the psychopathological disorders.

Evidence from many experimental and clinical studies has provided the basis for the hypothesis, as stated by Flor-Henry, that psychopathological disorders may be characterized by hemispheric-specific dysfunctions: this hypothesis relates schizophrenia to left-hemispheric overarousal and processing dysfunctions, while depression is considered a consequence of right-hemispheric malfunctioning and overactivation Flor-Henry and Koles, 1980; Tucker, 1982; Kemali, Vacca, Nolfe, Iorio & DeCarlo, 1980; a recent critical overview is provided by Cevey, 1984). This relationship has been strengthened by the finding of asymmetries in the EEG power spectra; for example, Serafetinides (1972, 1973) reported more desynchronization over the left hemisphere in schizophrenics (which was reduced under phenothiazine). Flor-Henry (1976, 1979) found more left-temporal power in the 13-20 Hz range in schizophrenic patients, while depressives demonstrated more right-temporal power in this range. As compared to normal control subjects, manic-depressives showed a significant reduction in alpha power over the right parietal area during verbal and visual-spatial tasks (see also Goldstein, 1979; Shaw, 1979). On the other hand, Abrams and Taylor (1979) found EEG-abnormalities only in 28% of the patient sample. Those patients showing lateralized abnormalities (8 out of 27 schizophrenics, 21 out of 132 affective psychotics) supported the hypothesis of a lateralized aberration as proposed by Flor-Henry.

----

* Address correspondence to:
Dr. Brigitte Rockstroh,
Arbeitsbereich Klinische und Physiologische Psychologie,
Gartenstr. 29,
D-4700 Tubingen, F.R. Germany.

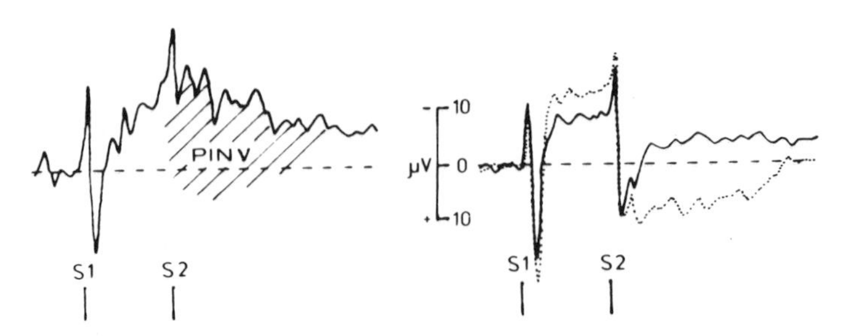

Fig. 1. Right-graph: Slow potential shift (CNV) during 2-sec interstimulus interval (S1-S2), as well as post-S2 CNV resolution in subjects with high (solid line) and low (dotted line) scores on the anhedonia scale. Left graph: SP in anticipation and following the S2 under two-stimulus reaction time conditions recorded from a schizophrenic patient. Notice the prolonged negativity (PINV). From: Elbert, 1985, with permission.

(Furthermore, the data of Perris & Monakhov (1979) point to a more complex relationship between symptoms in depression and hemisphere specific EEG patterns.)

The concept of anhedonia has a long history of association with theories of schizophrenia, from the psychoanalytic (Rado, 1956) to the biochemical (Stein & Wise, 1971). According to the concept of Meehl (1962), anhedonia is one of the four crucial source traits that characterize the schizotypic personality (together with cognitive slippage, ambivalence, and interpersonal aversiveness). Anhedonia is defined as "a marked, widespread and refractory defect in pleasure capacity" (Meehl, 1962, p.829) and is considered one of the most consistent and dramatic behavioral signs of schizophrenia. This personality trait is supposed to depict a potential risk factor for the development of a psychotic disorder. Meehl assumes a neural integrative defect as an underlying factor in schizophrenia, which he calls schizotaxia. The imposition of a social learning history upon schizotaxic individuals results in the personality organization of the schizotype. Although Meehl stresses that only a subset of schizotypic personalities decompensate into clinical schizophrenia, the investigation of subjects at risk in this respect (such as anhedonic individuals) may provide insight into characteristics of the schizophrenic disorder. This similarity was not only supported by psychological tests (anhedonics showed significant amounts of thought disorder on the Rorschach test (Edell & Chapman, 1979; Haberman, Chapman, Numbers & McFall, 1979) but also by psychophysiological responses (see Simons, 1981, 1982). With respect to event-related brain potentials, schizophrenic patients very often show smaller or altogther-lacking EP-components in the latency range between 50 and 350msec (Roth, 1977; for an overview of evoked potential results in schizophrenics see also Shagass, Ornitz, Sutton & Tueting, 1978; Shagass, 1979); this finding was also obtained in anhedonic subjects (see Simons, 1982; Miller, Simons & Lang, 1984).

Evidence for aberrant electrophysiological responses in anhedonic individuals was furthermore achieved in studies on slow brain potentials (SPs): whenever two stimuli are associated or contingent, in that a first or warning stimulus signals that a (cognitive or motor) response will be required by a subsequent imperative stimulus, then a slow negative DC-potential shift is recorded from the scalp during this anticipation or preparatory interval. This most prominent slow potential (SP) representative, the Contingent Negative Variation (CNV), was first reported by Walter, Cooper, Aldridge & McCallum (1964) and has been suggested to be a sign of priming and preparation. Much of the work (for summary see Rockstroh et al., 1982) suggests that negative SPs might be indicators of the preparation or mobilization of cerebral resources for the anticipated cerebral performance and overt response.

During or following the respective performance, the slow negative potentials usually decrease to baseline or to positive values. Under distinct conditions and in certain Ss, however, negativity is prolonged or even augmented. This prolonged negativity was originally observed in psychotic patients (Dongier, Dubrovsky & Engelsmann, 1976, 1977; Timsit-Berthier, Delaunoy & Rousseau, 1976) and has been called Postimperative Negative Variation (PINV) (see Fig. 1).

While it was first assumed that a PINV represents the underlying disorder, a PINV could also be induced in healthy Ss under conditions of sudden contingency change, uncontrollability, or disappointed expectations of contingencies or reinforcement (Rockstroh, Elbert, Lutzenberger & Birbaumer, 1979; Elbert, Rockstroh, Lutzenberger & Birbaumer, 1982; Delaunoy, Timsit-Berthier & Rousseau, 1975; Delaunoy, Gerono & Rousseau, 1978). This suggested to us to consider the PINV as a sign of prolonged contingency evaluation or preparedness for further information processing requirements (Birbaumer, Elbert, Rockstroh & Lutzenberger, 1986).

Psychiatric patients (schizophrenics and psychotic depressives) turned out to be more vulnerable to contingency change or contingency ambiguity, since they have already developed a PINV under conditions of less clearcut response-consequence contingencies, for example, if the response criterion was increased such that the imperative stimulus could be terminated by the motor respnse only in a certain proportion of trials (Giedke et al., 1980, 1982). This susceptibility was also observed in anhedonic Ss presumed to be at risk for the development of a psychotic disorder (see Lutzenberger, Elbert, Rockstroh, Birbaumer & Stegagno, 1981): Under conditions of sudden contingency change or loss of control over an aversive stimulus, anhedonic subjects developed a larger PINV than low-scoring subjects. When a response-irrelevant stimulus was introduced into some trials which had to be discriminated from the response-relevant imperative stimulus occurring 2 sec later, anhedonics exhibited larger anticipatory negativity prior to the possible additional stimulus, as well as a PINV tendency following the imperative stimulus (Lutzenberger, Birbaumer, Rockstroh & Elbert, 1983). Miller, Simons & Lang, 1984) report less anticipatory negativity (CNV) prior to hedonic stimuli (nude females) in anhedonic subjects as compared to controls which points at a reduced sensitivity for hedonic stimuli.

SUBJECTS

On the basis of Meehl's concept, Chapman and colleagues (Chapman, Chapman & Raulin, 1976, 1978) have developed a self-report questionnaire for the screening of subjects at high risk for schizophrenia. From a total of 750 university students who filled in the questionnaire a sample of 40 male Ss with extreme scores (high and low) on the scale was selected. The mean anhedonia score was 11.3+/-7. Ss with scores higher than 15 on the scale (that comprises 61 questions, i.e. 61 possible points)* were assigned to the experimental group, while Ss wih a mean score below 10 constituted the

---

*It has to be considered that subjects with low scores must not correspond to a control group in the sense of a normal distribution but may also include hedonic subjects with a tendency for mania.

control group. The anhedonia score of this sample of 40 subjects decreased across time: across a time period of by now six years, the subjects participated in several psychophysiological investigations and filled in the questionnaire four times; the mean anhedonia score decreased from 21.8 in 1978, to 15.1 in 1983. [The anhedonia-score in hospitalized schizophrenics exceed that of the present sample (15.9+/-8.7), although many schizophrenics do not exhibit anhedonic tendencies at all. On the other hand, 19 out of 20 psychotic depressives gave a strong anhedonic self-report (mean 22.9+/-6.9)].

In the present paper three studies are reported which investigate hemisphere specific SP-shifts in anhedonic subjects to gain insights into hemisphere specific functioning in individuals supposed to be at risk for the development of a psychotic disorder.

EXPERIMENT I

Method

In the first study, CNV and postimperative SPs were investigated in the standard two-stimulus reaction time paradigm. A click (white noise of 65 dB SPL, 100 msec duration) was presented via earphones as warning stimulus (S1). After an interstimulus interval of two seconds a light spot appeared on a tv-screen in front of the Ss as imperative stimulus (S2). Ss were requested to press a button as fast as possible in response to the S2 presentation. A total of 50 trials was given. Intertrial intervals varied between 9 and 17 sec around a mean of 13 sec.

The EEG was recorded monopolarly from F3, F4 and C3, C4 according to the International 10-20 system, with a time constant of 30 sec. Grass flat silver disc electrodes, chlorided prior to each experimental session, were used, Grass paste EC2 served as electrolyte. To achieve electrode impedance levels below 5 k Ohms, the skin below the electrodes was cleaned with alcohol and the outer layers of the skin were removed by gently scraping with a sterile scalpel. Reference electrodes were affixed to both earlobes and were shunted via a 10 k Ohm resistor. Eye movements were monitored for artifact control. Data were sampled at a rate of 100Hz and collapsed to 100 msec points by a digital filter without phase distortion for the slow potential analysis.

DEC laboratory computers controlled the timing and presentation of the experimental stimuli and stored the physiological data on magnetic tape. Pressing the microswitch closed the Schmitt trigger in the digital computer so that the response latency was stored to the nearest ms.

For data analysis, the digitized EEG and EOG data were averaged across 10 subjects and trials. Trials were rejected from further analysis, whenever the mean EOG shift exceeded 50 V, or whenever the mean slow potential (SP) shift during the stimulus interval exceeded 70 V relative to the baseline value. The mean of the second sec during the interstimulus interval was taken as measure of the CNV; the mean of the third half-second interval following the S2 referred to the CNV amplitude represented the postimperative shift, PINV. Experimental effects were statistically evaluated by an analysis of variance with the factors GROUPS (anhedonics vs controls) and LATERALITY (left vs right hemispheric recording). Furthermore, correlation coefficients determined the relationship between anhedonia score and the hemisphere-specific PINV-amplitudes.

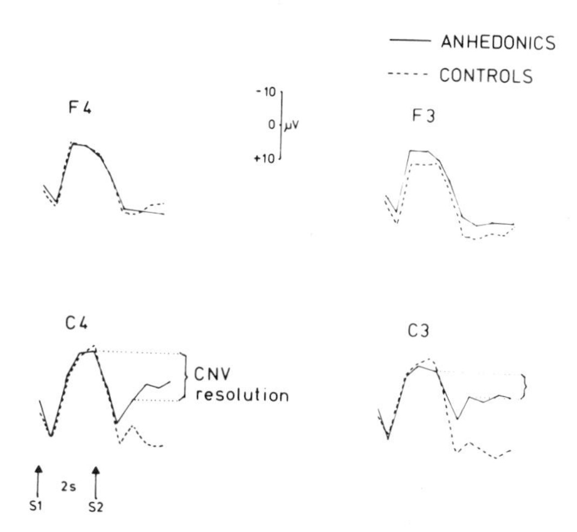

Fig. 2. CNV during the 2-sec anticipation interval and postimperative CNV-resolution in anhedonic (solid lines) and control subjects (dashed lines) recorded from the right (frontal: F4, precentral: C4) and the left (F3, C3) hemisphere. Less CNV-resolution indicates more pronounced postimperative negativity (PINV).

Results

Anhedonics demonstrate less postimperative negativity resolution following the S2/button press response, thus, a more pronounced relative PINV, than control Ss with low scores on the anhedonia scale. This is documented by a main effect of GROUPS for the right-precentral PINV-amplitude ($F(1,39)= 4.1$, $p < 0.05$), as well as for the left-precentral PINV-score ($F(1,39)= 6.7$, $p < 0.05$). Closer inspection of the hemisphere-specific SPs reveals the more pronounced postimperative negativity in anhedonics over the left precentral area (see Fig. 2). The ANOVA documents the group difference by an interaction of GROUP and LATERALITY with $F(1,39)= 6.7$, $p < 0.05$ for the PINV-amplitude. The group difference in hemisphere-specific PINV is supported by significant correlation coefficients between the anhedonia-score and the left-precentral PINV-amplitude, ($r= .40$, $p < 0.05$), as well as the left-right asymmetry ($r= .42$, $p < 0.05$).

This result suggests that increasing anhedonia-scores covary with more pronounced PINV, and with more pronounced lateral asymmetry, i.e. a left-hemispheric dominance of the PINV. Anhedonics seem espescially to activate the left hemisphere for their ongoing contingency evaluation.

EXPERIMENT II

Method

In the second study with the two-stimulus reaction time paradigm, the level of difficulty of the contingency evaluation was increased in that a response-irrelevant but similar stimulus had to be discriminated from the response-relevant imperative stimulus in half of the trials. Ss heard again a click (white noise of 80 db for 100 msec) as warning stimulus, which was followed after 6 sec by a hissing noise (white noise of 80 db) that had to be interrupted by an immediate button press. In half of the total of 50 trials, pseudorandomly interspersed, a hissing noise similar to the imperative one was presented after 4 sec, but Ss were instructed not to

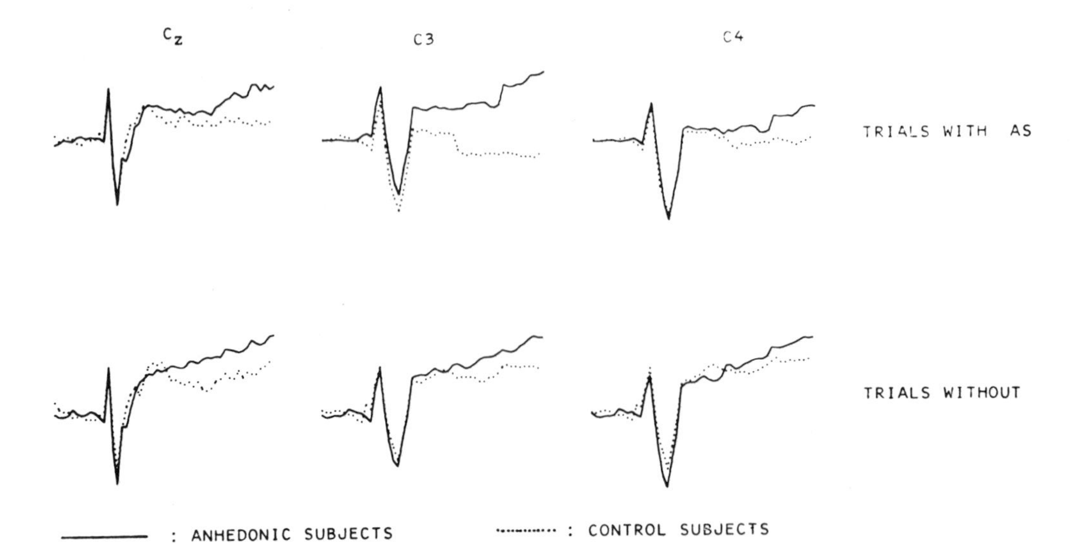

Fig. 3. SP-shifts during a 4-sec interval preceding the possible presentation of an additional stimulus (AS), recorded from the vertex (Cz), the left (C3) and the right (C4) precentral cortex in anhedonic (solid lines) and control subjects (dotted lines). Upper graphs: trials in which an AS occurred at the end of the 4-sec interval; lower graphs: trials without AS-presentation.

respond to this additional stimulus (AS) but only to the IS. Since subjects received delayed feedback about the adequacy of their discrimination only at the end of the trial, the stimulus-response contingency was assumed to be less clearcut or more difficult to evaluate.

The EEG was recorded along the midline (Fz, Cz, Pz), as well as from C3 and C4 according to the procedure described above. 38 Ss from the above described sample participated in the study.

For data analysis the digitized physiological data were averaged over trials separately for each subject x condition x recording cell. A varimaxed Principal Component Analysis (PCA) based on covariances was performed to decompose SP-waveforms. A weighted sum of principal components was fitted to each individual waveform to obtain component scores. These were submitted to an analysis of variance to evaluate statistical differences.

Results

In anhedonics, the negative shift prior to an expected AS* is significantly larger than in controls; this is supported by a main effect of

---

* If an AS is presented in 50% of the trials the conditional probabilities of two subsequent identical trials were 1 in 3, whereas the probabilities of a change in type of trial were 2 in 3. Subjects obviously have established subjective probabilities regarding the next trial on the basis of the preceding one. Following a trial with AS, a subsequent trial without AS is more likely to occur than another trial with AS, i.e. is expected. Consequently, subjects developed a more pronounced negative shift prior to the expected occurrence of an AS.

GROUPS for the late SP-component in the 4 sec interval between S1 and AS ($F(1,36) = 5.6$, $p < 0.05$).

As is illustrated by Fig. 3, this preparatory negativity in anhedonics, as well as the group difference, is most pronounced at C3, the left hemisphere, especially, if an AS was expected and presented. (see footnote on previous page).

Anhedonics also show less postimperative negativity resolution, thus, a more pronounced PINV, than controls, especially if no AS was presented during that trial (GROUP x AS: $F(1,36)=8.3$, $p < 0.01$, for an ANOVA of the recordings along the midline, Fz, Cz, Pz, and ($F1,36) = 3.9$, $p < 0.05$, for an ANOVA comprising the recordings Cz, C3, C4). However, this PINV did not show significant lateral asymmetry.

Furthermore, the N100 in the evoked potential turned out to be significantly larger at the left-precentral location (as compared to C4) in response to the AS (main effect of LATERALITY: $F(1,36) = 13.8$, $p < 0.01$) and in response to the imperative stimulus (S2)($F(1,36) = 16.2$, $p < 0.01$. The S2-elicited P200, too, shows left-hemispheric predominance ($F(1,36) = 9.3$, $p < 0.05$). This lateral asymmetry of the S2-elicited N100 is more pronounced in anhedonics, giving rise to a significant interaction between GROUPS and LATERALITY with $F(1,36) = 4.9$, $p < 0.05$).

Again, both results of this study, the larger preparatory negativity, as well as the more pronounced N100 to the response-relevant stimulus over the left hemisphere, indicated more pronounced left- than right-hemispheric phasic activation in anhedonics. The tonic activation, however, could also be lowered, which might give rise to higher phasic responses too. The N100 is mainly associated with selective attention (Hillyard, Hink, Schwent & Picton, 1973), which seems to require more effort in anhedonics, when they have to discriminate the S2 under the less clearcut contingency conditions of the present study. We may speculate that this increased activation may serve to compensate for left-hemispheric deficits, for example in the discrimination between the AS and the IS. The PINV then indicates longer lasting contingency evaluation in trials without AS, since following an AS the presentation of the IS verifies the discrimination and the adequacy of the response. Since the more pronounced PINV was observed in trials without AS, it cannot be argued that anhedonics take longer to recover from the distraction induced by the AS.

EXPERIMENT III

Finally, Ss participated in a SP-self-regulation training on the basis of a biofeedback paradigm. The perception and reward of electrical brain activity allows operant conditioning, so that negativity can be increased or reduced depending upon discriminative stimuli.

Method

The actual SP-shift is continuously fed back during intervals of 6 sec each by means of a little rocket-ship (Fig. 4), which moves across the tv-screen towards two goals. Ss are requested to direct the rocket into one of the goals depending upon which of two discriminative stimuli (a high- or a low-pitched tone) are presented simultaneously to the rocket's flight. The height of the rocket during its flight is a linear function of the integrated SP-shift at successive half-second averages, referred to the mean of a 4 sec baseline. Reaching the required goal, thus producing the correct SP-shift, is rewarded by a point signalling the gain of 1.-DM (40 U.S. Cents). Counterbalanced across Ss, one goal represents an SP-shift towards increased negativity, while reaching the other goal requires negativity

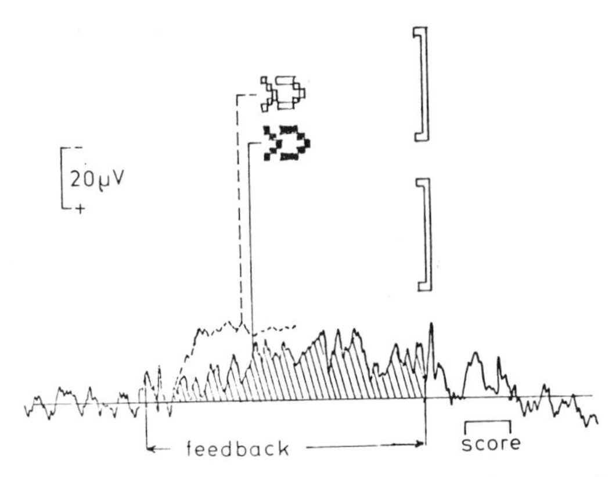

Fig. 4. The feedback display and the required changes in SP negativity. Dotted line and rocket-ship demonstrate that the more pronounced negative shift corresponds to the higher position of the rocket on the TV-screen as compared to less negativity (solid line and rocket). From: Rockstroh et al., 1984, with permission.

suppression. The EEG was recorded from Fz, Cz, Pz and from C3 and C4. The feedback stimulus was determined by the average of C3 + C4. (Various artifact control procedures prevent Ss from steering the rocket by eye movements or muscle tension/relaxation; for a detailed description of the feedback algorithm see Elbert, Rockstroh, Lutzenberger & Birbaumer, 1980; Rockstroh, Elbert, Birbaumer & Lutzenberger, 1982).

Many studies based on this paradigm have demonstrated that human Ss are able to achieve significant control, i.e., significant differentiation of SPs in the required polarity directions within 100 to 200 training trials. Control can also be maintained in transfer trials without feedback (see Elbert et al.; 1980, Birbaumer, Elbert, Rockstroh & Lutzenberger, 1981; Rockstroh, Elbert, Lutzenberger & Birbaumer, 1984).

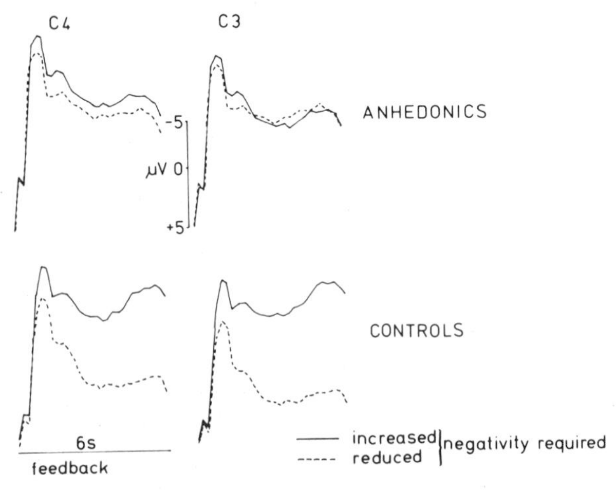

Fig. 5. SP-changes during the 6-sec feedback interval in anhedonics (upper graphs) and controls (lower graphs), averaged across trials with required increased negativity (solid lines) and required negativity suppression (dashed lines), and averaged separately for the right (C4) and left (C3) precentral recording.

Results

The SP-self-regulation ability in the control group of the present sample does not differ from that described for other samples of student volunteers, whereas anhedonics fail to achieve comparable SP-differentiation. Fig. 5 illustrates SP-regulation in anhedonics as compared to controls.

While control Ss achieve the most pronounced adequate differentiation especially at the left precentral recording, anhedonics fail to regulate their slow brain potentials, especially over the left hemisphere. This group difference is documented by a significant interaction of GROUPS x POLARITY x LATERALITY with $F(1,36) = 10.5$, $p < 0.01$.

Similar deficits in SP-self-regulation (however, not in a hemisphere-specific manner) have been observed in patients with bilateral lesions of the frontal cortex and in children with attentional disorders (see Rockstroh, Elbert, Lutzenberger & Birbaumer, 1984).

DISCUSSION

Taken together, the present results suggest a left-hemispheric SP-regulation deficit in anhedonics as compared to controls, which anhedonics might try to compensate by a left-hemispheric overactivation: they show more preparatory, more postimperative negativity, and an increased N100 amplitude over the left hemisphere. However, the present data do not allow a definite conclusion, whether the observed left-hemispheric overactivation results from overarousal or underarousal as the underlying deficit.

Do these results fit with findings as reported for schizophrenics (and probably also depressives)?

1.) Left-hemispheric overarousal was reported for schizophrenics, and if we consider anhedonics at risk for a schizophrenic career, then the present data suggest that this left-hemispheric overactivation is already evident in a pre-psychotic state. Increased left-hemispheric activity in schizophrenics is also supported by blood flow measurements, as described by Gur (this volume).

2.) Furthermore, left-hemispheric processing deficits were reported for schizophrenics. We may speculate that processing deficits become evident in the present sample in the impaired SP-self-regulation in the biofeedback paradigm (which, for example, involves attention and the processing of the feedback and reinforcement stimuli); they may also become manifest in the impaired or prolonged contingency processing as indicated by the postimperative negativity. This processing deficits might well be the consequence of a tonically underaroused state, which subjects try to compensate by increased effort, as indicated by increased (left-hemispheric) electrocortical activity. These results support the continuity between anhedonia and schizophrenia as proposed by Simons (1982).

3.) On the basis of EEG power spectra analyses, Flor-Henry and Koles (1980) as well as Kemali, Vacca, Nolfe, Iorio and Decario (1980) postulate less lateral asymmetry in schizophrenics than in controls. Although the present results are not based on power spectra analyses, they argue against a reduced lateralization in anhedonics (as compared to controls), but anhedonics certainly are not psychotic. We might speculate that a lasting left-hemispheric overarousal could turn into habituation or exhaustion, visible as reduced laterality in a later schizophrenic state.

On the one hand, it should be taken into account that anhedonia represents a disturbance of pleasure discrimination, which may characterize several psychopathological disorders, also depression. Although anhedonia is considered a source trait of the schizotypic personality, which may decompensate to a schizophrenic disorder, there is evidence that depressives score high on the anhedonic scale, so that the present results, too, may provide predictive evidence for a possible depressive development. The primarily left-hemispheric regulation deficit in anhedonics could contribute to their pleasure deficit, if we assume positive emotions to be represented in the left hemisphere (see Sackeim, Gur and Saucy, 1982; Flor-Henry and Koles, 1979; Tucker, Watson and Heilman, 1981). On the other hand, this hypothesis would predict that activation of the left hemisphere should induce positive emotions, which is not true for the present sample.

In the framework of our model about the behavioural meaning of slow cortical negativity (Rockstroh, Elbert, Birbaumer and Lutzenberger, 1982), the presently observed left-hemispheric activation might be interpreted as sign of increased effort to compensate for the (left-hemispheric) deficient regulation of attention and preparatory resources. Because of this regulation deficit, anhedonics (and psychotic patients, see Giedke, Bolz, Heimann and Straube, 1982) may be more susceptible to ambiguity of, or changes in contingencies.

REFERENCES

Abrams, R. and Taylor, M. (1979), Differential EEG patterns in affective disorder and schizophrenia. Archives of General Psychiatry, 36 1355-1358.

Birbaumer, N., Elbert, T., Rockstroh, B. and Lutzenberger, W. (1981). Biofeedback of event-related potentials of the brain. International Journal of Psychology, 16, 389-415.

Birbaumer, N., Elbert, T., Rockstroh B. and Lutzenberger, W. On the dynamics of the postimperative negative variation (PINV). Electroencephalography Clinical Neurophysiology Suppl. Amsterdam, Elsevier, 1986 (in press).

Cevey, B. (1984). Emotion und lateralisierte Aktivierung des Gehirns. Munchen, Profil-Verlag,

Chapman, L.J., Chapman, J.P., Raulin, M.L. (1976). Scales for physical and social anhedonia. Journal of Abnormal Pyschology, 85, 374-382.

Chapman, L.J., Chapman, J.P., Raulin, M.L. (1978). Body image aberration in schizophrenia. Journal Abnormal Psychology 87, 399-407.

Delaunoy, J., Timsit-Berthier, M., Rousseau, J. and Gerono, A. (1975). Experimental modification of the terminal phase of the CNV. Electroencephalography and Clinical Neurophysiology, 39, 551.

Delaunoy, J., Gerono, A. and Rousseau, J. (1978). Experimental production of postimperative negative variation in normal subjects. In D.A. Otto (Ed.) Multidisciplinary Perspectives in Event-Related Brain Potential Research. Washington: U.S. Environmental Protection Agency, 355-357.

Dongier, M., Dubrovsky, B. and Engelsmann, F. (1976). Event-related slow potentials: Recent data on clinical significance. Research Communities Psychology, Psychiatry and Behavior, 1, 91-104.

Dongier, M., Dubrovsky, B. and Engelsmann, F. (1977). Event-related slow potentials in psychiatry. In C. Shagass, S. Gershon and A. Friedhoff (Eds.) Psychopathology and Brain Dysfunction. New York: Raven Press, 339-352.

Edell, W. and Chapman, L. (1979). Anhedonia, perceptual aberration, and the Rorschach. Consulting and Clinical Psychology, 47, 37-384.

Elbert, T., Rockstroh, B., Lutzenberger, W. and Birbaumer, N. (1980). Biofeedback of slow cortical potentials. I.Electroencephalography and Clinical Neurophysiology, 48, 293-301.

Elbert, T., Rockstroh, B., Lutzenberger, W. and Birbaumer, N. (1982). Slow

brain potentials after withdrawal of control. Archives of Psychiatry and Neurological Sciences, 232, 201-214.
Flor-Henry, P. (1976). Lateralized temporal-limbic dysfunction and psychopathology. Annals of the New York Academy of Sciences, 280, 777-797.
Flor-Henry, P. (1979). Laterality, shifts of cerebral dominance, sinistrality and psychosis. In Gruzelier, J. & Flor-Henry, P. (Eds.) Hemisphere Asymmetries of Function in Psychopathology. Amsterdam: Elsevier, 3-20.
Flor-Henry, P. and Yeudall, L.T. (1979). Neuropsychological investigation of schizophrenia and manic-depressive psychoses. In Gruzelier, J. and Flor-Henry, P. (Eds.) Hemisphere Asymmetries of Function in Psychopathology. Amsterdam: Elsevier, 341-362.
Flor-Henry, P. and Koles, Z. J. (1980). EEG studies in depression, mania and normals: Evidence for partial shifts of laterality in the affective psychoses. Advances in Biological Psychiatry, 4, 21-43.
Flor-Henry, P., Koles, Z.J. and Tucker, D.M. (1982). Studies in EEG power and coherence (8-13 Hz) in depression, mania and schizophrenia compared to controls. Advances in Biological Psychiatry, 9, 1-17.
Giedke, H., Bolz, J. and Heimann, H. (1980). Pre- and postimperative negative variation (CNV and PINV) under different conditions of controllability in depressed patients and healthy controls. In H. Kornhuber and L. Deecke (Eds.) Motivation, Motor and Sensory Processes of the Brain. Amsterdam: Elsevier, 579-584.
Giedke, H., Heimann, H. and Straube, E. (1982). Vergleichende Ergebnisse psychophysiologischer Untersuchungen bei Schizophrenien und Depressionen. In G. Huber (Ed.) Endogene Psychosen: Diagnostik, Basissymptome und Biologische Parameter. Stuttgart: Schattauer Verlag, 295-312.
Goldstein, L. (1979). Some relationships between quantified hemispheric EEG and behavioral states in man. In Gruzelier, J. & Flor-Henry, P. (Eds.) Hemisphere Asymmetries of Function in Psychopathology. Amsterdam: Elsevier, 237-254.
Haberman, M., Chapman, L., Numbers, J. and McFall, R. (1979). Relation of social competence to scores on two scales of psychosis proneness. Journal of Abnormal Psychology, 88 675-677.
Hillyard, S., Hink, R., Schwent, V. and Picton, T. Electrical Signs of selective Attention in the Human Brain. Science, 1973, 182, 161-171.
Kemali, D., Vacca, L., Nolfe, G., Iorio, E. and DeCarlo, R. (1980). Hemispheric EEG quantitative asymmetries in schizophrenic and depressive patients. Advances in Biological Psychiatry, Vol.4, Basel: Karger, 14-20.
Lutzenberger, W., Elbert, T., Rockstroh, B., Birbaumer, N., Stegagno, L. (1981). Slow cortical potentials in subjects with high or low scores on a questionnaire measuring physical anhedonia and body image distortion. Psychophysiology, 18, 371-380.
Lutzenberger, W., Birbaumer, N., Rockstroh, B. and Elbert, T. (1983). Evaluation of contingencies and conditional probabilities - A psychophysiological approach to anhedonia. Archives of Psychology and Neurology Science, 233, 474-488.
Meehl, P. Schizotoxia, schizotypy, schizophrenia. American Psychology 1962, 17, 827-838.
Miller, G., Simons, R. and Lang, P. (1984). Electrocortical measures of information processing deficit in anhedonia. In Karrer, R., Cohen, J. and Tueting, P. (Eds.) Brain and Information. New York: The New York Academy of Sciences, 598-602.
Perris, C. and Monakhov, K. (1979). Depressive symptomatology and systemic structural analysis of the EEG. In Gruzelier, J. and Flor-Henry, P. (Eds.) Hemisphere Asymmetries of Function in Psychopathology. Amsterdam: Elsevier, 233-236.
Rado, S. (1956). Psychoanalysis of Behavior: Collected Papers. New York: Grune & Stratton.

Rockstroh, B., Elbert, T., Lutzenberger, W. and Birbaumer, N. (1982). Slow cortical potentials under conditions of uncontrollability. Psychophysiology, 16, 374-380.

Rockstroh, B., Elbert, T., Birbaumer, N. and Lutzenberger, W. Slow Brain Potentials and Behaviour, Baltimore: Urban and Schwarzenberg, 1982.

Rockstroh, B., Elbert, T., Lutzenberger, W. and Birbaumer, N. Operant Control of Slow Brain Potentials: A Tool in the Investigation of the Potential's Meaning and Its Relation to Attentional Dysfunction. (1984) In Elbert, T., Rockstroh, B., Lutzenberger, W. and Birbauber, N. (Eds.) Self Regulation of the Brain and Behavior. Heidelberg: Springer, 227-239.

Roth, W. T. (1977). Late event related potentials and psychopathology. Schizophrenia Bulletin, 3, 105-120.

Sackeim, H.A., Gur, R.C., Saucy, M.C. (1982). Functional brain asymmetry in the expression of positive and negative emotions: Lateralization of insult in cases of uncontrollable emotional outbursts. Archives of Neurology, 39, 210-218.

Serafetinides, E. (1972). Laterality and voltage in the EEG of psychiatric patients. Diseases of the Nervous System, 33, 622-623.

Serafetinides, E. (1973). Voltage laterality in the EEG of psychiatric patients. Diseases of the Nervous System, 34, 190-191.

Shagass, C., Ornitz, E.M., Sutton, S. and Tueting, P. (1978). Event-related potentials and psychopathology. In E. Callway, P. Tueting and S. Koslow (Eds.) Event-Related Brain Potentials in Man. New York: Academic Press, 443-496.

Shagass, C. (1979). Sensory evoked potentials in psychosis. In Begleiter, H. (Ed.) Evoked Brain Potentials and Behavior. New York, Plenum Press: 467-498.

Shaw, J. C. (1979). A comparison of schizophrenic and neurotic patients using EEG power and coherence spectra. In Gruzelier, J. & Flor-Henry, P. (Eds.) Hemispheric Asymmetries of Function in Psychopathology. Amsterdam: Elsevier, 257-284.

Simons, R. (1981). Electrodermal and cardiac orienting in psychometrically defined high-risk subjects. Psychiatry Research, 4, 347-357.

Simons, R. (1982). Physical anhedonia and future psychopathology: An electrocortical continuity. Psychophysiology, 19, 433-441.

Stein, L. and Wise, C. (1971). Possible etiology of schizophrenia: Progressive damage to the noradrenergic reward system by 6-hydroxydopamine. Science, 171, 1032-1036.

Tucker, D., Watson, R. and Heilman, K. (1981). Right frontal lobe activation and right hemisphere performance. Archives of General Psychiatry, 38, 169-174.

Timsit-Berthier, M., Delaunoy, J. and Rousseau, H. (1976). Some problems and tentative solutions to questions raised by slow potential changes in psychiatry. In W. McCallum and J. Knott (Eds.) The Responsive Brain. Bristol: Wright.

Walter, W.G., Cooper, R., Aldridge, V., McCallum, W.C. and Winter, A.L. (1964). Contingent negative variation: An electrical sign of sensorimotor association and expectancy in the human brain. Nature, 203, 380-384.

HEMISPHERIC DIFFERENCES IN RELATION TO SMOKING

Thomas Elbert* and Niels Birbaumer

Department of Clinical and Physiological Psychology
University of Tubingen
F.R.Germany

SUMMARY

The present investigations were designed to uncover differences in hemispheric interactions between tobacco smokers, smoking cigarettes with different nicotine concentrations, and non-smokers. In the first experiment, sensorimotor tasks, associated with hemisphere-specific processing, were presented either to the right hand (then stimuli had to be counted) or to the left hand (matching of stimulus patterns). A lever switch with the thumb of the stimulated hand was used to indicate the correct solution of the task. A warning stimulus presented to the same hand (6 sec prior to the task) primed these choice reaction time tasks. Non-smoking subjects, all right handers, performed faster in the right hand task. This difference, though more pronounced in sham-smoking subjects (smoking cigarettes with zero nicotine content) than in non-smokers vanished with increasing nicotine intake so that smokers responded faster in the left-hand task. Similarly, a difference in error rate between the two types of task was observed only in non-smokers or in smokers without recent nicotine intake. After nicotine intake, heart rate in smokers was higher for the left than for the right hand task and the task-dependent asymmetry of the slow brain potentials also interacted with smoking intake. Results may provide new evidence that nicotine in lower doses improves inter-hemispheric coordination or activates the right hemisphere of the brain.

The second experiment tested the self-regulation of the hemispheric asymmetry in slow brain potentials. Operant control was achieved by means of a biofeedback paradigm in non-smokers and smokers as well. Though only asymmetry over the central regions was required, smokers, unlike non-smokers, demonstrated an extended change in SP-distribution. Sham-smokers were less able to shift their slow negative brain potential to the left hemisphere. The results indicate that transmission of information between brain hemispheres might be impaired in deprived smokers and that nicotine helps to compensate for this regulation deficit. However, the finer tuning within limited brain regions is achieved by non-smokers only.

----

* Address correspondence to:
Dr. Thomas Elbert,
Department of Clinical & Physiological Psychology,
University of Tubingen, Gartenstr. 29, D-7400 Tubingen, West Germany.

INTRODUCTION

Based on psychological considerations, we have suggested that nicotine exerts its cortical actions primarily on the right hemisphere of the brain, or that it alters information interchange between the hemispheres (Birbaumer, 1981). On the one hand, smoking was found to reduce vigilance decrement (Mangan, 1982) and to increase performance in signal detection (Wesnes & Warburton, 1978; Williams, 1980), on the other hand it was reported to reduce recall and categorisation of written words (Gonzales & Harris, 1980). Such examples could indicate that smoking enhances performance for right rather than for left hemispheric tasks, although the alerting effects of nicotine generally confound this interpretation, and controversial results have been explained on the basis of a dose-response relationship and individual differences. (For a general review on smoking see Ashton & Stepney, 1982).

Nicotine is known to influence cholinergic (ACh) pathways. While facilitating synaptic transmission in low doses, it blocks the nicotine receptors at higher concentrations. The structural similarity (in electrical charge distribution) makes nicotine a perfect key to interact with ACh-receptors. The resulting combination, however, is much more enduring than the combination with ACh. ACh-receptors may still be occupied by nicotine when a subsequent nerve impulse arrives at the synapse. In this case, instead of being stimulated, the eventual response mediated at the synapse will be inhibited or suppressed. This explains, why nicotine exerts a biphasic effect on the arousal system (via the mesencephalic reticular formation), which might increase or lower electrocortical arousal (depending on dosage, deprivation time and actual arousal).

In contrast, nicotine has a generally inhibitory effect on the discharge of many cortical cells. While many cortical fibers are cholinergic, no cholinergic fibers were seen in the large tracts of the white matter (including corpus callosum and anterior commissure; Bindman & Lippold, 1980). By depressing cortico-cortical but not transcallosal connections, nicotine, therefore, would facilitate hemispheric integration, e.g. via presynaptic disinhibition of interhemispheric fibers. ACh is not symmetrically distributed in the human cerebral cortex, as it is in animals and this could also cause an asymmetrical action of nicotine. However, there is some evidence that nicotine receptors are not limited to cholinergic neurons (Abood et al., 1980), which makes predictions from psychopharmacological considerations very difficult.

The present investigations were designed to uncover differences in hemispheric coordination between smokers and non-smokers; during the experiment smokers smoked cigarettes of different nicotine content. Sensorimotor tasks with lateralized processing (Lutzenberger et al., 1985) were chosen for the first experiment, since the rate of acetylcholine release is greatest from the sensorimotor areas (Phillis, 1970). The primary projection areas are not connected by transcallosal fibers.

METHOD

Subject-sample. A total of 44 male right-handed students (as verified by the Edinburgh inventory of Oldfield, 1971) were paid for their participation in the two experiments. Age ranged from 18 to 32 years. 20 subjects were non-smokers, 24 smokers were selected according to the following criteria: at least 10 cigarettes per day with at least 0.8 mg nicotine per cigarette. The average consumption was rated to be 15.3 cigarettes per day. Based on the smoking pattern test (Taylor's version of Russell et al., 1974, published in Ashton & Stepney, 1982) subjects attributed their smoking to the categories INDULGENT (46+/-5%), SEDATIVE (38+/-7%) and SENSORIMOTOR

(40+/-3%) rather than STIMULATING (21+/-7%) or PSYCHOSOCIAL (18+/-4%).

Procedure. Subjects were instructed not to smoke for two hours before the experiment would start. After electrode attachment, a cigarette was offered. Subjects were informed that smoking might influence some of the measurements to be taken and, therefore, were asked to smoke a standardized cigarette type before each experiment. However, the smokers were assigned to one of four groups, each receiving cigarettes with a different nicotine content in a double blind setting. Concentrations were 0.0, 0.2, 0.4, 0.8 mg per cigarette.

Apparatus and Physiological Recordings. The EEG was recorded from C3-C4, P3-P4 and from Fz, Cz, C3, P3 in reference to shunted earlobes. A Beckman type R polygraph was modified to have a 30 sec time constant and high frequency cut-off of 30 Hz. The monopolar EEGs from C4 and P4 were obtained by off-line subtraction. Grass silver disc electrodes, chlorided before each experimental session, were used with Grass EC2 paste as a conducting agent. To keep electrode resistance below 5 k Ohms, the skin at the electrode location was cleaned with alcohol and the outer layers of the skin were removed by gently scraping with a sterile scalpel.

For the recording of vertical and lateral eye movements, (VEOG and LEOG), Beckman silver/silver chloride electrodes were attached 1 cm above and below the left eye and as near as possible to the outer canthi, respectively; Beckman electrode jelly served as electrolyte.

Data were sampled at a rate of 100 Hz and collapsed to 5 points/sec for slow potential analysis, using a phase-free digital filter.

The ECG recorded from the lower rib cage was fed into a cardiotachometer coupler to detect R-waves, which Schmitt-triggered the computer.

Timing and presentation of the experimental stimuli were enabled by means of digital computers. Computer programs were used to reject trials from averaging, when eye movements exceeded 70 μV, referred to prestimulus baseline, or when a shift of more than 50 μV/sec was detected in one of the EEG channels.

Data analysis. Varimaxed principal component analyses (PCA) were used to parameterize phasic physiological responses. The covariance matrices were normalized so that the trace of the matrix equalled its dimension, and two components were extracted to account for variances early and late in the anticipatory interval (early and late CNV). A linear combination of principal components was fitted to each individual physiological response. The component scores, as well as behavioural response measures were submitted to analyses of variance (ANOVAs). Separate ANOVAs were computed for the non-smokers (Non-S.) and for the smokers (Sm.), the latter included the between-subjects factor NICOTINE CONCENTRATION. Stars (*) indicate $p < 0.05$ or (**), $p < 0.01$. (Only effects relevant in the present context are reported.)

EXPERIMENT I

Design

Ss placed index and middle finger of each hand on a small surface, from which one to four solenoid driven pins would protrude for two sec. To maximize lateralized processing, the right hand task was a decision to be made according to the number of pins counted: either one or two pins, or more than two. The direction in which a lever was moved with the right thumb indicated the solution (see Fig. 1). The left hand task was to compare patterns of pins between the two fingers and to decide, whether those were equal or different (Fig. 1). In this case, a response with the left thumb was required. The different solutions of both tasks had equal probability. (The relationship between the direction of the lever switch and the solution of the tasks was balanced across Ss.)

The tactile task was preceded by a tactile warning stimulus S1 (all four pins protruded for one sec) on the hand to which the task would be given. For example, the perception of the pins at the right hand signalled that the right hand task (to count the number of pins) would be presented five sec after S1 offset.

Subjects experienced 32 trials with the right hand task and 32 trials with the left hand task in pseudorandom order. The total physical stimulation was equal for both types of task. Ss were asked to switch the lever as fast as possible to task presentation. If a wrong response or no response within a four sec interval following task onset was recorded, an error was counted. The intertrial interval varied between 10 and 25 sec.

During the experiment the subject sat in a reclining chair in a soundproof and dimly lit room. After the preparation for the physiological recordings, the subject read task instructions. Eight practice trials, supervised by the experimenter, introduced the task. Ss were advised to fixate a cross in front of them and to prevent blinks and eye movements.

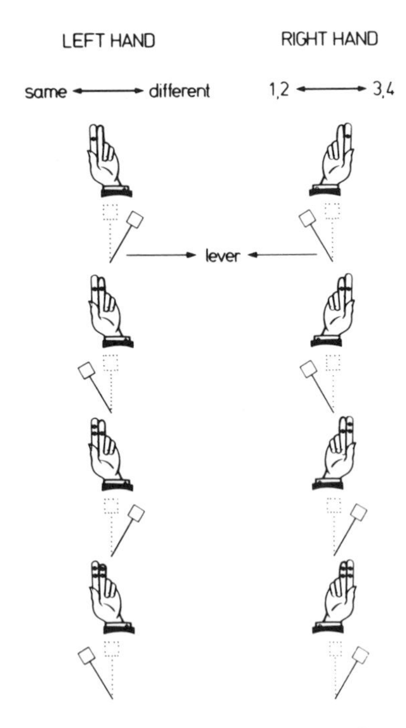

Fig. 1. Examples of the tactile stimulation and tasks for the left and right hand. Dots mark the number or configuration of protruded pins (S2). Below each hand, the required response, i.e., the correct direction of the lever switch is marked; dotted levers indicate the start position.

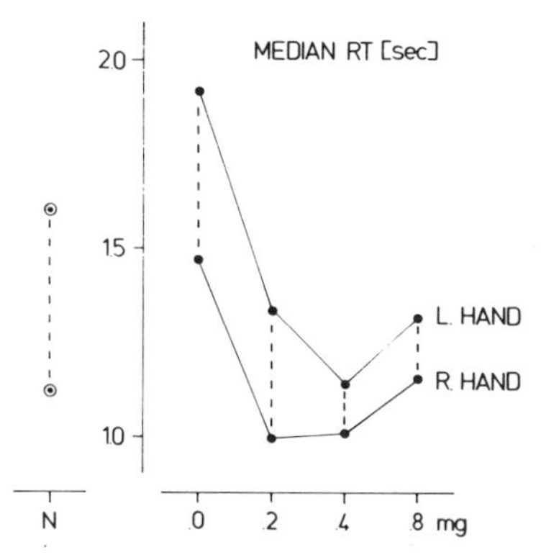

Fig. 2. Median of the reaction times (RT) for the left hand (l.) and right (r.) hand task. N: Non-smokers. Abscissa indicates nicotine content of smoked cigarettes.

Results

Reaction time and error rate. Subjects press the button faster for right-hand than for left-hand tasks (Non-S $F(1,19)=22.3^{**}$, Sm $F(1,20)=45.1^{**}$). For smokers, however, this difference vanishes with increasing nicotine consumption (NICOTINE CONCENTRATION X TASK: $F(3,20)=4.2^{*}$), as shown by the medians of the reaction time in Fig. 2.

Error rates, too, document superior task performance of the right (11%) as compared to the left hand (21%; Non-S $F(1,19)=38.3^{**}$, Sm $F(1,20)=6.5^{*}$). Across trials the error rate of the left-hand task decreases in non-smokers (TASK x SERIES, $F(1,19)=16.2^{**}$), as well as in sham-smokers, however, accompanied by an increase in the error rate for the right-hand task. With increasing nicotine uptake smokers, however, are able to reduce error rates for both tasks (NICOTINE x TASK x SERIES: $F(3,20)=4.1^{*}$) (see Fig. 3). This adaptive behaviour results in an overall improvement of performance under nicotine (NICOTINE, $F(3,20)=3.8^{*}$).

The Heart Rate (HR). The HR increment due to nicotine uptake (NICOTINE, $F(3,20)=3.4^{*}$) is illustrated in Fig. 4.

The HR in trials with the left-hand task is higher than during right-hand performance. This difference increases with nicotine uptake (NICOTINE x TASK: $F(3,20)=4.7^{*}$) (see Fig. 5).

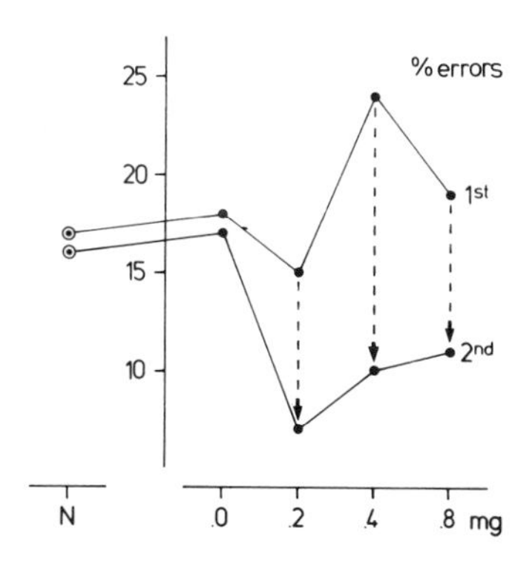

Fig. 3. Change in error rate from the first (1st) to the second (2nd) series. Abscissa as in Fig. 2.

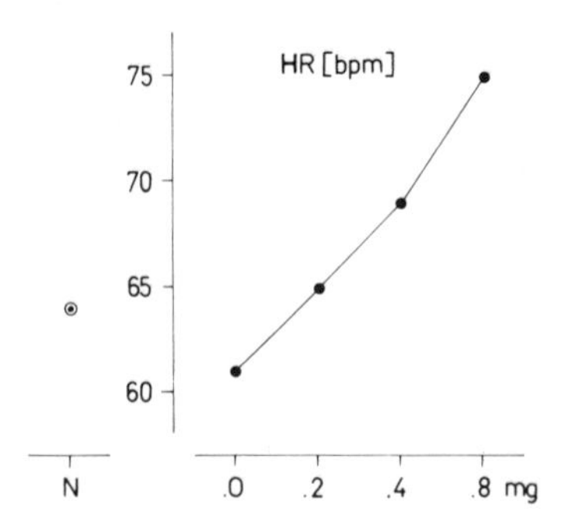

Fig. 4. Mean heart rate in beats per minute, averaged across the experimental session.

Slow Brain Potentials. The SPs as illustrated in Fig. 6, show a task-dependent asymmetry. The asymmetry develops immediately after S1 and increases in amplitude during the task presentation.

The ANOVA results in significant interactions for the early precentral component (TASK x LATERALITY: non-S $F(1,19)=21.2^{**}$; Sm $F(1,20)=22.4^{**}$). Principal components are depicted in Fig. 7.

The asymmetry can be described by the difference in component scores: C3-C4 (left hand) - C3-C4 (right hand). Fig. 8 shows these asymmetry scores for early and late components.

Average asymmetry is lower in sham-smokers than in non-smokers but increases with increasing nicotine inhalation to values comparable to the non-smoker level ($F(3,20)=4.2^{*}$ for the early and $F(3,20)=3.4^{*}$ for the late

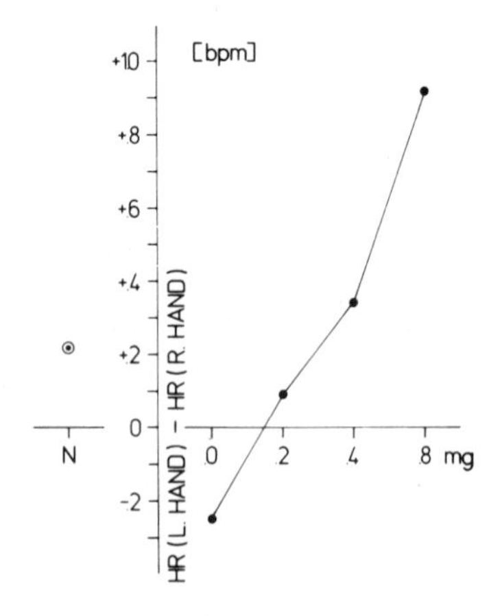

Fig. 5. Difference in heart rate between the two types of task. Positive values indicate a higher HR during left- than during right-hand tasks.

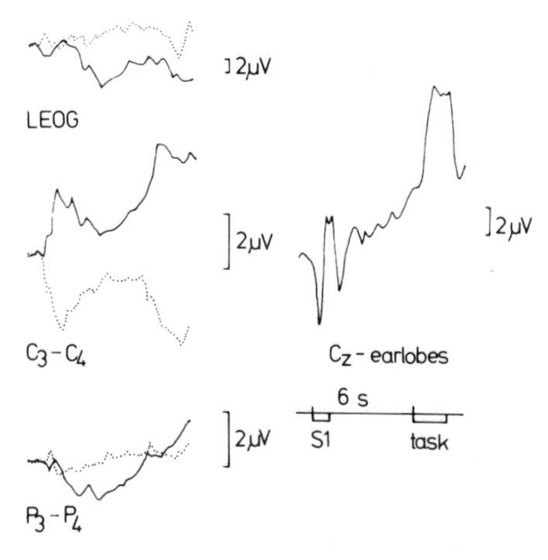

Fig. 6. Slow potentials averaged across groups in experiment I: Left column: Bipolar SPs are averaged separately for trials with left-hand - right hemispheric tasks (dotted lines) and right hand tasks (solid lines). Right column: Monopolar vertex potentials.

component). Thereby, the asymmetry develops earlier under nicotine. The amplitude of the late component decreases in parallel (Fig. 9), especially at the frontal recording site ($F(3,20)=3.5^*$).

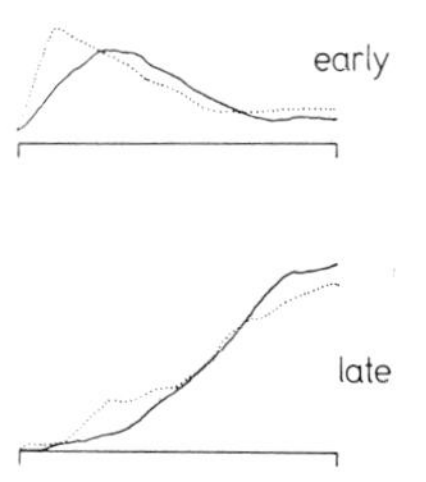

Fig. 7. Loadings for the three varimaxed principal components from PCA of the six sec anticipation intervals for experiment I (dotted lines) and experiment II (solid lines).

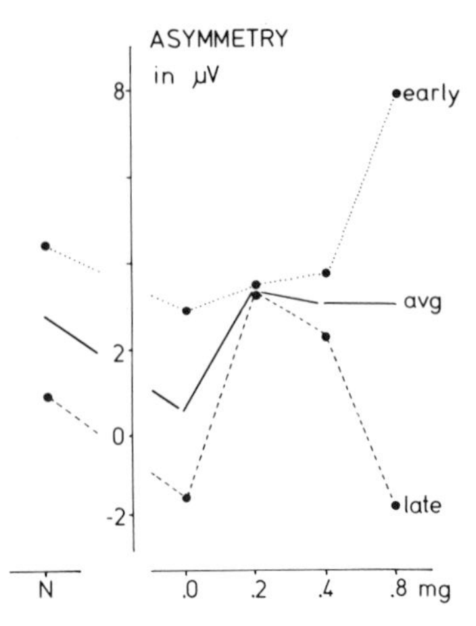

Fig. 8. Asymmetry scores for the early (dotted) and late (dashed) component. The solid line indicates the averaged SP-asymmetry.

Lateral and vertical eye movements are small and show no significant effects.

The EEG at the parietal sites was analyzed by means of a Fast Fourier Transformation (using a Tuckey-Hanning window) for 2.56 sec prior to S1- and 2.56 sec prior to S2- presentation (INTERVAL). During the anticipation interval a hemispheric asymmetry develops in a task-dependent manner for the 11-12 Hz frequency range: anticipation of the right-hand task reduces P3-power more than the left-hand-task expectancy (TASK x LATERALITY: Non-S $F(1,19)=5.3^{*}$, Sm $F(1,20)=5.4^{*}$). The desynchronisation in response to the S1 follows an U-shaped function with increasing nicotine intake, being lowest in the group with 0.2 mg (see Fig. 10, NICOTINE x INTERVAL: $F(3,20)=5.1^{**}$).

Discussion

The present tactile stimulation associated with different types of tasks for the left and right hand induces hemisphere-specific preparation and processing. This is confirmed by brain potential analysis (see Lutzenberger et al., 1985). The same physical stimulation was applied in both tasks, but input (contralateral hand), mode of processing (counting/pattern matching), and output (motor response with the contralateral hand) activate primarily one hemisphere.

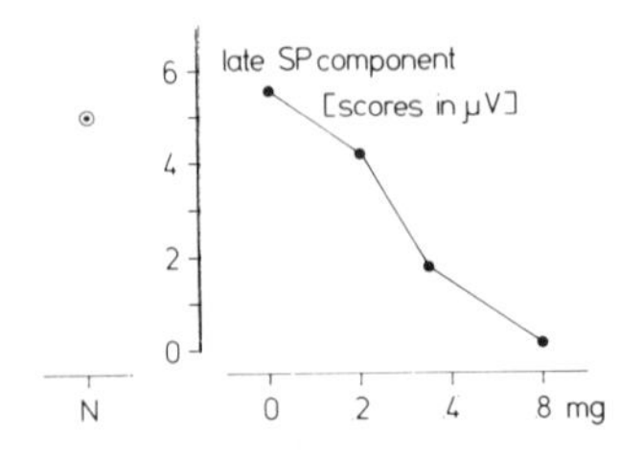

Fig. 9. Dependence of the late SP-component on nicotine intake.

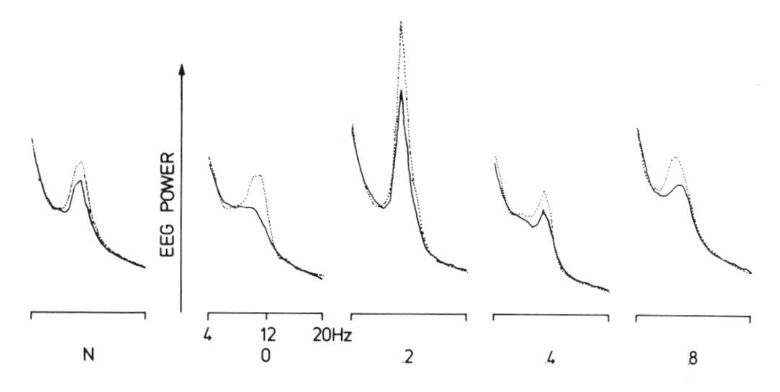

Fig. 10 EEG power spectra recorded from parietal locations. For experiment I (dotted lines) and experiment II (solid lines).

Obviously, the left- hand task (pattern matching), assumed to be processed in the right hemisphere, was harder for non-smokers (slower responses and more errors) than the presumed left-hemispheric task (stimulus counting).

The different results for smokers can be explained by the hypothesis that nicotine either interacts asymmetrically with the brain hemispheres (arousing the right side?) or facilitates the interchange between hemispheres. Smokers are better adapted to the presently required continuous switching between hemispheres. They react faster with less errors; their performance is more balanced between tasks and their brain potentials are well regulated, i.e. the anticipatory negativity is shifted to the site of processing. A higher heart rate during left-hand tasks also indicates a compensation for task difficulties. Smokers without nicotine, on the other hand, perform worse than non-smokers. The depressant action of nicotine on the cortex could result in a more extensive inhibition exerted through commissural fibers on the brain hemisphere, which is not processing the current task. Lower SP-amplitudes but increased SP-asymmetry, as presently observed, would be the consequence. Heart rate becomes generally higher with increasing nicotine consumption, indicating the well-known influence of nicotine on the autonomic nervous system and on general arousal (see Ashton & Stepney, 1982). These unspecific actions, however, cannot explain the reported specific effects. The second experiment was scheduled to pursue this hypothesis of an interaction between smoking and hemispheric integration.

EXPERIMENT II

Design

Regulation of CNV-like slow potentials (SPs) also may be achieved by means of the biofeedback paradigm, in which human subjects are rewarded for the generation of different SP-shifts. As shown previously also area-specific SP-patterns can be brought under operant control (Elbert, 1985; Rockstroh, 1986).

After a cigarette-break of 10 min the same subjects were tested for their ability to regulate brain electrical activity differentially between hemispheres. The bipolar C3-C4 SP difference was selected as the operant. During intervals of 6 sec each, the integral of C3-C4 EEG referred to pre-interval baseline was continuously fed back within a game presented on a

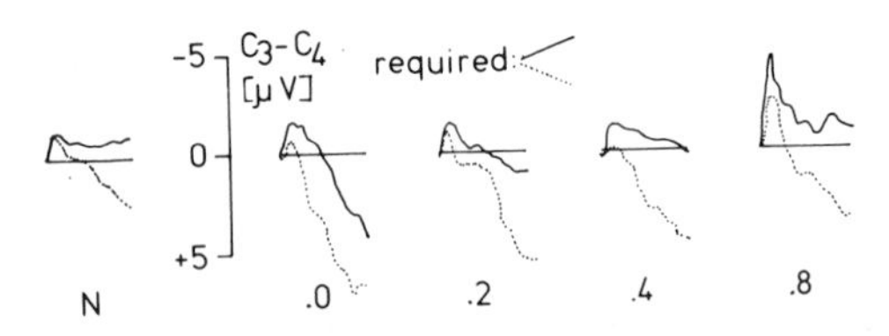

Fig. 11. Slow potentials recorded during biofeedback. Bipolar SPs are averaged separately for trials with required larger left (C3) than right (C4) precentral negativity referred to pre-feedback baseline (solid lines) and vice versa (dotted lines). The averaged C3-C4 differences meet the requirements of control.

video-screen. One of two discriminable visual stimuli (diffuse color light) required more negative left-, than right-precentral SP-shift, while the other color informed the subject that only SP-patterns which were less negative at C3 relative to C4 would be rewarded by winning points (50 c per point). A total of 80 trials (40 trials of each condition) was scheduled in two training blocks, each lasting about 20 min. Subjects smoked another cigarette during a 5 min break between the trial blocks. Simultaneous lateral EOG-shifts, which could have generated proportions of the recorded C3-C4 shift, prevented successful trials, but could result in a loss of points.

Results

Results are illustrated in Fig. 11. The data are averaged across artifact-free trials separately for the groups and the two self-regulation conditions. The significant C3-C4 differentiation of SPs documents that subjects have learned hemisphere-specific SP regulation ($p < 0.001$). The essential similarities among SPs of both experiments are confirmed by Principal Component Analysis (PCA). The separate PCAs disclosed in each case two nearly identical components (plotted as a function of time in Fig. 7).

The SP-differentiation shifts towards earlier segments of the 6 sec feedback interval in smokers (NICOTINE x CONTROL $P(3,20)=3.5^*$) for the early component. The differentiation is more widespread in smokers than in non-smokers (GROUPS x CONTROL x SCALP SITE, $F(1,42)=4.7^*$).

The EEG analysis reveals that asymmetry in desynchronisation covaries with SP-asymmetry: the hemisphere with higher negativity shows less alpha activity (CONTROL x LATERALITY, Non-sm $F(1,19)=4.7^*$). Again, power in the alpha-band follows an invertedly U-shaped function, being highest in the group with 0.2 mg (NICOTINE: $F(3,20)=4.5^*$; see Fig. 10).

In contrast to non-smokers an asymmetry develops in smokers with less right than left hemispheric synchronisation (GROUPS x LATERALITY: $F(1,42)=7.4^*$), but more beta-power (13-20 Hz) over P4 than over P3 (GROUPS x LATERALITY: $F(1,42)=10.4^{**}$). This asymmetry, evident in all smokers, becomes most pronounced after 0.4 mg nicotine cigarettes (NICOTINE x LATERALITY: $F(3,20)=4.7^*$).

Discussion

The results replicate previous findings that operant control of SP-asymmetry can be achieved (Elbert, 1985; Rockstroh, 1986). Since EEG-synchronisation depends on SP-control, significant artifactual influences can be excluded.

In line with the results of experiment I, the ability to regulate hemisphere-specific activation seems to be improved under nicotine, although all smokers regardless of the actual amount of nicotine intake exhibit a widespread SP-differentiation, which was not necessary to achieve reward. Possibly smokers lack the finer tuning observed in non-smokers who restrict SP-differentiation to the required brain regions.

We may even speculate that subjects with less efficient connections between hemispheres may compensate for the regulation deficit between hemispheres by taking advantage of the facilitating effect which nicotine might have on callosal transmission. A genetic predisposition of right-handedness (RS+) has been suggested by Annett (this volume). These genes might favour the developmental reduction in fibers crossing the corpus callosum (Witelson, this volume). This raises the question, whether subjects with RS++ are more prone to smoking than those with other genetic dispositions. Additionally, this would restrict the applicability of animal models, since the right shift and hemispheric organisation in general seems to be unique in humans (Annett, this volume).

Based on the assumption that cortical processing of emotions is lateralized, the speculative action of nicotine might, furthermore, contribute to an emotional balance (via the hemispheric balance) and control emotional arousal, as has been reported for tobacco smoking. Self-regulation of SPs affects various behavioural responses (Rockstroh et al., 1982). From these results we have concluded that SPs indicate the allocation of attentional resources in the brain. The superior regulation of the early SP-component suggests that nicotine can increase the flexibility of resource allocation. This adds to the possibility to use nicotine as an external regulator of various CNS responses explaining addictive behaviour.

Acknowledgement

We acknowledge the support and advice of Dr. Gosswein and Dr. Meloh. This paper is dedicated to Prof. Dr. Wolfgang Tunner on his 50th birthday.

REFERENCES

Abood, L., Reynolds, D., Bidlack, J. (1980). Stereospecific 3H-nicotine binding to intact and solubilized rat brain membranes and evidence for its noncholinergic nature. Life Sciences, 27, 1307-1314.

Annett, M. Implications of the right shift theory for individual differences in laterality. This volume.

Ashton, H. & Stepney, R. (1982). Smoking: Psychology and Pharmacology. London, Tavistock Publ.

Birbaumer, N. (1981) Zur Psychologie des Rauchens. Forschungsbericht an der Universitat Tubingen, University of Tubingen.

Bindman, L. & Lippold, O. (1981). The Neurophysiology of the Cerebral Cortex. London, W. Arnold.

Elbert, T. (1985). Externally and self-induced CNV-patterns of the brain hemispheres - a sign of task-specific preparation, Human Neurobiology (in press).

Gonzales, M. & Harris, M. (1980). Effects of cigarette smoking on recall and categorisation of written material. Perceptual and Motor Skills, 50, 407-410.

Lutzenberger, W., Elbert, T., Rockstroh, B., Birbaumer, N. (1985). Asymmetry of brain potentials related to sensorimotor tasks, International Journal of Psychophysiology, 2, 28-29.

Mangan, G. (1982). The effects of cigarette smoking on vigilance performance. Journal of General Psychology, 10C, 77-83.

Mangan, G. & Golding, J. (1978). An 'enhancement' model of smoking maintenance? In R.E. Thornton (Ed). Smoking Behaviour: Physiological and Psychological Influences. Edinburgh, Churchill Livingstone, 131-147.

Oldfield, R. (1971). The assessment and analysis of handedness: The Edinburgh inventory. Neuropsychologia, 9, 97-113.

Phillis, J. (1970). The Pharmacology of Synapses. Oxford, Pergamon Press,

Rockstroh, B., Elbert, T., Birbaumer, N., Lutzenberger, W. (1982). Slow Brain Potentials and Behaviour. Baltimore, Urban & Schwarzenberg.

Rockstroh, B. (1986). Operant control of slow brain potentials. In J. Hingtgen, D. Hellhammer & G. Huppman (Eds.) Advanced Design Methods in Psychobiology, Toronto, Hogrefe (in press).

Russell, M., Peto, J., Patel, U. (1974). The classification of smoking by a factorial structure of motives. Journal of the Royal Statistical Society, 137, 313-346.

Singh, M. M., Warburton, D.M., Lal, H. Central Cholinergic Mechanisms and Adaptive Dysfunctions. New York, Plenum Press.

Wesnes, K. & Warburton, D. (1978). The effects of cigarette smoking and nicotine tablets upon human attention. In R.E. Thornton (Ed.) Smoking Behaviour: Physiological and Psychological Influences. Edinburgh, Churchill Livingstone, 1978, 131-147.

Williams, A. (1980). Effect of cigarette smoking on immediate memory and performance in different kinds of smokers. British Journal of Psychology, 71, 83-90.

Witelson, S. Individual differences in anatomical asymmetry. (This volume).

INTERVAL HISTOGRAM ANALYSIS OF EEG HEMISPHERIC ACTIVITY IN SCHIZOPHRENIA

Antonio E. Puente and *Lelon J. Peacock

Department of Psychology
University of North Carolina
at Wilmington,
North Carolina, 28403

*Department of Psychology
University of Georgia
Athens,
Georgia 30602

While research into the psychophysiological correlates of maladaptive behavior has expanded during recent years, the investigation of the hemispheric processes related to psychopathology has experienced unprecedented growth during the last few years (e.g., Flor-Henry, 1979). One particular branch of this type of hemisphericity research which has received considerable attention and yielded results of heuristic value has been the study of hemispheric activity in schizophrenics, see Flor-Henry, Koles & Reddon (this volume); Flor-Henry & Koles (1984); Newlin, Carpenter and Golden (1981) and Walker, Hoppes and Emory (1981) for comprehensive reviews on this subject.

Within this general area we have developed a special interest in the electrophysiological (versus behavioral) asymmetries often observed in schizophrenics. While the literature generally provides strong support for left hemisphere abnormality, theories such as those exposed by Flor-Henry (1979) and Gur (1979), numerous disparities between studies are apparent. Most often, these disparities have been accounted for by the revision of existing theories (e.g., Walker & McGuire, 1982). Possibly a sounder approach to accounting for these discrepancies would be to examine methodological issues, rather than theoretical ones. Indeed, if methodological problems would account for the varied results reported in the literature, then the validity of existing theories would, in turn, be highly questionable.

While one could raise numerous methodological questions, one which has largely been ignored has been the measurement of dependent variables. It is not unusual for investigators using electroencephalographic (EEG) measures to use different electrode placements, filtering systems, and methods of analyses. While spectral or frequency analysis has been the most widely accepted method of EEG analysis, it is possible numerous statistical assumptions are typically violated when using this approach. Central to the assumptions underlying frequency methods of analysis is that the EEG wave is a periodic, unchanging variable (Burch, 1959). This assumption becomes especially critical since most investigators (due to the large amount of data obtained in EEG studies) tend to sample EEG for several seconds and then generalize to the entire period from which the sample was taken (usually minutes). In this study, we were interested in examining the hemispheric activation of schizophrenics using standard experimental paradigms (e.g., Galin & Ornstein, 1972) but analyzing the EEG from a time

Table 1. Mean Demographic Variables by Group

	Group		
Variable	Brain-damaged Schizophrenic	Non-Brain-Damaged Schizophrenic	Affective Disorders
Age	46.2	30.2	34.6
Year of Onset	31.5	21.4	29.3
Number of Hospitalizations	1.7	5.0	2.3
Current Length of Hospitalization (in days)	1163.1	113.7	57.5

rather than a frequency domain. Thus, the assumption of periodicity of the EEG would not be violated, in this technique.

METHOD

Subjects

From the in-patient population of a 1,200 bed state-supported psychiatric hospital, 57 volunteers participated in this study. To be included in the study patients had to be 18 years of age, have good corrected or uncorrected vision, be right-handed by report and observation, and had to read, comprehend, and sign an informed consent form. Additional biographical information is contained in Table 1.

From the pool of 57 subjects, three groups of 19 participants were formed. The first group was composed of schizophrenics without brain damage while a second group was composed of schizophrenics with brain damage. Diagnosis of schizophrenics with brain damage was ascertained using a standard diagnostic manual (DSM III) (American Psychiatric Association, 1980). Diagnoses were made independently by psychiatrists and psychologists using the history as well as the results of a clinical interview, the Shearn and Whitaker (1969) schizophrenic symptom checklist and the Whitaker Index of Schizophrenic Thinking (1973). The results of the scores are shown in Table 2. Brain lesions were determined by history of diffuse brain pathology as documented by medical examination and laboratory tests (e.g., CAT Scan) as well as by the Luria-Nebraska Neuropsychological Battery (Golden, Hammeke & Purish, 1978). A third group of affective disorders was included as a reference-control for the first two groups. Diagnoses were determined in the same manner as schizophrenia but using the criteria outlined in the DSM-III for affective disorders.

Experimental Procedure

Each subject participated in psychological and electroencephalographic (EEG) testing. Odd numbered subjects (e.g., 1, 3, 5, . . .) participated in the EEG part of the test in the morning and the psychological testing in the afternoon. Even numbered subjects (e.g., 2, 4, 6, . . .) did the reverse. Morning testing began at 9:00 am and was completed by 11:00 am while afternoon testing began at 1:00 pm and ended by 3:00 pm.

Table 2. Mean WIST and LNNB Scores by Group

		Group		
		Brain-Damaged	Non-Brain-Damaged	Affective
Test	Scale	Schizophrenic	Schizophrenic	Disorder
WIST;	Similarities	5.9	8.5	2.4**
	Word Pairs	2.7	3.4	1.3
	New Inventions	3.1	3.1	3.4
	Time	15.0	19.8	15.10
	Index	26.7	34.8	22.2
LNNB;				
	Motor	40.3	14.2**	
	Rhythm	12.0	6.6*	
	Tactile	13.5	8.9*	
	Visual	15.1	10.0	
	Receptive Speech	18.0	10.4*	
	Expressive Speech	26.3	15.2*	
	Writing	12.2	7.2*	
	Reading	9.3	6.5	
	Mathematics	17.8	8.4*	
	Memory	18.4	13.3*	
	Intellectual	38.5	30.0	
	Pathognomic	26.0	15.7	
	Left Hemisphere	11.8	5.4	
	Right Hemisphere	14.8	6.3	

* $p < .05$

** $p < .01$

Variables Measured.

Neuroleptic dosages were converted to chlopromazine equivalents according to the method suggested by Davis (1976). Dosages were altered during the course of the study. These conversions are found in Table 3.

Table 3. Chlorpromazine Equivalents (CPZE) by Groups

Group	CPZEmg		
	Mean	SD	Range
Brain-Damaged Schizophrenic	378.95	306.10	0-1200
Non-Brain-Damaged Schizophrenic	1628.95	2091.00	0-7500
Affective Disorder	531.59	656.85	0-2250

The EEG was recorded from the 0_1 and 0_2 sites referenced to respective ear-lobes (10-20 system) using a Grass Model 16 EEG. Signals were recorded graphically (for visual inspection) and on magnetic tape via an Ampex FM Recorder (for statistical analysis). Data from the tape was digitized using a PDP 8 by an individual unaware of group composition using the period analysis technique of Sharp, Smith and Surwillo (1975). Data was then coded "blind" by another technician to disc via an Apple II Plus.

The half-wave period was obtained using two manipulations. First, noise was subtracted using averaging of a specific number of sample waves. In addition, the peaks and troughs of a particular wave were selected relative to its opposite using a combination of criterion amplitude (measured in μV) and criterion duration (in msec). Once a half-wave had been selected, it was dumped into one of 250 possible bins, each having a duration of 1 msec. For the purposes of this study, 20 five msec bins were constructed to cover the entire spectrum of duration extending from 0 to 100 msec. A total of 50 half-wave were analyzed half way through the testing period for each period of the EEG testing.

There were five separate Test periods (each three minutes long and with the subject sitting facing away from the polygraph operator and towards a wall approximately six feet in front of the individual). During the first Test period, the participant sat quietly with eyes open while a resting baseline was obtained. A simple eye exercise involving fixating on a black circle and on a black square as well as performing a series of sequential eye movements followed. The next two Test periods were counter-balanced. For odd-numbered subjects, a visuo-spatial task involving the solution of both "easy" and "hard" (as determined by a pilot study with clinical subjects) problems from the Minnesota Paper Form Board Test followed the eye exercise. For even-numbered subjects, a task involving solving one and two digit (i.e., "easy" and "hard" as determined in a pilot study) multiplication problems followed the eye exercise. During presentation of the stimuli, the subjects were instructed to solve the respective problems in their head and that, after presentation of all the stimuli the subject would have the opportunity to provide the Experimenter with the solutions. (Indeed after completion of all problems, the polygraph was turned off and each stimulus was presented for five seconds once more and the verbal responses were then recorded). After the verbal reports, subjects were then presented with the alternate set of tasks; that is, for odd numbered subjects, the multiplication problems and the even numbered subjects, the geometric problems.

RESULTS

Two main sets of analyses were performed on the half-wave EEG data, using the statistical package developed by Steinmetz, Romano and Patterson (1981) on an Apple IIe.

The first set of analyses entailed examining within subject variables. Histograms were constructed for each subject, across Test periods and hemispheres, and were grouped according to clinical diagnosis. The grouped histograms were analyzed by means of correlated t-tests from Test period to Test period as well as within Test periods with respect to laterality. Thus, three sets of t-tests were performed, one for each group, in order to determine whether changes occurred in the last two Test periods relative to the baseline Test period and whether hemispheric measures differed (within Test periods). Regardless of the fact that a total of 30 t-tests were performed, no significant differences were observed. The lack of significance appears to be due to the wide dispersion of data within the histograms.

The second set of analyses employed the dominant interval recorded. Thus, instead of concentrating on the entire frequency distribution this set of analyses focused only on the EEG period during which time most waves were observed. An 3 x 2 analysis of variance, split-plot design was used. The design consisted of three between groups variables (clinical diagnosis) and two within group variables (hemispheres). In total, three 3 x 2 ANOVAs were performed, one from each major Test period of the experiment.

In Test period I, or "baseline", no significant interaction or main effects for group was noted. However, a main effect for hemisphere was observed ($F = 14.59$; $p < .01$). The means indicate that there is preponderance of right hemisphere activity for slower EEG periods (mean = 24. 31 msec) while faster intervals were observed in the left hemisphere (mean = 18.16 msec). For the geometric design Test period, a main effect of hemispheres was noted ($F = 22.76$; $p < .01$). A highly similar pattern to "baseline" was noted with regards to slow and fast activity across Hemispheres. The interaction indicated that this main effect appears to be due to the relatively slow right hemisphere (especially in brain-damaged schizophrenics) activity with the relatively fast duration-periods for the left noted for the two schizophrenic groups (especially the brain-damaged sample). During the last Test period, or the multiplication problem task, only a main effect for Hemisphere was noted ($F = 30.42$;$p < .01$). In general, the same pattern of right versus left hemisphere activity seen in the geometric Test period was observed with the largest left-right difference being noted for the brain-damaged schizophrenic group. The means for each group across Test periods and Hemispheres are found in Table 4.

DISCUSSION

The results indicate the hemispheric dysfunction in schizophrenia is a complicated phenomenon tempered by numerous variables. As Flor-Henry (1983) has suggested, hemispheric dysfunction is not a static but a dynamic situation.

Across the three Test periods, results indicate that faster EEG activity was observed for all three groups in the left hemisphere. Conversely, slower activity was noted in the non-dominant hemisphere. Baseline EEG indicated that largest hemispheric difference were found for the Affective group. Indeed, the non-brain damaged Schizophrenic group exhibited relatively similar activity for both hemispheres. Large left-right differences were observed in the geometric task Test period, this was especially true for the Brain-damaged sample. While slowing of right hemisphere activity was noted for the Brain-damaged Schizophrenic sample, slight increases in activity were noted for the two other groups. With regards to left hemisphere activation during this task, both Schizophrenic groups increased activity (especially the Non-damaged sample). Interestingly, the relatively large hemispheric differences noted for the affective disorders during the baseline Test period were eradicated during

Table 4. Group x Hemisphere x Group Interval (Half-wave) EEG Means

Test Period	Hemisphere	Group Brain-Damaged Schizophrenic	Non-Brain-Damaged Schizophrenic	Affective Disorder	Totals
	Left	16.6	21.4	16.4	18.1
Baseline "resting"	Right	22.1	25.3	25.6	24.3
	Difference	5.5	3.9	9.2	6.2
	Left	15.5	16.7	18.3	16.8
Geometric	Right	28.0	21.4	21.3	23.6
	Difference	12.5	9.7	3.0	6.8
	Left	16.4	15.8	15.9	16.0
Multi-plication	Right	30.5	24.9	24.6	26.7
	Difference	14.1	9.1	9.0	10.7

the geometric Test period. During the final Test period, similar hemispheric differences were noted between the non-brain-damaged schizophrenic and affective groups. Nevertheless, relative to the baseline Test period, the former group exhibited not only the fastest EEG activity but had a large increase in the activity generally. While slowing of the right hemisphere was not observed in the Non-brain-damaged Schizophrenic and Affective groups, significant decreases were noted for the Brain-damaged Schizophrenic sample.

Several conclusions can be reached. Foremost, period analyses of hemispheric EEG activity is a robust measure when considering group and task comparison. While wide variability of the data complicated analysis of frequency means, "dominant-wave" analysis provided a fruitful measure of EEG activity. Secondly, the use of brain-damaged schizophrenics as well as patients with affective disorders appear to be useful in defining the boundaries for the limits of laterality abnormalities in putatively CNS-intact schizophrenics. Indeed, the data derived from the subjects alone provide useful information into the role of brain dysfunction in schizophrenia as well as of affective pathology in hemisphericity issues. Finally, hemispheric differences in these groups appear to exist in response to task demands and, to a lesser degree, at rest.

Superficially these findings support those of Gur (1979) as well as interpretations of Newlin, Carpenter and Golden (1981), and Walker and McGuire (1982). While dominant hemisphere overactivation was noted, the data obtained indicate that sweeping generalizations about overactivation must be tempered by numerous factors including baseline-task Test period differences. Additionally, while changes were noted for left hemisphere activation care should be taken in considering right hemisphere activation (seen in the non-damaged schizophrenic group in response to geometric designs) and slowing (seen in the brain damaged-schizophrenic group in response to both tasks).

More definitive interpretations await replications and consideration of several limitations. Primarily, while dispersion of histogram data complicated the analysis of means, that in itself may be worthwhile to examine. As Surwillo (1978) indicated, variability (rather than central tendency) may be a more important factor in period analysis of the EEG. Secondly, "change score" analysis may have resulted in more accurate interpretation of the data. Differences were noted during the baseline Test period and while not significant, they do appear large enough potentially to mask group differences observed during task activity.

Period analysis provides a less restrictive approach to EEG interpretation of hemispheric activity. With further development, a more comprehensive interpretation of psychopathology and laterality should emerge as both frequency and time approaches to EEG analysis are considered.

REFERENCES

American Psychiatric Association (1980). Diagnostic and statistical manual (3rd ed.). Washington, D.C.: American Psychiatric Association.
Burch, N. R. (1959). Automatic analysis of the electroencephalogram: A review and classification of systems. Electroencephalography and Clinical Neurophysiology, 11, 827-829.
Galin, D. & Ornstein, R. (1972). Lateral specialization of cognitive mode: An EEG study. Psychophysiology, 9, 412-418.
Davis, J. (1976). Quantitative doses and costs of anti-psychotic medication. Archives of General Psychiatry, 33, 858-861.
Flor-Henry, P. (1979). Laterality, shifts of cerebral dominance, sinistrality and psychosis. In J. Gruzelier and P. Flor-Henry (Eds.) Hemispheric Asymmetries of Function in Psychopathology. Amsterdam: Elsevier/North Holland Biomedical Press, 3-19.
Flor-Henry, P. and Koles, Z.J. (1984). Statistical quantitative EEG studies of depression, mania, schizophrenia and normals. Biological Psychology, 19, 257-279.
Golden, C., Hammeke, T. & Purisch, A. (1978). Diagnostic validity of a standardized battery derived from Luria's Neuropsychological test. Journal of Consulting and Clinical Psychology, 46, 1258-1265.
Gur, R. E. (1979). Cognitive concomitants of hemispheric dysfunction in schizophrenia. Archives of General Psychiatry, 36, 269-274.
Newlin, D., Carpenter, B. & Golden, C. (1981). Hemispheric asymmetries in schizophrenia, Biological Psychiatry, 16, 561-582.
Sharp, F., Smith, G. & Surwillo, W. (1975). Period analysis of the electroencephalograms of EEG half-wave durations. Psychophysiology, 12, 471-475.
Shearn, C. & Whitaker, C. (1969). Selecting subjects in studies of schizophrenia. Schizophrenia, 1, 4-8.
Steinmetz, J., Romano, A. & Patterson, M. (1981). Statistical program for the Apple II microcomputer. Behavior Research Methods & Instrumentation, 13, 702.
Surwillo, W. (1978). Interval histogram analysis of period of the electroencephalogram in relation to age during growth and development of normal children. Psychophysiology, 12, 506-510.
Walker, E., Hoppes, E. & Emory, E. (1981). A reinterpretation of hemispheric dysfunction in schizophrenia. The Journal of Nervous and Mental Disease, 169, 378-380.
Walker, E., McGuire, A. (1982). Intra and interhemispheric information processing in schizophrenia. Psychological Bulletin, 92, 707-725.
Whitaker, L. (1973). The Whitaker Index of Schizophrenic Thinking. Los Angeles: Western Publishing Co.

AGE AND SEX DIFFERENCES IN LATERAL ASYMMETRIES TO VISUAL AND TACTILE STIMULI

Andrew W. Young, Pamela J. Bion and Kathryn H. McWeeny

Psychology Department
Lancaster University
Lancaster, LA1 4YF
England

INTRODUCTION

We have been studying the development of right hemisphere abilities in normal children for a number of years. We began this research with the strong expectation that right hemisphere superiorities would change across age, and an equally strong distaste for the idea of sex differences in cerebral organisation. Our findings have consistently tended to undermine these preconceptions.

We will describe three series of experiments. The first series is concerned with age and sex differences in recognising faces, the second series with age and sex differences in lateral asymmetries obtained to complex visual stimuli, and the third series with age and sex differences in lateral asymmetries obtained to tactile stimuli. Most of the studies of complex visual and tactile stimuli that we will describe involve the enumeration of collections of dots.

Face recognition and dot enumeration are tasks which had already been found to be associated with right hemisphere superiority in both the normal and the clinical literature when we began the research (De Renzi and Spinnler, 1966; Kimura, 1966; Warrington and James, 1967a, 1967b; Rizzolatti, Umilta and Berlucchi, 1971; Hilliard, 1973). We chose to concentrate on them in order to look at abilities acquired at different times during the first years of life. The abilities involved in face recognition are present in early infancy (Schaffer, 1971; Young, 1985) and hence are well established in the age range most suitable for investigation using the techniques we had developed (five years to adult). In contrast, although infants are now known to be sensitive to some aspects of number (Starkey and Cooper, 1980), explicit enumeration is not learnt until an age much closer to the bottom end of the range we could use (Klahr and Wallace, 1973: Schaeffer, Eggleston and Scott, 1974).

EXPERIMENTS

All of our experiments involved right-handed subjects. Children's handedness was determined by a combination of hand used in writing, teacher's report, and self report.

The methods used in the experiments involve lateral presentation of stimuli in the left or right visual hemifields (LVF or RVF), and to the left and right hands. The basic principles involved in such studies have been reviewed by Young (1982a) and Beaumont (1983), and their application to developmental studies is discussed by Witelson (1977), Beaumont (1982), and Young (1982b, 1983, 1985).

A. Face recognition

Experiment 1: Recognition of small sets of faces by children

(Young and Bion, 1980, Experiment 1)

Rationale: In this experiment, we studied differences in ability to recognise the members of small sets of faces of unfamiliar people presented in the LVF or in the RVF. Both upright and inverted faces were used as stimuli. Previous research (Leehey, Carey, Diamond and Cahn, 1978) had shown that the LVF superiority for face recognition is reduced by inversion of stimulus faces. This seemed a useful technique because it can eliminate the possibility that any LVF superiority obtained only to upright faces derives solely from the complexity of faces as visual stimuli, since faces are equally complex whether upright or inverted.

Method: Faces were bilaterally presented (one in the LVF, one in the RVF) for 150 ms; the task involved identifying the LVF face and the RVF face from a display of four faces. The faces used were all of people who were not known by the children studied. Each face subtended a horizontal visual angle of $3^{o}20'$, and the centres of the LVF and RVF faces were offset from the midline of each stimulus card by $3^{o}48'$. During the course of the experiment only one set of four faces was used for each subject, but for half the subjects these faces were easy to discriminate and for the other half they were difficult to discriminate. The orientation of the presented stimulus faces and response cards was upright for half the subjects and inverted for the other half.

The presence of central fixation before each stimulus presentation was determined by monitoring the subject's eyes with a video camera; this is a simple and accurate technique (Young, 1982a). Order of report of the bilateral faces was counterbalanced by cuing the face to be reported first.

Sixteen right-handed boys and sixteen right-handed girls of each of the ages seven, ten and thirteen years acted as subjects. Data for each subject consisted of LVF and RVF accuracy scores out of a possible maximum of 40 correct trials. Each subject was given practice trials with appropriate stimuli before the experimental trials commenced.

Results: Proportions of upright and inverted faces correctly identified from the LVF and RVF by children of each age are shown in Fig. 1.

A five-factor analysis of variance of the accuracy scores was carried out, to determine the effects of Age, Sex, Orientation (upright or inverted stimuli), Difficulty (use of the easy or difficult to discriminate set of faces), and Visual hemifield (repeated measure).

The principal findings were that faces were more accurately identified from the LVF than from the RVF (Visual hemifield, $F = 29.13$, d.f.1,72, $p < 0.001$), but this result held only for the upright stimuli (Orientation x Visual hemifield, $F=16.27$, d.f.1,72, $p < 0.001$). There were also main effects of Age ($F=13.09$, d.f.2,72, $p < 0.001$), Orientation ($F=26.18$, d.f.1,72, $p < 0.001$), and Difficulty ($F=34.67$, d.f.1,72, $p < 0.001$). No other statistically significant effects were found.

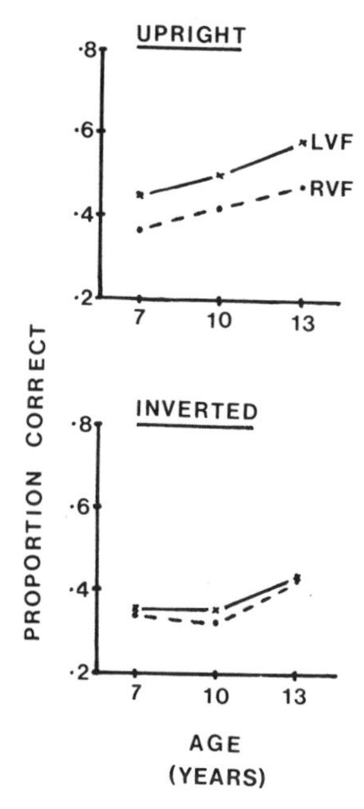

Fig. 1. Proportions of upright and inverted faces correctly identified from the LVF and RVF by right-handed seven-, ten- and thirteen-year-old children in Experiment 1.

The results of the experiment thus showed LVF superiority to upright but not to inverted faces, with no change in the size of this LVF superiority across age. The absence of visual hemifield differences in the case of inverted faces shows that the LVF superiority to upright faces derives from a right hemisphere superiority for face processing *per se*, rather than for any complex visual stimulus. No evidence was found to indicate that this right hemisphere superiority changes in any way across the age range studied. LVF superiority for upright faces was found for both boys and girls.

Experiment 2: Recognition of larger sets of faces by children

(Young and Bion, 1980, Experiment 2)

Rationale: In this experiment a larger pool of stimulus faces was used than in Experiment 1. This gives subjects less opportunity to learn the faces that can appear on each trial.

Method: The design and procedure were the same as for Experiment 1, except that we increased the number of stimulus faces by changing the set of four possible faces at regular intervals. This was done after blocks of four trials throughout the experiment, so that subjects now encountered a total of forty different faces instead of the total of four faces seen by each subject in Experiment 1. Sixteen right-handed boys and sixteen right-handed girls of each of the ages seven, ten and thirteen years acted as subjects.

Results: Proportions of upright and inverted faces correctly identified from the LVF and RVF by boys and girls of each age are shown in Fig. 2.

A four-factor analysis of variance of the accuracy scores was carried out, to determine the effects of Age, Sex, Orientation (upright or inverted), and Visual hemifield (repeated measure).

The principal findings were that faces were more accurately identified from the LVF only by boys, with no visual hemifield difference in girls (Sex x Visual hemifield, $F=4.83$, d.f.1,84, $p < 0.05$); this was again only the case for upright faces (Sex x Orientation x Visual hemifield, $F=4.23$, d.f.1,84, $p < 0.05$). There was also a main effect of Age ($F=29.95$, d.f.2,84, $p < 0.001$), and an Age x Orientation interaction ($F=18.98$, d.f.2,84, $p < 0.001$) in which only older children performed better to upright than inverted faces. No other statistically significant effects were found.

Results of Experiment 2 showed LVF superiority to upright faces for boys only, with no change in the degree of LVF superiority across age. The principal difference between these findings and those of Experiment 1 thus lies in the absence of any LVF superiority for girls in Experiment 2. This unexpected sex difference was stable across the range of ages studied.

FACE RECOGNITION:
LARGER SETS OF STIMULI

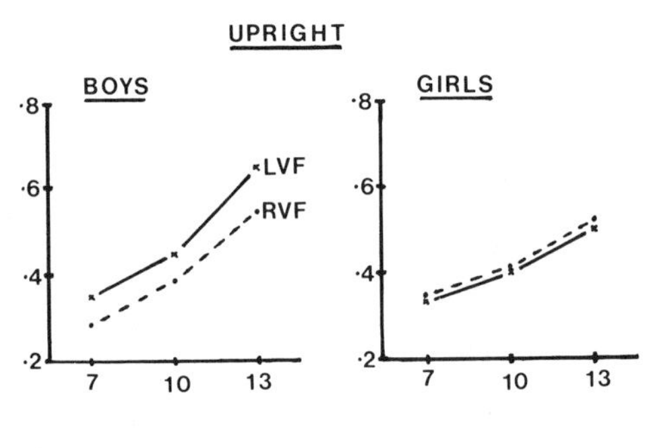

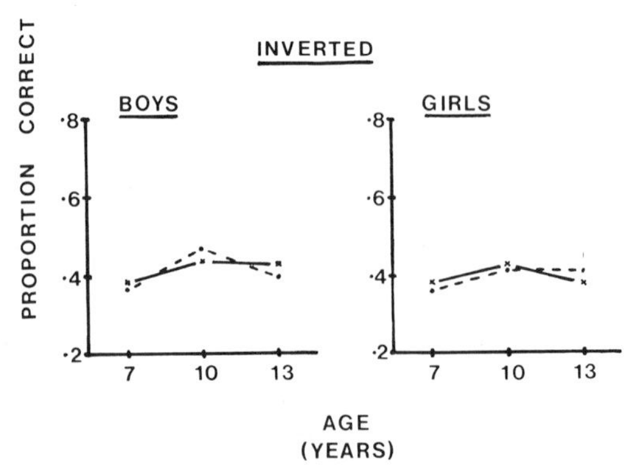

Fig. 2. Proportions of upright and inverted faces correctly identified from the LVF and RVF by right-handed seven-, ten- and thirteen-year-old boys and girls in Experiment 2.

Experiment 3: Effect of different stimuli to trials ratios on adults (Young and Bion, 1983, Experiment 1)

Rationale: Experiment 2 of the present series had shown that increasing the number of faces used in the experiment can eliminate LVF superiority for female subjects. We decided to explore the nature of this sex difference by increasing the number of stimuli still further (to the maximum possible), to see whether this would have any effect on the performance of male subjects. The experiment was carried out with adult subjects in order to check that the sex difference found for children in Experiments 1 and 2 would also be found in adults.

Method: This experiment used adult subjects, and upright faces only. The number of stimulus faces was systematically increased across the levels used in Experiments 1 and 2 to a third level at which a new set of faces was introduced on every experimental trial. Thus the ratio of stimuli to trials was manipulated across three experimental conditions (low, medium or high ratio of stimuli to trials); these conditions were given to subjects in a counterbalanced order.

Twenty-four right-handed male and twenty-four right-handed female adults acted as subjects, with accuracy scores out of a possible maximum of 24 correct being recorded for LVF and for RVF performance in each experimental condition. Other features of design and procedure followed those used in Experiments 1 and 2.

Results: Proportions of faces correctly identified from the LVF and from the RVF by male and female subjects in each experimental condition are shown in Fig. 3.

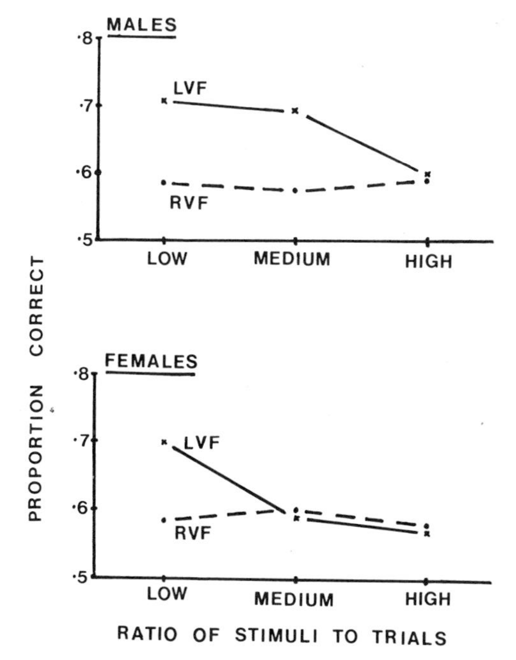

Fig. 3 Proportions of upright faces correctly identified from the LVF and RVF by male and female right-handed adult subjects in each condition of Experiment 3.

A three-factor analysis of variance of the accuracy scores was carried out, to determine the effects of Sex, Visual hemifield (repeated measure), and experimental Condition (repeated measure). There was an overall LVF superiority (Visual hemifield, $F=32.12$, d.f.1,46, $p < 0.001$), and a Visual hemifield x Condition interaction ($F=9.64$, d.f.2,92, $p < 0.001$). The principal finding, however, was the Sex x Visual hemifield x Condition interaction ($F=6.58$, d.f.2,92, $p < 0.01$), in which LVF superiority was lost for male subjects with the high ratio of stimuli to trials but for female subjects with the medium and high ratios of stimuli to trials. No other statistically significant effects were found.

Results thus showed LVF superiority for both males and females with small numbers of stimulus faces (as in Experiment 1), LVF superiority for males only with an increased number of stimulus faces (as in Experiment 2), and no visual hemifield differences for male or female subjects with the large set of faces (high ratio of stimuli to trials). This demonstrates that the finding of a sex difference made with children in Experiments 1 and 2 also holds in the adult range, and extends the finding by showing that sex differences are only obtained at certain points along the continuum of stimuli to trials ratios. Thus the sex difference is clearly not due to any absence of right hemisphere superiority in females, since right hemisphere superiority can be either found or not found for both male and female subjects according to the conditions used. It was also notable that changes in the ratio of stimuli to trials affected LVF performance for both male and female subjects, but had no effect on RVF performance.

B. Complex visual stimuli

Experiment 4: Enumeration of random dots by children
(Young and Bion, 1979)

Rationale: The aim of this experiment was to explore visual hemifield differences for dot enumeration in children with a technique approximating to that used by Kimura (1966) and McGlone and Davidson (1973) with adult subjects.

Method: Collections of 2-6 randomly arranged dots were presented in the LVF or in the RVF to five, seven and eleven-year-old children, who were asked to enumerate them as quickly and accurately as possible. Each collection of dots fell within a circular area having a diameter subtending a visual angle of $3^{o}48'$, and the centres of the LVF and RVF collections were offset by $5^{o}4'$ from the midline of each stimulus card. Accuracy of LVF and RVF performance out of a possible maximum of 20 correct trials per visual hemifield and vocal reaction times (timed from stimulus onset) for correct responses were recorded. Trials with central stimuli were also used to control fixation on the argument that performance with central stimuli should be equal to or better than both LVF and RVF performance if central fixation is occurring (Young and H. Ellis, 1976). Using this criterion, groups of ten right-handed boys and ten right-handed girls at each of the ages five, seven and eleven years were built up. Presentation times of 80 ms for each stimulus were used for seven and eleven-year-olds, and 100 ms for five-year-olds. Each subject was given practice trials with appropriate stimuli before the experimental trials commenced.

Results: Accuracies and vocal reaction times for correct responses by boys and girls of each age are shown in Fig.4.

A three-factor analysis of variance of the accuracy scores was carried out, to determine the effects of Age, Sex and Visual hemifield (repeated measure). A Sex x Visual hemifield interaction was found ($F=7.45$, d.f.1,54, $p < 0.01$), with only boys showing greater LVF accuracy. There were also

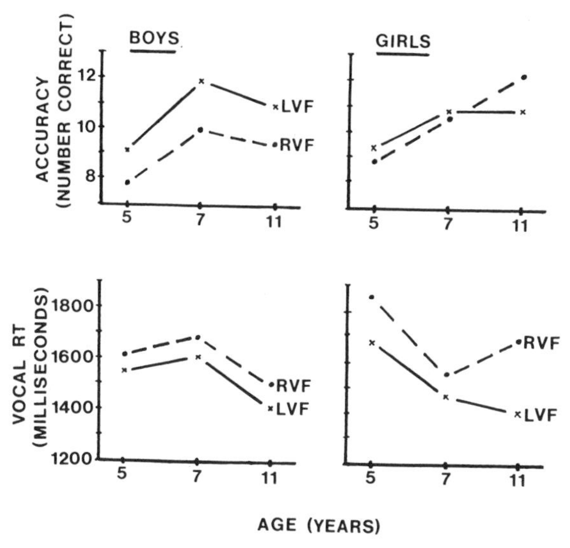

Fig. 4. Accuracies and vocal reaction times for enumeration of LVF and RVF collections of randomly arranged dots by right-handed boys and girls aged five, seven and eleven years in Experiment 4.

main effects of Age (F=5.66, d.f.2,54, $p < 0.05$) and of visual hemifield (F=5.30, d.f.1,54, $p < 0.05$). No other statistically significant effects were found.

A three-factor analysis of variance of the vocal reaction times to correct responses was also carried out. The only statistically significant effect involved the faster RT's to LVF stimuli (Visual hemifield, F=5.63, d.f.1,54, $p < 0.05$).

In this experiment, then, LVF superiority for accuracy was found for boys only. This finding of an LVF accuracy superiority for right-handed male subjects only is consistent with a previous accuracy study using adult subjects reported by McGlone and Davidson (1973). However, vocal reaction times for correct responses in Experiment 4 were faster to LVF presentations for both boys and girls. No changes were found in the size of visual hemifield differences across age.

Experiment 5: Enumeration of random dots by adults

Rationale: Although the vocal reaction times recorded in Experiment 4 produced an interesting pattern of findings, they were collected in an experiment that was primarily intended to examine accuracy. We decided to look more carefully at vocal RT's in a study designed explicitly for this purpose; this study used adult subjects.

Method: Collections of 1-4 randomly arranged dots were presented in the LVF or in the RVF to adults, who were asked to enumerate them as quickly and accurately as possible. Trials on which errors occurred were replaced later in the series, until 20 correct LVF (5 with each size of collection) and 20 correct RVF reaction times had been achieved. Thirty-six right-handed adults acted as subjects, with six males and six females being run at each of the stimulus presentation times 40 ms, 80 ms and 150 ms. Central fixation before each trial was checked by monitoring the subject's eyes using a video camera. Stimulus visual angles were as for Experiment 4. Each subject was given practice trials with appropriate stimuli before the experimental trials commenced.

Results: Vocal reaction times for enumeration of LVF and RVF dot collections by male and female subjects are shown in Fig. 5.

A three-factor analysis of variance of the vocal RT's was carried out, to determine the effects of Sex, stimulus presentation Time (40 ms, 80 ms, or 150 ms), and Visual hemifield (repeated measure). Although the number of items to be enumerated is also plotted in Fig.5, this was not included as a factor in the analysis because of the small number of trials (5) in each cell of the design at this level. The only statistically significant effect was the Sex x Visual hemifield interaction (F=8.84, d.f.1,30, $p < 0.01$), with male subjects showing faster responses to RVF than LVF stimuli and female subjects showing faster responses to LVF than RVF stimuli.

A three-factor analysis of variance of the number of trials that needed to be repeated showed that more trials had to be repeated at the shorter presentation times (Time, F=8.96, d.f.2,30, $p < 0.01$). With 40 ms presentations 18% of trials had to be repeated, with 80ms presentations 15%, and with 150ms presentations 7%. There was also a main effect of Visual hemifield (F=5.74, d.f.1,30, $p < 0.05$), in which more RVF than LVF trials needed to be repeated. The main finding, however, was a Sex x Visual hemifield interaction (F=10.69, d.f.1,30, $p < 0.01$), which showed that only male subjects needed more RVF than LVF trials to be repeated. No other statistically significant effects were found.

Results of this experiment show faster vocal RT's for LVF stimuli in females, but faster vocal RT's to RVF stimuli in males. However, in order to achieve an equal number of correct RT's in each cell of the design more RVF than LVF trials had to be repeated for male subjects. It is thus clear that despite the instruction to be both fast and accurate, male subjects trade RVF accuracy against speed (though we do not mean to imply that they do this consciously and deliberately). The need to repeat more RVF trials for male subjects parallels the LVF accuracy superiority seen for males in Experiment 5, but that experiment did not reveal the extent of the trade-off of accuracy against speed.

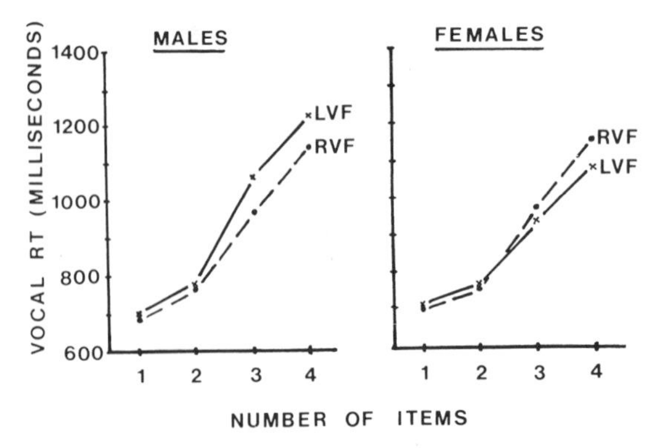

Fig. 5. Vocal reaction times for enumeration of LVF and RVF collections of randomly arranged dots by male and female right-handed adult subjects in Experiment 5.

Experiment 6: Enumeration of linear dots by children

Rationale: This experiment was intended to explore the possibility that the sex difference found in the accuracy data in Experiment 4 (LVF accuracy superiority for boys only) was linked to the particular quantification strategies adopted. Thus we arranged the dots in straight lines; this makes it difficult to use strategies based on grouping the dots into identifiable visual pattern configurations before enumerating them.

Method: The general design and procedure were the same as for Experiment 4, but the collections of 2-6 dots were arranged along regularly spaced horizontal or vertical straight lines. Presentation times of 150 ms were used with seven- and eleven-year-olds, and 200 ms with five-year-olds.

Results: Accuracies and vocal reaction times for correct enumerations by right-handed five, seven and eleven-year-old children are shown in Fig. 6.

A three-factor analysis of variance of the accuracy scores was carried out, to determine the effects of Age, Sex, and Visual hemifield (repeated measure). Older children were found to be more accurate (Age, $F=4.93$, d.f.2,54, $p < 0.05$), and this was especially true for girls (Age x Sex, $F=4.37$, d.f.2,54, $p < 0.05$). No other statistically significant effects were found. The main effect of Visual hemifield did not reach statistical significance ($F=2.73$, d.f.1,54, $p > 0.1$).

A three-factor analysis of variance of the vocal reaction times to correct responses was also carried out. Vocal RT's were faster to LVF stimuli (Visual hemifield, $F=15.31$, d.f.1,54, $p < 0.001$), and vocal RT's were also faster for older children (Age, $F=10.99$, d.f.2,54, $p < 0.01$). No other statistically significant effects were found.

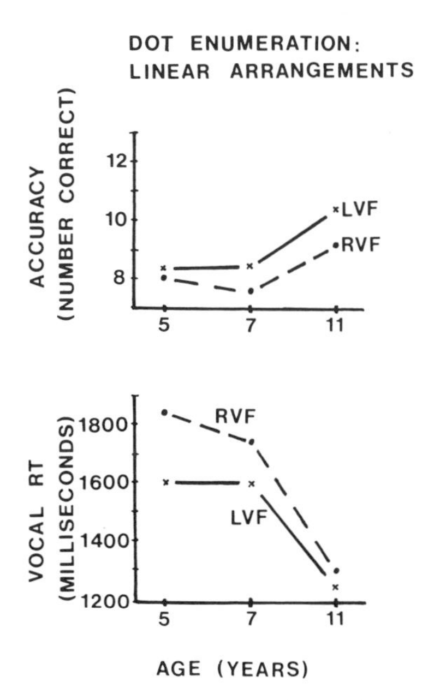

Fig. 6. Accuracies and vocal reaction times for enumeration of LVF and RVF linear collections of dots by right-handed five-, seven- and eleven-year-olds in Experiment 6.

The principal aim of this experiment was to study accuracy differences in the enumeration of LVF and RVF stimuli. However, no visual hemifield differences in accuracy were found, but the experiment did reveal faster vocal reaction times for LVF stimuli. There were no age differences in this LVF superiority in vocal reaction time.

Experiment 7: Enumeration of linear dots by adults

Rationale: This experiment was intended to explore the faster vocal reaction times to LVF presentations of linear collections of dots found in Experiment 6, using a design explicitly suited to this purpose.

Method: Collections of 2-4 dots in regularly spaced horizontal or vertical straight line arrangements were presented for 150ms in the LVF or in the RVF. Adult subjects were asked to enumerate them as quickly and as accurately as possible; trials on which errors occurred were replaced later in the series. Twelve right-handed male and twelve right-handed female adults acted as subjects. Other features of design and procedure were as for Experiment 5.

Results: Vocal reaction times for enumeration of linear arrangements of dots presented in the LVF and in the RVF are shown in Fig. 7.

A two-factor analysis of variance of the vocal reaction times was carried out, to determine the effects of Sex and Visual hemifield (repeated measure). Although the number of items to be enumerated is also plotted in Figure 7, this was not included as a factor because of the small number of trials (5) in each cell of the design at this level. Reaction times were found to be faster to stimuli presented in the LVF (Visual hemifield, $F=4.62$, d.f.1,22, $p < 0.05$). No other statistically significant effects were found.

A two-factor analysis of variance of the number of trials needing to be repeated to obtain enough correct vocal RT's was also carried out. No statistically significant effects were found, but a tendency for more RVF trials to need to be repeated approached significance (Visual hemifield, $F=3.14$, d.f.1,22, $0.1 > p > 0.05$). For LVF trials 4% had to be repeated, and for RVF trials 9%.

The findings of Experiment 7 thus show faster enumeration of stimuli presented in the LVF by both male and female adult subjects. This shows that the findings obtained from children in Experiment 6 also hold for adults, and that they hold when an experimental design explicitly suited to the investigation of vocal reaction times is used.

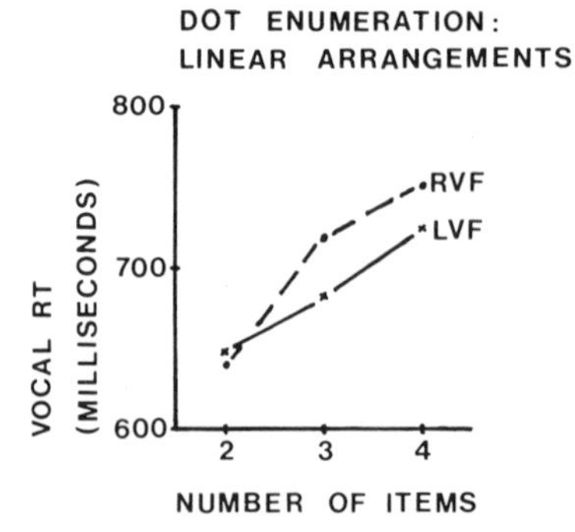

Fig. 7. Vocal reaction times for enumeration of LVF and RVF linear collections of dots by right-handed adult subjects in Experiment 7.

C. Complex tactile stimuli

Experiment 8: Tactile enumeration of random dots

Rationale: This experiment examined differences between the left and right hands in the tactile enumeration of collections of randomly arranged raised dots. Adults and seven-year-old children were studied. The technique used was adapted from one found to produce left hand superiorities in right-handed male and female adults by Young and A. Ellis (1979). The principal aim of the present experiment was thus to see whether this finding would also hold for children.

Method: Right-handed people aged seven years or adult felt collections of 2-4 raised dots with the middle fingers of their left or right hands. There were ten male and ten female subjects of each age. The dots were approximately 1.5 mm in diameter, and raised by 1 mm. They were formed by pressing heavily on the back of a piece of card with a ball-point pen. Each of the raised dots was located at random in one of the nine positions in a 1.5 cm x 1.5 cm grid. Subjects felt the raised dots with their hands placed out of their own sight behind a vertical black cloth screen; a circular wooden surround was used to define the area within which the raised dots might be found. The subject's task was to enumerate the dots as accurately as possible. Trials on which errors occurred were replaced later in the series, until eighteen correct left hand trials (six with each size of collection) and eighteen correct right hand trials had been achieved. Reaction times for correct enumerations were recorded with a stopwatch.

Left hand and right hand trials were separated into a number of blocks given in counterbalanced order. During left hand trials a large felt-tip pen was grasped in the right hand, and during right hand trials the same pen was grasped in the left hand. This was done to provide some degree of bilateral tactile input (see Witelson, 1974). Each subject was given practice trials with appropriate stimuli before the experimental trials commenced.

Results: Reaction times for correct tactile enumeration of random arrangements of dots by right-handed seven-year-olds and adults using their left or right hands are shown in Fig. 8.

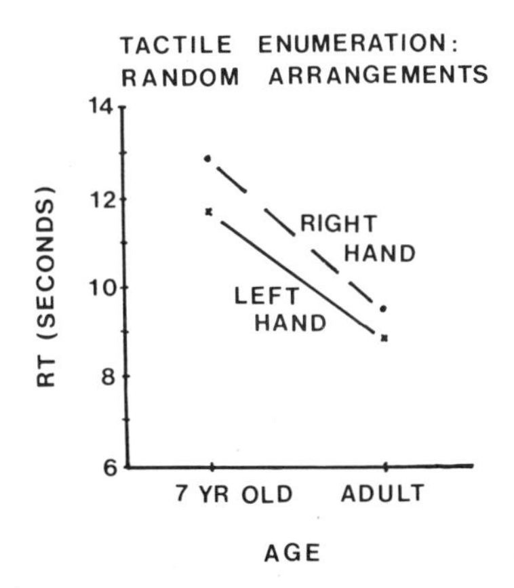

Fig. 8. Reaction times for correct left and right hand tactile enumeration of randomly arranged raised dots by right-handed seven-year-olds and adults in Experiment 8.

A three-factor analysis of variance of the reaction times was carried out, to determine the effects of Age, Sex, and use of the left or right Hand (repeated measure). This showed faster reaction times for the left Hand ($F=13.58$, d.f.1,36, $p < 0.001$) and faster reaction times by adult subjects (Age, $F=25.92$, d.f.1,36, $p < 0.001$). No other statistically significant effects were found.

A three-factor analysis of variance of the number of trials that needed to be repeated showed that more trials needed repeating for the seven-year-olds (Age, $F=6.99$, d.f.1,36, $p < 0.05$). For seven-year-olds 18% of trials had to be repeated, and for adults 6%. No other statistically significant effects were found.

The results of the experiment thus show that right-handed people were faster at enumerating the randomly arranged raised dots when using their left hands. There were no sex differences, and no accuracy differences between left and right hands. No change in the size of the left hand superiority was found across age.

Experiment 9: Tactile enumeration of linear dots

Rationale: Young and A. Ellis (1979) found that left hand superiorities for tactile enumeration arose to randomly arranged dots but were eliminated when the dots were placed along regularly spaced straight lines (which makes their potential spatial locations much more predictable during tactile exploration). The aim of the present experiment was to see whether this finding would also hold for children.

Method: Design and procedure were as for Experiment 8, but the collections of dots were arranged along regularly spaced straight lines.

Results: Reaction times for correct tactile enumeration of linear arrangements of raised dots by seven-year-olds and adults using their left or right hands are shown in Fig. 9.

A three-factor analysis of variance of the reaction times was carried out, to determine the effects of Age, Sex, and use of the left or right Hand (repeated measure). The only statistically significant effect was the tendency of adults to be faster (Age, $F=25.48$, d.f.1,36, $p < 0.001$). A similar analysis of the number of trials that needed to be repeated also showed only.

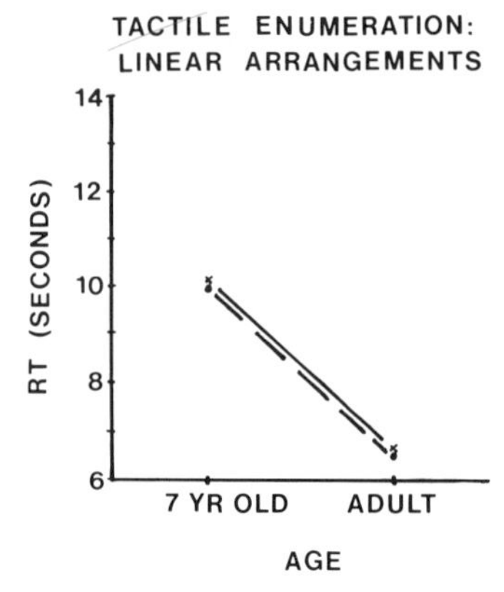

Fig. 9. Reaction times for correct left and right hand tactile enumeration of linear dots by right-handed seven-year-olds and adults in Experiment 9.

that fewer trials needed to be repeated for adults (Age, F=13.59, d.f.1,36, $p < 0.001$). For seven-year-olds 8% of trials had to be repeated, and for adults 4%. Thus no hand differences in RT or accuracy were found at either age or for either sex in Experiment 9.

Experiments 8 and 9 confirm the findings made by Young and A. Ellis (1979), who used a similar method with adult subjects, and show that the results also hold for seven-year-old children.

Experiment 10: Tactile form matching

Rationale: Unlike our studies using visually presented stimuli (Experiments 1-7), Experiments 8 and 9 had not found any sex differences. However, we doubted that the absence of sex differences was due to the tactile modality of presentation, because sex differences in the tactile modality had previously been reported (e.g. Witelson, 1976). Witelson's (1976) study used a tactile form matching task, and found left hand superiority for boys only in the age range six to thirteen years. Experiment 10 was thus intended to examine hand differences in tactile form recognition in adults and children.

Method: Subjects were required to decide whether or not pairs of plastic shapes were the same as or different from each other. The plastic shapes were all 0.5 cm thick and had a surface area of 6 cm^2. They were mounted on 10 x 15cm pieces of card. All of the shapes used are shown in Figure 10. On each trial the subject first felt one of the shapes for 5 seconds, using her or his left or right hand. A second shape was then presented to the same hand, and the subject had to decide as quickly and as accurately as possible whether or not it was the same as the original shape. Reaction times were timed from the presentation of the second shape.

Six right-handed males and six right-handed females of each of the ages six-years-old and adult acted as subjects. Incorrectly answered trials were repeated later in the series. Other features of design and procedure were as for Experiments 8 and 9.

Results: Reaction times for correct tactile form matching by male and female six-year-olds and adults using their left or right hands are shown in Fig. 11.

A three-factor analysis of variance of the reaction times was carried out, to determine the effects of Age, Sex and use of the left or right Hand

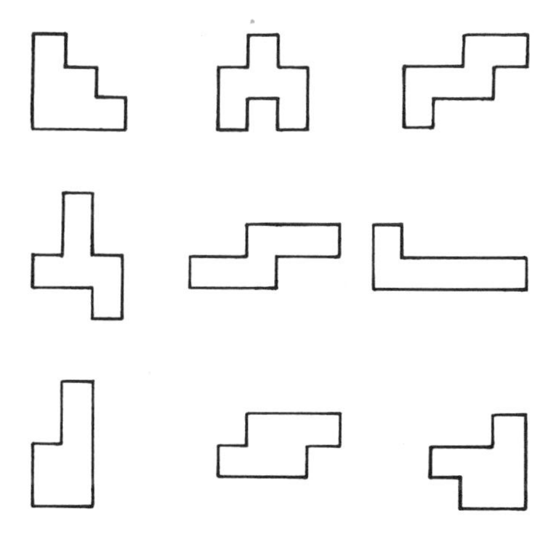

Fig. 10. Shapes used for tactile form matching in Experiment 10.

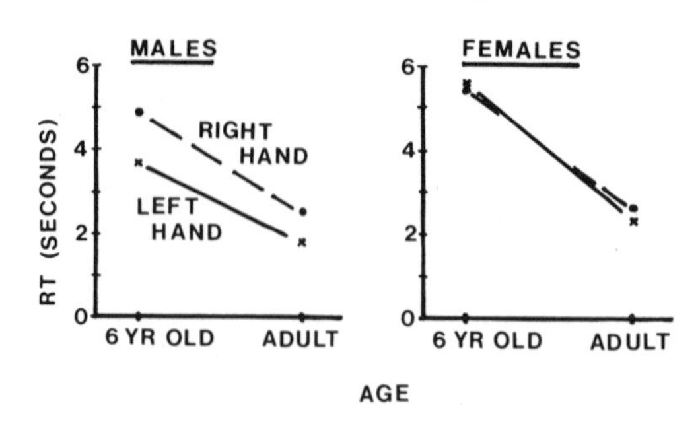

Fig. 11. Reaction times for correct left and right hand tactile form matching by male and female right-handed six-year-olds and adults in Experiment 10.

(repeated measure). A Sex x Hand interaction ($F=4.63$, d.f.1,20, $p < 0.05$) showed that left hand RT's were faster for male subjects only. There were also main effects of Age ($F=52.87$, d.f.1,20, $p < 0.001$) and Hand ($F=6.65$, d.f.1,20, $p < 0.05$). No other statistically significant effects were found.

Thus the left hand was faster for male subjects only with the tactile form matching task used in Experiment 10. There was no change across age in the size of the left hand superiority.

SUMMARY OF FINDINGS

A. Face recognition (Experiments 1-3)

The demonstration of LVF superiority for upright faces was affected by how well subjects got to know the faces used. With a small set of faces (that were in consequence often repeated) LVF superiority was found for both male and female subjects; with a large set of faces (and hence little opportunity to learn them) no visual hemifield differences were obtained. Females switched from LVF superiority to no visual hemifield difference at an earlier point along the continuum of decreasing stimulus familiarity than did males. No developmental changes in LVF superiority were found.

B. Complex visual stimuli (Experiments 4-7)

These experiments produced the most complicated pattern of findings. It was possible to show LVF superiority in both male and female subjects, but this was related both to the type of stimulus arrangement (random or linear dot collections) and the measure of performance adopted (accuracy or vocal RT); males sometimes produced consistently fast but inaccurate RVF responses. No developmental changes in LVF superiority were found.

C. Complex tactile stimuli (Experiments 8-10)

Left hand superiorities were found for both male and female subjects in the enumeration of random collections of raised dots, with no hand differences for collections of raised dots arranged in straight lines. These findings replicate those made by Young and A. Ellis (1979) with adult subjects, and extend them to seven-year old children. The use of a form recognition task (Experiment 10) led to a finding of left hand superiority for male subjects only. No developmental changes in left hand superiority were found.

IMPLICATIONS

We have not found any evidence of developmental changes in the size of lateral asymmetries. Thus it would seem that the underlying cerebral functional asymmetries do not change in the age range studied (five years to adult). This conclusion holds regardless of whether we have studied abilities at ages relatively close to or distant from those at which they are first acquired. Our findings are in line with those of other studies of lateral asymmetries at different ages, but we will not review these here because they have already been extensively examined elsewhere (e.g. Witelson, 1977; Beaumont, 1982; De Renzi, 1982, (Chapter 2); Young, 1982b; Bradshaw and Nettleton, 1983, (Chapter 12); Young, 1983, 1985).

We have, however, often (but by no means always) encountered sex differences in lateral asymmetries. These sex differences have always proved to be stable across age in our experiments, and demand explanation. Their stability across age is clearly inconsistent with theories (reviewed by Harris, 1978) that try to link sex differences to a postulated development of cerebral hemispheric lateralisation of abilities.

Other studies of lateral asymmetries in the processing of visual and tactile stimuli have also produced findings of sex differences (see Harris, 1978; McGlone, 1980; for reviews). Two general types of explanation of these sex differences in lateral asymmetries have been advanced. The first type of explanation is in terms of sex differences in the degree of functional asymmetry of the cerebral hemispheres, the second type is in terms of sex differences in the cognitive processes used. To some extent these explanations can be seen as existing at different levels, so that they are not necessarily incompatible. Despite this, however, we think that certain points can be made concerning the suitability of each type of explanation for explaining the sex differences in lateral asymmetries found in studies of normal subjects.

We do not think that an account in terms of a greater or lesser degree of functional asymmetry in the male or female brain will alone prove adequate in explaining sex differences in lateral asymmetries. We have consistently found that sex differences can be produced or eliminated by what at first sight appear to be relatively small procedural changes, and that they can also depend on the measure used. The general trend of our findings is in the direction of it being easier to demonstrate right hemisphere superiority for male subjects, but there is no question that right hemisphere superiority can be shown in females under appropriate conditions and that it may occasionally be possible to reverse the usual sex difference and find right hemisphere superiority in females but not males. The key problem is thus to specify the conditions that do or do not allow right hemisphere superiority to be demonstrated by male and female subjects.

The second type of explanation, in terms of sex differences in the cognitive processes used, seems to us more likely to be of use (see also Marshall, 1973; Bryden, 1979). In essence, the proposal is that males and females may rely on lateralised or on non-lateralised cognitive processes under different conditions. However, it is important to distinguish two variants of this type of explanation. The first variant sees the differences in the cognitive processes used as reflecting strategy differences, i.e., a voluntary choice from a repertoire of alternatives. We were originally very attracted to this idea, and sought to test its prediction that findings of sex differences could be altered or even reversed by instructing subjects to use different strategies. Unfortunately we have never managed to produce any evidence for this from our own studies (e.g. Young and Bion, 1983, Experiment 2). We have also found the strategy explanation to be intellectually unsatisfying because a clear and independently validated description of what the different strategies might be is seldom offered by researchers using this notion in attempting to account for changes in laterality effects. However, advances in cognitive psychology may help to overcome this particular problem (e.g. Cooper, 1982).

The second variant would see sex differences in cognitive processes as reflecting factors that can be affected by task and procedural demands, but over which the person has little voluntary control. Because of our lack of success in manipulating subject strategies by mere instruction we are now inclined to favour this alternative. However, we are also only too well aware that it still remains necessary to determine what these different cognitive processes might be, and why males and females rely on them to differing degrees under certain conditions.

Acknowledgement

Most of this research was carried out with the support of SSRC Grant HR5078.

REFERENCES

Beaumont, J.G. (1982). Developmental aspects. In J.G. Beaumont (Ed). Divided Visual Field Studies of Cerebral Organisation. London: Academic Press, 113-128.

Beaumont, J.G. (1983). Methods for studying cerebral hemispheric function. In A.W. Young (Ed). Functions of the Right Cerebral Hemisphere. London: Academic Press, 113-145.

Bradshaw, J.L. and Nettleton, N.C. (1983). Human Cerebral Asymmetry. New Jersey: Prentice-Hall.

Bryden, M.P. (1979). Evidence for sex differences in cerebral organization. In M. Wittig and A. Peterson (Eds). Determinants of Sex-Related Differences in Cognitive Functioning. New York: Academic Press, 121-143.

Cooper, L.A. (1982). Strategies for visual comparison and representation: individual differences. In R.J. Sternberg (Ed). Advances in the Psychology of Human Intelligence, Vol 1. New Jersey: Erlbaum, 77-124.

De Renzi, E. (1982). Disorders of Space Exploration and Cognition. Chichester: Wiley.

De Renzi, E. and Spinnler, H. (1966). Facial recognition in brain-damaged patients. Neurology, 16, 145-152.

Harris, L.J. (1978). Sex differences and spatial ability: possible environmental, genetic and neurological factors. In M. Kinsbourne (Ed). Asymmetrical Function of the Brain. London: Cambridge University Press, 405-522.

Hilliard, R.D. (1973). Hemispheric laterality effects on a facial recognition task in normal subjects. Cortex, 9, 246-258.

Kimura, D. (1966). Dual functional asymmetry of the brain in visual perception. Neuropsychologia, 4, 275-285.

Klahr, D. and Wallace, J.G. (1973). The role of quantification operators in the development of conservation of quantity. Cognitive Psychology, 4, 301-327.

Leehey, S.C., Carey, S., Diamond, R. and Cahn, A. (1978). Upright and inverted faces: the right hemisphere knows the difference. Cortex, 14, 411-419.

McGlone, J. (1980). Sex differences in human brain asymmetry: a critical survey. The Behavioral and Brain Sciences, 3, 215-263.

McGlone, J. and Davidson, W. (1973). The relationship between cerebral speech laterality and spatial ability with special reference to sex and hand preference. Neuropsychologia, 11, 105-113.

Marshall, J.C. (1973). Some problems and paradoxes associated with recent accounts of hemispheric specialization. Neuropsychologia, 11, 463-470.

Rizzolatti, G., Umilta, C. and Berlucchi, G. (1971). Opposite superiorities of the right and left cerebral hemispheres in discriminative reaction time to physiognomical and alphabetical material. Brain, 94, 431-442.

Schaeffer,B., Eggleston, V.H. and Scott, J.L. (1974). Number development in young children. Cognitive Psychology, 6, 357-379.

Schaffer, H.R. (1971). The Growth of Sociability. Harmondsworth: Penguin.

Starkey, P. and Cooper, R. G. (1980). Numerosity perception in human infants. Science, 210, 1033-1035.

Warrington, E.K. and James, M. (1967a). An experimental investigation of facial recognition in patients with unilateral cerebral lesions. Cortex, 3, 317-326.

Warrington,E.K. and James,M. (1967b). Tachistoscopic number estimation in patients with unilateral cerebral lesions. Journal of Neurology, Neurosurgery and Psychiatry, 30, 468-474.

Witelson, S.F. (1974). Hemispheric specialization for linguistic and nonlinguistic tactual perception using a dichotomous stimulation technique. Cortex, 10, 3-17.

Witelson, S.F. (1976). Sex and the single hemisphere: specialization of the right hemisphere for spatial processing. Science, 193, 425-427.

Witelson, S.F. (1977). Early hemisphere specialization and interhemisphere plasticity: an empirical and theoretical review. In S.J. Segalowitz and F.A. Gruber (Eds). Language Development and Neurological Theory. New York: Academic Press, 213-287.

Young, A.W. (1982a). Methodological and theoretical bases of visual hemifield studies. In J.G. Beaumont (Ed.) Divided Visual Field Studies of Cerebral Organisation. London: Academic Press, 11-27.

Young,A.W. (1982b). Asymmetry of cerebral hemispheric function during development. In J.W.T. Dickerson and H. McGurk (Eds). Brain and Behavioural Development. Glasgow: Blackie, 168-202.

Young,A.W.(1983). The development of right hemisphere abilities. In A.W. Young (Ed). Functions of the Right Cerebral Hemisphere. London: Academic Press, 147-169.

Young, A.W.(1985). Subject characteristics in lateral differences for face processing by normals: age. In R. Bruyer (Ed). The Neuropsychology of Face Perception and Facial Expression. New Jersey: Erlbaum.

Young, A.W. and Bion, P.J. (1979). Hemispheric laterality effects in the enumeration of visually presented collections of dots by children. Neuropsychologia, 17, 99-102.

Young, A.W. and Bion, P.J. (1980). Absence of any developmental trend in right hemisphere superiority for face recognition. Cortex, 16, 213-221.

Young, A.W. and Bion, P.J.(1983). The nature of the sex difference in right hemisphere superiority for face recognition. Cortex, 19, 215-225.

Young, A.W. and Ellis,A.W. (1979). Perception of numerical stimuli felt by fingers of the left and right hands. Quarterly Journal of Experimental Psychology, 31, 263-272.

Young, A.W. and Ellis, H.D. (1976). An experimental investigation of developmental differences in ability to recognise faces presented to the left and right cerebral hemispheres. Neuropsychologia, 14, 495-498.

ESTIMATING CEREBRAL ASYMMETRY AND INTERHEMISPHERIC TRANSMISSION TIME FROM INDIVIDUAL DIFFERENCES IN BIMANUAL RESPONSE TO LATERALIZED STIMULI

W. N. Schofield

Department of Experimental Psychology
University of Cambridge
Downing Street
Cambridge, England, CB2 3EB

ABSTRACT

This chapter is partly a critical review of selected divided visual field studies of cerebral lateralisation using manual reaction time mostly to simple stimuli such as dots or flashes of light, and of studies purporting to estimate interhemispheric transmission times from these procedures. It is suggested that inconsistencies in the findings reviewed rest on failure to take into account individual differences in processing strategy, amongst other things. A method for overcoming these difficulties is proposed, and support for this is found by re-analysis of right and left hand RTs from an early study. The remainder of the chapter reports a simple reaction time experiment in which bimanual response to bilateral stimulus presentation is related to bimanual response to unilateral presentation. It is shown that sex differences in lateralization following unilateral presentation reduce to difference in between hand response speeds as established for bilateral presentation, and it is suggested that this represents individual hemispheric differences in processing strategy. An influence from sighting dominance is shown both in initial analyses and in multivariate regression on factor scores. Estimates of interhemispheric transmission time are provided, but with suggestions for improved control. It is argued that hand differences in reaction time will vary in relation not to the hemisphere of initial projection of the visual stimulus but to the hemisphere in which the response command is issued, and that this could well be a better indicator of processing hemisphere than visual field differences.

INTRODUCTION

This paper reports an experiment in which the dependent variable was reaction time (RT) to an identical stimulus - a plain circular light source - on all trials, the only variation being that presentation was in the right visual field, the left visual field, or the light source was duplicated and appeared simultaneously in both visual fields. The presentation procedure was thus similar to many other RT studies for which the example of Poffenberger (1912) is often cited and which have been reviewed by Johnson (1923), Woodworth (1938), Teichner (1954) and Swanson et al. (1978). In the present experiment subjects always made the same response, they raised both hands no matter which stimulus condition was presented. These presentation and response procedures were employed in an endeavour to resolve conflicting findings reported for simple reaction time (SRT) studies of lateralised hemispheric function, and also to resolve the very evident confusion in the literature on using RT paradigms to estimate interhemispheric transmission time (IHTT). The design of the study was in effect for two experiments run simultaneously with the same subjects. Both experiments involved bimanual response, but for one stimulus presentation was always bilateral and for the other it was always unilateral. The purpose of the bilateral presentation was to provide control information on hypothesized individual differences in processing hemisphere relevant to analysis and interpretation of the results for lateralised presentation. The assumption of the study was that lateralisation of function, if any, would be consistent within individual subjects for a comparatively briefly tested task, and that information on this from the bilateral procedure could be tested in the unilateral data. The trials for the experiments were integrated and presented randomly in one block and the whole procedure can be considered as one experiment with two conditions and will be reported in those terms. Sex was controlled, and sex differences will be reported and will be related to individual differences in bimanual response time.

Research on response to simple laterally presented visual stimuli has included what have been called cross education and dominance studies (Herlitzka, 1908; Terashi, 1920; Metfessel and Warren, 1934; Cook, 1934; Hellebrandt, 1951); it has included studies of stimulus-response compatibility (Fitts and Seeger, 1953; Rabbit, 1967; Simon, 1968; Craft and Simon, 1970; Wallace, 1971; Brebner et al., 1972; Bertera et al., 1975; Anzola et al., 1977; Berlucchi et al., 1977; Bowers and Heilman, 1980; Cotton et al., 1980); and many studies of the lateralization of cerebral function (Kimura, 1969; Jeeves, 1969; Jeeves and Dixon, 1970; Jeeves, 1972; Umilta, et al., 1974; Pohl et al., 1972; Bryden, 1976; Allard and Bryden, 1979; Levy and Reid, 1978; McKeever and Huling, 1970; Berlucchi, 1974), including measurement of IHTT (Poffenberger, 1912; Jeeves, 1969; Bradshaw and Perriment, 1970; Berlucchi et al., 1971; Rizzolatti et al., 1971; Berlucchi et al., 1977; Anzola et al., 1977; Swanson et al., 1978; Harvey, 1978; Rizzolatti, 1979; McKeever and Hof, 1979; Di Stefano et al., 1980). Basic to this work has been the assumption that mediational differences (i.e. visual and/or motor pathway differences; hemispheric processing differences; or associative influence in cognitive processes) can be inferred from differences in response related to the lateralization of input and output.

The present experiment parallels to some extent published work in all of these areas but attention will be concentrated on studies of hemispheric functional differences in dot detection or localisation and on the estimation of IHTT from lateralized RT paradigms since the findings to be reported have specific relevance in both areas.

Hemispheric Specialisation for Dot Localisation or Detection

Among straightforward studies of hemispheric functional differences Kimura's (1969) early experiments on dot detection and localisation have been influential, but unfortunately are open to criticism on a number of grounds, and it is not surprising that attempts to replicate have produced both supportive and conflicting results (Pohl, Butters and Goodglass, 1972; McKeever et al., 1975; Bryden, 1976; Allard and Bryden, 1979; Davidoff, 1977; Levy and Reid, 1978; Pitblado, 1979a & b). For dot localisation Kimura (1969) reported what she called a "stronger" left visual field (LVF) advantage for males than for females which was her interpretation of four localisation experiments where in every case an LVF advantage was found for the male subjects ($P<0.02$, $P<0.10$, $P<0.01$) but in only two instances was significance reached, or approached, for the females ($P<0.10$, $P<0.01$). Kimura argued (from the lack of a finding) that for a task which could be performed by either right or left hemisphere males would employ right hemisphere systems whereas females would be more likely to use a left hemisphere (verbal) mechanism. It should be noted that none of Kimura's localisation experiments produced a left hemisphere right visual field, (RVF) advantage, either for males or for females, and her proposals apparently are based on inspection of unreported non-significant effects.

For dot detection Kimura (1969) reported that in three threshold experiments there were no significant RVF/LVF differences, neither overall nor for males and females separately. The first experiment, however, was grossly imprecise as measurement was made in 10 millisecond bands, whereas any difference was likely to be in the range of perhaps 2-4 msec. The second was also imprecise since the ceiling for detection was approached by both males and females for both visual fields (subjects were asked to respond "yes" if they saw a 5 msec. dot) and possible interaction from response modality was not considered. In the third experiment, detection thresholds were established by presenting a stimulus for 1 msec. and if both its occurrence and position (RVF or LVF) were not correctly reported presentation time was increased in 1 msec. steps to a maximum of 5 msec. This improved measurement to an appropriate interval size and range, but contaminated the detection task with a localization requirement. Consequently the third experiment might be seen as a failure to replicate the localization experiments when the localization task was simplified, rather than as replication of the non-significant findings for dot detection.

Apart from measurement, design and analysis problems the main defect in Kimura's (1969) detection experiments is the use of threshold differences to operationalise functional differences in hemispheric processing, since these might not register a difference which had effect subsequent to initial representation of the stimulus in the hemisphere tested. Although this particular criticism does not apply to the findings for the localization experiments it does apply to their interpretation, since it cannot be accepted that Kimura has succeeded in her intention of ruling out functional differences at the level of detecting occurrence of a stimulus.

Pohl et al. (1972), using adult male subjects, found no visual field differences for dot localization when the array of dots appeared in a square frame with its centre marked by a cross. But when the frame and the cross were omitted, a significant difference was found in the opposite direction to that reported by Kimura for males, i.e., there was a significant left hemisphere superiority. One defect in the Pohl et al., (1972) study was the use after each error response of repeated testing with increased stimulus presentation times. Each subject made on average more than 160 errors in the 65 trials; yet no information is given on the distribution of the retesting, nor is there information on variability, or on individual results.

Bryden (1973) also presented dots in the RVF and LVF but found no visual field differences. Bryden (1976) conducted four experiments similar to those of Pohl et al. (1972) and Kimura (1969) but with design improvements, and in which the dependent variable was a count of correct report for localization, or for detection and localization separately. These failed to resolve the conflicting findings of Kimura and Pohl et al., as there were no significant visual field differences in any of the experiments.

Allard and Bryden (1979), using identical stimuli to Bryden (1976), supported Kimura (1969) in so far as finding and LVF advantage, but this was for detection where Kimura had found no difference. Further, Allard and Bryden (1979) found no difference for localization, whereas for this Kimura did report a significant difference. Worse, in Bryden and Allard's data the LVF detection advantage in their control condition was significantly greater for females than for males, again turning around Kimura's findings.

One explanation of these inconsistent findings is the possibility that when analysis is made of mean scores then visual half-field (VHF) differences present at an individual level either cancel, or produce results in one direction or another as an artifact of subgroup composition. Evidence supporting this contention can be found in Bryden's 1976 experiments. Certainly none of the studies reviewed has established that dot detection is not lateralized, and the inconsistency in the reported findings would be neatly explained if individual differences could be related to the reported overall outcome of the experiments.

Using RT Paradigms to Estimate Interhemispheric Transmission Time

All three of Donders' (1868, tr. 1969) RT paradigms have been used in VHF lateralization experiments to determine IHTTs The basic assumptions have been that, at least for a simple reaction time (SRT) procedure, (1) processing and response initiation take place equally well in either hemisphere, and (2) the difference between conditions employing crossed and uncrossed pathways is the time taken to cross the corpus callosum once. Obviously the pathways actually involved in initiating and controlling a manual response are more complicated than this (Rolls, 1983) but a further assumption is that they will be balanced in the relevant comparisons. Bashore (1981) has reviewed research designed to measure IHTT. The widely differing estimates of IHTTs ranging from about 2-3 msec. (Jeeves, 1969; Berlucchi et al., 1971; Anzola et al., 1977; Berlucchi et al., 1977; McKeever and Hof, 1979; Rizzolatti, 1979; Di Stefano et al., 1980) to values of 16 msec. or much greater (Bradshaw and Perriment, 1970; Swanson et al., 1978; Harvey, 1978; Filbey and Gazzaniga, 1969; Rizzolatti, 1979) have been accounted for by the fact that the shorter times are from SRT studies, whereas the greater times are from choice reaction time (CRT) studies (Rizzolatti, 1979; Bashore, 1981; Bradshaw and Nettleton, 1983). Since the shorter 2-3 msec. times approximate to what might be expected for transmission over the neuropathways involved (but see Swadlow et al., 1979, on variability in the speed of conduction of callosal axons) it has been taken for granted (e.g. Rizzolatti, 1979; Bradshaw and Nettleton, 1983) that these are the best estimates of IHTT and that the larger figures are inflated by processes necessary for the discriminative components of the CRT task (including possibly stimulus-response contiguity effects). This reasoning ignores the fact that both figures would be incorrect if hemispheric differences were confounded in their measurement, and from the findings which have just been reviewed this has not been ruled out.

Rizzolatti et al. (1979) have suggested that findings of hemispheric differences in SRT experiments is a consequence of interference from a concomitant task, however in an adequately controlled experiment such effects, if present, should be unsystematic, i.e., they should be taken care of by the randomisation of error variation assumed in the statistical models applied. Presumably with grounds such as these in mind, in another paper, Rizzolatti (1979) has argued that in SRT studies of IHTT the estimates should reflect the conduction velocity of callosal fibres and should remain constant irrespective of the absolute value of the responses. If all else were balanced such constancy could be expected, but to assert (as he does) that fairly constant small values confirm that IHTTs are being measured in specious. Both the large and the small difference could be consistently preserved if the difference at mean score level reflected a misbalance of differences at individual subject level for classes of response differing greatly, but consistently, in mean and dispersion. A systematic source of uncontrolled error variation could be concealed equally well behind consistent small or consistent large differences.

Stimulus-response compatibility (or contiguity) effects (Fitts and Seeger, 1953, Broadbent and Gregory, 1962, 1965) are a problem for IHTT and other lateralization studies. Basically this work has shown that response to a laterally presented visual stimulus is faster if the hand making the response is on the same body side as the stimulus, even if the hand is held across the body to achieve this (Wallace, 1971; Brebner et al., 1972; Brebner, 1973). Thus the right hand responds faster than the left hand to a flash of light in the RVF and left hand responds faster to a flash of light in the LVF, but if the hands are held across the body midline so that the right hand operates the response key on the left body side and the left hand operates the key on the right body side then these effects are reversed. Response is fastest to the response position (not the hand in that position) nearest to the stimulus position, irrespective of whether the contralateral or ipsilateral pathways are being tested. Clearly, if these findings are sound, then the difference between the uncrossed and crossed visual field and response hand connections is not a measure of IHTT, but of the global time taken for cognitive requirements related to the stimulus-response configuration, or to a mixture of these and the pathway times. If correct, this not only destroys the rationale of manual RT studies of hemispheric transfer time, but points to a source of bias overlooked in many manual RT studies of laterality.

These findings have been challenged in recent work comparing SRT and CRT results. Rizzolati (1979) describes the work of his colleagues Berlucchi et al. (1977) and Anzola et al. (1977). The former reported that both for a SRT experiment and for a go-no/go design response speed was related to hand used rather than to the body side on which the hand was held during response. The latter confirmed the SRT finding and further reported that for a CRT task a strong S-R compatibility effect swamped any attempt to measure IHTT. Similar results have been reported previously by Callan et al. (1974) for manual response to auditory stimuli, and by Bertera et al. (1975) for eye movement RTs. Taken at face value these appear to be convenient findings for researchers working on transfer time, provided they use SRT procedures, and this work will be reviewed as it is highly relevant to the present findings.

Berlucchi et al. (1971), using a SRT paradigm and two separate groups of subjects, estimated IHTTs of 3.3 and 2.1 msec. They reasoned that the right hand would respond faster to stimuli in the RVF (uncrossed reaction) and that the contrary would be the case for the left hand (crossed reaction) because "the neural pathways serving crossed reactions should contain at least one more synaptic link than the pathways mediating uncrossed reactions" (1971, p419). This follows Poffenberger (1912). Subsequent work by these researchers, and others at the same laboratories in Pisa, Parma and Bologna, has failed to provide unequivocal support for the contention that the time cost of one synaptic link can be measured in the context of even a "simple" RT study, although the contrary is claimed (Rizzolatti, 1979; Di Stefano et al., 1980). For a start the procedure assumes that the hemispheres are equal in their ability to deal with a SRT task (Rizzolatti, et al., 1971). Surprisingly this was not the case even in Berlucchi et al's. (1971) SRT experiment where a significant right hemisphere advantage can be seen for one group of subjects. Consider the implications of this. Interhemispheric transmission time was found by subtracting "uncrossed" from "crossed" RTs: uncrossed being when the visual stimulus is projected to the same hemisphere as would initiate the manual response, and crossed being when initial projection was to the hemisphere contralateral to that controlling the manual response. Thus for Berlucchi et al's. (1971) procedure the uncrossed responses were the average of the RVF-right hand response and the LVF-left hand response, and the crossed responses were the average of RVF-left hand and LVF-right hand. The claim was that this procedure cancelled out unwanted differences in motor mechanism, and in retinal hemifield/visual half field effects (these were confounded as subjects viewed solely with the right eye); and that for both components of the crossed condition the manual response would require one additional transcallosal cross, and, further, that although these would be in opposite directions it was plausible to consider they would cost the same time. But if stimulus detection/processing takes place in the right hemisphere (as was found for one group of subjects) then LVF-L would require no interhemispheric crosses; RVF-R would require two crosses, one for detection and one for output; RVF-L would require one for detection, but none for output, and finally LVF-R would require a cross solely for output. In this complex situation, it is hard to say what the source of a significant difference between uncrossed and crossed averages might be. There are, however, clues. The authors reported that 9 of their 14 Ss responded faster with the left hand than with the right hand and presumably assumed that these differences were balanced in the averaging. The validity of this assumption depends on what the individual differences represent. If, for example, they reflect hemispheric processing differences then the supposed estimates of IHTT would be nothing more than bias in an unbalanced mix of confounded effects.

Rizzolatti et al. (1971) extended this work to a CRT paradigm reasoning that this would involve hemispheric functional asymmetry, and that the difference between response following stimulus presentation to the hemisphere preferred for the task and to the nonpreferred hemisphere would reflect IHTT. This experiment found a significant RVF advantage for a task involving letter recognition (18.5 msec.) and a significant LVF advantage for a human face recognition task (15.5 msec.). The results were unusually consistent for a laterality study, with every subject except one for each task having VHF differences in the same direction as the mean scores. Surprisingly, this consistency was not maintained in between hand comparisons where it might have been expected that the group showing a left hemisphere preference would respond faster with the right hand, and that the group showing a right hemisphere preference would respond faster with the left hand. The mean scores, although not significant ($F=2.02$, $Df=1,20$, $P=0.171$, calculated from the published values) went in the opposite direction, and there was a high level of inconsistency for individual subjects. Since this was an experiment where the authors predicted and found a preferred hemisphere effect it is interesting to apply the same method as was used by Berlucchi et al. (1971) when they estimated IHTT in an experiment where no preferred hemisphere effect was predicted, but where one was found (see above). From the published mean scores it was found that the difference between what in an SRT design would have been called the crossed and uncrossed pathways was 2.5 msec. The same difference calculated for the face recognition task was 7.5 msec. These differences are not significant in the data given, but are similar in size to those found in Berlucchi et al's. (1971) SRT study. The possibility remains in both cases that something other than straightforward IHTT was represented.

Berlucchi (1974) reported that the findings of these experiments, (Rizzolatti et al., 1971), held when the letter and face stimuli were presented in random sequence, and that a square patch of light included in the same sequence produced no VHF differences, but gives no further details. If substantiated, these further findings would be evidence against attentional bias due to block administration of stimuli (Kinsbourne, 1970). Berlucchi (1974) suggested further that a RVF superiority was found in the first experiment because the letter stimuli were easily verbalised, but that this would be more difficult when faces were discriminated.

In a further series of experiments a group including the same researchers (Umilta et al., 1974) trained right-handed Ss to discriminate rectangles in varying orientations. In a go-no/go RT task Ss pressed a key with right or left hand if there was a match in orientation. In the first experiment RVF response was 12 msec. faster than LVF response ($P<0.025$) and there were no other significant effects. In a second experiment requiring more difficult discrimination there were no significant effects, but in the third and final experiment with even more difficult discriminations the significant advantage was for the LVF ($p<0.01$) with no other significant effects or interactions. Response hand differences were not significant in any of the experiments nor were there significant interactions between response hand and visual field. The authors account for the significant RVF effect in the first experiment by proposing left hemisphere mediation for a task which they felt was more easily verbalised than that of the second and third experiments. If the reasoning of the earlier work is to be followed then the difference between the RTs for the preferred and nonpreferred hemisphere should estimate IHTT. By this method the estimate from the first experiment would be 12 msec. As there was a left hemisphere advantage a further estimate of IHTT should have been given by the difference between right and left hand RT since the latter would require one extra callosal cross, but the results were in the opposite direction and were not significant. Once again the interesting point is unexplained individual subject differences. The mean scores used in the analyses are based on: 6 subjects faster with the right hand 6 subjects faster with the left hand, suggesting that the VHF and hand difference scores for these subgroups might usefully have been examined. There is similar disparity among subjects for the third experiment. Clearly it would be inappropriate to estimate IHTT by the crossed versus uncrossed pathway method in view of the hemispheric difference in two of the three experiments. If it is done, however, then the estimate would be 7.0 msec. for the first experiment and 4.0 msec. for the second, both in the direction of the expected uncrossed pathway advantage, showing that plausible results can be produced by invalid comparisons. For the third experiment the difference is in the wrong direction and the hypothetically longer pathway is 5 msec. faster than the shorter pathway. Of the 12 Ss in this experiment 8 were fast with their right hand and 4 with the left hand. If IHTT paradigm assumptions are correct then a normal distribution of left hand advantage difference RTs might have been expected with possibly a preferred hand practice advantage applying a counter bias to shift the distribution without altering its shape, but no information is provided on this in the published findings. These experiments, also, did not take into account stimulus-response compatibility effects.

Berlucchi et al. (1977) repeated the work of Berlucchi et al. (1971) but with binocular viewing and 50% of the trials given with hands and response keys on the same body side as stimulus presentation and with 50% with hands crossed to operate keys on the opposite body side. This work was planned as a rejoinder criticism from to Broadbent (1974) and to test S-R compatibility effects (reference given earlier). In discussing their findings the authors adhere to the belief that "the difference between ipsilateral and contralateral responses in simple reaction time to lateralized stimuli is presumably a pure measure of interhemispheric transmission time" (1977, p511), although their previous findings, and also the findings they now report, include evidence against this assumption. Firstly for one condition they report a significant LVF advantage, and from a figure showing the mean scores (values are not reported) it can be seen that for most conditions the VHF differences are consistent in direction with this. Secondly, although the mean difference between the uncrossed and the crossed pathways is significant (2.5 msec. $P<0.01$) the results for 25% of subjects went in the opposite direction when keys and hands were on the same body side, and this increased to 31% when hands were crossed. No information is given on the size of the differences for the reversed cases, but the percentages are much greater than would be expected it these were cases of reversed lateralization of function (Zangwill, 1960), and even if they were they would form a highly unwelcome subgroup, unless they could be treated separately, considering the objectives of the research. Thirdly, there were similar inconsistencies in response hand times. For example almost one third of subjects were faster for the left hand for the stimulus presented in the RVF, whereas both IHTT and S-R compatibility hypotheses would predict that the right hand should have been faster. In some circumstances the presence of error to this degree might be expected and attention would only be given to estimates of central tendency and the standard deviations from these. But in the present experiments subjects have undertaken many trials and the analyses are made on the means of medians from these, and it is reasonable to expect consistent individual results in the summary statistics. The size and distribution of the wayward differences cannot be calculated from the results given, but if IHTT methods are valid these should be in the tail of a unimodal distribution, and any suggestion of bimodality would cast doubt on validity.

An important finding from this experiment was the complete lack of S-R compatibility effects. It would be helpful to know however, if there was evidence of this effect in the data from the early test sessions and if this was lost in calculating the overall mean scores. Each subject had one practice session followed by 8 test sessions on different days and a total of 960 RTs was recorded per subject. Wallace's (1971) demonstration of stimulus-response compatibility effect was based on just one test session with 8 practice and 128 test trials. Brebner et al., (1972) used a large number of test sessions and highly experienced subjects, and when their data were re-examined Brebner (1973) found that the last trial of the main experimental sessions gave an RT pattern which in the case of each subject differed from the overall pattern, and which suggested change with practice. Brebner (1973) interpreted findings from 10 sessions for 10 subjects as showing the gradual acquisition of a spatial frame of reference in which identification of the position of the hand in space becomes independent of its normal body side. The figures published by Berlucchi et al. (1977) do seem to suggest that individual comparisons might establish more differences than those reported. Callan et al. (1974) also reported no S-R compatibility effect for a SRT design, but although there were 504 trials in 8 separate sessions no information is given on the possible effect of practice.

In their second experiment, Berlucchi et al. (1977) further examined stimulus-response compatibility effects in a go-no/go design and, as in the first (SRT) experiment, found none. This finding rests on a significant interaction between response hand and VHF ($P<0.005$), together with a non-significant three-way interaction when response hand position (same or opposite body side) was introduced ($P<0.20$). Once again, however, no information is given on practice. The effect of this could have been considerable as there were 1280 trials per subject, plus those of an initial training session. The point of interest is whether the three-way interaction remained stable across these. Also of interest was the finding that response with hands in the normal positions was faster to LVF than to RVF stimuli and also that for this condition response was faster with the left hand than with the right. This implied right hemisphere mediation. Both visual field and hand effects were in the opposite direction for response with hands held across the body, implying left hemisphere mediation. These effects suggest to this reviewer a change in strategy for the different conditions of the experiment. The possibility of changing strategy is also supported by individual subject differences. For uncrossed hands the finding in the mean scores of an LVF advantage rested on 11 subjects faster in that field, but 5 were faster in the RVF; and further, only 9 were faster with their left hands than with their right hands. There were similar individual differences for the crossed hands RVF advantage. Undeterred by these inconsistencies the authors provided estimates of IHTT of 3.4 msec. from the uncrossed hands condition and 2.7 msec. from the hands crossed condition with a mean estimate of 3.0 msec. ($P<0.01$). An amazing feature of these predictions of IHTT is that a large subgroup of subjects were faster for the supposedly longer cerebral pathways than for the shorter pathways (44% for the hands across the body condition, 19% for the normal condition). It is hard to accept Berlucchi et al's., conclusion that the significant difference which remains when these individual differences are averaged out are a valid estimate of IHTT. An alternative possibility is that the mean value represents not IHTT, but lack of balance in strategy preference and processing hemisphere.

The 1971 study of Rizzolatti et al. was repeated by Rizzolatti and Buchtel (1977) using the same go-no/go design for face recognition, but with both male and female subjects. The LVF advantage of the 1971 study was confirmed for the males, but there were no significant VHF differences for the females. The experiment was repeated with fresh groups of men and women but with presentation time reduced from 100 msec. to 20 msec. It was found that the LVF advantage for the males was greatly increased, and the finding for females was unchanged. The significant finding in the analysis of variance is largely because the men show a deficit when initial projection is to the left hemisphere, however this is interpreted as showing that for brief exposure and immediate judgment, "a lateralised mechanism specialized for faces can be activated only in males" (1974, p304). The published report (Rizzolatti and Buchtel, 1977) has nothing to say on IHTT, but the findings are relevant to earlier work by Rizzolatti and his colleagues. The RVF advantage for the males was about 40 msec. when stimulus exposure was 100 msec. and 140 msec. when stimulus exposure was 20 msec. This can be compared to the value of 15.5 msec. reported for the 1971 experiment. Just why the more difficult discrimination should substantially increase a supposed indicator or IHTT (and for men but not for women) and not simply overall processing time (as was the case for women) remains unanswered.

Anzola et al. (1977) followed similar objectives to Berlucchi et al. (1977) but with separate subgroups for the crossed and uncrossed conditions. Again there were two experiments. The first was a SRT design and the second a CRT design. Thus according to the logic of Rizzolatti (1979) and the whole rationale of IHTT methods there should have been significant VHF differences in the second experiment but not in the first. Unfortunately, it was for the first experiment, the SRT design that an (unpredicted) LVF advantage was found; this once again undermines the sense of estimating IHTT by subtracting uncrossed from crossed response times. The authors express surprise at this finding but totally overlook its implication for their work and report an overall "uncrossed pathway" advantage with a probability from the analysis of variance of $P<0.025$. Clearly there is a systematic effect from some source or another, but it cannot be what the authors claim.

In the second, CRT, experiment reported by Anzola et al. (1977) subjects responded with the hand on the same, or on the opposite, body side as the stimulus; and had crossed, or uncrossed arms, for separate blocks of trials. There was no main effect for VF, but a strong stimulus-response compatibility effect ($P<0.005$) and main effects for hand (the right hand was faster ($P<0.025$). In both experiments 450 trials were given across 9 test sessions, i.e., about half the number of trial as in the go-no/go experiment (Berlucchi et al., 1977) in which there was no effect in the overall means. The difference between the ipsilateral and contralateral pathways was 10 msec. but this was not significant, and in any case was contaminated by the significant VHF asymmetry. No information is given on individual data.

The experiments reviewed in the preceding paragraphs rest on the assumption that more time will be required for a process which crosses the corpus callosum than for a process which is identical in all other respects except the need to cross. None of the work reviewed is satisfactory because of the inability of the researchers to achieve balance of all factors other than transcallosal-cross time, irrespective of whether a simple or CRT design is used, and because of their failure to take into account conflicting evidence visible even in their own data. Although S-R compatibility effects remain a major problem, it appears to have been shown that either they are not present in some SRT designs, or else that they decline with practice and thus can be brought under control. The findings of hemispheric asymmetry even in SRT tasks is a more fundamental difficulty especially if this varies between or within subjects depending on strategy used for performance. The firm stance on interhemispheric transfer times adopted by Rizzolatti (1979) on the basis of this work (much of it authored or co-authored by him) is without justification. His statements that "the main factor determining which hand is faster in responding to a lateralized stimulus is the basic organization of sensory and motor pathways" (1979, p395), and that in a SRT experiments "both hemispheres are virtually equal in their ability to analyze the stimulus (1979, p396) and that in "SRT experiments since the ´transfer time´ reflects the conduction velocity of callosal fibres, it should remain constant regardless of the absolute value of the responses" (1971, p396) are contradicted by findings evident in papers published by himself or which he has co-authored. Wider acceptance (Di Stefano, 1980; Biseach et al., 1982; Bradshaw and Nettleton, 1983) is unfortunate.

THE PRESENT EXPERIMENT

The present experiment provided an opportunity to examine IHTTs in a SRT procedure and at least to test some possible sources of uncontrolled variation. It further provided the opportunity to test statistical procedures which might increase control and sensitivity in IHTT or lateralization studies. The presentation of the identical stimulus in either the RVF or LVF allowed comparison between these fields, but with problems similar to some of those in the studies reviewed. The introduction of a randomly presented condition in which the stimulus appeared simultaneously in both visual fields provided control data to aid interpretation of the results for unilateral presentation.

Further control was introduced by use of a method of response which also was identical on all trials. Subjects had both hands resting on keys and raised them both as quickly as possible thus making the same response no matter what the stimulus presentation condition. The action trained was at the wrist, and there was nothing to prevent additional movement either at the fingers or elbows. Thus there was the possibility that both distal and proximal musculature was used to some extent. Di Stefano et al. (1980) used key pressing and a lever-pulling task in a SRT design and reported that the two methods produced similar results after unilateral presention of the visual stimulus, but that ipsilateral-contralateral time differences were reduced for bilateral key presses, and were completely removed for bilateral lever pulling. Studies of bilateral movement time as reviewed by Kerr (1978) are not directly relevant to the present experiment since the response was simply to raise both hands, and this terminated the trial, i.e., there is no requirement of subsequent movement to different targets, or to any target at all. It has been reported (Salow, 1913;Jeeves, 1969; Jeeves and Dixon, 1970; Nakamura and Saito, 1974) that bilateral responses are slightly but systematically slower than unilateral responses.

Separate right and left hand response have been used in many manual RT studies and comparisons between hands have also been made (Metfessel and Warren 1934; Kerr et al., 1963). The published results have usually been reduced to differences between hands, or to the mean of left and right hand, in an attempt to balance uncontrolled effects. For unimanual response Anzola et al. (1977) reported an overall superiority for the right hand in a CRT experiment. Berlucchi et al. (1977) reported no overall difference between hands in a SRT experiment, as did Rizzolatti et al. (1971) for a Donders' type C design and Umilta et al. (1974) for a line orientation discrimination task. Bradshaw and Perriment (1970) also reported no overall difference between hands. Poffenberger (1912) published data for 4 subjects but only two were tested for all retinal conditions used. One of these was right-handed, and one was left-handed and it is interesting to note that for foveal presentation the left-handed subject invariably responded faster with his left hand both for left and for right presentation. The right-handed subject showed a less consistent advantage for the right hand.

Metfessel and Warren (1934) have published data giving separate results for right and left hand for a bimanual SRT task in which subjects were trained to press response keys simultaneously when a centrally positioned light was briefly flashed. It seems likely that subjects responded rhythmically to the expected flash, but even so the report of a clear left hand advantage for the right-handers, and a clear right hand advantage for the left-handers is of interest. Groups of dextro-sinistrals and stutterers (who were mostly right-handed) also had a left hand advantage. These findings were consistent both for manual RTs and for action potentials taken from the two forearms immediately prior to hand response - which was with the first two fingers of each hand. Metfessel and Warren (1934) interpreted these findings as due to over-compensation of the less used hand in an attempt at simultaneous movement. However an alternative explanation could be the intervention of hemispheric functional differences which, considering the repetitive rhythmical nature of the task may have been consistently maintained once established, but differentially between subjects.

Fortunately these authors have published their raw data and from this it can be seen that for a sample of right-handed subjects 17 were faster with the left hand at times ranging from 1.2 to 18.3 msec. and 6 were faster with the right hand at times ranging from 1.7 to 12.4 msec. It is at least possible that these times identify separate groups, the former mediating responses via a right hemisphere non-linguistic spatial strategy, and the latter via a left hemisphere verbal strategy. Using Metfessel and Warren's data it was calculated that the mean left minus the mean right hand RT for the first group was 7.01 msec. with a standard error of only 1.22 and that for the second group the mean was -5.68 msec. with a standard error of 2.07. Thus when considered separately the two groups are clearly placed either side of zero with the suggestion of a systematic source of variation influencing the differences. Recalculation produced similar results for the other samples. The left-handed sample had 11 subjects faster with the right hand (mean 6.71 msec. standard error 1.55) and 10 faster with the left hand (mean -6.98, standard error 1.57). A sample of dextro-sinistrals (lefthanded writers who had been corrected in schooling to righthand writing or righthanded writers who were lefthanded for most other purposes) had 16 faster with the left hand (mean 6.93, SE 1.22) and 5 with the right hand (mean -4.94, SE 1.45). In a sample of 23 students (21 righthanded and 2 lefthanded), 19 were faster with the lefthand (mean 8.72, SE 1.18), one of whom was left handed, and 4 were faster with the right hand (mean -6.20, SE 2.17). When data for all 89 subjects were pooled and the histogram for the difference scores inspected there was a clear bimodal distribution both for the action potentials and manual RTs. In the RT data the peaks were at 7.43 and -6.15 msec. and since the standard error was only 0.87 the separation seemed decisive. The comparable figures for the action potential times were 8.95, -6.78 and 1.08 respectively on a sample reduced to 83. Thus in each subset of separate hand data whether for the sample as a whole or the hand preference/clinical subgroups, there is a suggestion not of overcompensation, i.e., bias in one population towards a particular tail, but of separate distributions. In a footnote, Metfessel and Warren (1934) mention results for a further group of 46 normal righthanded subjects. Thirty of these were consistently faster with left hand; 12 with the right hand, and 4 had no advantage one way or the other.

In his review of IHTT studies Bashore (1981) concluded that responses made with one hand were faster than responses made with both hands simultaneously. In support of this he cited Jeeves, 1969; Jeeves and Dixon 1970; and Di Stefano et al. 1980. Bashore (1981) further reported that speed of onset of bilateral movements involving different muscles (measured by latency to electromyograph activation following an imperative stimulus) varied as a function of response side and handedness (Hongo et al., 1976). That different cortical mechanisms mediate the production of unimanual and bimanual finger movements is indicated by the fact that components of movement related potentials measured at the scalp vary as a function of which movement is being executed (Kristeva et al., 1979). In the present study <u>all</u> responses are bimanual and the only variation is whether the signal to respond is presented in the RVF, the LVF or in both visual fields simultaneously. The possibility of an interaction between mixed use of both unimanual and bimanual response within subjects is eliminated.

In the experiment to be reported stimulus presentation took place on every trial, i.e., there were no blank cards to detect anticipatory responding (Bryden, 1976). The reason for this was that use of blank cards would have transformed the procedure from that of a SRT to a go-no/go task. Subjects viewed the dark interior of the tachistoscope fixating a low illumination dot of light. The stimulus occurred after a random delay of between 1 to 3 sec. and this should have been a perfectly adequate method of reducing anticipation to a few randomly distributed occurrences. But if blank cards had been introduced subjects would have been set to expect one of two stimuli: the flash of light from the blank card (or whatever), or the dot of light, and would have to decide which had been presented before responding. Bashore (1981) correctly describes this as introducing a go-no/go component to the initial SRT procedure, but he is entirely incorrect in his assumption that the former involves detection and that the latter does not. He writes: "In an SRT task, stimulus detection is not actually required because the subject knows that on every trial a stimulus will be presented as soon as the fixation stimulus leaves the central visual field; the subject need only prepare and execute the task" (p361). It would, however, be a unfortunate defect in an SRT experiment if the fixation dot disappeared at the onset of a laterally presented stimulus, since there would be no way of knowing whether the response followed the lateral or the central information. Since subjects are trained to maintain fixation in the central position it seems highly probable that the information signalling response would be found there.

Predictions

Predictions derived from the work reviewed included that there would be an overall significant LVF (right hemisphere) advantage since despite inconsistencies, and denial of what appear to be positive findings, these have been previously found. Again, there have been inconsistent findings of sex differences; even direction of prediction seemed uncertain, and the safest prediction was that there would be none. Both the IHTT and S-R compatibility research support a VHF and response hand interaction and it was predicted that the right hand would be faster in the RVF and that the left hand would be faster in the LVF at the level of mean scores. In general these predictions follow the balance of reported findings, but despite the volume of literature are certainly not strong in expectation.

The study included original predictions aimed at accounting for the inconsistency in the research reviewed and at exploring the possibility that current RT methods of measuring IHTT are defective. These predictions were tested in data from random presentation of unilateral and bilateral identical stimuli (the dot of light in the RVF, LVF or both VFs) with uniform bimanual response to all stimuli. This ensured that S-R compatibility effects were balanced since response was always with a pair of hands one of which was on the same body side as the stimulus and one of which was not. The most important innovation, however, was an attempt to use the data from bimanual RT to bilateral stimuli as a control for comparison with data from the bimanual RT to unilateral stimuli. If a hemispheric preference were shown for the former (bimanual RT, bilateral stimuli) then hypotheses could be tested in the latter (bimanual RT, unilateral stimulus). It was reasoned that if there were hemispheric differences for the bilateral condition these would be revealed by faster response for the hand on the opposite body side to one specific hemisphere and that the other hand would be slower by a difference commensurate with the need for one IHTT. Consequently the general research predictions were that subjects faster in the both VF condition with the right hand would respond on average faster in the RVF than in the LVF for unilateral presentation, and that for these subjects response would be faster with the right hand than with the left hand irrespective of visual field of initial projection by a difference similar in extent to the RVF/LVF difference since in both cases this would represent one cross of the corpus callosum. For subjects faster in the both VF condition with the left hand it was predicted that response would be faster in the LVF than RVF for the unilateral condition; that the left hand would be faster than the right hand no matter which VHF received the stimulus, and again that the RT difference between hands would be similar in extent to the RVF/LVF difference.

The present experiment differs from most of those reviewed in that the subjects were 9 year old children rather than adults (usually students, or the researchers themselves). This should in no way be a methodological defect. If there were a lack of hemispheric specialisation for a SRT task at 9 years of age then findings would follow those predicted for the IHTT research reviewed, although the innovative hypotheses would remain untested. A further difference is the comparatively small amount of data collected from each subject - only 120 RTs each, whereas 1000 or more is not unusual. This was seen as a necessary condition for minimising possible changes in response strategy (and processing hemisphere) within subjects and was matched by an increase in sample size.

Method

Subjects. The subjects were 11 girls and 9 boys drawn at random from two parallel mixed ability classes at an inner city school in the south of England. In addition 3 boys and 3 girls were tested when the procedure was being established, and the data from these pilot Ss were included in some of the analyses made. Mean age (for 26 Ss) was 8.91 years (SD 0.16 years). All subjects were London born; had normal vision in both eyes; and were without physical or neurological handicap. Information on hand preference and visual dominance was not collected until after the conclusion of the experiment. None of the children had previously taken part in research of any sort. Procedure and objectives were not discussed with children or teachers.

Stimuli. The stimuli were two circular sources of light each with a luminance of 8.6 cd/m^2 appearing in a blackground of approximately 0.14 cd/m^2. Each stimulus subtended approximately 1 degree of visual angle, as established by tachistoscopic presentation. They were arranged 2 degrees either side of fixation and could be presented singly or simultaneously.

Stimulus duration was 60 msec. throughout the experiment. The fixation dot subtended approximately 0.3 degrees of visual angle and had a luminance 4.3 cd/m^2 in a black background of 0.14 cd/m^2.

Equipment and Procedure. Stimuli were presented using two fields of a three field tachistoscope (Electronic Developments Ltd), fitted with a long viewing mask. A dull black 7cm x 12cm tachistoscope card with a 0.30cm diameter hole at its centre was prepared from heavy mounting board. White translucent cine leader film was mounted on the back of the card completely covering this hole so that the appearance from the front was of a white opaque dot at the centre of a black card.

The card so prepared was sealed with adhesive to the internal black mask provided by the manufacturers within the tachistoscope approximately 8cm in front of the usual card presentation position. Thus the card was located between the field lamps and the S's eye and the only light which could pass was through the central hole. A plain grey filter, fitted within the tachistoscope viewing mask, further reduced the amount of light which could pass to the subject's eye. This field was on throughout the experiment.

A similar card was mounted in the second field of the tachistoscope but with two holes of 1 cm diameter centred on the horizontal midline of the card but 2 cm to each side of the card centre. This card was also covered on the rear side to give the same effect as for the fixation card. An assembly was fastened to the rear surface of this card containing two covers (one for each aperture) moveable in fixed horizontal tracks. These could be operated from outside the tachistoscope. With either or both covers in the open position light could pass through the apertures to the viewer's eye. The covers were moved between every trial, whether necessary or not. The plain grey filter at the viewing mask also reduced the amount of light reaching the viewer's eye from this field.

After these internal alterations had been made the rear of the tachistoscope was completely shielded from external light and the fidelity of the arrangements were checked with both child and adult subjects not involved in the experiment.

The effect to a viewer when both fields were off was of total darkness, nothing was visible within the tachistoscope. When the first field was on (and it remained on throughout the experiment) the viewer saw what appeared to be a small light in the central position. What the viewer saw when the second field was flashed depended on the position of the two covers: in this experiment he saw either two lights evenly spaced either side of the fixation light, or he saw one or other of these lights together with the fixation light.

Subject's were individually trained and tested in one session. After chair height had been adjusted S was seated at the tachistoscope and trained to fixate and to raise both hands from the surface of the table whenever a stimulus was presented. Micro switches linked to millisecond counters were inset beneath switch covers level with the table top. The switch covers were centred 75cm on either side of the S's midline. Table height was adjusted so that S's forearm was parallel to the surface. The keys, wiring and timers were alternated between hands for each subject. Calibration checks confirmed that the difference between timers was less than one msec. for measurement up to one sec. Timing began when a stimulus was presented and ceased if a hand movement broke the circuit to the millisecond counter associated with that hand. There were 6 practice trials (2 for each condition) followed immediately by 60 trials (giving 120 RTs), 20 for each condition (40 RTs) presented in a randomly determined sequence. Ss were not told how fast they had responded. Less than 0.5% of data was lost through procedural error on the part of Ss.

When all Ss had been tested each was again seen separately, to obtain data on hand preference and visual dominance, which in this report is taken to mean sighting dominance (Crider, 1944; Porac and Coren, 1976) or preference. A binocular rivalry task, tachistoscopically presented, was used to assess sensory dominance (Coren and Kaplan, 1973), but as this was found to be unrelated to sighting dominance, and made no significant contribution either as a main effect or as an interaction with other variables, the data have not been used in the analyses reported. Subjects were asked to write their names and the hand used was recorded. Ss stood near a table and a rectangular card 7cm by 12cm with a single 0.5cm hole at its centre was placed on the table at the S's midline. The S was asked to pick the card up and to look through the hole at the experimenter. The eye used was recorded. Eye used was similarly recorded for sighting a toy gun and for looking into a kaleidoscope. During normal class activities all Ss were observed writing, cutting with scissors and painting and it was found that two Ss used their left hand for all three of these activities, whereas eighteen always used the right hand.

The design of the experiment and the available measures permitted a series of multifactor analyses of variance involving both between and with S comparisons, and also the introduction of covariates. Factor analysis was used to display relationships between within Ss variables. Multivariate regression and canonical analyses were used to explore the possibility of predicting visual field differences from the factors.

Results

The raw data were screened and examined in histograms and values outside a range of 100 to 750 msec. were arbitrarily excluded. Winsorized mean scores were then calculated and 95% confidence limits established for each variable (Dixon, 1983). Less than 1% of the data were lost by these procedures, and it was clear that outlying values were sparsely and randomly distributed. The remaining data were logtransformed, to correct the typical RT skew, and the mean scores were recalculated. Medians and other estimates of central tendency were calculated, including Hampel's, Andrew's and Tukey's estimates of location (Andrews et al., 1972). These followed a similar pattern to the mean scores from the winsorized log transformed data on which all analyses were made. Analysis was also made of RTs for individual trials following logarithmic transformation with no exclusions other than values outside the 100 to 750 msec. range, and of means computed from this untrimmed data. These precautions were taken because the mean scores were estimated from fewer observations than is customary.

Preliminary analysis for fixation control. To establish that fixation was maintained, a three-way analysis of variance was made in the form, sex (male or female) x presentation position (right or left or both) x response hand (right or left). The only significant main effect was for visual field position ($F=3.94$, $Df=2,32$, $P<0.028$) and post-analysis of variance comparisons established that the mean score for the Both condition was significantly faster than for RVF and LVF presentation grouped ($t=2.054$, $Df=36$, $P<0.01$), and it was concluded that Ss maintained fixation. The mean RTs in msec. were 269.35, 266.43 and 261.31 for the RVF, LVF and Both positions respectively. The 2.92 msec. difference between RVF and LVF was not significant ($t=0.952$, $Df=36$).

Analysis with visual pathway and contiguity effects balanced. To establish which hand responded fastest when pathway and visual contiguity effects were balanced, a two-way analysis of variance was made of data for the bilateral VF condition in the form, sex (male or female) x hand (right or left) in which levels for the final factor were repeated measures. Although in the overall mean scores for this analysis, the left hand was, as predicted, significantly faster than the right hand ($F=5.32$, $Df=1,18$, $P<0.033$) this finding was complicated by a significant interaction with sex ($F=4.51$, $Df=1,18$, $P<0.05$). For the males the left hand advantage of about 8 milliseconds, was highly significant ($t=2.415$, $Df=18$, $P<0.01$); but for the females response times for the separate hands were almost identical ($t=0.134$, $Df=18$). This effect is illustrated in Figure 1 where it can also be seen that the males and females did not differ when comparison was made between their left hands ($t=0.178$, $Df=18$), but they did differ significantly when the comparison was between right hands ($t=2.937$, $Df=18$, $P<0.0001$).

The possibility that these results were artifacts of sighting dominance was tested. Since most Ss achieved either maximum or minimum score on the sighting dominance indicators the scores were dichotomised at 1.5. This gave 7 girls dominant in the right eye and 4 dominant in the left eye, against 3 boys dominant in the right and 6 boys dominant in the left eyes (Phi=0.302, corrected Chi-square=0.808, $Df=1$, NS).

The analysis (which was by a regression method) was repeated with sighting dominance introduced as a second grouping variable. Sighting dominance was not significant either as a main effect ($F=1.06$, $Df=16$) or in interaction with other variables.

For each subject RT for the left hand in the bilateral VF condition was subtracted from RT for the right hand in the same condition. A histogram of mean scores for all 20 subjects on the variable thus created was distinctly bimodal with peaks either side of zero, suggesting two separate groups of observations (see Figure 2). From the statistics given below Figure 2 it can be seen that the mean for each group shown is more than four times its standard error from zero, and that only one observation is more than three standard errors from its mean in the direction of the other group mean.

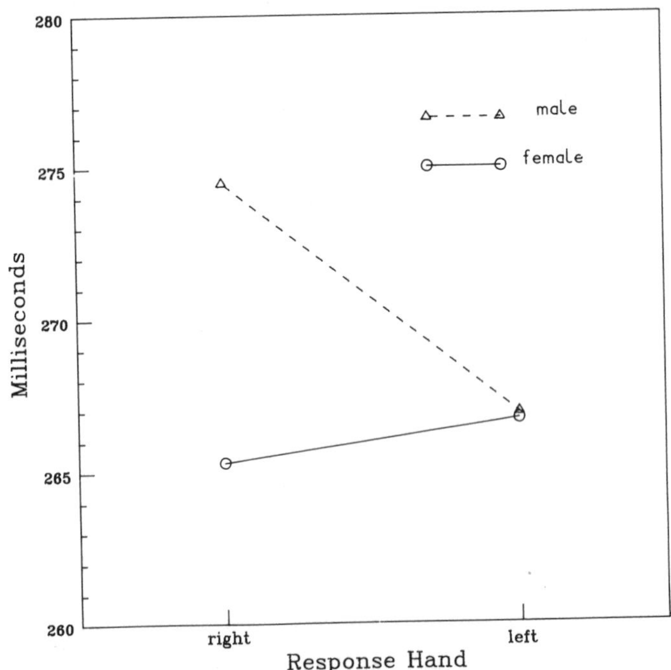

Fig. 1. The highly significant interaction between sex and response hand for the stimulus presented simultaneously in both visual fields ($F=4.51$, $Df=18$, $P<0.05$.

Histograms were also computed displaying the difference scores for each S individually and none of these were bimodal. In each case the distribution of difference scores was centred on one side or the other of zero, with mostly only a few tail values crossing to the other side. Data for the 6 Ss tested during pilot work were similar.

Subjects were divided into two groups as shown in Figure 2. Males and females were unevenly distributed within these groups. There were 6 females but only 1 male in the group which was faster with the right hand, as against 5 females and 8 males in the other group (Phi=0.453, corrected Chi-square=2.428, Df=1, NS). To simplify presentation the two groups will be referred to as the fasthand right and the fasthand left groups, which should be understood to mean "the group of Ss individually faster with the right hand than with the left hand in the bilateral condition" and "the group of Ss individually faster with the left hand than with the right hand in the bilateral condition", respectively.

```
                          Group 1                Group 2
                   (right hand faster)      (left hand faster)

    Milliseconds

        32 )
        28 )
        24 )                                    *
        20 )                                    *
        16 )                                    **
        12 )                                    M*
         8 )                                    ****++
         4 )                                    **
         0 )                                    *
       - 0 )
       - 4 )             +
       - 8 )             M*****++
       -12 )             *
       -14 )             +
       -16 )
       -20 )
```

Group means shown by "M" which also represents one data value.
Statistics (with pilot Ss included given in brackets):

	Group 1	Group 2
MEAN	-5.853 (-6.56)	9.997 (9.91)
S.E.	1.231 (1.09)	2.158 (1.88)
MAXIMUM	-3.721 (-1.65)	30.486 (30.49)
MINIMUM	-7.652 (-16.2)	0.365 (0.31)
N	7 (11)	13 (15)

Fig. 2. Histograms of the bilateral difference RTs (right hand - left hand) for the two groups. Each asterisk represents mean data for 60 trials from one subject. Pilot Ss are indicated by the symbol "+".

Analysis of RT for RVF and LVF Presentation

A three-way analysis of variance was made comparing response times after RVF and LVF stimulus presentation. The factors in the analysis were sex (male or female), visual field (RVF or LVF), and response hand (right or left). The two final factors were repeated measures. There was a highly significant two-way interaction between sex and visual field (F=8.28, Df=1,18, P<0.01) and the three-way interaction between sex, visual field and hand approached significance (F=3.04, Df=1,18, P<0.098). The two-way interaction is illustrated in Figure 3. Although the hand x sex interaction looked very similar to the significant effect reported for the both hand condition it was not significant (F=2.74, Df=1,18, P<0.115), and there were no other significant effects in the analysis. Post-analysis of variance comparisons established that the males responded significantly faster in the LVF than in the RVF (t=2.912, Df=18, P<0.01) but that there was no difference between visual fields for the females (t=1.062, Df=18). The females were significantly faster than the males for RVF presentation (t=3.519, Df=18, P<0.005), but not following LVF presentation (t=0.544, Df=18).

Taken at face value these findings would appear to suggest that:

(1) the males responded significantly faster when initial projection was to the right hemisphere than when initial projection was to the left hemisphere.

(2) although there was no RVF/LVF difference for the females, they were significantly faster than the males when projection was to the left, language dominant, hemisphere; but not when initial projection was to the right hemisphere.

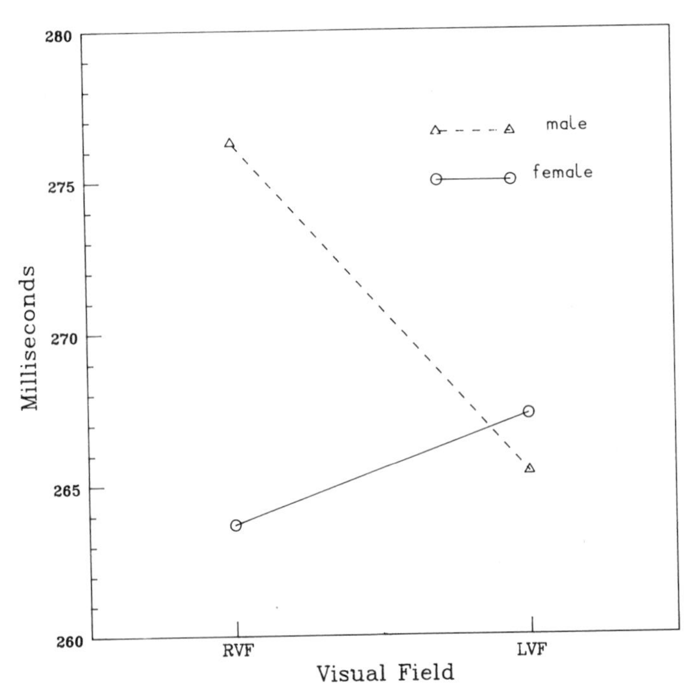

Fig. 3. The significant interaction between sex and visual field (F=8.28, Df=1,18, P<0.01).

Further analysis took advantage of the findings for the bilateral condition and established that these VHF by sex differences could not be taken at face value.

Re-analysis taking into account individual bilateral findings. The analysis of the previous section was repeated with the introduction of fasthand (as explained previously) as a second grouping variable. The factors in this analysis were sex (male or female); fasthand (right or left); VHF (right or left); and response hand (right or left). The only significant effect was an interaction between fasthand and response hand ($F=15.58$, $Df=1,16$, $P<0.0012$), but as there was only one male in the right fasthand group the analysis was remade with the data for the 6 pilot subjects included. Results were almost identical to those obtained for the 20 subjects. The only significant effect was the interaction between fasthand and response hand ($F=16.61$, $Df=1,22$, $P<0.0005$), and there were no significant, or near significant, effects involving sex.

Taking into account difference in response hand speed from the bilateral presentation condition removed all suggestion of sex differences related to visual field of presentation and initial projection to different hemispheres. The apparent VHF difference between males and females reduced to the finding that in the augmented sample of 14 females 9 responded faster with their left hand, and 5 were faster with their right hand; whereas of the 12 males only 2 were faster with their right hand, against 10 who were faster with the left hand. This distribution of frequencies was significant ($Phi=0.481$, corrected chi-square$=4.211$, $Df=1$, $P<0.05$).

As sex was not significant, and to improve the quality of the analysis as affected by subgroup size, it was dropped from the design. The effect of the fasthand grouping variable on VHF differences was tested in an analysis of variance in which the factors were fasthand (right or left), VHF (right or left) and response hand (right or left), and into which a covariate was introduced representing overall reaction time. The purpose of the covariate was to remove from the analysis individual differences in response speed since a difference of say six msec. might represent different functional magnitudes for relatively fast and relatively slow responding subjects. The estimate of overall RT was obtained from the bilateral condition data which was not being otherwise used in the analysis except to provide the grouping variable. The correlation between mean RT for the both visual fields condition and the overall mean RT was 0.99. Slope for the covariate did not differ significantly for the subgroups. In the event the covariate was highly significant ($F=86.82$, $Df=1,17$, $P<0.0001$), but the main effect for fasthand remained non-significant. There was a highly significant interaction between fasthand and response hand ($F=34.94$, $Df=1,18$, $P<0.0001$), which was an expected consequence of the design. In addition the interaction between VHF, hand and fasthand closely approached significance ($F=4.01$, $Df=1,18$, $P<0.061$).

This interaction is illustrated in Figure 4 and is of considerable interest, despite the border-line probability level. Since both the visual field and hand differences were predicted, straightforward post-analysis of covariance comparisons remained appropriate. Only the comparisons relevant to the predicted within group effects have been tested and are given in Table 1. To provide real time information the comparisons are given for the unadjusted mean scores, but significance levels were confirmed for adjusted comparisons. It can be seen from Table 1 that for both fasthand groups the differences between right and left hand were significant at $P<0.001$, both for the RVF and for the LVF, and were in the expected direction. It can further be seen that the VHF differences also all go in the predicted directions. For the fasthand left group the VHF differences are significant

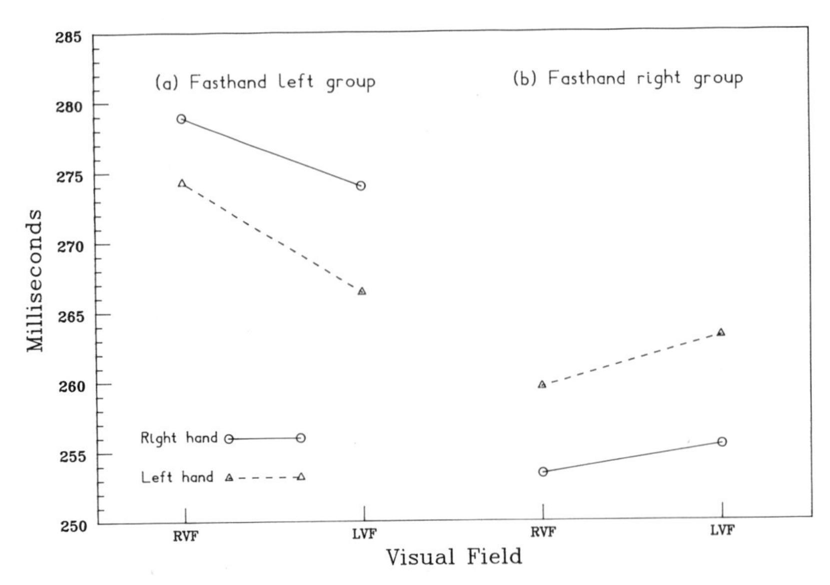

Fig. 4. The interaction between visual half field, response hand and fasthand (F=4.01, Df=1,18, P<0.061).

at P<0.001, both for the right and left hands tested separately. But for the smaller fasthand right group this reduced to P<0.02 for the left hand, and the right hand VHF comparison was not significant.

Two cells in Table 1 have larger differences than any other within subject comparison. These compare right hand in the RVF to left hand in the LVF, separately for each fasthand group. For both groups, the level of significance is considerably greater than the value of P<0.001 tabled. According to the assumptions of IHTT studies these comparisons should not be significant because both involve uncrossed relationships, but according to the predictions of the present experiment they should be larger than any other within group comparison because they compare a mean from a condition which requires no inter-hemispheric transfer to a condition which requires a transfer for detection and a transfer back for manual response.

Analyses including Sighting Dominance. The analyses of the preceding section were repeated with the inclusion of sighting dominance as a grouping variable. Sighting dominance was not significant as a main effect but approached significance in interaction with visual field (F=3.45, Df=1,16, P<0.082). All the analyses reported were also made on untrimmed logtransformed mean RTs and although this produced the same results as earlier, for this particular interaction there was an increase in significance to P<0.039, and there was a mean LVF advantage attributable to sighting dominance of 5.99 msec. for the group dominant with the left eye, and a mean RVF advantage of 2.90 msec. for the group dominant with the right eye. Also, when analysis was made of the logtransformed individual RTs here again the sighting dominance x visual field interaction was significant (P<0.04). Finally, for an analysis grouping by sex and sighting dominance the interaction was significant (P<0.029) and also when the 6 pilot Ss were introduced (P<0.019).

Exploratory factor analysis. Further descriptive, rather than hypothesis testing, analyses were made aimed partly at parsimonious description of the relationships already described, and partly at the general issue of measurement in lateralization studies (Marshall, Caplan and Holmes, 1975; Repp, 1977; Bryden and Sprott, 1981; Richardson, 1976).

Table 1. Individual comparisons for the visual field (RVF or LVF) x response hand (right or left) x response group (right or left hand faster in the bilateral condition) interaction from the analysis of data for the unilateral condition (F=4.01, Df=1,18, P<0.06).

	FASTHAND RIGHT (N=7)				FASTHAND LEFT (N=13)			
	RVF		LVF		RVF		LVF	
	right hand	left hand	right hand	left hand	right hand	left hand	right hand	left hand
	(1)	(2)	(3)	(4)	(5)	(6)	(7)	(8)
MEAN	253.41	259.56	255.39	263.70	278.89	274.31	274.37	266.45
(1)	-							
(2)	-6.15**	-						
(3)	-1.98	4.17*	-					
(4)	-10.29**	-4.15*	-8.31**	-				
(5)					-			
(6)					4.58**	-		
(7)					4.52**	0.06	-	
(8)					12.44**	7.86**	7.92**	-

* = P<0.02; ** = P<0.001, one-tailed tests

The RT data for the three visual field positions (RVF, LVF and both VF) for each hand; the difference between hand scores, and the three indicators of sighting dominance were factor analysed. Sex and age were also included in this analysis. Factor extraction by the principal components method followed by orthogonal rotation produced four distinct factors together accounting for 90.9% of the variance. Communalities were high for all variables, confirming the relevance of the variables included to the analysis made. Figure 5 illustrates the absolute values of the correlations in sorted shaded form. From these the constitution and the distinctiveness of the factors can be seen.

The first factor, which took up 50% of the variation loaded heavily on the six RT variables, with negligible contribution from other variables. This would appear to be a plain RT factor representing response speed. The second factor (about 18.5% of the variation) loaded heavily on the three hand difference variables with a small contribution from sex, and more importantly, very little from the sighting dominance measures. It represents the difference between right and left hand in response time, and confirms that all three image representation positions provide almost completely the same information for the hand difference measure. The third factor (14.7% variation) represents sighting dominance. In a preliminary analysis hand preference also loaded well on this third factor, but a larger sample with more left handed Ss would be needed for inclusion of this variable to be meaningful. The fourth factor (8% variation) loaded heavily on age and sex alone. The conjunction of these two variables would seem to

Variables

5 LVFL	■													
3 RVFL	■	■												
4 LVFR	■	■	■											
2 RVFR	■	■	■	■										
7 BOTHL	■	■	■	■	■									
6 BOTHR	■	■	■	■	■	■								
8 RVFDIF	.	.		-			■							
10 BOTHDIF			.	.		.	■	■						
9 LVFDIF			-	.		-	■	■	■					
28 EYEGUN									.	■				
25 EYEKAL				.			-	-	-	■	■			
21 EYECARD	.		.	.	.	.				■	■	■		
16 AGE	.	-	-	-	.	.	.	-	-		.		■	
17 SEX								-	+	+	-	.	X	■

The absolute values of the matrix entries have been shown above in shaded form according to the following scheme:

	less than or equal to	0.123
.	0.123 to and including	0.246
-	0.246 to and including	0.369
+	0.369 to and including	0.492
X	0.492 to and including	0.615
■	0.738 to and including	0.861
■	greater than	0.861

KEY: LVFL = left visual field, left-hand response, RVFR = right visual field, right-hand response, etc. DIF = difference; EYEGUN, EYEKAL, EYECARD = eye used for aligning gun, or card, or for looking into kaleidoscope.

Fig. 5. In this table the factor structure is illustrated by representing the absolute values of the correlations between variables in sorted and shaded form.

have been fortuitous, since although in a one-way analysis of variance the males and females did not differ significantly on age, there were more females than males at the upper end of the six months age range of the sample. The interesting point was that sex could largely be separated into a factor orthogonal to those representing RT and hand differences.

A similar analysis including the 6 pilot Ss produced the same 4 initial factors, loading on the same variables, and accounting for 46%, 23%, 13% and 7% of the variance respectively - a total of 90%. There were similar findings for a maximum likelihood factor analysis followed by both orthogonal and oblique rotations, and for analysis using Kaiser's second generation little jiffy (Kaiser, 1970) with orthoblique rotation. The loadings from these methods differed in no important way from those of the orthogonal methods. Mostly correlation was introduced between factor 4 and the other factors, but also between factor 2 (hand differences) and factor 3 (sighting dominance) and between factors 1 (reaction speed) and 2 (hand difference).

Analysis of Factor Scores. Two-way analysis of variance of the form fasthand (right or left) x eye dominance (right or left) were made of the factor scores. For the first factor there were no significant main effects for either fasthand or sighting dominance and the interaction was not significant, which is consistent with the interpretation of this factor as representing general speed of RT. For the second factor, representing differences between left and right hand in response time, there was a highly significant effect on fasthand ($F=41.36$, $Df=1,16$, $P<0.0001$). There was also a significant interaction between this variable and sighting dominance ($F=4.85$, $Df=1,16$, $P<0.042$), hinting at construct validity problems in the sighting dominance measure connected possibly with hand preference. (It seems likely that a kaleidoscope will be put to the eye on the same body side as the hand which picks it up). For the third factor, loading largely on the sighting dominance indicators, the only significant effect was for sighting dominance ($F=93.01$, $Df=1,16$, $P<0.0001$); and for the fourth factor, as would be expected, there were no significant effects.

The question of interest was whether the factors gave a clearer definition of visual differences than the initial unfactored variables, and that was why the visual field differences were not included in the set of variables factor analysed. Canonical correlation established that there was a high level of dependence between the 4 factors and the right and left hand VHF differences considered jointly, and that about 59% of the difference score variation could be expressed by the first canonical variable extracted (Chi-square=22.34, $Df=6$, $P<0.0001$). However this loaded more or less equally on all three factors, and the second variable extracted, which loaded mostly on the hand difference factor, did not provide significant prediction.

In a multivariate regression analysis, the VF difference scores were predicted from the first three factors. The results differed between hands. For the left hand VF differences the RT speed factor approached significance ($t=1.91$, $Df=22$, $P<0.07$), the hand difference factor predicted significantly ($t=2.62$, $Df=22$, $P<0.01$), as did sighting dominance ($t=3.14$, $Df=22$, $P<0.005$). The results were in the same direction in the right hand data, but only the contribution from sighting dominance was significant.

The analyses presented in this report have been weakened by the procedure of applying separate analyses of variance to the data studied in piecemeal fashion. This, however, has allowed exploration of complex effects more directly than would have been possible in one global analysis where the effects of interest would have appeared as complex interactions. It is also in accord with the basic design of the work as two separate experiments run in an integrated framework. The analyses reported included two subjects who were lefthanded. Each analysis was repeated with these subjects excluded with no effect on the findings except to reduce the degrees of freedom and thus to a small extent probabilities.

DISCUSSION

In the experiment reported, a plain circular light source subtending approximately one degree of visual angle was used as a stimulus on all trials, and on all trials Ss responded by raising both hands as quickly as possible from micro-switches built into the table top. The treatment of the experiment was that the light was randomly presented either to the RVF or to the LVF. Randomly interspersed with this treatment was a further condition in which the stimulus was duplicated and presented simultaneously in both visual half fields. This bilateral presentation condition, in which stimulus and response parameters were balanced, provided control information to aid the analysis and interpretation of results for the unilateral condition.

To place the experiment in the context of previous work using reaction times in divided VF studies of cerebral lateralization the data were first analysed without taking into account the information from the bilateral presentation condition, except to establish that fixation had been maintained.

In these analyses the prediction that response would be faster when initial projection of the stimulus was to the right hemisphere failed in the data overall, as did the prediction that response would be faster with the left hand than the right hand. However this was due to a cancelling of differences between male and female subjects. Both predictions held for males, but in neither case was there a significant difference for females. Female Ss were, however, significantly faster than males when initial projection was to the left, language dominant, hemisphere but were not faster when initial projection was to the right hemisphere.

Taken at face value these results appear to be yet another addition to the inconsistent findings reviewed in the introduction. Thus there was a VHF difference for detection, following Allard and Bryden (1979) but not Kimura (1969); and a right hemisphere advantage was shown by males but not by females, a finding in the same direction to that of Kimura (1969) but opposite to Allard and Bryden (1979). It now seems clear that the only valid conclusion from studies such as these is that different results will be produced on different occasions, and that replication with improved procedures and analyses such as undertaken by Bryden and Allard will not in itself be sufficient to resolve inconsistency. A possible reason for this is suggested by Bryden's 1976 report of no VHF difference for detection or localisation in 4 replications of Kimura's experiments, yet in the individual data 88% of subjects were faster in one VHF than in the other. Of these 60% were faster in the LVF and 40% were faster in the RVF. Bryden interpreted this as corroborating the finding from the analysis of variance made on mean data of no significant differences between visual fields. But a more plausible interpretation would have been that the individual differences were evidence of an uncontrolled effect confounded in the mean scores, and their distribution should have at least been inspected. It is unfortunate that Allard and Bryden also failed to provide information on the individual distribution of error counts in their 1979 data.

Jeeves and Dixon (1970) also mention, but do not describe or take into account, individual differences inconsistent with the mean scores on which their analyses were made. The point is that some inconsistent subjects might be expected as a consequence of measurement or experimental error, but overall the difference RTs should be normally distributed with the inconsistent subjects in the tail of a distribution which crosses zero difference. The fact that these authors test their results with a distribution-free test suggests that perhaps this was not the case. They propose that the motor responding area in the left hemisphere is faster at initiating a response than the corresponding area in the right hemisphere, since in their mean scores LVF/left hand response was slower than for LVF/right hand, i.e., the condition held to require a callosal cross was faster than the condition which did not. This interpretation overlooks the possibility of confounded individual differences within the mean RTs. Jeeves (1972) repeated this work with children; once again there was inconsistency in the individual data, and on this occasion also in the mean scores. No information is given on the distribution of the difference RTs and the possibility that the individual inconsistencies affect <u>all</u> the mean scores is not considered. It seems possible that the direction of the mean scores was an artifact of group composition and would depend on the incomplete balancing of confounded individual difference and not on cerebral processing or pathway differences.

The bilateral presentation condition was introduced into the design in an attempt to provide a method of predicting in which direction the results would go. This method was suggested by previous work (in preparation); but there also seemed to be evidence in the early SRT literature of unnoticed hemispheric differences when response was bimanual. The use of simultaneous response with both right and left hand has been common in RT studies, but usually only the average time for both hands has been analysed in the mistaken belief that this provided control - which it did to some extent, but at the cost of confounding central processing differences.

As has been seen, one study which did separately report right and left hand RTs was that of Metfessel and Warren (1934). Their finding was that subjects responded on average faster with the non-preferred than with the preferred hand, and they accounted for this by supposing that Ss asked to respond as quickly as possible with both hands "overcompensated" with the non-preferred hand in an attempt to produce simultaneous movement. However re-analysis of their data (see Introduction) established that the hand difference scores were distinctively bimodal in distribution. For example the 21 lefthanded subjects were not all faster with their right hands, nor was there a normal distribution of difference scores centered on one side from zero, to represent an advantage on the part of one hand (the right) with one tail crossing zero in the direction of a left hand advantage for some subjects, as would be expected if "overcompensation" were biasing a preferred-hand advantage. Rather 11 of the 21 lefthanded subjects were faster with the right hand, but 10 were faster with the left hand, and the difference scores were distributed as if from two separate normal distributions either side of zero with means of 6.71 and -6.98 msec. respectively with standard errors of less than 1.6 msec. in each case.

The prediction that the data for the bilateral component of the present experiment would follow a similar bimodal distribution held. Distinct separation of Ss into two groups can be seen in Figure 2. The mean difference scores of 9.99 and -5.85 msec. are each separated by more than 4 standard errors from zero in opposite directions, and are comparable in size to those found in the data of Metfessel and Warren (1934).

It had been proposed that this effect, if found, would be consistent with hemispheric processing differences; and it was predicted that if subjects were separated into two groups, those faster with the right hand and those faster with the left hand to bilaterally presented stimuli, then the former group would respond faster in the LVF and the latter group in the RVF to unilateral presentation of the same stimuli. In the event 65% of the sample were faster with the left hand and 35% with the right hand, and when this classification was taken into account in the analysis of the unilateral data the results were as predicted. The group which was faster in the bilateral condition with the right hand responded faster to LVF presentation than to RVF presentation in the unilateral condition, whereas the group which was faster in the bilateral condition with the left hand responded faster to RVF presentation than to LVF presentation in the unilateral condition. These findings applied equally to males and females.

The results for these important findings were presented in detail in Table 1 and some comment on the significance levels is required. Firstly, the overall effect in the analysis of variance reached only $P<0.06$, but this need cause no reservation as the paired comparisons and their direction were predicted (Winer, 1971). Twelve comparisons in Table 1 were relevant to the predictions of the experiment, and although this increased the possibility of a type 1 error this seemed preferable to an increase of risk in the opposite direction, by use of a conservative post hoc method. The use of Scheffe´s method would have been indefensible considering the large number of paired and grouped comparisons of no interest. Tukey´s HSD method may have been a preferable alternative, but even so 16 of the possible paired comparisons were not relevant. A cautious reader will not search for alternative methods of testing comparisons; but might tentatively hold as interesting the effects which can be seen in Table 1, pending replication.

The VHF results as listed in Table 1 are all in the predicted direction and for Ss who were faster with the left hand in the bilateral condition are highly significant ($P<0.001$ or less), both for the right hand and for the left hand. This is strong evidence for right hemisphere processing, and for consistency between hand difference and VHF difference as measures of functional cerebral asymmetry. The results are also in the expected direction for Ss who were faster with the right hand in the bilateral condition; but although the VHF difference for left hand response at 4.14 milliseconds was similar in size to the highly significant difference for the other group, the greater variability for the smaller group reduced significance to $P<0.02$; and for the right hand the VHF difference, although in the predicted direction, was not significant.

The 4 cells in Table 1 where both VHF and hand are varied are of special interest. These are RVF right hand compared to LVF left hand; and RVF left hand compared to LVF right hand, separately for each of the two response groups. If the assumptions quoted from Rizzolatti (1979) are correct then there should have been no significant difference for any of these 4 comparisons. This is because the first compares uncrossed pathways via one visual field to uncrossed pathways via the other visual field, and the second compares crossed pathways by one visual field to crossed pathways by the other visual field, i.e., in both cases both sides of the comparison have the same advantages and disadvantages. But if the interpretation offered for the present findings is correct, then in the case of the fasthand right group the RVF right hand/LVF left hand comparison will have one component requiring no crosses (the first) and one component requiring a cross for detection plus a cross (or use of the less efficient ipsilateral motor pathways) for response and accordingly the difference should be twice the estimated IHTT for that group. From Table 1 it can be seen that the difference of 10.29 msec. was indeed very close to twice the time estimated as the mean for the other 4 conditions (5.15 msec.). For the fasthand left group the same comparison will also involve one condition with no crosses (LVF, left hand) and one with two crosses (RVF, right hand) and here the difference score (12.44 msec.) is precisely twice the IHTT as estimated by the mean for the other 4 comparisons for the same group. The remaining two cells for within response group comparisons in Table 1 are not predictable. For both components there is a single cross, but for one component this is for detection whereas for the other it is for response. The outcome consequently is uncertain, and in the event for the larger group there was no difference, but for the smaller group the condition when the cross was for detection was significantly faster than when the cross was for response.

The research reviewed in the introduction on estimating IHTT needs very little further discussion. If the present findings are correct then the assumptions on which this work is based are defective, and the times given for example by Berlucchi et al. (1971, 1977), Rizzolatti et al. (1971, 1979) and Anzola et al. (1980) using the SRT paradigm are not estimates of IHTT but are small residual differences left after the balancing of various confounding effects. This in itself requires no further discussion in the present context since it is a conclusion which can be drawn by anyone who looks closely at these researchers own published findings.

The present findings do provide estimates of IHTT where at least relevant processing differences are not confounded, but have also shown that sighting dominance should be taken into account, and group size was too small for this to be usefully done. There is, however, a fair measure of consistency, and IHTT estimates based on hand differences were 6.15, 8.31, 4.58 and 7.92 msec. giving a mean of 6.74 msec. which compares closely with the results extracted by a similarly regrouped analysis of Metfessel and Warren's (1934) data. Except for one non-significant value the estimate based on VHF comparisons with response hand held constant, including the non-significant value, are 1.98, 4.15, 4.52 and 7.86 msec. Thus the mean for the estimates from the hand difference comparisons is 6.74, and the mean for the estimates from the VHF comparisons is 4.63. The overall mean from the estimates from both sources is 5.68, and if the weighted times for the comparisons with two IHTT crosses are introduced then a mean rounding to 5.68 msec. is produced as the best estimate of IHTT using the present data, with the important qualification that this estimate is known to include significant uncontrolled bias due to sighting dominance and possibly also from a preferred hand practice effect to be discussed later. Also, there is no reason to suppose that results for 9-year-old children either will or will not, match those for adult subjects.

Very little need be said about the additional analyses presented in the results section. These tested alternate explanations for the findings, and a certain amount of discussion has been necessary in their presentation. Subjects were sampled randomly from their school year groups to avoid the possibility of reactivity from sampling procedures or overt selection of any sort apparent to teachers or children. This led to the testing of two left-handed children, but it was shown that inclusion or deletion of these Ss in no way affected the findings. Since a method was being used to infer processing hemisphere, any relationship of this with handedness might not have seemed important. However, any practice advantage for the preferred hand was not been controlled.

The inclusion of a sighting dominance measure was important. Although sighting dominance was not significant as a main effect and did not influence the VHF interaction with response group and response hand, there were other grounds for concluding that visual dominance would need to be taken into account in a more finely tuned experiment. The additional analyses established that visual dominance was significant in interaction with other variables in a changing pattern of responses across the 120 RTs for each subject. It was clearly separated from the hand difference variables and from the indicators of overall RT in the factor analysis, and when visual field difference were regressed on the factor scores the sighting dominance factor was the best predictor of these differences.

Kimura (1969) also tested sighting dominance, and reported a significant bias on the part of right-handed females, but not right-handed males, ($P<0.02$, $N=192$) towards left eye sighting dominance. An effect such as this might conceivably provide an alternative explanation of sex differences in a lateralization study using a VHF method. But the value of Kimura's finding has to be balanced by the fact that the relationship accounted for only 2.7% of the variation and also against the lack of significance in any of the 7 data sets pooled to give the sample of 192 (calculated from the published results). There were no sex differences by sighting dominance in the overall analyses made of the present data, nor were the frequencies by sex significant but sample size was small. About 58% of the sample were right eye dominant; 42% were left eye dominant, and none were classified as mixed. These figures are within the range of values for eleven studies reviewed by Porac and Coren (1976), but on average show a lower incidence of right dominance contributed mostly by the male subjects.

From the mean scores it could be seen that the general direction of the visual dominance effect was for faster response when presentation was to the visual field on the same body side as the dominant eye, i.e., when the image was presented to the nasal hemiretina of the dominant eye and to the temporal hemiretina of the non-dominant eye. This observation is speculative, and is based on inspection of mean scores and not on significance testing, but it suggests that it would be wise to control both sighting dominance and use of the crossed and uncrossed visual pathways if IHTT is to be reliably estimated by a lateralization study. Young (1982) has remarked that although it was once customary to measure eye dominance of Ss in VHF studies, it was now accepted that this was of no use as an indicator of cerebral organisation. However the importance of sighting dominance in the present context is not whether it is a useful indicator of cerebral lateralization, but whether or not it is confounded with the VHF differences used to operationalise predicted differences in cerebral organisation. Bryden (1982) has suggested that the relationship between sighting dominance and lateralization should be examined in a dichotic listening task, since in such a task sighting dominance would not have a peripheral sensory effect, and any relation observed would be a truer assessment of the relation between sighting dominance and cerebral specialisation. However, if no effect were found this would say nothing about the influence, if any, of sighting dominance on the results of VHF studies. The only way to gain such information would be by careful VHF experiments with sighting dominance, for one thing, controlled.

It was notable that the introduction of the fasthand factor into the analysis of variance removed the significant sex differences. This in itself supports the interpretation of the hand difference measure as an indicator of hemispheric differences, since it seems improbable to suppose that males and females differ as to which side of the body produces the fastest movement quite independently from central processing differences. Put another way, one interpretation of the findings could be that some people are faster with their right hands and others are faster with their left hands for reasons which have nothing to do with VHF or hemispheric differences. It is hard to see what these reasons could be other than hand preference, which could not be applicable in the present case. It is even harder to see what reasons could be given compatible with the finding of sex differences, apart from strategy/processing differences.

The procedure followed removed the sex differences from the analysis of variance and showed that if Ss were grouped in terms of a proposed indicator of hemispheric asymmetry for the set task then it was this and not sex which accounted for visual field differences. However this merely moved the sex differences back one level. A substantial majority of the males was faster with the left hand, suggesting right hemisphere processing, whereas the females were more evenly balanced. Kimura's (1969) suggestion that females would be more likely to use a verbal, left hemisphere, procedure and males a spatial, right hemisphere procedure (see also Witelson, 1976; McGlone, 1980) rested on, as has been shown, very shaky ground but is supported by the present findings. The apparent sex and VHF interactions in the initial analyses are a consequence of this difference. Marshall (1973) on the basis of a review concluded that many reports of sex differences in hemispheric specialisation could be accounted for as differences in strategy preference. Segalowitz and Stewart (1979) have suggested that in a letter matching paradigm sex differences might be accounted for as strategy differences, males tending to use nonverbal strategies and females being more variable in their use of either a verbal or nonverbal strategy. It is one thing to find that males and females choose, possibly as set by unnoticed situational influence, prior experience, or experimenter reactivity, different task strategies, either linguistic or non-linguistic, and quite another to propose that there are functional brain differences between males and females. Further when group size is small a significant lack of balance on an uncontrolled difference is not unlikely, and could go by chance in either direction. Allard and Bryden (1979) provide one instance when a sex difference for a comparable task is in the opposite direction to the present finding. It remains to be seen whether or not further use of the present design and analyses will show that sex differences in response group composition consistently go in one direction or not.

CONCLUSION

In conclusion, it has been shown that even in a SRT procedure in which Ss made identical response on every trial to unilateral and bilateral presentation of a dot of light, there are VHF differences which are confounded, unless Ss are grouped by response strategy. It has been shown that sex differences in VHF effects evaporate when response strategy is taken into account; or rather that the VHF differences reduce to strategy differences, which is something different as these seem more likely to have an experimental rather than a structural origin. An approach for correcting deficiencies in IHTT studies has been outlined, and it has been argued that none of the previous work on IHTT is satisfactory. It is suggested that if a SRT procedure is to be used for reasonably accurate estimations of IHTT then not only hand preference, sighting dominance, and hemiretina of stimulus presentation will need to be controlled, but also response strategy - however difficult that may be. The alternative is that individual hemispheric differences will be confounded, either to the extent of cancelling all evidence of VHF differences, or to the extent of throwing VHF differences in one direction or the other as an artifact of subgroup composition. It also seems clear that the habit of collecting a large amount of data from individual subjects may well provide narrower confidence limits for the means estimated, but the question has to be asked, means of what? Control of response strategy is more likely to be achieved in small amounts of data from many subjects, since the likelihood of a change in strategy on the part of individual Ss will be reduced. Finally the experiment has suggested an important method for the operationalisation of predicted asymmetries in hemispheric function. This involves the analysis of hand difference RTs since the direction of these will vary in relation to the hemisphere not of initial projection of the visual stimulus but to the hemisphere in which the response command is initiated, and this could well be a better indicator of processing hemisphere than VHF differences. Certainly if this technique can be reliably developed many interesting possibilities for lateralization experiments and their analysis will follow.

REFERENCES

Allard, F. and Bryden, M. (1979). The effect of concurrent activity on hemispheric asymmetries. Cortex, 15, 5-17.

Andrews, D.F., Bickel, P.J., Hampel, F.R., Huber, P.J., Rogers, W.H. and Tuckey, J.W. (1972). Robust Estimates of Location: Survey and Advances. Princeton: Princeton UP.

Anzola, G.P., Bertoloni, G., Buchtel, H.A., Rizzolatti, G. (1977). Spatial compatibility and anatomical factors in simple and choice reaction times. Neuropsychologia, 15, 295-302.

Bashmore, T.R. (1981). Vocal and manual reaction time estimates of interhemispheric transmission time. Psychological Bulletin, 89, 352-368.

Berlucchi, G. (1974). Cerebral dominance and interhemispheric communication in normal man. In Smith, F.O. and Worden, F.G. (Eds.) The Neurosciences: Third study program. Cambridge, Mass: MIT Press.

Berlucchi, G., Crea, F., Di Stefano, M. and Tassinari, G. (1977). The influence of spatial stimulus-response compatibility on reaction time of ipsilateral and contralateral hand to lateralized light stimuli. Journal of Experimental Psychology: Human Perception and Performance, 3, 505-517.

Berlucchi, G., Heron, W., Hyman, R., Rizzolatti, G. and Umilta, C. (1971). Simple reaction times of ipsilateral and contralateral hand to

lateralized visual stimuli. Brain, 94, 419-430.
Berlucchi, C., Marzi, C., Rizzolatti, G. and Umilta, C. (1976). Functional hemispheric asymmetries in normals: influence of sex and practice. Paper presented at the XII International Congress of Psychology, Paris.
Bertera, J.H., Callan, J.R., Parsons, O.A. and Pishkin, V. (1975). Lateral stimulus-response compatibility effects in the oculomotor system. Acta Psychologica, 39, 175-181.
Bisiach, E., Mini, M., Sterzi, R. and Vallar, G. (1982). Hemispheric lateralization of the decisional stage in choice reaction times to visual unstructured stimuli. Cortex, 18, 191-198.
Bowers, D. and Heilman, K.M. (1980). Pseudoneglect: effects of hemispace on a tactile line bisection task. Neuropsychologia, 18, 491-498.
Bradshaw, J.J. and Nettleton, N.C. (1983). Human Cerebral Asymmetry. Englewood Cliffs, NJ: Prentice Hall.
Bradshaw, J.J., Nettleton, N. and Spehr, K. (1982). Sinistral inverters do not possess an anomalous visuomotor organisation. Neuropsychologia, 20, 605-609.
Bradshaw, J.L. and Perriment, A.D. (1970). Laterality effects and choice reaction time in a unimanual two-finger task. Perception and Psychophysics, 7, 185-187.
Brebner, J. (1973). S-R compatibility and changes in RT with practice. Acta Psychologica, 37, 93-106.
Brebner, J., Shephard, M. and Cairney, P. (1972). Spatial relationships and S-R compatibility. Acta Psychologica, 36, 1-15.
Broadbent, D.E. (1974). Division of function and integration of behavior. In Smith, F.O. and Worden, F.G. (Eds.) The Neurosciences: Third Study Program. Cambridge, Mass: MIT Press.
Broadbent, D.E. and Gregory, M. (1962). Donders' B- and C-reactions and S-R compatibility. Journal of Experimental Psychology, 6, 575-578.
Broadbent, D.E. and Gregory, M. (1965). On the interaction of S-R compatibility with other variables affecting reaction time. British Journal of Psychology, 56, 61-67.
Bryden, M. (1973). Perceptual asymmetry in vision: relation to handedness, eyedness and speech lateralization. Cortex, 9, 419-435.
Bryden, M. (1976). Response bias and hemispheric differences in dot localisation. Perception and Psychophysics, 19, 23-28.
Bryden, M.P. (1982). Laterality: Functional Asymmetry in the Intact Brain. New York: Academic Press.
Bryden, M.P. and Sprott, D.A. (1981). Statistical determination of degree of laterality. Neuropsychologica, 19, 571-581.
Callan, J., Klisz, D. and Parsons, O.A. (1974). Strength of auditory stimulus-response compatibility as a function of task complexity. Journal of Experimental Psychology, 102, 1039-1045.
Cook, T.W. (1934). Studies in cross education. Journal of Experimental Psychology, 17, 749-762.
Coren, S. and Kaplan, C.P. (1973). Patterns of ocular dominance. American Journal of Optometry, 50, 283-292.
Cotton, B., Tzeng, O.J.L. and Hardyck, C. (1980). Role of cerebral hemispheric processing in the visual half-field stimulus-response compatibility effect. Journal of Experimental Psychology, 6, 13-23.
Craft, J.L. and Simon, J.R. (1970). Processing symbolic information from a visual display: Interference from an irrelevant directional cue. Journal of Experimental Psychology, 83, 415-420.
Crider, B. (1944). A battery of tests for the dominant eye. Journal of General Psychology, 31, 179-190.
Davidoff, J. (1977). Hemispheric differences in dot detection. Cortex, 13, 434-444.
Di Stefano, M., Morelli, M., Marzi, C.A. and Berlucchi, G. (1980). Hemispheric control of unilateral and bilateral movements of proximal and distal parts of the arm as inferred from simple reaction time to lateralized light stimuli in man. Experimental Brain Research, 38, 197-204.

Dixon, W.J. (1983). Biochemical Computer Programs. Berkeley: UCP.
Donders, F.C. (1969). On the speed of mental processes. Acta Psychologia, 30, Attention and Performance 11, 412-431. (Translation of Donders, 1868).
Donders, F.C. (1868). Over de snelheid van psychische processen. Nederlandsch Archief voor Genees - en Naruurkunde, 4, 117-145.
Filbey, R.A. and Gazzaniga, M.S. (1969). Splitting the normal brain with reaction time. Psychonomic Science, 17, 335-336.
Fitts, P.M. and Seeger, C.M. (1953). S-R compatibility: spatial characteristics of stimulus and response codes. Journal of Experimental Psychology, 46, 199-210.
Harvey, L.O. (1978). Single representation of the visual midline in humans. Neuropsychologia, 17, 653-660.
Hellebrandt, F.A. (1951). Cross education: ipsilateral and contralateral effects of unimanual training. Journal of Applied Physiology, 4, 136-144.
Herlitzka, A. (1908). Ricerche cronografiche sui movimenti volontari bilaterali. Archivio di Fisiologia, 5, 277-284.
Hongo, T., Nakamura, R., Narabayashi, H., and Oshima, T. (1976). Reaction times and their left to right differences in bilateral symmetrical movements. Physiology and Behaviour, 16, 477-482.
Jeeves, M.A. (1969). A comparison of interhemispheric transmission times in acallosals and normals. Psychonomic Science, 16, 245-246.
Jeeves, M.A. (1972). Hemispheric differences in response rates to visual stimuli in children. Psychonomic Science, 27, 200-203.
Jeeves, M.A. and Dixon, N.F. (1970). Hemisphere differences in response rates to visual stimuli. Psychoneurological Science, 20, 249.
Johnson, H.M. (1923). Reaction time measurements. Psychological Bulletin, 20, 562-589.
Kaiser, H.F. (1970). A second generation little jiffy. Psychometrica, 35, 401-415.
Kerr, B. (1978). Task factors that influence selection and preparation for voluntary movements. In G.E. Stelmach, (Ed.) Information Processing in Motor Control and Learning, New York: Academic Press. 55-67
Kerr, M., Mingay, R. and Elithorn, A. (1963). Cerebral dominance in reaction time responses. British Journal of Psychology, 54, 325-336.
Kimura, D. (1969). Spatial localisation in the left and right visual fields. Canadian Journal of Psychology, 23, 445-458.
Kinsbourne, M. (1970). The cerebral basis of asymmetries in attention. Acta Psychologica, 33, 193-201.
Kristeva, R., Keller, E., Deecke, L. and Kornhuber, H.H. (1979). Cerebral potentials preceding unilateral and simultaneous bilateral finger movements. Electroencephalography and Clinical Neurophysiology, 47, 229-238.
Levy, J. and Reid, M. (1978). Variations in cerebral organization as a function of handedness, hand posture in writing and sex. Journal of Experimental Psychology: General, 2, 119-144.
Marshall, J.C. (1973). Some problems and paradoxes associated with recent accounts of hemispheric specialisation. Neuropsychologia, 11, 463-470.
Marshall, J.C., Caplan, D., and Holmes, J.M. (1975). The measure of laterality. Neuropsychologia, 13, 315-321.
McGlone, J. (1980). Sex differences in human brain asymmetry: a critical review. The Behavioural and Brain Sciences, 3, 215-263.
McKeever, W.F. and Hoff, A. (1979). Evidence of a possible isolation of left hemisphere visual and motor areas in sinistrals employing an inverted handwritting posture. Neuropsychologia, 17, 445-455.
McKeever, W.F., and Huling, M.D. (1970). Left-cerebral hemisphere superiority in tachistoscopic word-recognition performances. Perceptual and Motor Skills, 30, 763-766.

McKeever, W.F., Gill, K.M. and Van Deventer, A.D. (1975). Letter versus dot stimuli as tools for splitting the normal brain with reaction time. Quarterly Journal of Experimental Psychology, 27, 363-374.
Metfessel, M. and Warren, Neil D. (1934). Over-compensation by the non-preferred hand in an action-current study of simultaneous movements of the fingers. Experimental Psychology, 17, 246-256.
Nakamura, R. and Saito, H. (1974). Preferred hand and reaction time different movement patterns. Perceptual and Motor Skills, 39, 1275-1281.
Pitblado, C. (1979a). Visual field differences in perception of the vertical with and without a visible frame of reference. Neuropsychologia, 17, 381-392.
Pitablado, C. (1979b). Cerebral asymmetries in random-dot stereopsis: reversal of direction with change in dot size. Perception, 8, 683-690.
Poffenberger, A.T. (1912). Reaction time to retinal stimulation with special reference to effects of hemispace on a tactile line bisection task. Neuropsychologia, 18, 491-498.
Pohl, W., Butters, N. and Goodglass, H. (1972). Spatial discrimination systems and cerebral lateralisation. Cortex, 8, 305-314.
Porac, C. and Coren, S. (1976). The dominant eye. Psychological Bulletin, 83, 880-897.
Rabbit, P.M. (1967). Signal discriminability, S-R compatibility, and choice reaction time. Psychonomic Science, 7, 419-420.
Repp, B.H. (1977). Measuring laterality effects in dichotic listening. Journal of the Acoustical Society of America, 62, 720-737.
Rizzolatti, G. (1979). Interfield differences in reaction times to lateralised visual stimuli in normal subjects In I. Steele Russell, M.W. Hof and G. Berlucchi (Eds.) Structure and Function of Cerebral Commissures, London: Macmillan. 390-399.
Rizzolatti, G., Bertoloni, G. and Buchtel, H. (1979). Interference of concomitant motor and verbal tasks on simple reaction time: a hemispheric difference. Neuropsychologia, 17, 323-330.
Rizzolatti, G. and Buchtel, H. (1977). Hemispheric superiority in reaction time to faces: a sex difference. Cortex, 13, 300-305.
Rizzolatti, G., Umilta, C. and Berlucchi, G. (1971). Opposite superiorities of the right and left cerebral hemispheres in discriminative reaction time to physiognomical and alphabetical material. Brain, 94, 431-442.
Rolls, E.T. (1983). The initiation of movements. Experimental Brain Research, Suppl., 7, 97-1130.
Salow, P. (1913). Untersuchungen zu uni- und bilateralen Reaktionen Eigene Versuche am Chronographen. Psychol. Studieren, volume 8, 506-540.
SAS Institute Inc., User's Guide, 1982 Edition. Cary, NC: SAS, NC.
Segalowitz, S.J. and Stewart, C. (1979). Left and right lateralization for letter matching: strategy and sex differences. Neuropsychologia, 17, 521-525.
Simon, J.R. (1968). Effect of ear stimulated on reaction time and movement time. Journal of Experimental Psychology, 78, 344-346.
Swadlow, H.A., Geschwind, N. and Waxman, S.G. (1979). Commissural transmission in humans. Science, 204, 530-531.
Swanson, J., Ledlow, A. and Kinsbourne, M. (1978). Lateral asymmetries revealed by simple reaction time. In M. Kinsbourne, (Ed) Asymmetrical Function of the Brain, Cambridge: CUP, 274-291.
Teichner, W.H. (1954). Recent studies of simple reaction time. Psychological Bulletin, 51, 128-149.
Terashi, Y. (1920). Ricerche cronografiche sui movementi bilaterali di reazione. Giornale Accademia Mediche Torino, 83, 90-95.
Umilta, R., Rizzolatti, G., Marzi, C.A., Zamboni, G., Franzini, C., Camarda, R. and Berlucchi, G. (1974). Hemispheric differences in the

discrimination of line orientation. Neuropsychologia, 12, 165-174.
Wallace, R.J. (1971). S-R compatibility and the idea of a response code. Journal of Experimental Psychology, 88, 354-360.
Winer, B.J. (1971). Statistical Principles in Experimental Design. New York: McGraw-Hill.
Witelson, S. (1976). Sex and the single hemisphere: right hemisphere specialisation for spatial processing. Science, 193, 425-427.
Woodworth, R.S. (1938). Experimental Psychology. New York: Holt.
Young, A. (1982). Methodological theoretical bases. In J.G. Beaumont, (Ed.) Divided Visual Field Studies of Cerebral Organisation. London: Academic Press.
Zangwill, O. (1960). Cerebral Dominance and its Relation to Psychological Function. Edinburgh: Oliver Boyd.

SELF-REPORT OF NEUROPSYCHOLOGICAL DIMENSIONS OF SELF-CONTROL

Thomas R. O'Connell, *Don M. Tucker and Thomas B. Scott

Counseling Department
University of North Dakota

*Psychology Department
University of Oregon

ABSTRACT

A central notion in research on individual differences in hemispheric specialization is that some people emphasize one hemisphere's cognitive processes as a general mode of functioning. We begin this chapter by considering this construct of a lateralized cognitive style, and the methodological and theoretical problems it has raised. If the construct is to be scientifically useful, it must be supported by empirical assessment of cognitive performance - or at least cognitive strategy. For the neuropsychological approach to contribute uniquely to personality research, it will be important to relate the psychological traits to physiological measures of brain activity. In addition, an important observational method is self-report: given the appropriate questionnaire, we argue that most people can describe their cognitive and emotional functioning in ways that are meaningful to neuropsychological constructs of personality. We present preliminary data on the O'Connell Cognitive and Affective Style Scale (OCASS), a self-report scale designed to assess cognitive and emotional dimensions of self-control.

INTRODUCTION

The traditional neuropsychological study of brain lesions showed that damage to certain areas may impair specific psychological functions. The commissurotomy studies of the 1960s (Sperry, 1966; Sperry, Gazzaniga and Bogen, 1969) provided the more dramatic demonstration that an isolated hemisphere could perform the fundamental cognitive operations we consider necessary for an intelligent mind. As the characteristics of each hemisphere were described, it became clear that a hemisphere's specialization involved not just specific functions, but a general mode of approach to problem solving; the left hemisphere seemed to analyze tasks into parts, while the right hemisphere was better able to integrate parts into wholes (Bogen and Bogen, 1969; Levy, 1969). It was not necessarily the case that a hemisphere's involvement in a task would be appropriate to the task requirements. Levy's (1977) studies suggested that activation of a hemisphere during a task may or may not be appropriate to the hemisphere's competence to perform the task.

Thus, it has become apparent that hemispheric function during cognitive activity is a dynamic process, and it is insufficient to consider brain laterality only in terms of a static localization of mental abilities. The progression from being concerned with fixed abilities to considering dynamic strategies is reminiscent of the development of individual difference approaches to intelligence. Whereas the issue at the beginning of the century was differences among people in intellectual ability, by the middle of the century the popular issue had become differences in strategy or cognitive style (e.g., Witkin, Dyk, Faterson, Goodenough and Karp, 1962).

For many readers of the hemispheric specialization studies, the particular styles shown by the two hemispheres seemed intuitively appealing as descriptions of differences among people in cognitive style. The left hemisphere's analytic, verbally-mediated cognition was contrasted with the right hemisphere's more holistic, spatially-organized cognition; certainly some people seemed to draw on one of these cognitive strategies more than the other. The problem in researching how these hemispheric modes could be related to normal cognitive style was one of measurement; how is differential hemispheric function to be assessed in an intact brain?

A convenient method suggested itself in the form of eye movements to the left or right during reflective thought. Teitlebaum (1954) had suggested that a person is likely to move the eyes laterally while thinking deeply about something. Day (1961; 1964) observed that an individual's characteristic direction of movement was related to emotional and personality characteristics. A right-looker was described as experiencing anxiety from external causes, whereas a left-looker was described as experiencing anxiety in a more internal, subjective fashion.

The important link to lateral brain function was made by Bakan (1969). He proposed that a person's direction of eye movement during thought indicates use of the contralateral hemisphere; thus the more vivid imagery reported by left- than right-lookers shows a greater use of the right hemisphere by those subjects. Bakan (1969) reported that left-lookers were more likely to choose college majors in humanistic, as opposed to scientific or quantitative, fields. Similar results were found by Combs, Hoblick, Czarnecki and Kamler (1977). Bakan (1969) and Weiten and Etaugh (1974) reported left-lookers to have lower quantitative SAT scores than right-lookers. Greater usage of right than left hemisphere processing by left-lookers would be compatible with the finding by Crouch (1976) that they are more responsive to facial than verbal cues. Consistent with Day's original findings, other studies found left-lookers to be more inner-attentive (Meskin and Singer, 1974) and more artistically diverse and creative (Harnad, 1972; Hines and Martindale, 1974). On the Allport-Vernon-Lindzey Scale of Values, left-lookers scored higher on aesthetic and social scales, while right-movers scored higher on theoretical and economic scales (Weitan and Etaugh, 1973).

Bakan (1969) found that left-lookers were more hypnotizable than right-lookers. Similar results were reported by Gur and Gur (1974) and Gur and Reyher (1973). Day (1967) had reported higher amplitude, lower frequency EEGs in left-lookers. Consistent with this suggestion of a lower level of brain activation, Bakan (1969) found higher resting alpha in left- than right-lookers. Bakan and Svorad (1969) replicated the finding of higher alpha in left-lookers, and observed that higher hypnotizability was also related to higher resting alpha.

The general pattern of results in this research seemed encouraging for a hemispheric model of cognitive style. Yet, as is often the case with personality measures, replicability has often been a problem (Ehrlichman, 1977; Etaugh, 1972; Etaugh and Rose, 1973; Hartlage and Tollison, 1979).

Furthermore, the rationale for linking an eye-movement to engagement of the opposite hemisphere has been shown to be less compelling than at first thought (Ehrlichman and Weinberger, 1978). Efforts to validate lateral eye movements with measures of ability in verbal and spatial cognition were largely unsuccessful (Ehrlichman, 1977; Galin and Ornstein, 1974; Hiscock, 1977b).

There have been attempts to examine cognitive style with measures of hemispheric activation other than lateral eye movements. Galin and Ornstein (1972) evaluated the EEGs of lawyers and potters, exemplars of left and right hemispheric cognitive styles, respectively. Because there may be differences between left- and right-lookers in overall EEG activity (Bakan, 1969; Day, 1967), EEG studies must consider not only lateral asymmetries but the problems of assessing characteristic overall activation levels that have arisen in the introversion-extraversion literature (see Gale, 1973).

The validity of lateral eye movements as measures of individual differences in characteristic hemispheric activation has received some support from studies of event-related potentials (Shevrin, Smokler and Kooi, 1980) and regional cerebral blood flow (Gur and Gur, 1980). Remarkably little attention has been given the question of why activating a given hemisphere would cause subjects to move their eyes in the opposite direction. Bakan (1969) and Kinsbourne (1972) suggested that motor activity in the more activated hemisphere would overflow into the frontal eye fields. This ocular-motor synkinesia would not seem to be a particularly adaptive way for the brain to work. Another interpretation considers the function of eye movements while thinking: to avoid the visual distraction of the questioner. The direction of the eye movements is selective; a right movement, for example, shields the right visual half fields and left hemisphere from the distraction facing the subject. Considered in this way, lateral eye movement behaviour could reflect not only characteristic hemispheric balance, but an individual's susceptibility to interfering visual input (Yutrzenka, 198; Tucker, Yutrzenka and Heck, 1984).

Just as the measurement of "hemispheric activation" has proved more complex than initially thought, the theoretical aspects of this notion have also become complex. Levy (1983) has emphasized the need to consider hemispheric activation as a dynamic process in research with normal subjects as well as with commisurotomy patients. Although people may vary in their tendency to activate one hemisphere over the other characteristically, the two hemispheres are variably activated according to task demands, and this must be considered when assessing hemispheric contributions to task performance. In their research with normals, Levy, Heller, Banich and Burton (1983) assessed the tendency to activate a preferred hemisphere by measuring each subject's visual field asymmetries on a syllable-identification task. Those with a left hemisphere approach (i.e., a right field advantage) showed poorer performance when tested on a face recognition task, consistent with the proposal that these subjects show less left and greater right hemisphere activation in a variety of situations.

Caplan and Kinsbourne (1981) examined the preference of children for a verbal rather than a visual style of problem-solving. Those preferring a verbal approach were better readers, suggesting that cognitive style is related to skill development. The preference for a verbal strategy also showed a small but significant correlation with a right ear advantage, and thus left hemisphere processing, on a dichotic listening task. Caplan and Kinsbourne emphasized the importance of differentiating between brain lateralization - the stable, anatomically-based functional differences between the hemispheres - and cognitive style - the dynamic usage of the two hemispheres in problem-solving.

A similar distinction has been drawn recently by Sackeim, Weiman and Gregia (1984). They suggest that activation asymmetry, a person's tendency to engage a given hemisphere preferentially, should not be assumed to be identical with brain lateralization. They assessed LEM (Lateral Eye Movement) measures of preferred hemispheric activation in groups known to differ in brain lateralization - left handers, right-handers and those with left-handers in the immediate family. Interestingly, significantly more right LEMs were found in left-handers and those with left-handedness in the family. However, there was no simple congruence between lateralization group and LEMs, supporting Sackeim et al.'s contention that activation asymmetry, which is analogous to lateralized cognitive style in the context of the present discussion, should not be seen as identical with basic functional lateralization.

It is simple enough to think that people vary in their tendency to use one hemisphere more than the other. It is more complex to consider a hemispheric function in a dynamic sense. In addition, the evidence is growing that not only cognitive but emotional factors influence hemispheric functioning. In reviewing this evidence, Tucker (1981) concluded that in both normals and psychiatric patients certain forms of emotional arousal have been associated with asymmetries of hemispheric activation. High trait-anxious normals showed a right-ear auditory attentional bias and a detail-oriented perceptual style suggesting left hemisphere activation (Tucker, Antes, Stenslie and Barnhardt, 1978; Tyler and Tucker, 1982). For both normal students and depressed patients, changes in mood level seemed to influence the right hemisphere's function particularly (Tucker, 1981; Tucker, Stenslie, Roth and Shearer, 1981). The implication of these observations for cognitive style is that emotional factors - either traits or states - may be important determinants of asymmetric brain activation and thus hemispheric contributions to cognitive function.

Self-regulation Through Activation and Arousal

Perhaps the more important implication of these observations is that the processes regulating emotional arousal in the brain seem to be inherently asymmetric. To explore this implication, Tucker and Williamson (1984) examined the literature on the control systems determining brain activity. They drew on Pribram and McGuinness's (1975) distinction between activation and arousal systems, then went on to formulate the differential control properties of these systems by reviewing the relevant literature on the neurochemical substrates of activation and arousal in animals. The activation system augments motor readiness. It produces a tonic increase in brain activity, supporting vigilant attention. When overactivated pharmacologically in animals or humans, this system produces stereotyped, repetitive behaviour. Tucker and Williamson (1984) conclude that this reflects the _redundancy bias_ which is the elementary cybernetic mode of the activation system. In contrast, the most elementary attentional control on the perceptual systems seems to operate in a different fashion, with a phasic increment in brain activity in response to novel input. This arousal system seems to operate under a _habituation bias_, a tendency not to respond to perceptual input unless it is novel.

As McGuinness and Pribram (1980) emphasized, these are fundamentally motivational and emotional systems, yet they are essential to controlling attention during higher cognitive processing. Tucker and Williamson (1984) review evidence that the neurotransmitter substrates of activation and arousal are asymmetric in the human brain. This evidence could help explain the hemispheric asymmetries observed in the psychopathological disorders that are thought to be associated with neurotransmitter dysfunctions. Furthermore, it is possible to formulate a theoretical model of how the elementary controls of activation and arousal are differentially important in regulating the cognitive operations of the left and right hemispheres.

The redundancy bias of the activation system is important to focusing attention; this may be integral to the analytic cognitive operations of the left hemisphere. Many features of higher motor control seem to depend on left hemisphere processing. In contrast, the right hemisphere seems to show an affinity for integrating the perceptual input processes of the posterior brain. The cybernetics of the arousal system may facilitate this directly. By selecting for novel input, the arousal system fills representational capacity with a broad range of unique information. This may be the attentional control underlying the right hemisphere's skill in holistic perception.

The major division of the brain found when considering activation and arousal is not right/left but anterior/posterior: arousal facilitates input while activation provides the sequential control for motor organization. Tucker and Williamson theorized that the lateral specialization of the brain occurred to capitalize on each of the cybernetic modes of activation and arousal within a hemisphere, the left hemisphere elaborating its cortical organization around control by activation, the right hemisphere elaborating an entire hemisphere around control by arousal.

Pribram (1981) emphasized the distinction between anterior and posterior cortical systems in self-regulation. The posterior intrinsic cortex facilitates representation of the external context, and thus tends to make the organism context-sensitive. Pribram terms this the aesthetic mode. The frontal cortex offers inhibitory control over this tendency, thus allowing the organism greater internal control over behaviour. This is the ethical mode. Pribram points to early psychometric studies of brain lesions suggesting that extraversion is increased by frontal lesions (Petrie, 1952). Tucker and Williamson (1984) propose that the specific cybernetic effects of activation and arousal can explain many of the differences between introverts and extraverts. Whereas Eysenck (1967) formulated this personality dimension around the notion of reticular activating system function, this was a unidimensional concept of brain arousal; clearer formulations of the self-control patterns of introversion and extraversion are possible by considering the qualitatively-specific cybernetics of activation and arousal. The redundancy bias of activation affords constancy that facilitates internal control over behavior. In an extreme form, this can produce pathological introversion. Animals and humans whose activation is pathologically augmented by dopaminergic agonist drugs show social withdrawal (Ellinwood, 1967; Kokkinidis and Anisman, 1980). The habituation bias of the arousal system causes novel environmental events to gain control over the extravert's behavior, leading to a responsive emotionality, a need for novel excitement, and a limited attention span.

Although they will undoubtedly require modification and revision, these formulations offer ways of thinking about neurocybernetics - how an elementary neural control process can be important to higher-order attentional control. In the current cognitive psychology literature, an important distinction is that between "controlled" cognitive processes, which require attention and volitional effort, and "automatic" processes, which seem to organize themselves spontaneously (Neisser, 1967; Posner, 1978; Schneider and Shiffrin, 1977). The unique control properties of redundancy and habituation biases would seem integral to these different kinds of cognitive control.

Thus there are specific predictions available for experimental test on how alterations in motivational and emotional control systems in the brain lead to changes in information processing. Neuropsychology may allow a more integrated approach to the study of personality, wherein emotional self-control and cognitive self-control are described within a single framework.

A Scale for Self-Report

If the neuropsychological approach is to provide more than theory, it will be necessary to examine measures of brain activity and relate these to the hypothesized processes of cognitive and emotional self-control. Performance measures are required to find if there are attentional effects of the hypothesized control processes. These empirical assessments are predicted directly by the neuropsychological constructs, and could provide new experimental directions for the study of personality.

In addition, we suggest that there is an important place in this research for self-report measures of cognitive style. Any laboratory assessment of a person's cognitive approach or emotional responsiveness can only cover a few representative situations at best. The subject has some degree of awareness of the approach he or she takes to handling the tasks of daily living in innumerable specific contexts. We propose that at least the major dimensions of cognitive style are sufficiently available to everyday introspection that, given the appropriate self-report questions, most subjects will be able to describe where they fall on these dimensions.

In asserting the utility of self-report, we are aware of the evidence that self-report abilities are remarkably limited, and have been traditionally overrated. When asked to introspect on a variety of processes, subjects often show little evidence of being able to access their internal psychological processes directly, but rather make up plausible explanations for their behaviour *post hoc* (Nesbitt and Wilson, 1977; Bem, 1972). Furthermore, when considering a lateralized cognitive style, there may be intrinsic biases in self-report depending on which hemisphere is most important to the person's style. Bear and Fedio (1977) observed that left-temporal-lobe epileptics, whose intellectualized and ideational style suggested an exaggeration of left hemisphere functioning, were more critical of their own behaviour than were observers. Right-temporal-lobe epileptics, whose increased emotionality suggested an exaggerated right hemisphere contribution, tended to deny problems and bias their self-report positively in comparison to observers. In research with normal university students, Swenson and Tucker (1983) found that those reporting a more right hemispheric cognitive style on the Preference Questionnaire (Zenhausern and Gebhardt, 1979) not only reported less timidity and depression on the Emotions Profile Inventory (Plutchik and Kellerman, 1974) but also showed a greater bias toward positive self-description than those with a left hemispheric style.

Thus, it must be considered an open question whether subjects are sufficiently aware of their cognitive processes to describe their characteristic styles accurately (assuming, of course, that these styles exist), and whether there will be intrinsic biases in self-report associated with one style or the other. Our intention was to devise a scale that would present a number of alternatives that exemplify major dimensions of emotional and cognitive style, and allow the subjects to choose which style seemed most characteristic of their behaviour.

The psychometric method of this study was thus a rational, rather than empirical one. In developing psychological assessment instruments, there has been a tradition of attempting to avoid problems of self-report by ignoring the content of the questionnaire items and blindly selecting those items which relate empirically to some external criterion, such as a diagnostic category. Recent psychometric thinking has questioned whether this approach leads to meaningful assessment (Burisch, 1984; Jackson, 1976). The recent trend seems to be to accept the limits of self-report, and to develop a questionnaire which targets well-defined psychological constructs, with items that are clearly representative of those constructs (Goldberg, 1972; Jackson, 1976).

The constructs targeted by the present scale were cognitive and emotional self-control styles, as described by Tucker and Williamson (1984). Although the theory proposes that cognitive and emotional self-control are parallel manifestations of the same underlying neural systems, our primary objective was to assess cognitive control as a separate psychological dimension from emotional control. Thus, any correspondence between the cognitive and emotional dimensions would occur empirically, rather than being created artifactually by including these within the same scale. This has been a cause of concern with the Preference Test of hemispheric style, which Zenhausern and associates (Coleman and Zenhausern, 1979; Zenhausern and Gebhardt, 1979) developed from Torrance's "Your Style of Thinking" test (Torrance, Reynolds, Ball and Riegel, 1978). Some of the items, such as those dealing with verbal versus nonverbal cognition, seem clearly related to hemispheric function. Other items, such as one questioning the subject's preference to read while sitting upright, are related only to an intuitive notion of what comprises right and left hemispheric styles. For the present self-report instrument, the intention was to develop subscales for assessing cognitive and emotional control, each of which was related explicitly to a major theoretical construct of the Tucker and Williamson (1984) model, and each of which was comprised of homogenous items rationally consistent with the construct targeted by that subscale.

Construction of the Scale

O'Connell reviewed the Tucker and Williamson (1984) formulation to determine those aspects of cognitive and emotional self-control that were central to the model and that would be reasonable targets for self-report. In addition, he drew on Shapiro's (1965) characterization of neurotic personality styles, which takes an ego psychology approach in describing attentional processes as integral to a person's personality style. Tucker and Williamson pointed to Shapiro's clinical descriptions in suggesting how exaggerations of normal self-regulatory processes may produce the cognitive dysfunctions of psychiatric patients.

The emotional control dimension was termed "Control-Spontaneity", and was characterized as follows (O'Connell, 1985):

"Controlled individuals show restraint when expressing emotions, interacting with others, and making decisions. Their emotional reactions are subdued and controlled. They come across to others as calm and rational. They can be described as cautious, private, dependable, disciplined, patient and serious. They can tolerate boredom, and they use routines to structure and control their lives. They are goal-oriented, and they use self-control to attain their goals. They base decisions on rational considerations; they do not act on impulse. Because of their need to think before acting, they are sometimes slow to make decisions and to act. In the extreme, control shades into inhibition, where feelings and impulses are rigorously guarded against; these highly controlled individuals might be described as "obsessive-compulsive" personalities: they are hyper-rational to the point of "isolating" themselves from their emotions.

Spontaneous individuals are emotionally expressive and open. They tend to make decisions on the basis of their immediate feelings and impulses. They have a low threshold for boredom and frustration, and they have a need for excitement and emotional stimulation. They come across to others as emotional, open, lively, playful, enthusiastic. They make decisions relatively quickly. Just as control shades into inhibition, spontaneity shades into impulsivity. Lacking adequate self-control, highly spontaneous-impulsive people may act unpredictable and undependable; there is an element of immediacy to their behaviour and decision-making which may get them into trouble".

The characterization of the primary cognitive control dimension, termed "Deliberate-Impressionistic" is as follows (O'Connell, 1985):

"The deliberate cognitive mode may be described as reflective, rational and logical. Individuals who rely on this tend to think in ways that are practical, systematic and organized, and realistic. They are able to narrowly focus their attention on their ideas or the task at hand; they are able to concentrate intensely. Their style of thinking is suited for paying attention to details. Sometimes their thinking is repetitive; they keep processing the same ideas over and over in greater detail. They prefer conversation and group discussions that are structured and do not wander off from the main topic. Because they pay attention to details, their memory is exact and precise; they also express their ideas with precision. Their attention to details leads them to think in complex or complicated ways. They like to use their minds for intense, focused and concentrated thinking. To some extent, this style of cognition is similar to Freud's concept of secondary process thinking.

The impressionistic cognitive mode involves the use of imagination, fantasy and intuition. Individuals who rely on this cognitive mode like to daydream, fantasize, engage in romantic thinking. Their thoughts are often loosely organized, drifting. They do not impose organization and structure on their ideas and thoughts. They think in a free-flowing, unstructured way. They have difficulty focusing their attention, and their attention span is short. They value original and "creative" thinking, rather than realistic and pragmatic thinking. They get "hunches" and act on them. They do not require evidence before accepting something as true. Ideas tend to race quickly and disconnectedly through their minds. Their mode of thinking is suited for focusing on the "whole" or the general picture; they lack the focused attentional mode needed for attention to details. They lack an interest in intellectual matters or other interests that require sustained concentration. Their lack of detailed, analytical thinking leads them to be open to getting hunches, being surprised by things that happen, and remembering dreams. Decisions are formed quickly on the basis of hunches and impressions, and may be changed just as quickly. This cognitive mode incorporates Freud's notion of primary process thinking".

In addition to these primary subscales, two additional subscales were developed. Because anxiety has been an integral issue in both clinical (Shapiro, 1965) and neuropsychological (Tucker, 1981) thinking, an Anxiety affective subscale was developed. As implemented within the present scale this included two facets: a tendency to be tense and worrisome, rather than relaxed and easy-going, and a tendency to be self-critical, rather than confident and optimistic. A Verbal-Nonverbal cognitive subscale was developed, since this aspect of cognitive preference is not necessarily aligned with the Deliberate-Impressionistic dimension, and is important to the notion that these cognitive styles are related to hemispheric function. A verbal mode involves a preference for dealing with words or numbers, such as shown by writers or accountants. A nonverbal preference includes dealing with physical-sensory data (colour, music, movement) and thinking with mental pictures and images, and would be more characteristic of the activities of painters, photographers or dancers.

A pool of 700 simple declarative statements was written by the first author to exemplify these psychological dimensions. These were edited to

377 items that were evaluated for clarity and content relevance by five judges with expertise in psychology and psychiatry. In addition, Tucker reviewed the items for their reasonableness as exemplars of the Tucker and Williamson formulation. At this point, 257 remaining items were administered to 183 university students and evaluated for the students' ratings of item clarity and for statistical evaluation of internal consistency within each subscale. The Marlowe-Crowne Social Desirability Scale (Crowne and Marlowe, 1960) was also administered to the first sample to assess response bias; these results were used to assure that selected items correlated more highly with their subscale on the OCASS than with the Social Desirability Scale. A set of 129 items was then selected and cross-validated for internal consistency with a second sample of 99 students.

One noteworthy feature of the item selection process was that items relating to analytic, detail-oriented thinking did not show good internal consistency within the cognitive control subscale, and thus tended not to be included on further revisions. Apparently, the analytic/holistic aspect is not integral to the Deliberate-Impressionistic dimension, or it is not as accessible to self-report as the more central issue of tight versus loose cognitive control.

From the second administration, 99 items were selected to create a preliminary version of the O'Connell Cognitive and Affective Style Scale (c. Thomas R. O'Connell, 1984). Internal consistency indicated by coefficient alpha for the subscales of the OCASS was: Emotional Control-Spontaneity, .86; Anxiety, .87; Deliberate-Impressionistic Cognition, .84; Verbal-Nonverbal Cognition, .81. On each scale there was a sufficient range of responses to differentiate among individuals, and the scores were symmetrically distributed (the measures of central tendency were similar). The only significant difference between males ($n=53$) and females ($n=46$) was the higher score by females on the Anxiety subscale ($p < .02$). Handedness and presence of sinistrality in first-degree relatives were assessed. Although a significant effect of familial sinistrality on the Deliberate Cognition dimension was observed ($p < .04$), with subjects who reported familial sinistrality showing a more impressionistic cognitive mode, no effect of subject handedness was found on this or any other subscale. However, it is of interest that seven of the eight effects of familial sinistrality on the four OCASS subscales differed in the same direction, i.e., left handedness and familial sinistrality tended to be related to higher scores on dimensions of right hemisphere control. A complete description of the psychometric issues and the procedures for test construction and item selection will be provided elsewhere (O'Connell and Scott, in preparation).

Subscale Intercorrelations and Concurrent Validity

An important theoretical question was whether or not the cognitive and emotional control dimensions, assessed independently by the two sub-scales would correlate as predicted. Table 1 presents the intecorrelations among the OCASS subscales, and the correlations of the subscales with the total OCASS score. These data suggest that if there is a "core" subscale to the OCASS, it is Emotional Control-Spontaneity; this was the most highly correlated with the other subscales and with the total score. Consistent with the theoretical model, greater emotional control was associated with more deliberate cognition, with verbal cognition, and with higher anxiety. The cognitive subscales (Deliberate and Verbal) were more highly correlated with each other (.44) than with the emotional subscales, and the emotional subscales (Control and Anxiety) were more highly correlated with each other (.55) than with the cognitive subscales, suggesting that, although the

Table 1. Correlations of the OCASS with itself, Trait Anxiety, the EPI, and the HPQ

	OCASS Subscales				OCASS Total
	Emot. Control	Anxiety	Cog. Control	Verbal	
OCASS					
Emot. Control	--	.55**	.34**	.29**	.81**
Anxiety	.55**	__	.11	.03	.65**
Cog. Control	.34**	.11	--	.44**	.68**
Verbal	.29**	.03	.44**	--	.61**
Trait Anxiety	.27**	.63**	-.29**	-.12	.20*
EPI					
Introversion	.75**	.45**	.41**	.24**	.69**
Impulsivity	-.53**	-.21**	-.51**	-.37**	-.59**
Sociability	-.69**	-.48**	-.17**	-.04	-.50**
Preference Test					
Left hem. score	.24**	.16	.57**	.29**	.45**
Right hem. score	-.31**	.20*	-.22*	-.53**	-.45**

* $p < .05$ ** $P < .01$

emotional and cognitive control dimensions are interrelated, they may be considered as separate dimensions. Although it was correlated with Emotional Control, the Anxiety subscale of the OCASS was not correlated with the cognitive dimensions (Deliberate and Verbal) as was expected.

In order to assess the concurrent validity of the OCASS, several theoretically relevant self-report scales were selected and administered with the OCASS to the second sample of students. The correlations of these scores with the OCASS are also presented in Table 1. The Speilberger (1968) Trait Anxiety measure showed fairly strong correlation with the OCASS Anxiety subscale (.63) and a positive correlation with Emotional Control (.27). Consistent with the findings reported above, Speilberger Trait Anxiety also failed to correlate with the OCASS cognitive subscales as predicted, and actually correlated with the Deliberate subscale (-.29) in the direction opposite to the theoretical predictions.

The correlations of the OCASS with the Eysenck Personality Inventory Introversion-Extraversion (I-E) scale were in the predicted direction of introverts reporting greater Emotional Control (.75), Anxiety (.45), Deliberate Cognition (.41) and Verbal Cognition (.24). As with the internal correlations for the OCASS, Emotional Control had the strongest correlations with I-E and was most representative of the total OCASS score correlation with I-E. The subdivisions of the I-E dimension, impulsivity and sociability, showed an interesting divergence in their relations to the OCASS subscales. Impulsivity showed the expected negative correlation with all 4 OCASS subscales, but sociability was more specific in its relation to the affective scales of the OCASS.

Table 1 also shows the correlations of OCASS subscales with Zenhausern's Preference Test. The items from this test were summed separately to provide scores for both the right and left hemisphere style dimensions. These data are generally consistent with the notion that both these scales are related to similar cognitive styles (although the question of actual hemispheric usage, of course, requires independent verification).

There was a remarkable symmetry in the correlations of the Preference Test Left Hemisphere and Right Hemisphere scores with the OCASS total score. The cognitive subscales of the OCASS showed the stronger relations to the Preference Test, as might be expected, with a Left Hemisphere style on the Preference Test more strongly related to Deliberate Cognition on the OCASS (.57) and a Right Hemisphere preference negatively related to OCASS Verbal Cognition negatively (-.53). Again, the Emotional Control dimension of the OCASS showed a relation to the cognitive dimensions of the Preference test in the expected direction.

Scores on the Marlow-Crowne Social Desirability Scale were correlated with the OCASS Anxiety subscale as would be expected (-.22), but did not show significant correlations with other OCASS subscales.

DISCUSSION

The aim of this work was to create a self-report scale that would have reasonable psychometric properties and that would accurately target the constructs of emotional and cognitive self-control. The OCASS contains items that the subjects found understandable and that the judges found consistent with the appropriate constructs. The subscales showed high internal consistency and enough variance to serve as a sensitive individual difference measure with normal subjects. Furthermore, the pattern of correlations among the subscales and between the OCASS subscales and other relevant instruments was generally as expected, further indicating the validity of the items as exemplars of the desired constructs.

In the scale construction process, the items written to indicate an analytic, detail-orientated cognitive approach did not show high internal consistency with other items on the Deliberate-Impressionistic Cognition subscale. One possibility is that this aspect of cognitive structure - analytic versus holistic - must be considered separately from the dimension of cognitive control assessed by the Deliberate Cognition subscale. Construction of a specific subscale for this purpose could be an objective for further research. Alternatively, it may be that subjects are insufficiently familiar with notions of cognitive structure to be able to recognize and report on their own preferred organizational tendencies. We have observed individual differences in preference for local versus global perceptual organization in laboratory tasks (Tyler and Tucker, 1982). Actual performance measures, rather than self-report, were used in the traditional assessment of analytic versus global cognitive styles (Witkin et al., 1962). Because the analytic/holistic dimension is so central to neuropsychological theory and to the question of cognitive style, further examination of this issue is required.

We expected that higher scores on the Anxiety scale would be associated not only with greater Emotional Control, but with higher scores on the Deliberate and Verbal Cognition subscales. Instead, the OCASS Anxiety subscale did not relate to the cognitive subscales, and the Speilberger Trait Anxiety measure actually showed a small negative correlation with the Deliberate Cognition scale.

Based on a number of previous findings (Tucker, 1981; Tucker, Antes, Stenslie and Barnhardt, 1978; Tyler and Tucker, 1982) we expected that greater anxiety would be associated with a more tightly controlled, verbal cognitive style, consistent with greater left hemisphere activation. It could be argued that anxious subjects often experience left hemisphere overactivation, and thus a more dysfunctional Deliberate cognitive mode. Anxious subjects do show deficits such as speech anxiety that suggest impaired function on verbal, focused attentional tasks. Yet this _post-hoc_ explanation is unsatisfactory: even if their cognitive performance is impaired in the Deliberate Cognition domain, it might be expected that anxious subjects would still show a cognitive style with these qualities. Further research is required to clarify the conditions under which activation or overactivation of controlled, left-hemispheric processing is found in high anxious subjects. A related issue is whether there may be dissociations between self-report and cognitive performance in anxious persons.

One problem in research on anxiety may be that this single construct is used to mean too many things. It often refers to general emotionality or to autonomic arousal, and these may be quite different phenomena than the hypervigilant attention and overly routinized cognition and behaviour characteristically observed in the chronic anxiety of obsessive-compulsive or schizophrenic persons. It may be that better theoretical differentiation of the construct of anxiety will be necessary for greater precision in measurement. In the OCASS Anxiety subscale, two features were indicated by the items in the final version of the subscale: a tense, worrisome attitude and a tendency toward self-criticism. Although the high internal consistency of the Anxiety subscale argues that these dimensions were closely related, it might be important for theoretical purposes to assess them separately.

Related to self-criticism is the question of the influence of social desirability on the OCASS. Because items were selected to minimize social desirability, the OCASS will not be helpful in evaluating what may be an intrinsic relation between self-report bias and lateralized cognitive style (Bear and Fedio, 1977; Swenson and Tucker, 1983). This should not be a major drawback; it is probably not the best approach to relate self-report bias to a self-report measure of cognitive style. For example, it could be argued that the Swenson and Tucker (1983) findings simply reflected a positive bias toward the right hemispheric items written by Zenhausern and Torrance for the Preference test. Further research on self-report biases might better examine cognitive performance or brain activity measures of individual differences in neuropsychological function rather than a self-report measure of cognitive and affective style.

The major purpose of constructing the OCASS was to assess social and emotional aspects of self-control separately from cognitive aspects of self-control, then to allow the relatedness of these domains to be assessed empirically. The present results did indicate these dimensions to be related. The Emotional Control-Spontaneity subscale was not only significantly related to the Deliberate and the Verbal Cognition subscales, but appeared to be the core subscale of the OCASS. Furthermore, this dimension showed the predicted correlation with the Introversion-Extraversion measure that was high enough to be within the range of the reliability for many self-report tests.

This relation is consistent with Tucker and Williamson's (1984) formulation of the differential effects of self-regulation with activation versus arousal systems on the individual's degree of internal versus external control. It is also consistent with Pribram's (1981) account of the shift between internal and external control provided by the relative balance between anterior and posterior cortical systems. Eysenck (1967) proposed that it is differences in characteristic neural arousal that leads to the personality differences between introverts and extraverts. By considering how the cybernetic effects of a redundancy bias differ qualitatively from those of a habituation bias, Tucker and Williamson (1984) could explain how one form of neural control augments a tonic internal control over behaviour, while an increase in a different neural control system augments a phasic responsivity to external influences.

Increasing our understanding of the self-regulation of neural systems, thus provides new ways of theorizing about self-control in the personality domain. Although the interest in neuropsychological models of individual differences in recent years has centered around concepts of differential hemispheric function, and these concepts remain important to the present approach, it is important to recognize that the notions of activation and arousal are more fundamentally related to the anterior/posterior than the left/right dimension of brain organization. Thus we seem to be returning to earlier neuropsychological concepts of personality, considered in terms of frontal versus posterior systems (Petrie, 1952; see Pribram, 1981). This emphasis could pose some interesting measurement problems for attempts to relate psychological issues to neural function. The laterality approach has been relatively easy to implement in research with normal persons because the function of the two hemispheres is roughly parallel. Thus, stimulus materials can be presented to one hemisphere or the other, or the EEG responses of one hemisphere can be compared with those of the other during a cognitive or emotional process. These fairly sensitive right/left functional comparisons may not be as easy when attempting to assess the degree to which individuals show a tendency to engage frontal versus posterior brain systems as general cognitive mode. Certainly the left/right dimension will continue to be important; it seems to be closely interdependent with the anterior/posterior dimension (Tucker and Williamson, 1984). But considering this dimension presents new challenges to the assessment of individual differences in a neuropsychological framework. It will be important to meet these challenges if we are to move beyond self-report scales of cognitive and emotional styles and show how these styles are related to brain activity, emotional behaviour, and cognitive performance.

REFERENCES

Bakan, P. (1969). Hypnotizability, laterality of eye movement and functional brain asymmetry. Perceptual and Motor Skills, 28, 927-932

Bakan, P. and Svorad, D. (1969). Resting EEG alpha and asymmetry of reflective lateral eye movements. Nature, 223, 975-976.

Bem, D.J. (1972). Self-perception theory. In L. Berkowitz (Ed.) Advances in Experimental Social Psychology (Vol. 6). New York: Academic Press.

Bogen, J. E. and Bogen, G. M. (1969). The other side of the brain III: The corpus callosum and creativity. Bulletin of the Los Angeles Neurological Society, 34, 191-220.

Burisch, M. (1984). Approaches to personality inventory construction: A comparison of merits. American Psychologist, 39, 214-227.

Caplan, B. and Kinsbourne, M. (1981). Cerebral lateralization, preferred cognitive mode, and reading ability in normal children. Brain and Imagery, 14, 349-370.

Coleman, S. and Zenhausern, R. (1979). Processing speed, laterality patterns, and memory encoding as a function of hemispheric dominance. Bulletin of the Psychonomic Society, 14, 357-360.

Combs, A. L., Hoblick, P. J., Czarnecki, M. J. and Kamler, P. (1977). Relationship of lateral eye movement to cognitive mode, hemispheric interaction and choice of college major. Perceptual and Motor Skills, 45, 983-990.

Crouch, W. (1976). Dominant direction of conjugate eye movements and responsiveness to facial and verbal cues. Perceptual and Motor Skills, 42, 167-174.

Crowne, D.P. and Marlowe, O.A. (1960). A new scale of social desirability independent of psychopathology. Journal of Consulting Psychology, 24, 349-354.

Day, M. E. (1961). An eye movement phenomenon relating to attention, thought and anxiety. Perceptual and Motor Skills, 19, 443-446.

Day, M. E. (1961). An eye movement indicator of type and level of anxiety:

some clinical observations. Perceptual and Motor Skills, 19, 443-446.
Day, M. E. (1967). An eye movement indicator of individual differences in the physiological organization of attentional processes and anxiety. Journal of Psychology, 66, 51-62.
Ehrlichman, H. (1977). Field dependence-independence and lateral eye movements following verbal and spatial questions. Perceptual and Motor Skills, 44, 1229-1230.
Ehrlichman, H. and Weinberger, A. (1978). Lateral eye movements and hemispheric asymmetry: A critical review. Psychological Bulletin, 85, 1080-1101.
Ellinwood, E.H. (1967). Amphetamine psychosis: I. Description of the individuals and process. Journal of Nervous and Mental Disease, 144, 273-283.
Etaugh, C. F. (1972). Personality correlates of lateral eye movement and handedness. Perceptual and Motor Skills, 34, 741-754.
Etaugh, C. F. and Rose, M. (1973). Lateral eye movements: Elusive personality correlates and moderate stability estimates. Perceptual and Motor Skills, 37, 211-217.
Eysenck, H. J. (1967). The biological basis of personality. Springfield: Thomas.
Gale, A. (1973). The psychophysiology of individual differences: Studies in extraversion and the EEG. In P. Kline (Ed.) New Approaches to psychological Measurement. New York: John Wiley, 211-256.
Galin, D. and Ornstein, R. (1972). Lateral specialization and cognitive mode: An EEG study. Psychophysiology, 9, 412-418.
Galin, D. and Ornstein, R. (1974). Individual differences in cognitive style: I. Reflective eye movements. Neuropsychologia, 12, 367-376.
Goldberg, L. R. (1972). Some recent trends in personality assessment. Journal of Personality Assessment, 36, 547-560.
Gur, R. C. and Gur, R. E. (1974). Handedness, sex and eyedness as moderating variables in the relation between hypnotic suscetibility and functional brain asymmetry. Journal of Abnormal Psychology, 83(6), 635-643.
Gur, R. C. and Gur, R. E. (1980). Handedness and individual differences in hemispheric activation. In J. Herron (Ed.) Neuropsychology of Left-Handers. New York, Academic Press, 211-231.
Gur, R.E. and Reyher, J. (1973). Relationship between style of hypnotic induction and direction of lateral eye movements. Journal of Abnormal Psychology, 82, 499-505.
Harnad, S. (1972). Creativity, lateral saccades and the nondominant hemisphere. Perceptual and Motor Skills, 34, 653-654.
Hartlage, L.C. and Tollison, C. D. (1979). MMPI correlates of looking left or right during mental tasks. Journal of Clinical Psychology, 35, 92-94.
Hines, D. and Martindale, C. (1974). Induced lateral eye movements and creative and intellectual performance. Perceptual and Motor Skills, 39, 153-154.
Hiscock, M. (1977). Eye movement asymmetry and hemispheric function: An examination of individual differences. Journal of Psychology, 97, 49-52.
Jackson, D. N. (1976). Jackson Personality Inventory Manual. Goshen, New York: Research Psychologist Press.
Kinsbourne, M. (1972). Eye and head turning indicates cerebral lateralization. Science, 176, 539-541.
Kokkinidis, L. & Anisman, H. (1980). Amphetamine models of paranoid schizophrenia: An overview and elaboration of animal experimentation. Psychological Bulletin, 88, 551-578.
Levy, J. (1969). Possible basis for the evolution of lateral specialization of the human brain. Nature, 224, 614-615.
Levy, J. (1983). Individual difference in cerebral hemisphere asymmetry: Theoretical issues and experimental considerations. In J. Hellidge

(Ed.) Cerebral hemisphere asymmetry: Method, Theory and application. Praeger Scientific Publishers.

Levy, J. (1977). Manifestations and implications of shifting hemi-inattention in commisorotomy patients. In E. A. Weinstein and R. P. Freidland (Eds.) Hemi-inattention and Hemisphere Specialization. New York: Raven Press, pp. 83-92.

Levy, J. (1983). Is cerebral asymmetry of function a dynamic process? Implications for specifying degree of lateral differentiation. Neuropsychologia, 21, 3-11.

Levy, J., Heller, W., Banich, N.T., & Burton, L. A. (1983). Are variations among right-handed individuals and perceptual asymmetries caused by characteristic arousal differences between the hemispheres? Journal of Experimental Psychology: Human Perception and Performance, 9, 329-359.

McGuinness, D. & Pribram, K. (1980). The neuropsychology of attention: Emotional and motivational controls. In M.C. Wittrock (Ed.) The Brain and Psychology. New York: Academic Press, 95-131.

Meskin, B. B. and Singer, J. L. (1974). Daydreaming, reflective thought and laterality of eye movements. Journal of Personality and Social Psychology, 30, 64-71.

Neisser, U. (1967). Cognitive Psychology. New York: Appleton-Century-Crofts.

Nesbitt, R. E. and Wilson, T. D. (1977). Telling more than we can know: Verbal reports on mental processes. Psychological Review, 84, 231-259.

O'Connell, T. R. (1985). Development and preliminary validation of an instrument measuring four cognitive and affective psychological dimensions. Unpublished doctoral dissertation, University of North Dakota.

O'Connell, T. R., and Scott, T. (in preparation). Construction of the O'Connell Cognitive and Affective Style Scale.

Petrie, A. (1952). Personality and the Frontal lobes. London: Routledge and Kegan.

Plutchik, R., and Kellerman, H. (1974). Emotions Profile Index Manual. Western Psychological Services, Los Angeles.

Posner, M. I. (1978). Chronometric Explorations of Mind. Lawrence Erlbaum. Hillsdale, New Jersey.

Pribram, K. H. and McGuinness, D. (1975). Arousal, activation and effort in the control of attention. Psychological Review, 82, 116-149.

Pribram, K. H. (1981). Emotions. In S.K. Filskov and T.J. Boll (Eds.) Handbook of clinical Neurophysiology. New York: Wiley - Interscience, 102-134.

Sackeim, H. A., Weiman, A. L., and Gregia, D. M. (1984). Effects of preditors of hemispheric specialization on individual differences in hemispheric activation. Neuropsychologia, 22, 55-64.

Schneider, W., and Shiffrin, R. M. (1977). Controlled and automatic human information processing: I. Detection search and attention. Psychological Review, 84, 1-66.

Shapiro, D. (1965). Neurotic Styles. New York: Basic Books.

Shevrin, H., Smokler, I., and Kooi, K. A. (1980). An empirical link between lateral eye movements and lateralized event-related brain potentials. Biological Psychiatry, 15, 691-697.

Sperry, R. W. Gazzaniga, M. S. & Bogen, J. E. (1969). Interhemispheric relationships: The neocortical commisures; syndromes of hemisphere disconnections. In P.Vinken & C. Bruyn (Eds.) Handbook of Clinical Neurology. Amsterdam: North Holland.

Sperry, R. W. (1966). Brain bisection and consciousness. In J.C. Eccles (Eds.) Brain and Conscious Experience. New York: Springer - Verlag, 298-313.

Speilberger, C. D. (1975). Anxiety: State-trait-process. In C. D. Speilberger & I. G. Sarason (Eds.) Stress and Anxiety, Vol 1. New York: Wiley, 115-143.

Swenson, R. A. & Tucker, D. M. (1983). Lateralized cognitive style and self-description. International Journal of Neuroscience, 21, 91-100.

Teitelbaum, H. A. (1954). Spontaneous rhythmic ocular movements. Their possible relationship to mental activity. Neurology, 4, 350-354.

Torrance, E., Reynolds, C., Riegel, T. and Ball, D. (1978). "Revised norms-Technical Manual for Your Style of Learning and Thinking", University of Athens, Georgia.

Tucker, D. M., & Williamson, P.A. (1984). Asymmetric Neural Control Systems in Human Self-Regulation. Psychological Review, 91, 185-215.

Tucker, D. M., Yutrzenka, B. A., and Heck, D. G. (1984). Emotion and eye movements. Paper presented at the symposium, "Current status of lateral eye movement research, American Psychological Association meetings, Toronto, August.

Tucker, D.M., Antes, J.R., Stenslie, C.E., and Barnhardt, T.N. (1978). Anxiety and lateral cerebral function. Journal of Abnormal Psychology, 87, 380-383.

Tucker, D.M., Stenslie, C.E., Roth, R.S., and Shearer, S. (1981). Right frontal lobe activation and right hemisphere performance decrement during a depressed mood. Archives of General Psychiatry, 38, 169-174.

Tyler, S.K., and Tucker, D. M. (1982). Anxiety and perceptual structure: Individual differences in neuropsychological function. Journal of Abnormal Psychology, 91, 210-220.

Weiten, W., and Etaugh, C. F. (1974). Lateral eye movement as a function of cognitive mode, question sequence, and sex of subject. Perceptual and Motor Skills, 38, 439-444.

Weiten, W., and Etaugh, C. F. (1973). Lateral eye movement as related to verbal and perceptual-motor skills and values. Perceptual and Motor Skills, 36, 423-428.

Witkin, H.A., Dyk, R.B., Faterson, H. F., Goodenough, D. R., and Karp, S. A. (1962). Psychological Differentiation: Studies of Development. Wiley, New York.

Yutrzenka, B. A. (1981). An introduction to an initial examination of the visual half field shut down theory of reflective lateral eye movements. Unpublished doctoral dissertation, University of North Dakota.

Zenhausern, R., and Geghardt, M. (1979). Hemispheric dominance in recall and recognition. Bulletin of the Psychonomic Society, 14, 71-73.

HANDWRITING POSTURE AND CEREBRAL ORGANIZATION

A.M. Weber and *J.L. Bradshaw

Department of Psychology
University of Victoria
P.O. Box 1700
Victoria, B.C., Canada
V8W 2Y2D

* Department of Psychology
Monash University
Clayton, Vic., Australia 3168

ABSTRACT

There seems little support for Levy's (1982) revised hypothesis that writing hand/posture indexes lateralization of selected aspects of verbal or spatial functions. There is limited support from certain reading studies and dichotic listening studies using speech sounds, but even this evidence is at best equivocal. While there is some suggestion that motor control may be more bilaterally organized in inverters, the evidence so far is inconclusive and requires further clarification. Moreover, the notion of ipsilateral pyramidal control of writing movements in inverters is generally contradicted by the available data. Finally, there is evidence that callosal agenesis is not associated with inverted writing posture. In conclusion, we do not regard the modifications offered by Levy (1982) as adding substantially to her earlier position.

INTRODUCTION

Levy and Reid (1976, 1978) proposed that cerebral lateralization of verbal and spatial functions could be predicted from a knowledge of the subject's writing hand (left or right) and posture adopted in writing (inverted or normal). Their hypothesis received only very limited support from other empirical studies, and these findings together with Levy and Reid's theoretical rationale were critically evaluated by Weber and Bradshaw (1981). Levy (1982) responded to such criticism with a modification of her original hypothesis; instead of assuming that verbal (or spatial) functions are lateralized to one (or the other) hemisphere in left-handers, she now (p. 592) proposes that in such subjects language (or spatial) functions may be bilaterally represented with different aspects of a given function occurring in separate hemispheres. Levy also contests Weber and Bradshaw's interpretation of her earlier position and of the relevant empirical literature. The present paper evaluates her new proposals and also offers some clarification concerning the conflict between her interpretation and our own.

CONFLICT OF INTERPRETATION

To avoid burdening the reader at this stage with a detailed refutation of Levy's criticisms of our 1981 paper, point by point, this refutation has been relegated to the appendix. However, it might be helpful to outline briefly the method by which we arrived at our interpretation of Levy's ideas. Her publications concerning writing hand/posture (Levy, 1972, 1974;

Levy and Reid, 1976, 1978), were carefully examined and interrelated proposals were grouped together. Where inconsistencies or ambiguities apparently arose in Levy's statements, these were evaluated in the context of the overall trend and logic of related statements in order to establish the predominant theme of such statements. The results of the analysis were the four postulates presented and evaluated by Weber and Bradshaw (1981). Such a clarificatory formulation of Levy's proposals was a necessary prerequisite to their evaluation. We believe that if the interested reader performs the same painstaking analysis, he or she will come to the same conclusion.

LEVY'S MODIFIED HYPOTHESIS - PRELIMINARY COMMENTS

Levy (1982) stresses the non-unitary nature of functional cerebral asymmetry in left-handers. Certainly, many research findings have supported the notion that left-handers show less marked lateralization of verbal function than do right-handers (see review by Bradshaw, 1980). There is also considerable evidence of at least some bilateral representation of certain aspects of language in right-handers (Gainotti, Caltagirone, Miceli, & Masullo, 1981; Knopman, Rubens, Klassan & Meyer, 1982; Patterson, 1981; Searlman, 1977; Urcuioli, Klein & Day, 1981; Wapner, Hamby & Gardner, 1981; Zaidel & Peters, 1981). Prediction of functional asymmetry must indeed take into account such bilateral representation and, as Levy (1982) points out, the exact nature of the associations and dissociations of hemispheric representation of the different aspects of language (or spatial) functions is a matter for empirical investigation.

However, Levy's realization of the well-known "bilateralization" factor is surely too general and speculative to constitute a substantial new contribution to our theoretical understanding of cerebral asymmetry or to be of practical use at a clinical or applied research level. Furthermore, although Levy has modified her original hypothesis, much of her reply is devoted to defending this earlier hypothesis. The resulting confusion and vagueness may encourage people to continue referring to the original more clearly delineated hypothesis. There is particular risk that busy clinicians without the time for extensive review of the research literature will continue to accept the original hypothesis as a quick and simple index of language lateralization as, for example, in the recommendation of Eastwood and Stiasny (1978).

Before proceeding to evaluate the new Levy hypothesis, it seems desirable to list briefly all recent empirical evidence not previously discussed by either Levy (1982) or Weber and Bradshaw (1981) and which pertains to the original hypothesis. Such findings will be dealt with in more detail subsequently. Of the sixteen known studies, none give clear support for the old hypothesis, eight give at best a little tenuous or partial support (Dabbs & Choo, 1980; Miller, 1983; Natale, Gur & Gur, 1984; Reuter-Lorenz, Givis, & Moscovitch, 1983; Searleman, 1980; Springer & Searleman, 1980; Tapley & Bryden, 1983; Wellman & Allen, 1983), and eight offer strongly contrary findings (Ajersch & Milner, 1983; Bashore, Nydegger, & Miller, 1982; Beaumont & McCarthy, 1981; Ellis & Miller, 1981; Jones, 1980; Strauss, Wada, & Kosaka, 1984; Volpe, Sidtis, & Gazzaniga, 1981; Warrington & Pratt, 1981). These findings are consistent with those previously reviewed (Weber & Bradshaw, 1981) in their rejection of the notion that writing hand/posture is a reliable index of cerebral functional asymmetry as formulated in the original hypothesis.

MODIFIED HYPOTHESIS

Levy (1982, p. 593) modified the Levy and Reid conclusions regarding the relationship of hand/posture to functional asymmetry by substituting "reading of tachistoscopically presented nonsense syllables and location of tachistoscopically presented dots" for the general terms "verbal" and "spatial" functions. She argues (p. 598) that many research findings support the notion that there is some difference in the asymmetry patterns of left-inverted and left-normal subjects, and that left-inverted and right-normal subjects are similar and both are different from left-normal subjects. Later (p. 605), she concludes that "my interpretation of the currently available data is that these (left-inverted and left-normal subjects) are neuropsychologically distinct populations, that the hand-posture dimension is an important correlate of neurological organization" (parentheses added).

If writing posture does indeed distinguish two neurologically different groups with respect to cerebral asymmetry, the crucial question is how. Levy's evaluation of the literature is not at all clear on this point. Given that various aspects of language (or other) function may be differentially lateralized, the following analysis seeks some pattern in the current data.

Replication Studies

As Levy and Reid's (1976, 1978) results give strong support to the idea that lateralized function is predictable given a knowledge of hand/posture, the appropriate starting point would seem to be an examination of similar studies. Wellman and Allen (1983) administered the Levy and Reid tachistoscopic tasks to 7-9 year old children but did not replicate the hand/posture differentiation findings. Levy (1982, p. 593) mentions Smith and Moscovitch (1979) as one such study and points out that they have confirmed Levy and Reid's findings on their syllable task. On their dot-location task, however, while right-normal and left-inverted subjects showed the expected left visual field (LVF) advantage, left-normals showed no significant asymmetry. Levy points out that the Smith and Moscovitch dot-location task may not have indexed the same underlying function as that of Levy and Reid because they used a wider visual angle of dot-presentation (up to 7° lateral to fixation) than Levy and Reid (up to 2.7^{o}). Levy refers to Kimura (1969) as finding larger visual-field asymmetries for more medial dots when compared with relatively more lateral dots. In fact, Kimura states that the most marked field differences occurred along the middle horizontal row and apparently found field asymmetries using a range of visual angles similar in lateral extent to that of Smith and Moscovitch. Thus, visual angle differences do not seem to explain the discrepant findings of the two studies with respect to dot-location.

A more recent study of dextral inverters and normals by Tapley and Bryden (1983) found a right visual field (RVF) superiority on the syllable task and a LVF superiority on the dot-location task but no writing posture effects. If, therefore, there is a neurological differentiation of inverters from normals with respect to the two tasks, it would seem only to apply to sinistrals and not to dextrals. If confirmed, such a neurological phenomenon would require explanation.

Even if sinistrals were to show the task asymmetry differences specified by Levy and Reid, and were to do so consistently, the question would remain as to what functions were being measured. The response mode together with the use of a fixation number (see below) may well introduce possible contaminants, and render difficult any simple intepretation of the findings in terms of phonological processing or spatial location of visually presented material. Thus, Smith and Moscovitch's subjects were required to respond in the dot-location task by calling out the identification number of the judged location. Although Levy and Reid (1978) do not explicitly specify this response mode, they do say (p. 127) that location was reported from a card mounted below the tachistoscope. As Levy (1982) does not mention response mode as a methodological difference between the Levy and Reid study and that of Smith and Moscovitch, and as she and Reid made no mention of any counterbalancing of response hand in their methodology, it would seem probable that Levy and Reid used a similar response mode to that of Smith and Moscovitch. Such a response mode is perhaps rather surprising in view of the fact that it was Levy, together with Trevarthen and Sperry (1972), who demonstrated that response mode may influence hemispheric lateralization of stimulus processing. An oral response in which subjects name the identification number of the dot positions is likely to invoke the hemisphere dominant for speech. A counterbalanced left-right manual response might have been more appropriate. Tapley and Bryden (1983) also appear to have used the number method of identifying dot positions.

In an attempt to ensure that subjects maintained central fixation during the stimulus display, Levy and Reid (1976, 1978) used the McKeever and Huling (1970) procedure of requiring subjects to report a central fixation digit presented within the stimulus display. Smith and Moscovitch (1979) and Tapley and Bryden (1983) also used this method. This procedure can be criticized in that such digits may constitute a form of verbal stimulus (Bryden, 1970; Gordon & Carmon, 1976; Kimura, 1961; Levy & Bowers, 1974; Satz, Achenbach, & Fennell, 1967; Zurif & Bryden, 1969). Thus the central digit may enhance the verbal nature of the syllable task and add a verbal element to the dot-location task. Findings concerning the possible effects of the central stimulus upon the nature of the stimulus display are currently equivocal. Kershner, Thomae, and Callaway's (1977) study demonstrated that the verbal or nonverbal nature of the central fixation check influenced the direction of field superiorities for verbal lateral stimuli. Similar conclusions were reached by Carter and Kinsbourne (1979), Hellige and Cox (1976), and Mancuso, Lawrence, Hintze, and White (1979). However, in other studies, Duda and Kirby (1980) and Hines (1978) found no effect on field asymmetries from the nature of the central check. Indeed, this procedure may not necessarily even guarantee fixation. If a subject can fixate centrally and still see the lateral stimulus, there is no reason why the subject may not also look at the lateral area and still be able to see the central digit. In any case, there is evidence (Geffen, Bradshaw, & Nettleton, 1972; Harcum, 1978, p.149; Jones & Santi, 1978) that such an attempt to control fixation is unnecessary and that such deviations from fixation as occur are either very rare or are likely to be evenly distributed between the fields.

Weber (1980) omitted the central digit from the syllable and dot-location tasks and also used a manual response mode (hands counterbalanced) for the dot-location task. With such modifications in the interests of improved experimental controls, neither task was found to be associated with hand/posture. An overall RVF advantage was found for males on both tasks and females showed a RVF advantage on the syllable task and a LVF advantage on dot-location.

Although each of the four studies have used very slightly differing hand/posture criteria, they would appear to be generally similar and deviation from Levy and Reid's criteria as such does not appear to account for the extent to which the other studies do or do not support the original finding.

Other Studies

Other studies concerned with the hand/posture effect are now examined to see whether there are any clues as to what aspects of cerebral functioning might be associated with hand/posture.

Spatial and nonverbal material. Bradshaw, Nettleton, and Taylor (1980) in a face discrimination task found no hand/posture effect. This finding is in keeping with the face-processing data of Heller and Levy (1981, cited in

Levy, 1982), Levy, Heller, Banich, and Burton, (manuscript, cited in Levy, 1982), Jones (1980), but not those of Natale, Gur and Gur (1983) who found that left-inverters were lateralized at a level intermediate between dextrals and left-normals, or Reuter-Lorenz, Givis, and Moscovitch (1983) who found that left-normals and left-inverters were oppositely lateralized. A person's own facial expression does not appear to be lateralized in such a way as to discriminate between inverters and non-inverters (Campbell, 1979; Moscovitch and Olds, 1982). Lawson (1978) did find the expected hand/posture effect for males but the reverse pattern for females. Levy (1982) raises the possibility of field dependence and independence effects to explain Lawson's findings. It is not entirely clear why such a factor should account for Lawson's data more than that of other studies. However, the trend of findings at this stage indicates that there is no hand/posture effect in relation to face-processing. Ellis and Miller (1981) found only very weak evidence that there may be some difference between right-normals and left-normals with respect to which side of an advertisement picture subjects preferred the written message. In summary, there seems to be little evidence of hand/posture effects in association with particular facets of spatial or nonverbal functioning so far, and these negative findings seem to apply across the various hand/posture criteria used by different studies.

Verbal material. Examination of visual field studies does not reveal any clear association between aspects of language function and hand/posture: Bashore, Nydegger, and Miller (1982) using letter naming, Bradshaw, Nettleton, and Taylor (1980) for lexical decisions, Bradshaw and Taylor (1979) for reading words and nonwords, McKeever (1979) for word reading, and McKeever and VanDeventer (1980) for letter masking. McKeever (1979) found weak support for a hand/posture effect for color naming, left-normals showing a weaker RVF advantage than left-inverters. Miller (1983) reported a nonsignificant trend with respect to a hand/posture effect for word recognition.

Dichotic listening studies using speech sounds provide weak or no evidence of hand/posture effects (Searleman, 1980; Smith & Moscovitch, 1979; Tapley & Bryden, 1983; Volpe, Sidtis, & Gazzaniga, 1981). Studies using digits show no hand/posture effects (Beaumont & McCarthy, 1981; McKeever & VanDeventer, 1980; Warrington & Pratt, 1981). Other dichotic listening studies whose material is unspecified were Thistle's (1976) study which found no hand/posture effect, and Springer and Searleman's (1980) study which found some hand/posture differences for dichotic listening in twins but not in singletons.

Other relevant studies include Allen and Wellman's (1980) finding that the closer dextral children came to using normal posture, the better their reading level. But Wellman and Allen (1983) found that this association of writing posture and reading skill in dextral children was not so much related to hemispheric lateralization of verbal or spatial skills as to maturational factors. They suggested that nonnormal posture indicates maturational lag which is also associated with left-right directionality problems which in turn lead to lower reading scores. Bryson and McDonald (1984) found that among grade 5 sinistral children, it was boys who used parallel posture (neither inverted nor normal) who showed poor reading. Younger boys (grades 1 & 3) and all girls (grades 1, 3, & 5) showed no such association between reading and posture. Gregory and Paul (1980) showed a variety of cognitive deficits in inverted adults relative to normals. Perhaps these various reading and cognitive deficits are related to maturational lag factors. Herron, Galin, Johnstone, and Ornstein (1979) found evidence of more right occipital involvement in reading and writing tasks for left-normals than for left-inverters or dextrals. Perhaps the most substantial evidence against the argument that hand/posture can index the lateralization of speech processes comes from the recent unilateral ECT and amytal-ablation studies. Thus Ajersch and Milner (1983), Strauss, Wada and Kosaka (1984), and Volpe, Sidtis, and Gazzaniga (1981) found that inverters could not be distinguished from normals with respect to speech lateralization. Exactly similar findings were reported by Warrington and Pratt (1981) on the basis of transient dysphasia following unilateral ECT in non-neurological (depressive) patients.

Generally the findings with respect to verbal material offer no really clear indications that lateralization of specific aspects of language function are predictable on the basis of hand/posture. Although Herron et al.'s findings suggest some link with reading, the exact nature of this link is unclear and would seem to be a very specific one, possibly related to maturational lag factors, given the generally negative findings of the visual field studies involving reading. Levy and Reid's (1976, 1978) and Smith and Moscovitch's (1979) syllable task findings constitute the only strong evidence, and it is difficult to see what their tasks and those of Allen and Wellman and Herron et al. might have in common to distinguish them from the other visual field tasks. Dichotic listening studies are hardly more encouraging although there is some weak evidence for the hand/posture hypothesis from studies of speech sound discrimination. Warrington and Pratt (1981), using transient dysphasia after unilateral ECT as a measure of speech lateralization, concluded that dichotic listening was not a reliable index of speech lateralization. Chiarello (1980) also points out the unreliability of dichotic listening in subjects over time, and that it has never been shown that size of ear effect is proportionate to the degree of hemispheric dominance for language. Halsey, Blauenstein, Wilson, and Wills (1980) claim a relationship between lesion side, aphasia and hand/posture but unfortunately give no details. These generally negative findings with respect to verbal material occur across a variety of hand/posture criteria and do not seem to be associated with such variations.

It is not feasible, either, to relate all these findings to the question of a possible interaction of hand/posture with familial handedness in predicting cerebral lateralization. However, the findings with respect to a possible relation between familial handedness and functional cerebral asymmetry are in any case equivocal. Some authors have claimed that right hemisphere language is more prevalent in people with left-handed relatives than in people without such relatives (Andrews, 1977; Hecaen & Sauguet, 1971; Zurif & Bryden, 1969) while others have made the reverse claim (Newcombe & Ratcliff, 1973; Warrington & Pratt, 1973).

Levy (1976, 1978) suggests that Bradshaw and colleague's negative results with respect to hand/posture and cerebral lateralization differ from those of American researchers. The above review of findings suggests that in fact the Australian findings are closely similar to many American ones. In considering Levy's reference (1982, p.599) to strong confirmation for some form of relationship between hand/posture and cerebral asymmetry, the reader should also note that Levy makes only brief reference (p. 598) to the fact that "not all investigators have found a distinction between the two hand-posture groups with respect to asymmetry." Any objective appraisal of the status of Levy's proposal must of course include such negative findings as well as the positive ones. We believe that Levy (1982) rather neglects to do this.

However, in the above context of possible cultural differences, Levy (1982) speculates that perhaps Australian school children are strongly pressured to write with a normal posture, and that obviously "the hand-posture dimension cannot represent the same underlying factors in the two continents". The guidelines for teaching writing specified by the Education Department of Victoria (Australia), and the senior author's extensive professional contact with children and teachers regarding educational problems, together with anecdotal reports from some inverters about how they changed to this posture in imitation of peers after some years of schooling, all indicate that children are not typically pressured to use a normal posture, or for that matter an inverted one. More importantly, if hand/posture is a cultural phenomenon as Levy (1982) suggests, then this would detract from the probability of its also being "an important correlate of neurological organization" which is the major thrust of Levy's hypothesis (and see 1982, p.605). The varying incidences of inverted posture across age groups suggest the involvement of some non-neurological factor, as discussed by Weber and Bradshaw (1981, pp.80-81) and as has also been suggested by Shanon (1978), Peters (1983), and Guiard and Millerat (1984). Searleman, Porac, and Coren (1982) found some association between inverted posture and birth stress in left-handed males but not in females or right-handed males and the birth-stress factor did not unambiguously discriminate inverters from noninverters even among left-handed males (94% stressed and 41% no stress left-handed male subjects were inverters). Searleman, Porac, and Coren (1984) also found that among left-handed males, unlike other groups, inverted posture was associated with left-sided preference with respect to hand, eye, foot, and ear. They suggest that left-handed male inverters would be the most likely candidates to display leftward lateral preferences owing to an excess of pathological left-sidedness. However, although this association between hand/posture and lateral preference is statistically significant, it is far from reflecting a perfectly correlated association. Porac, Coren, and Searleman (1983) report a relationship between hand/posture in parents and offspring which could be indicative of a genetic factor but not a simple, straight forward one and cultural influences are not ruled out. Hence, although these data suggest the possibility of a neurological basis for inverted posture, they are far from conclusive in this respect. There is in fact some suggestion of a link between writing hand/posture and possible brain damage or defect (Allen & Wellman, 1980; Gregory & Paul, 1980; McKeever & Hoff, 1979; Strauss et al., 1984; Wellman & Allen, 1983). The exact nature of such a link remains to be defined.

Motor function. Parlow's (1978) data indicate some relation between motor control and hand/posture, but also that the neural mechanism for hand preference is not based on an asymmetry in control of fine movement. The exact nature of this phenomenon requires further clarification.

A number of studies have suggested that perhaps motor control is a relatively more bilateral affair for left-inverters than for left-normals (Halsey, Blauenstein, Wilson, & Wills, 1980; Parlow & Kinsbourne, 1981; Warshal & Spirduso, 1981; and perhaps Todor, 1980, and Wilke & Sheeley, 1979). This line of enquiry might be worth pursuing further, but it should be noted that Peters and Durding (1979) did not find such a tendency among their female subjects and Peters (1983) and Guiard and Millerat (1984) were unable to find any difference between left-inverters and left-normals in fine motor control. Note also that this possible "bilaterality" applies to motor phenomena and not, as in the original Levy and Reid hypothesis, to verbal or spatial functions.

The neurological mechanisms of hand control are discussed in the next section.

Mechanisms of Hand Control

In their 1978 paper, Levy and Reid suggested that left-inverters controlled their writing hand from the ipsilateral hemisphere and raised the question as to whether this control was exerted via transcommissural pathways or via uncrossed pyramidal tracts. While briefly expressing some uncertainty as to which alternative was the more likely, by far the main thrust of Levy and Reid's discussion was directed to supporting the idea of uncrossed motor tracts. Levy (1982) has modified her speculations with respect to hand control: ipsilateral hand control in inverters is now said to be restricted to pathways for visuomotor reactions, and it is allowed that some left-handers with left hemisphere specialization for writing would use transcommissural pathways while others (for such visuomotor reactions) would use ipsilateral motor pathways.

The question remains, however, as to whether ipsilateral control for visuomotor responses does characterize a substantial group of inverters. Some support for this notion comes from the findings of Moscovitch and Smith (1979) and Smith and Moscovitch (1979). In these and other related studies, lateralization of motor control is inferred by assuming that manual responses will be faster to stimuli initially entering the same hemisphere that controls the motor response than to stimuli entering the other hemisphere. Levy (1982) also refers to McKeever and Hoff's (1979) data as supporting this notion of ipsilateral control in inverters, but in fact only 8/15 of McKeever and Hoff's subjects showed the appropriate performance - a random result. A similar finding was also reported by McKeever and Hoff for their 1983 study. It might perhaps be argued that given the bias of possible spatial compatibility effects to favor a "contralateral" pattern, even this result is suggestive. However, Bradshaw, Nettleton, and Spehr (1982) investigated response times to visual stimuli with controls for spatial compatibility effects and found no difference in the performance of inverters and normals. Although Levy (1984) disputed Bradshaw et al.'s procedure, her criticisms were answered (Bradshaw & Umilta, 1984) and further reaction-time evidence has been found (Guiard, 1983) which negates the notion of ipsilateral motor control for inverters with respect to visuomotor tasks. Bashore, McCarthy, Heffley, Clapman, and Donchin (1982) found that with respect to readiness potentials over the sensorimotor area prior to writing, inverters showed the same contralateral control as did normals. They further concluded that although movement may be ipsilaterally controlled in a small number of left-handers, handwriting posture did not seem to index this difference, and that perhaps both left- and right-handers may vary in neurological control of hand movements. Herron, Galin, Johnstone and Ornstein (1979) also concluded on the basis of EEG measures that the right central region of the brain is involved in the control of left-hand writing for both inverters and normals. For visuomotor activities such as writing, it seems that the evidence pertaining to this modified speculation concerning ipsilateral control in inverters is at best equivocal. Consideration of hand/posture variations does not seem to account for the different findings of these studies.

A further problem regarding the idea that motor control is ipsilateral for inverters when responding manually to visual stimuli, is the question of the nature of such control. Evarts (1980) states that motor cortex activity prior to a given movement is the same regardless of whether the movement is triggered by a visual or an auditory stimulus and studies such as that of Haaxma and Kuypers (1975) suggest that intrahemispheric fibers subserve visual guidance of the premotor cortex. It would seem, on the basis of present knowledge, that the only neurological basis for the Moscovitch and Smith results is an abnormality of the visual system and/or its connections to the motor system. McKeever and Hoff (1979) suggest that isolation of visual from motor areas within the left hemisphere of left-inverters may necessitate a double transcallosal transmission for a right-hand response to the RVF stimuli. This idea is similar to that proposed and refuted by Levy (1982, p.600), namely, that left-inverters have a single commissural relay for the heterolateral condition (condition referring to visual field - response hand) but a double such relay for the homolateral condition. Levy argues that were this the case, one could expect mean reaction time to be longer in inverters and mean variabilities to be larger, which is not the case for Moscovitch and Smith's (1979) and McKeever and Hoff's (1979) findings where inverters showed shorter and less variable reaction times than normals. However, if McKeever and Hoff's hypothesis is correct, we would expect that for left-inverters, the right-hand/RVF condition would show the longest reaction time, the left-hand/LVF the shortest, and the two heterolateral conditions in between. This was indeed the case for McKeever and Hoff's (1979, Table 2, p. 449) data, and it remains so even if 3 msec is added to the two LVF conditions to eliminate right-hemisphere sensory advantages. However, Moscovitch and Smith's (1979) findings do not follow this pattern and so the hypothesis remains a very tentative one. McKeever and Hoff's later (1983) investigation indicates that disconnection of motor and visual areas does not characterize left-inverters after all.

Levy (1982, p. 600) refers to Jensen's (1980, cited in Levy, 1982) finding that reaction time variation distinguishes retarded from normal people more reliably than mean reaction time. Levy seems to be suggesting that perhaps the more efficient pathways (ipsilateral for inverters, contralateral for normals) will show less variability in reaction time than other pathways. This is born out to some extent by the findings of Moscovitch and Smith (1979) but not by those of McKeever and Hoff (1979). Inverters in both studies also showed less variability under homolateral conditions than did normals - a fact which tends to contradict the idea of variability distinguishing between the two writing posture groups with respect to the efficiency of pathways.

In further support of her contention regarding ipsilateral motor control, Levy (1982, p. 599) argues against the contralateral transcommissural hypothesis by referring to Hecaen and Sauguet's (1971) finding that agraphia for left-handers with right-hemisphere lesions was much less common than in either left- or right-handers with left-hemisphere lesions. Levy argues that such facts are difficult to reconcile with the postulate that the right hemisphere is crucial for control of left-hand writing, either as a central programmer or as a relay station. Although the left-handers of Hecaen and Sauguet's study did show a higher tendency to agraphia following right-hemisphere lesions than did dextrals, this tendency was not statistically significant. However, when interpreting Hecaen and Sauguet's findings it is very important to note that their subjects were heterogeneous with respect to location of lesion within the hemisphere. As Hecaen, Angelergues and Douzenis (1963) pointed out, there are at least three different sorts of agraphia, each related to different lesion locations (i.e. frontorolandic, temporal, and parietal). Any evaluation of agraphia findings should consider the type and degree of agraphia, the location of the lesion, and if possible the writing hand/posture of the person prior to the lesion occurring. Such analysis could be very helpful to our understanding of the neurological basis of motor control.

The findings of Brinkman and Kuypers (1973) and of Geschwind (1975) indicate that not only is there a distinction between motor skill programming areas and those controlling specific segments of muscular activity, but there is also a distinction between control of fine and gross muscle-limb segments, both with respect to brain areas and brain-to-muscle-pathways. With respect to limb segments, this model indicates that fine, distal movements which are important in writing are controlled primarily from the precentral gyrus via contralateral pyramidal pathways (and to a lesser extent extrapyramidally via contralateral pathways). Writing also involves some gross movements of the wrist, elbow, and shoulder and these segments are controlled via ipsilateral pathways. Such gross movements could produce writing although it is likely to be a cruder or slower form of writing than that which also includes pyramidal fine motor control. The location of the motor skill programming area remains unknown. Evarts (1980) has suggested that such programming may occur in the cerebellum and basal ganglia which strongly influence the motor cortex. However Phillips and Porter (1977) argue that it is not possible to attribute motor skill programming in this way, even though the interaction between the cerebellum, basal ganglia, and motor cortex is an important one.

If inverters in fact possess ipsilateral control of writing movements, one might expect them to be able to use the inverted posture of the nonpreferred hand more easily (also controlled via ipsilateral tracts) than to use the normal posture for either hand (requiring contralateral tracts). Weber (1983) timed writers using each of the four hand/postures in turn and had independent judges rank the coordination of the written product. Contrary to the expectation deriving from the ipsilateral control hypothesis, it was found that left inverters showed more skill using left-normal posture than in using either right-normal or right-inverted posture. Normal writers also adapted better to using the preferred hand with nonpreferred posture compared to using either posture with the nonpreferred hand. Such findings suggest that neural control of hand preference for writing may be more important than any factors relating to posture and thus are compatible with the model described in the previous paragraph. Furthermore, when forced to write with their right hand, both left inverters and left normals showed greater skill using normal posture than inverted. The latter fact suggests that perhaps inverted writing is largely an adaptation to left-hand writing rather than a reflection of ipsilateral pathways.

Anomalies of Midline Development

Although Levy (1982) stresses that the idea of callosal agenesis being associated with inverted posture is highly speculative, she still devotes some space to defending this notion. The simplest solution to this question seems to be to determine whether known acallosal or partially acallosal patients do in fact use inverted posture. No known studies of acallosal patients mention inversion, including the recent review by Chiarello (1980). Given the detailed investigations that have been done on such patients, surely if inverted posture were a consistent feature it would have been noted by now. Personal communication from one of the leading investigators of callosal agenesis (M. A. Jeeves, 5 November 1982) states that such patients do not invert. Using an indirect measure of callosal functioning, Miller (1983) inferred normal interhemispheric communication on a word recognition task from lowering of recognition scores when words were presented in two visual fields rather than one. Such lowering of scores was considered to reflect callosally mediated interference between the hemispheres. The lack of such lowering of scores was interpreted as indicating lack of callosally mediated communication. Findings suggested that there is less callosally mediated communication in left- than right-handers. There was, however, no clear difference in this respect between left-normals and left-inverters and, as Miller himself points out, any interpretation of this measure as indicating midline anomaly is highly speculative at this stage.

CONCLUSIONS

Some research data do seem to support the hypothesis of an association between writing hand/posture and cerebral lateralization of function, but these positive findings are like tiny islands in a sea of otherwise unsupportive results. Editorial policy of reporting more unusual or positive findings rather than unsuccessful replications or inability to confirm, would seem to give extra significance to the generally negative or weak evidence concerning any relationship between hand/posture and cerebral asymmetry.

There are other factors which may help to account for the equivocal findings which pervade the cerebral laterality field in general, not least the Levy hypothesis. Sackeim, Weiman and Grega (1984) and Levy, Heller, Banich and Burton (1983) report that arousal-level differences between hemispheres should be taken into account and Harshman, Hampson and Berenbaum (1983) claim that level of reasoning ability is relevant. Perhaps the most succinct way of taking such factors into account is that of considering subject strategies. Informal post-task questioning of subjects by Weber (1980) revealed that subjects were using a variety of strategies on the syllable and dot-location tasks. These strategies not only varied between subjects but also from trial to trial for any one subject, and sometimes combined strategies were used on a single trial. It was noted that most subjects reported using both verbal and nonverbal strategies on each task.

Clarification of the contribution of strategies to laterality effects perhaps even requires a totally different experimental paradigm, one that goes beyond the usual specification of stimulus, response, and instructional set factors. It needs to establish the subjects' actual mode of performing the task, trial by trial. Ways could be devised for subjects to report their strategies. The main practical problems would be limits in the subjects' awareness of their strategies and their capacity to describe such strategies. Overcoming such problems might require training subjects in such self-awareness and self-report. It might also be necessary to investigate and perhaps control for the possible influence of the mode of reporting such strategies, upon the direction and extent of lateral asymmetries.

In summary, there seems to be little clear and consistent evidence that would indicate a relationship between writing hand/posture and particular aspects of cerebral functional asymmetry, even when considered with respect to Levy's (1982) modifications. At best there are some suggestions of relatively more symmetrical motor function in inverters than in normal writers, but even this is tentative and requires confirmation and more precise elaboration. Such relative symmetry would also have to be accommodate the apparent fact of contralateral pyramidal motor control in such writers. In view of the above arguments, and the fact that the two hand/postures cannot be easily dichotomized (Buchtel & Rueckert, 1984; Peters, 1983) and have been subjected to a variety of distinguishing criteria (Fudin & Lembessis, 1982), we should perhaps concur with Guiard and Millerat (1984) that an inverted hand/posture may be largely an adaptive strategy for producing an adductive movement.

APPENDIX

All page references to Levy's work refer to the 1982 publication.

Levy (p. 589) says that Weber and Bradshaw give a misleading picture of the experimental findings, fail to take into account relevant clinical observations, and do not distinguish between descriptions of observations and interpretations of those observations.

Reply: Weber and Bradshaw have received unsolicited support from other researchers in the field concerning their 1981 paper. This support, together with the careful reviewing system prior to acceptance for publication, would seem to indicate that Levy's comments are unfounded. If readers follow Weber and Bradshaw's methodology for analysing Levy and Reid's work and look at the relevant literature they can check on the validity of the conclusions reached in Weber and Bradshaw's paper themselves.

Levy (p. 591) attributes to Weber and Bradshaw the belief that any reliable and valid index of lateralized function in left-handers reflects an underlying and general pattern of cerebral asymmetry, regardless of the specific process within the verbal or nonverbal domain that a task assesses.

Reply: This statement is publically contradicted by the existence of Bradshaw's (1980) review, and both authors were well aware of the evidence relating to bilateral representation of language functions in both left- and right-handers. Our focus on the more general language and spatial asymmetry reflected our concern to evaluate Levy and Reid's proposals which were in fact general.

Levy (p. 599) disagrees with Weber and Bradshaw's conclusion that the relation between hand/posture and cerebral asymmetry is "almost certainly not as predicted by Levy and Reid."

Reply: Such disagreement by Levy is surprising in view of her current modification which seems implicitly to endorse Weber and Bradshaw's statement.

Levy (p. 599) says that Weber and Bradshaw were incorrect in attributing to Levy and Reid (1978) the claim that inverters control writing via ipsilateral pathways and that Levy and Reid actually considered both this notion and that of transcommissural control.

Reply: Weber and Bradshaw agree that Levy and Reid raised both possibilities but note that the great weight of Levy and Reid's argument was in support of the ipsilateral pathway proposal, as indeed is also largely true of Levy's (1982) discussion, although perhaps to a slightly lesser degree. Thus, evaluation of Levy and Reid's proposals necessarily involved focus on their ipsilateral control notion.

Levy (p. 599) states that Levy and Reid (1978) did not discount the case described by Heilman, Coyle, Gonyea and Geschwind (1973) as asserted by Weber and Bradshaw, and that Levy and Reid cited this case to show that the transcommissural hypothesis could not be ruled out.

Reply: Certainly Levy and Reid (1978, p. 136) do mention that the Heilman et al. case suggests that the transcommissural hypothesis cannot be ruled out. However, they then proceed to discount its significance (p. 137) by suggesting that long-standing brain damage might account for this unusual case and continue to argue in favor of ipsilateral control. Thus Weber and Bradshaw's (1981, p. 78) statement is an accurate representation of the position of Levy and Reid.

Levy (p. 601) says that Weber and Bradshaw question Levy and Reid's citations re the similarity in performance patterns between left-inverters and callosal agenesic patients.

Reply: In fact Weber and Bradshaw (p. 83) agreed with Levy and Reid re this similarity.

Levy (p. 602 and elsewhere) says that Weber and Bradshaw converted Levy and Reid's speculative notions into claims.

Reply: Weber and Bradshaw's evaluation focussed on the main thrust of Levy and Reid's arguments. However, we are pleased to note that, as a possible consequence of this evaluation, Levy is now differentiating between fact and speculation.

Levy (p. 602) states that Francois, Eggermont, Evens, Logghe and

DeBock (1973) report a variety of anomalies of midline development in association with callosal dysgenesis. She also suggests that Weber and Bradshaw dispute this statement.

Reply: Weber and Bradshaw did not dispute this statement but rather commented that this study by Francois et al. made no mention of pyramidal nondecussation.

Levy (pp. 602-603) states that Levy and Reid's citing of Tennyson (1970) was to point out that, apparently, axogenesis of presumptive callosal fibers occurs in callosal agenesis, but the axons that develop fail to cross the midline and not (as implied by Weber and Bradshaw) as evidence supporting generalized anomalies of midline development in association with callosal agenesis.

Reply: Again, Levy is attacking a straw man. What Weber and Bradshaw actually said (p. 84) was that Tennyson's paper was not at all concerned with the corpus callosum or its pathology.

Levy (p. 603) suggests that Dennis' (1976) partially acallosal patient (whose intentional finger movements were accompanied by unintended, although not always homologous, movement in the fingers of the other hand) shows that synkinesis does occur in such patients, and that this somehow contradicts Weber and Bradshaw's challenge to Levy and Reid's logic concerning synkinesis in acallosal cases.

Reply: This argument does not seem at all clear and the reader is referred to pages 83 to 85 of Weber and Bradshaw's (1981) paper for the appropriate context in which they referred to Dennis's study.

REFERENCES

Ajersch, M.K., and Milner, B. 1983. Handwriting posture as related to cerebral speech lateralization, sex, and writing hand, Human Neurobiology, 2, 143-145.

Allen, M., and Wellman, M. M. 1980, Hand position during writing, cerebral laterality and reading: age and sex differences, Neuropsychologia, 18, 33-40.

Andrews, R.J. 1977, Aspects of language lateralization correlated with family handedness, Neuropsychologia, 15, 769-778.

Bashore, T.R., McCarthy, G., Heffley, E.F., Clapman, R.M., and Donchin, E. 1982, Is handwriting posture associated with differences in motor control?: an analysis of asymmetries in readiness potential, Neuropsychologia, 20, 327-346.

Bashore, T. R., Nydegger, R. V., and Miller, H. 1982, Left visual field superiority in a letter naming task for both left- and right-handers, Cortex, 18, 245-256.

Beaumont, J. G., and McCarthy, R. 1981, Dichotic ear asymmetry and writing posture, Neuropsychologia, 19, 469-472.

Bradshaw, J. L. 1980. Right-hemisphere language: familial and nonfamilial sinistrals, cognitive deficits and writing hand position in sinistrals, and concrete-abstract, imageable-nonimageable dimensions in word recognition. A review of interrelated issues, Brain and Language, 10, 172-188.

Bradshaw, J. L., and Taylor, M. J. 1979, A word-naming deficit in nonfamilial sinistrals? Laterality effects of vocal responses to tachistoscopically presented letter strings, Neuropsychologia, 17, 21-32.

Bradshaw, J. L., and Umilta, C. 1984. A reaction time paradigm can simultaneously index spatial compatibility and neural pathway effects: a reply to Levy, Neuropsychologia, 22, 99-101.

Bradshaw, J. L., Nettleton, N., and Spehr, K. 1982, Sinistral inverters do not possess an anomalous visuomotor organization, Neuropsychologia, 20, 605-609.
Bradshaw, J. L., Nettleton, N., and Taylor, M. J. 1980. "Right Hemisphere Language and Cognitive Deficit in Sinistrals?" Unpublished manuscript. Available from J. L. Bradshaw, Department of Psychology, Monash University, Clayton, Victoria, Australia 3168.
Brinkman, J., and Kuypers, H. G. J. M. 1973, Cerebral control of contralateral and ipsilateral arm, hand and finger movements in the split-brain rhesus monkey, Brain, 96, 653-674.
Bryden, M. P. 1970. Laterality effects in dichotic listening: relations with handedness and reading ability in children, Neuropsychologia, 8, 443450.
Bryson, S. E., and MacDonald, V. 1984. The development of writing posture in left-handed children and its relation to sex and reading skills, Neuropsychologia, 22, 91-94.
Buchtel, H. A., and Rueckert, L. 1984. Hand posture in writing: possible artifacts from self report, Cortex, 20, 435-439.
Campbell, R. 1979. Left-hander's smiles: asymmetries in the projection of a posed expression, Cortex, 15, 571-579.
Carter, G. L. and Kinsbourne, M. 1979. The ontogeny of right cerebral lateralization of spatial mental set, Developmental Psychology, 15, 241-245.
Chiarello, C. 1980. A house divided? Cognitive functioning with callosal agenesis, Brain and Language, 11, 128-158.
Dabbs, J. M., and Choo, G. 1980. Left-right carotid blood flow predicts specialized mental ability Neuropsychologia, 18, 711-713.
Dennis, M. 1976. Impaired sensory and motor differentiation with corpus callosum agenesis: a lack of callosal inhibition during ontogeny?, Neuropsychologia, 14, 455-469.
Duda, P. D., and Kirby, H. W. 1980. Effects of eye-movement controls and frequency levels on accuracy of word recognition, Perceptual and Motor Skills, 50, 979-985.
Eastwood, M. R., and Stiasny, S. 1978. Cerebral dominance and ECT, American Journal of Psychiatry, 135, 4.
Education Department, Victoria, Australia. 1964, "Course of Study for Primary Schools: Handwriting".
Ellis, A. W., and Miller, D. 1981, Left and wrong in adverts: neuropsychological correlates of aesthetic preference, British Journal of Psychology, 72, 225-229.
Evarts, E. V. 1980, Brain mechanisms in voluntary movement in: "Neural Mechanisms of Behavior," D. McFadden, ed., Springer-Verlag, New York. 223-259.
Francois, J., Eggermont, E., Evens, L., Logghe, N., and DeBock, F. 1973, Agenesis of the corpus callosum in the median facial cleft syndrome and associated ocular malformations, American Journal of Ophthalmology, 76, 241-245.
Fudin, R., and Lembessis, E. 1982. Note on criteria for writing posture used to test Levy and Reid's cerebral organization hypotheses, Perceptual and Motor Skills, 54, 551-556.
Gainotti, G., Caltagirone, C., Miceli, G., and Masullo, C. 1981, Selective semantic-lexical impairment of language comprehension in right brain-damaged patients. Brain and Language, 13, 201-211.
Geffen, G., Bradshaw, J. L., and Nettleton, N. C. 1972, Hemispheric asymmetry: verbal and spatial encoding of visual stimuli, Journal of Experimental Psychology, 95, 25-31.
Geschwind, N. 1975, The apraxias: neural mechanisms of disorders of learned movement, American Scientist, 63, 188-195.
Gordon, H. W., and Carmon, A. 1976, Transfer of dominance in speed of verbal response to visually presented stimuli from right to left hemisphere, Perceptual and Motor Skills, 42, 1091-1100.

Gregory, R., and Paul, J. 1980, The effects of handedness and writing posture on neuropsychological test results, Neuropsychologia, 18, 231-235.

Guiard, Y. 1984, Spatial compatibility effects in the writing page: a comparison of left-handed inverters and noninverters, Acta Psychologica, 57, 17-28.

Guiard, Y, and Millerat, F. 1984, Writing postures in left-handers: inverters and hand-crossers, Neuropsychologia, 22, 535-538.

Haaxma, R., and Kuypers, H. G. J. M. 1975, Intrahemispheric cortical connexions and visual guidance of hand and finger movements in the rhesus monkey, Brain, 98, 239-260.

Halsey, J. H., Blauenstein, U. W., Wilson, E. M., and Wills, E. L. 1980, Brain activation in the presence of brain damage, Brain and Language, 9, 47-60.

Harcum, E. R. 1978, Lateral dominance as a determinant of temporal order of responding in: "Asymmetrical Function of the Brain", M. Kinsbourne, ed., Cambridge University Press, Cambridge, pp 141-226.

Harshman, R. A., Hampson, E., and Berenbaum, S. A. 1983. Individual differences in cognitive abilities and brain organization, part 1: sex and handedness differences in ability, Canadian Journal of Psychology, 37, 144-192.

Hecaen, H., and Sauguet, J. 1971. Cerebral dominance in left-handed subjects, Cortex, 7, 19-48.

Hecaen, H., Angelergues, R., and Douzenis, J. A. 1963, Les agraphies, Neuropsychologia, 1, 179-208.

Heilman, K. M., Coyle, J. M., Gonyea, E. F., and Geschwind, N. 1973, Apraxia and agraphia in a left-hander, Brain, 96, 21-28.

Hellige, J. B., and Cox, P. J. 1976, Effects of concurrent verbal memory on recognition of stimuli from left and right visual fields, Journal of Experimental Psychology: Human Perception & Performance, 2, 210-221.

Herron, J., Galin, D., Johnstone, J., and Ornstein, R. E. 1979. Cerebral specialization, writing posture, and motor control of writing in left-handers, Science, 205, 1285-1289.

Hines, D. 1978, Visual information processing in the left and right hemispheres, Neuropsychologia, 16, 593-600.

Jones, B. 1980, Sex and handedness as factors in visual - field organization for a categorization task, Journal of Experimental Psychology: Human Perception & Performance, 6, 494-500.

Jones, B., and Santi, A. 1978, Lateral asymmetries in visual perception with and without eye movements, Cortex, 14, 164-168.

Kershner, J., Thomae, R., and Callaway, R. 1977, Nonverbal fixation control in young children induces a left-field advantage in digit recall, Neuropsychologia, 15, 569-576.

Kimura, D. 1961, Cerebral dominance and the perception of verbal stimuli, Canadian Journal of Psychology, 15, 166-171.

Kimura, D. 1969, Spatial localization in left and right visual fields, Canadian Journal of Psychology, 23, 445-458.

Knopman, D. S., Rubens, A. B., Klassen, A. C., and Meyer, M. W. 1982, Regional cerebral blood flow correlates of auditory processing, Archives of Neurology, 39, 487-493.

Lawson, N. C. 1978, Inverted writing in right- and left-handers in relation to lateralizaton of face recognition, Cortex, 14, 207-211.

Levy, C. M., and Bowers, D. 1974, Hemispheric asymmetry of reaction time in a dichotic discrimination task, Cortex, 10, 18-25.

Levy, J. 1972, Lateral specialization of the human brain: behavioral manifestations and possible evolutionary basis in: "The Biology of Behavior," J. Kiger, ed., Oregon State University Press, Corvallis, pp.158-180.

Levy, J. 1974, Psychobiological implications of bilateral asymmetry in: "Hemisphere Function in the Human Brain", S. J. Dimond and J. G. Beaumont, eds., Elek Science, London, pp. 121-183.

Levy, J. 1982, Handwriting posture and cerebral organization: how are they related? Psychological Bulletin, 91, 589-608.
Levy, J. 1984, Can a reaction time paradigm simultaneously index arm position effects, spatial compatibility effects and neural pathway effects? Comments on Bradshaw, Nettleton and Spehr, Neuropsychologia, 22, 95-97.
Levy, J., Heller, W., Banich, M. T., and Burton, L. A. 1983, Are variations among right-handed individuals in perceptual asymmetry caused by characteristic arousal differences between hemispheres, Journal of Experimental Psychology:Human Perception & Performance, 9, 329-359.
Levy, J., and Reid, M. 1976, Variations in writing posture and cerebral organization, Science, 194, 337-339.
Levy, J., and Reid, M. 1978, Variations in cerebral organization as a function of handedness, hand posture in writing, and sex, Journal of Experimental Psychology: General, 107, 119-144.
Levy, J., Trevarthen, C., and Sperry, R. W. 1972, Perception of bilateral chimeric figures following hemispheric deconnexion, Brain, 95, 61-78.
Mancuso, R. P., Lawrence, A. F., Hintze, R. W., and White, C. T. 1979, Effect of altered central and peripheral visual field stimulation on correct recognition and visual evoked response, International Journal of Neuroscience, 9, 113-122.
McKeever, W. F. 1979, Handwriting posture in left-handers: sex, family sinistrality and language laterality correlates, Neuropsychologia, 17, 429-444.
McKeever, W. F., and Hoff, A. L. 1979, Evidence of a possible isolation of left hemisphere visual and motor areas in sinistrals employing an inverted handwriting posture, Neuropsychologia, 17, 445-455.
McKeever, W. F., and Hoff, A. L. 1983, Further evidence of measurable interhemispheric transfer time in left-handers who employ an inverted handwriting posture, Bulletin of the Psychonomic Society, 21, 255-258.
McKeever, W. F., and Huling, M. D. 1970, Left cerebral hemisphere superiority in tachistoscopic word-recognition performance, Perceptual and Motor Skills, 30, 763-766.
McKeever, W. F., and VanDeventer, A. D. 1980, Inverted handwriting position, language laterality, and the Levy-Nagylaki genetic model of handedness and cerebral organization, Neuropsychologia, 18, 99-102.
Miller, L. K. 1983, Hemifield independence in the left-handed, Brain and Language, 20, 33-43.
Moscovitch, M., and Olds, J. 1982, Asymmetries in spontaneous facial expressions and their possible relation to hemispheric specialization, Neuropsychologia, 20, 71-81.
Moscovitch, M. and Smith, L. C. 1979, Differences in neural organization between individuals with inverted and noninverted handwriting postures, Science, 205, 710-713.
Natale, M., Gur, R. E., and Gur, R. C. 1983, Hemispheric asymmetries in processing emotional expressions, Neuropsychologia, 21, 555-565.
Newcombe, F., and Ratcliff, G. 1973, Handedness, speech lateralization and ability, Neuropsychologia, 11, 399-407.
Parlow, S. 1978, Differential finger movements and hand preference, Cortex, 14, 608-611.
Parlow, S. E., and Kinsbourne, M. 1981, Handwriting posture and manual motor asymmetry in sinistrals, Neuropsychologia, 19, 687-696.
Patterson, K. 1981, Neuropsychological approaches to the study of reading, British Journal of Psychology, 72, 151-174.
Peters, M. 1983, Inverted and noninverted lefthanders compared on the basis of motor performance and measures related to the act of writing, Australian Journal of Psychology, 35, 405-416.
Peters, M., and Durding, B. 1979, Left-handers and right-handers compared on a motor task, Journal of Motor Behavior, 11, 103-111.
Phillips, C. G., and Porter, R. 1977, "Corticospinal Neurones. Their Role in Movement", Academic Press, London.

Porac, C., Coren, S., and Searleman, A. 1983, Inverted versus straight handwriting posture: a family study, Behavioral Genetics, 13, 311-320.
Reuter-Lorenz, P. A., Givis, R. P., and Moscovitch, M. 1983, Hemispheric specialization and the perception of emotion: evidence from right-handers and from inverted and noninverted left-handers, Neuropsychologia, 21, 687-692.
Satz, P., Achenbach, K., and Fennell, E. 1967, Correlations between assessed manual laterality and predicted speech laterality in a normal population, Neuropsychologia, 5, 295-310.
Sackeim, H. A., Weiman, A. L., and Grega, D. M. 1984, Effects of predictors of hemispheric specialization on individual differences in hemispheric activation, Neuropsychologia, 22, 55-64.
Searleman, A. 1977, A review of right hemisphere linguistic capabilities, Psychological Bulletin, 84, 503-528.
Searleman, A. 1980, Subject variables and cerebral organization for language, Cortex, 16, 239-245.
Searleman, A., Porac, C., and Coren, S. 1982, The relationship between birth stress and writing hand posture, Brain and Cognition, 1, 158-164.
Searleman, A., Porac, C., and Coren, S. 1984. Writing hand posture and four indexes of lateral preference, Brain and Cognition, 3, 86-93.
Shanon, B. 1978, Writing positions in Americans and Israelis, Neuropsychologia, 16, 587-591.
Smith, L. C., and Moscovitch, M. 1979, Writing posture, hemispheric control of movement and cerebral dominance in individuals with inverted and noninverted hand postures during writing, Neuropsychologia, 17, 637-644.
Springer, S. P., and Searleman, A. 1980, Left-handedness in twins: implications for the mechanisms underlying cerebral asymmetry of function in: "Neuropsychology of Left-Handedness," J. Herron, ed., Academic Press, New York, pp. 139-158.
Strauss, E., Wada, J., and Kosaka, B. 1984, Writing hand posture and cerebral dominance for speech, Cortex, 20, 143-147.
Tapley, S. M., and Bryden, M. P. 1983, Handwriting position and hemispheric asymmetry in right-handers, Neuropsychologia, 21, 129-138.
Tennyson, V. M. 1970, The fine structure of the developing nervous systen in: "Developmental Neurobiology", W. A. Himwich, ed., Charles C. Thomas, Springfield, Illinois, pp. 47-116.
Thistle, A. B. 1976, Dichotic listening and orientation of the writing hand in left-handers, Journal of Acoustic Society of America, 59, 1,s6.
Todor, J. I. 1980, Sequential motor ability of left-handed inverted and non-inverted writers. Acta Psychologica, 44, 165-173.
Urcuioli, R. J., Klein, R. M., and Day, J. 1981, Hemispheric differences in semantic processing: category matching is not the same as category membership, Perception and Psychophysics, 29, 343-351.
Volpe, B. T., Sidtis, J. J., and Gazzaniga, M. S. 1981, Can left-handed writing posture predict cerebral language laterality?, Archives of Neurology, 38, 637-638.
Wapner, W., Hamby, S., and Gardner, H. 1981, The role of the right hemisphere in the apprehension of complex linguistic materials, Brain and Language, 14, 15-33.
Warshal, D., and Spirduso, W. W. 1981, Concurrent verbal-manual performance in inverted and non-inverted writers, Perceptual and Motor Skills, 53, 123-126.
Warrington, E. K., and Pratt, R. T. C. 1973, Language laterality in left-handers assessed by unilateral ECT, Neuropsychologia, 11, 423-428.
Warrington, E. K., and Pratt, R. T. C. 1981, The significance of laterality effects, Journal of Neurology, Neurosurgery and Psychiatry, 44, 193-196.
Weber, A. M. 1980, "Writing Hand/Posture and Cerebral Lateralization", Unpublished master's thesis, University of Melbourne.
Weber, A. M. 1983, Capacity to vary writing hand/posture in relation to the Levy and Reid model for the control of writing, Journal of Motor Behavior, 15, 19-28.

Weber, A. M., and Bradshaw, J. L. 1981, Levy and Reid's neurological model in relation to writing hand/posture: an evaluation, Psychological Bulletin, 90, 74-88.
Wellman, M. M., and Allen, M. 1983, Variations in hand position, cerebral lateralization, and reading ability among right-handed children, Brain and Language, 18, 277-292.
Wilke, J. T. and Sheeley, E. M. 1979, Muscular or directional preferences in finger movement as a function of handedness, Cortex, 15, 561-569.
Zaidel, E., and Peters, A. M. 1981, Phonological encoding and ideographic reading by the disconnected right hemisphere: two case studies, Brain and Language, 14, 205-234.
Zurif, E. B. and Bryden, M. P. 1969, Familial handedness and left-right differences in auditory and visual perception, Neuropsychologia, 7, 179-187.

INDIVIDUAL DIFFERENCES IN DYNAMIC PROCESS ASYMMETRIES IN THE NORMAL AND PATHOLOGICAL BRAIN*

John H. Gruzelier

Department of Psychiatry
Charing Cross and Westminster Medical School
Fulham Palace Road
London, U.K. W6 8RF

STRUCTURAL VERSUS DYNAMIC PROCESS ASYMMETRIES

Dynamic concepts of brain function have played an important role in the history of neuropsychology. They were integral to Holism, a nineteenth century school of thought which had its origins in the empirical studies of Flourens (1794-1869), who deduced from extirpation experiments with birds that individual areas had specific effects ('action propre') but also generalised influences, such that the removal of any part affects every other part ('action commune'). This dynamic view was upheld by Lashley (1929) who proposed principles of mass action and equipotentiality from evidence that learning impairments in rats were inversely related to the extent of the brain tissue destroyed (mass action), but the precise location of the destruction was unimportant because of an equivalence of function (equipotentiality). Later Luria (1973) introduced the concept of a functional system whereby complex processes involve integrative participation of widespread regions, the involvement of which could alter with maturation (Vygotsky, 1963).

Strict localisationism has been the more influential school of thought on Western neuropsychology. Its empirical foundations arose from discoveries such as those of Broca in 1861 and Wernicke in 1874 of the relationship between focal lesions of the left hemisphere and specific language disorders. Investigations of cerebral asymmetry have followed in the tradition of strict localisationism, initially attempting to delineate absolute differences in function between the hemispheres but now giving way largely to a search for relative differences in function, in view of evidence of recovery of function after gross unilateral lesions, and the failure to find clear-cut asymmetries in the intact brain. Neuroanatomical asymmetries, thought to provide a plausible foundation for language lateralisation, do not support the high estimates of left hemispheric dominance based on inferences from hand dominance; around 30% of dextrals do not show structural asymmetries in favour of the left hemisphere. Hemispheric equivalence of function is gaining popularity to account for language lateralisation in a substantial proportion of sinistrals (Satz, 1979), and may contribute to gender differences as an account of language

----

*This research was supported by the Wellcome Trust, The Science and Engineering Research Council and by the MRC in the form of student fellowships

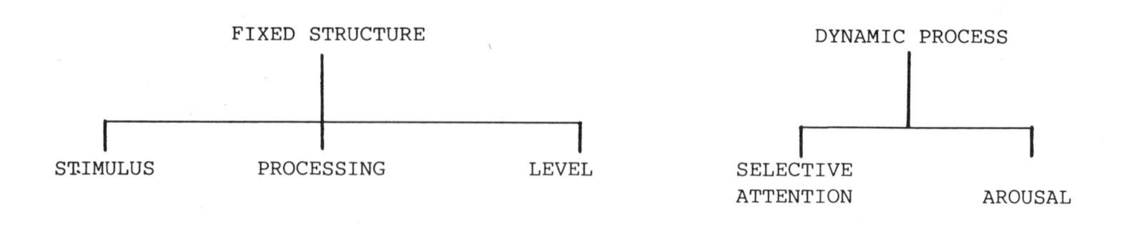

Fig. 1. The two classes of cerebral asymmetry (see text for details).

language lateralisation in females. Gender together with handedness form the two widely accepted individual differences in hemispheric specialisation.

Disquiet at exclusively structural determinants of lateralisation has arisen from consideration of the temporal variation in asymmetries seen across trials or sessions in the same subject revealed by laterality measures such as dichotic listening, divided visual field presentation, haptic sorting and so forth (Cohen, 1982). These are attributed usually on a post-hoc basis to factors such as unfamiliarity, fatigue, stress or shifts in attention or cognitive strategy. Cohen (1982) p.104 posits a combined structural-dynamic model. "To extend the power of the structural model so as to explain this kind of variability, it is necessary to postulate some form of dynamic mechanism which controls or influences the functioning of fixed structures. Dynamic models therefore represent an extension or amplification of structural models, rather than a distinct alternative. Although an increase in explanatory power is achieved by marrying structural and dynamic models, there is a corresponding loss of predictive power because the principles which govern the dynamic mechanism are not at present sufficiently well specified." Here preliminary attempts are made at the specification of dynamic process asymmetries.

An adaptation of Cohen's model is shown in Fig. 1. On the one hand there are cerebral asymmetries presumed to have structural determinants such as verbal versus visuospatial and analytic versus gestalt processes, etc. On the other hand, there are asymmetries in dynamic processes such as arousal and attention. One modification of the schema is to suggest that instead of subsuming arousal under attention, that arousal and attention may involve different mechanisms. An example of this independence has arisen from dichotic listening studies of paranoid schizophrenic patients where a presumed left hemispheric bias in activation seen in free recall has placed a constraint on ear advantages in cued recall when attention was directed to the left ear (Gruzelier, this volume Fig. 2).

The most developed theory of a dynamic process in competition with a structural one belongs to Kinsbourne's (1970, 1975) model of attention. Kinsbourne posits that hemispheric activation patterns are reciprocally related in a finely tuned balance. Moment-to-moment imbalances determine the direction of orientational responses in the lateral plane. Through incompatible expectancies as to the nature of the task, priming carried over from preceding tasks, or distraction from concurrent tasks, attention may be diverted away from the structural process best suited for analysis. One of the reasons we suggest why tests of the model have not always been confirmatory is because activational imbalances may be out of keeping not only with structural process requirements but also with attention. Dynamic processes themselves may be in conflict.

Activational asymmetries independent of attention have received scant attention in normal subjects, though there is an extensive literature in psychiatric patients (e.g., Gruzelier & Flor-Henry, 1979; Flor-Henry & Gruzelier, 1983). One exception is found in the work of Tucker (1981) who has proposed that anxiety through over-arousal places a processing load on the left hemisphere giving rise to a right hemisphere processing dominance. Our investigation of this proposition with a within-subject design will be outlined because it exemplifies how a dynamic process can reverse structural asymmetries. We examined the effect of anxiety on a divided visual-field task using a life-stressor to manipulate level of anxiety (Gruzelier & Phelan, 1986). The effect of the stressor was validated with psychophysiological measures of sympathetic reactivity and self-report scales of anxiety. Thirty three medical students, of whom nine were female, were tested before an examination and again at a relatively unstressful time at least four weeks before or after, with order of testing counterbalanced. Sympathetic activity was monitored with electrodermal activity throughout a stimulus habituation paradigm consisting of 22 tones of 70dB SPL, 1 second duration, 1000 Hz, with controlled rise and decay times, and 20-40 second interstimulus intervals. The 15th and 19th tones were of 690 Hz and 2160 Hz to measure dishabituation. Three other measures were obtained; the rate of habituation of orienting responses to non-signal tones, the number of non-specific responses between tones, and levels of tonic activity. It has been shown that electrodermal responses increase with anxiety, that tonic levels rise and rate of habituation is retarded (Stern & Janes, 1973); studies which have included the effects of examination stress on students (Maltzman, Smith & Cantor, 1971). The IPAT anxiety questionnaire was administered in each condition to a sub-group (N=18); seven had scores above 40 which was in the psychopathological range (Krug, Scheier & Cattell, 1976).

Convincing evidence of increased stress prior to an examination was provided by the electrodermal measures. Rate of habituation of orienting

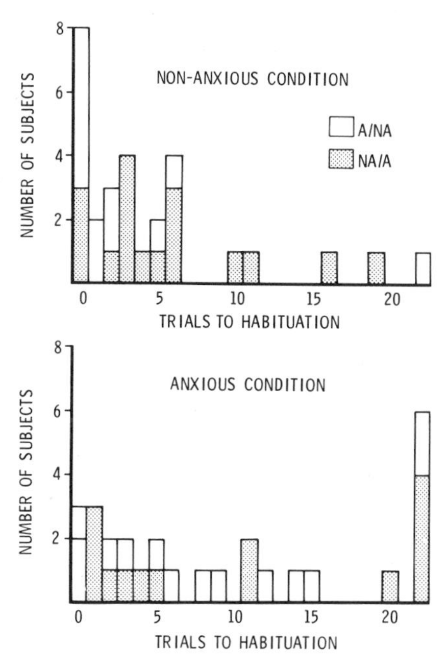

Fig. 2. Rate of habituation in students in anxious and non-anxious conditions.

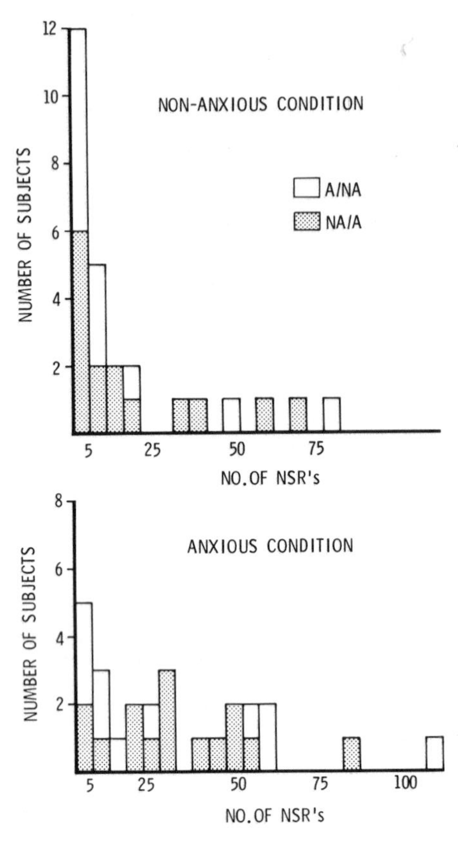

Fig. 3. Number of electrodermal nonspecific responses.

responses is shown in Fig. 2. Habituation was retarded under stress ($p<0.006$). Non-specific responses, shown in Fig. 3, were more frequent prior to exams ($p<0.0007$), as were orienting responses ($p<0.0009$), shown Fig. 4. Skin conductance levels rose from a mean of 21.56 mhos to 26.03 mhos ($p<0.03$, one tailed). Self-report anxiety increased from 28.4 to 31.1 ($p<0.05$, one tailed), particularly the frustration-tension sub-scale which increased from 7.00 to 8.37 ($p<0.03$, one tailed).

Hemispheric asymmetries were examined with a divided visual-field task shown to produce a left hemisphere advantage in normal subjects (Connolly, Gruzelier & Manchanda, 1983). The task required discrimination between consonants and vowels with a manual reaction time recorded to target stimuli (vowels). Stimuli were presented randomly in either visual field for 240 trials, 81 of which were targets. Stimuli were of 50 msec duration and subtended an angle of 4 degrees on either side of a centrally positioned 500 msec warning light which occurred 400 msecs before the stimulus. The interstimulus interval was 4 seconds.

In the non-anxious condition there was a nine msec right visual-field advantage whereas in the anxious condition the left visual-field showed a six msec advantage ($p<0.02$). The results are shown in Fig. 5. Between hemisphere differences were both significant ($p<0.05$). Within hemisphere post-hoc comparisons revealed that the right hemisphere showed the greater variation between conditions ($p<0.01$). Anxiety significantly improved right hemisphere performance to an extent that it surpassed the performance of the

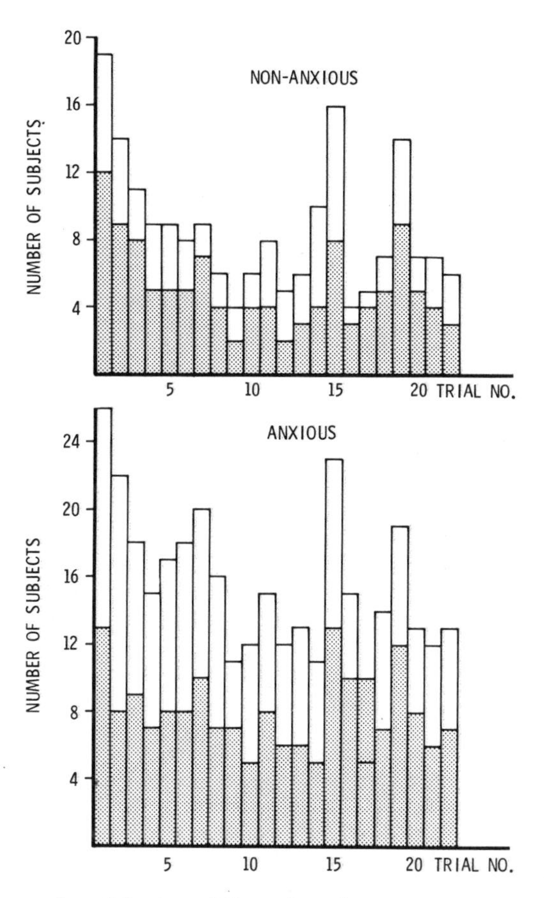

Fig. 4. Number of subjects with orienting responses onn each trial.

left hemisphere on either occasion. Left hemisphere performance did not vary.

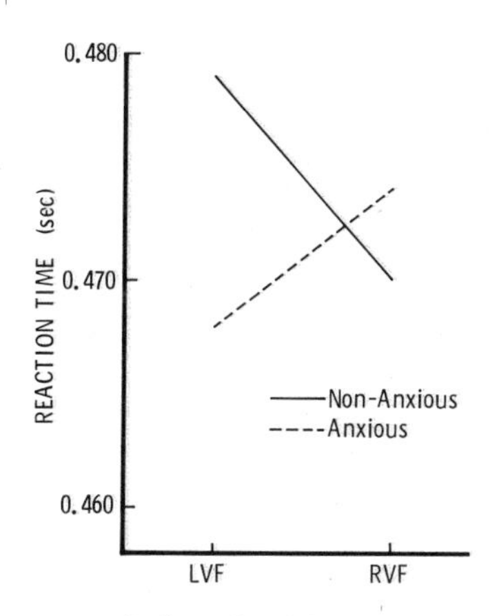

Fig. 5. Reaction times to verbal stimuli presented in the left and right visual fields.

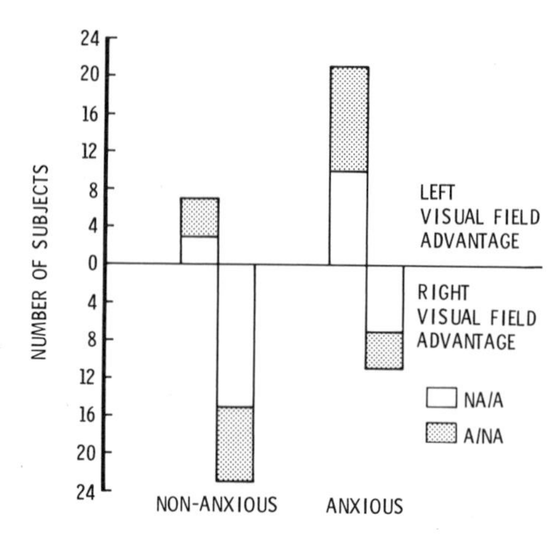

Fig. 6. The number of subjects showing left or right visual field advantages when anxious and non-anxious.

Considering individual subjects Fig. 6, 23/31 subjects showed a left hemisphere advantage in the non-anxious condition while 21/31 subjects showed a right hemisphere advantage in the anxious condition ($p<0.001$). 22/31 subjects showed a shift between conditions towards superior right hemisphere performance in the anxious condition. Thus, more than two-thirds of subjects showed a reversal of a left hemisphere advantage for verbal processing under anxiety, an effect largely due to an enhancement in right hemisphere processing. Thus, there was support for the prediction that the relative difference between the hemispheres under anxiety would favour the right hemisphere, but there was no evidence of a deterioration in performance in the left hemisphere as might be supposed if anxiety placed a processing load on the left hemisphere. Nor would the model predict an improvement in right hemispheric performance. This will be returned to in a later section.

RATE OF HABITUATION AND INDIVIDUAL DIFFERENCES IN CEREBRAL ASYMMETRY

A further implication from the study of examination stress was that rate of habituation of electrodermal activity, which varied with anxiety, may itself represent an individual difference with a bearing on hemispheric activational balance. In support of this contention, correlations have been found between rate of habituation of the orienting response to tones and the lateral asymmetry in response amplitudes, an asymmetry which is presumed to provide one reflection of hemispheric activation. This was discovered in three experiments (Gruzelier, Eves & Connolly, 1981a) involving, in the first, patients awaiting minor surgery in the Copenhagen General Hospital; in the second, first year medical students at Charing Cross Hospital and, in the third, hospital staff at the latter hospital. The tone series was similar but shorter to the one above involving 13 70 dB tones in experiments 1 and 2 and 10 tones in experiment 3. For the hospital staff this was followed by a 90 dB SPL series, with stimuli of five seconds duration and with the 10th of the 11 tones of ten seconds duration. Correlations between rate of habituation to a criterion of three successive failures to respond and lateral asymmetries in orienting responses, calculated as a laterality index, 2(left-right/left+right), are shown in Table 1 for the three experiments for the total number of subjects, for male and female subjects considered separately, and for sinistrals. In all three studies significant

Table 1 Correlations between trials to habituation and lateral asymmetries in orienting response amplitudes.

GROUP	SEX	N	Rho	P
			70-dB Tone Intensity	
Surgical Patients	M	16	.53	.05
Hospital Staff	M	18	.56	.02
	F	10	.52	.10
	M + F	28	.46	.01
Students	M	37	.46	.01
	F	25	.13	n.s.
	M + F	62	.39	.002
All Groups Combined	M	71	.49	.001
	F	35	.24	n.s.
	M + F	106	.41	.001
Sinistrals	M + F	9	.42	n.s.
			90-dB Tone Intensity	
Hospital Staff	M	18	.46	.06
	F	13	.18	n.s.
	M + F	31	.53	.01

correlations were obtained. Fast habituation was associated with larger left than right hand response amplitudes and slow habituation with larger right than left hand response amplitudes. The relationship was found to be true of males whereas in females it was weaker, falling short of significance in the hospital staff and absent altogether in female students. The relationship also fell short of significance in sinistrals.

It is well documented that non-specific responses are typically more frequent in subjects who are slow habituators; here correlations ranged between 0.76 and 0.84 for the total samples. The same relationship held for non-specific responses between lateral asymmetries in response amplitude and the number of responses. A scatter plot is shown in Fig. 7. Correlations ranged between 0.45 and 0.73 for the total samples, all significant at the $p<0.001$ level, and were significant for both men and women.

This cerebral asymmetry has an important bearing on phasic reactivity in the electrodermal system whether responses are stimulus specific or non-specific. However, this relationship does not extend to tonic levels of skin conductance activity. Correlations between asymmetries of orienting responses and tonic levels ranged between 0.06 - 0.32 (ns). Furthermore, the partialling-out of the influence of levels on amplitudes still gave correlations of 0.39 and 0.49 ($p<0.06$ - $p<0.002$) between the asymmetry in responses and rate of habituation. An independance of lateral asymmetries in levels and responses has been a consistent finding in our studies, as can be seen in two illustrative cases (Figs. 8a + b). This dissociation deserves further neurophysiological investigation and may imply a more focal distribution of influences on phasic responses.

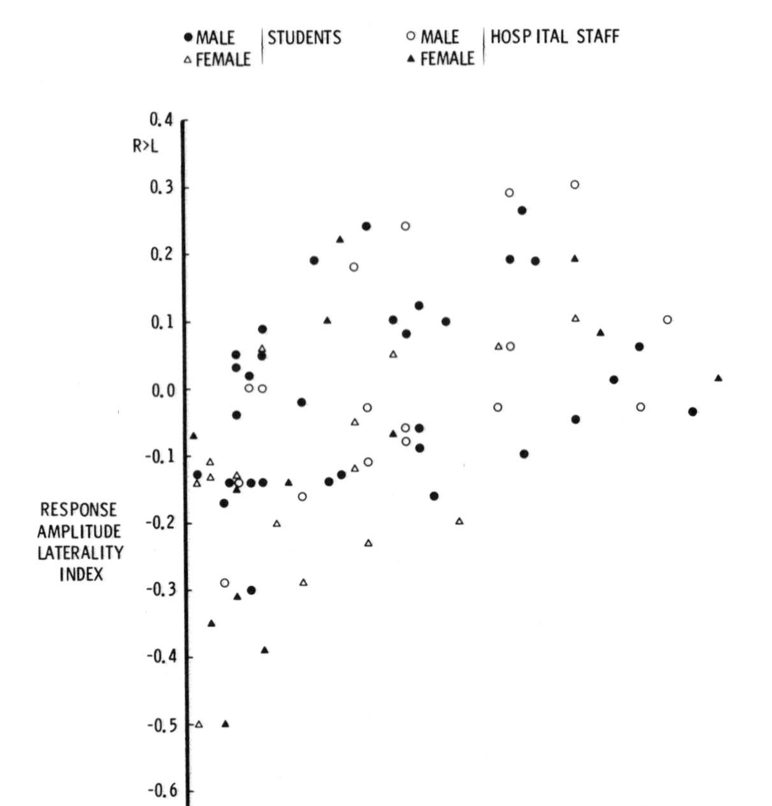

Fig. 7. Scatterplot of the number of non-specific responses as a function of the lateral asymmetry in reponses amplitudes.

Given that there was a reversal in lateral asymmetries, two types of hemispheric influence may underly the effect on response amplitudes as rate of habituation decreased. Either there is a reciprocal relationship between the hemispheres, as Kinsbourne's theory might suppose, such that when the balance is tipped one way habituation is fast and when tipped the other habituation is slow, or alternatively control of habituation may be unilateral such that when control is strengthened or weakened, amplitudes will show variation predominantly on the hand contralateral to the controlling hemisphere. We reasoned that reciprocal control would lead to correlations between rate of habituation and unilateral amplitudes that were opposite in sign on one hand compared with the other, whereas unilateral hemispheric control should provide significant correlations on one hand only. The obtained correlations supported reciprocal hemispheric control. They were negative (-0.25 to -0.45) with the left hand and positive (+0.28 to +0.69) with the right hand.

The importance for cerebral asymmetry of individual differences in rate of habituation is underscored by the finding that rate of habituation of

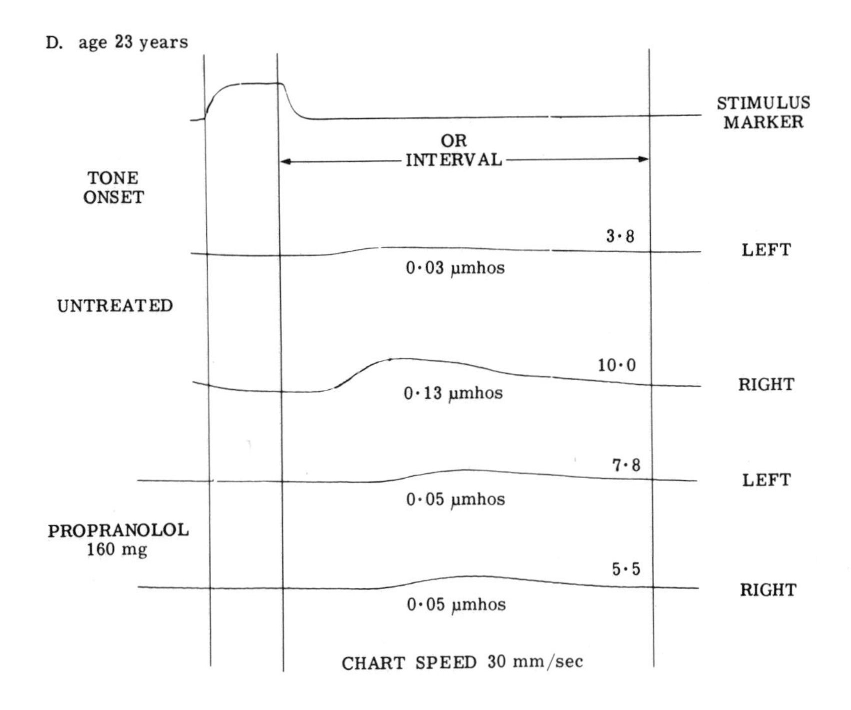

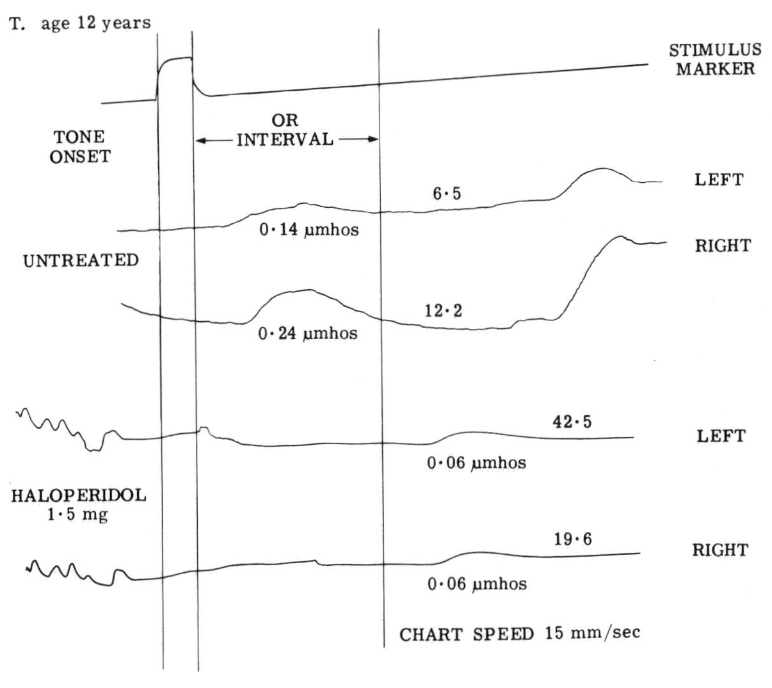

Fig. 8a & b. Bilateral electrodermal polygraph tracings in two patients during presentation of a one second tone.

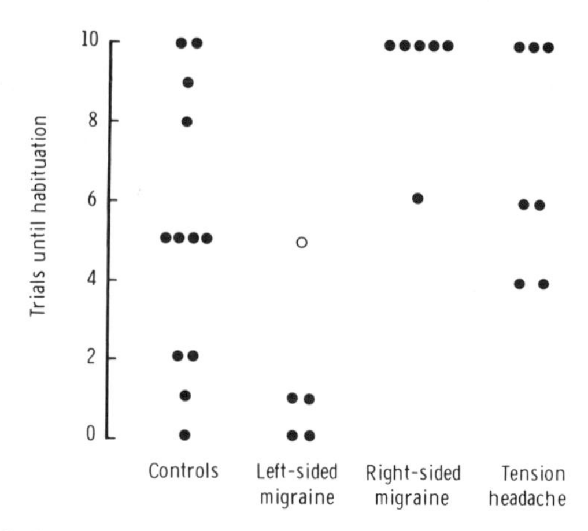

Fig. 9. Rate of habituation to flashes. The unfilled circle represents the sinistral migraineur.

electrodermal orienting responses is predictive of the laterality of pain in migraine in patients where the headache is consistently lateralised (Gruzelier et al., 1986). Two experiments were undertaken recording rate of habituation between headaches. In the first experiment stimuli were flashes of light and, in the second, moderate intensity (70 dB) tones in a series similar to the ones used in the experiments above. It was hypothesised, that, as patients often report sensory discomfort preceding and during attacks, central mechanisms regulating sensory imput may be deficient and this may have a hemispheric basis. In the first experiment, 5 left-sided migraineurs were compared with 6 right-sided cases, 7 with bilateral tension headache and 12 medically fit controls. Rate of habituation is shown in Fig. 9. There was no overlap between left and right-sided migraineurs and the one sinistral was midway between the groups. Patients with left-sided headache were on average faster to habituate than all other groups whereas patients with right-sided headache were the slowest to habituate. Extremes of habituation in migraineurs were replicated in the second experiment, where seven left-sided cases were compared with six right-sided cases, see

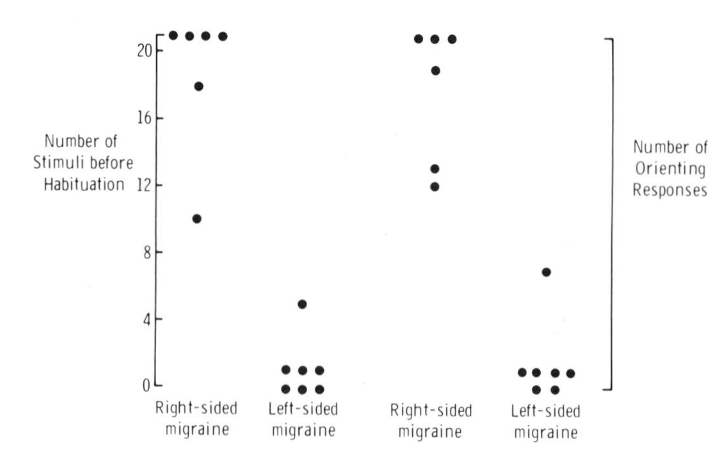

Fig. 10. Rate of habituation and number of orienting responses to tones in migraineurs.

Fig. 10. All were dextral and free of medication. Here the extremes of habituation found parallels in the number of non-specific responses and in tonic electrodermal responsiveness measured as the difference between the highest and lowest skin conductance level. In both experiments, the patients were all severe cases having migraine attacks as often as one a week. The laterality of headache was the only clinical feature that distinguished them. The association of extremes of habituation of the orienting response with the laterality of pain makes it tempting to conclude that hemispheric imbalance is a predispositional factor in migraine. Of relevance here was the link between cerebral asymmetry and individual differences in rate of habituation.

A study of neurological cases with unilateral lesions had earlier shown opposite extremes of electrodermal responsivity according to the side of the lesion. Heilman, Schwartz & Watson (1978) examined seven right-sided patients, five of whom had infarctions in the distribution of the middle cerebral artery, one with a right temporoparietal glioma and one a putamenal haemorrhage. All manifested flattened affect and contralateral behavioural neglect. Four had participated in an earlier experiment revealing impairments in the comprehension, discrimination and expression of affectively intoned speech. They were compared with six left-sided patients with comparable lesions, five having infarctions in the distribution of the left middle cerebral artery and one a left-sided putamenal haemorrhage. All patients exhibited dysphasia, three with Broca's aphasia and three with anomic and conduction aphasia. These groups, matched for the site of the lesion, were in turn compared with a group of patients of similar age but without known structural brain lesions. Electrodermal activity (skin resistance) was recorded from the hand ipsilateral to the lesion to two blocks of five presentations of brief electrical pulses (10 sec ISI) delivered to the forearm, the intensity of which was individually gauged to present discomfort but non-pain:- $\bar{X}$:0.44 in mA (right), $\bar{X}$:0.28 mA (left), $\bar{X}$:0.53 mA (controls). There were no statistical differences in the amount of current delivered. Levels of electrodermal activity were significantly reduced in the right-sided group (right $<$ left $<$ controls), and 5/7 patient were non-responders. In contrast, on average, the magnitude of responses of the left-sided group was approximately four times that of the controls.

It is unlikely that the differences in reactivity could be attributed to recording from the left hand in one group and the right hand in the other because though hemispheric influences are asymmetric they are nevertheless bilaterally mediated; while the evidence so far is sketchy, structural lesions including surgical ablation produce asymmetric responsivity in the direction of larger responses on the contralateral hand (Gruzelier, 1979b for review). Furthermore, the authors note that when recordings were obtained in right-sided cases from the contralateral hand, three continued to be non-responsive over a range of stimulus intensities. Nevertheless, replication of the study with bilateral recording and a standardised stimulus is desirable, for there was a broad range of individual variation in the levels of the current delivered. These incidental findings do not detract from the major difference in responsivity between the left and right hemisphere lesion groups. Moreover, support is found from other reports showing that aphasic patients maintain an ability to show differential responses to emotionally loaded versus neutral stimuli whereas patients with right-sided lesions show no such discrimination (Boller et al., 1978; Morrow et al., 1981).

Morrow et al., (1981) compared electrodermal responses in three groups of 14 patients classified according to whether they had dominant or non-dominant hemisphere neuropathology or no known nervous system pathology; one of the dominant hemisphere group had a right-sided lesion and demonstrated right-hemisphere speech dominance. With two exceptions, the patients all had vascular lesions and the dominant hemisphere group were all dysphasic. Electrodermal responses were recorded from the hand ipsilateral to the lesion to a randomized sequence of emotional and neutral slides. Regarding responsivity to the slides on which all groups differed significantly the groups were ordered as follows:- non-dominant, < dominant, < controls. In addition emotional slides produced larger responses than neutral slides in all but the non-dominant group. Thus the difference between the unilateral lesion groups was in keeping with the results of Heilman et al. (1978) despite the different nature of the stimuli. There was a discrepancy between the reports in the degree of responsiveness in controls. However, in view of the wide individual variation in electrodermal responsiveness this may have affected differences in sampling. Leaving aside this discrepancy, it is clear that comparable left and right-sided lesions do have differential effects on phasic responsivity in the electrodermal system. With left-sided lesions responsiveness is retained, and may in some instances be augmented, whereas with right-sided lesions patients become hypo-responsive. If we assume the opposite polarity of influences operates in the normal compared with the damaged brain, then the left hemisphere has inhibitory influences and the right hemisphere excitatory influences on electrodermal activity. Accordingly the evidence with neurological patients is consistent with the results above in normal subjects and with migraineurs, in whom increased cortical excitability has been hypothesised (Wilkins et al., 1984).

The association of fast habituation with the left hemisphere and slow habituation with the right hemisphere is also consistent with theories of hemispheric specialisation in attention and with empirical evidence linking electrodermal responsiveness and habituation with vigilance and selective attention. The electrodermal response when elicited to a repetitive stimulus is regarded as one manifestation of orienting behaviour which reflects investment of attention in the stimulus, while non-specific reactivity is thought to be related to cortical excitability and alertness. In vigilance experiments parallel declines in vigilance and electrodermal responses have been reported (Ross, Dardano & Hackman, 1959; Stern, 1966). In a visual detection study examination of the temporal relationship between electrodermal responses and correct detections and misses revealed that responses (skin potential) were more frequent before hits than misses (Surwillo & Quilter, 1965). Blakeslee (1979) found a similar relationship when measuring the amplitude of responses in time epochs before and after the stimulus. Furthermore Krupski, Ruskin & Bakan (1971) found that errors of commission were negatively related to reactivity.

The measurement of habituation rates prior to vigilance tasks has shown an association between fast habituation and poor vigilance (Coles & Gale, 1971; Siddle, 1972; Crider & Augenbraun, 1975). On the other hand prior habituation to a stimulus later employed as a distractor was shown to minimise its potential for distraction (Waters, MacDonald & Karenko 1977). On the basis of this evidence we have argued that extremes in electrodermal reactivity are associated with differences in attention such that fast habituation is associated with focussed, selective attention, and slow habituation with sustained, broad, vigilant attention (Gruzelier, Eves & Connolly, 1981a). Focussed selective attention is a prerequisite of analytic, sequential processing of the left hemisphere while broad vigilant attention is compatible with the parallel, holistic processing of the right hemisphere. In support of this, Dimond & Beaumont (1973) have found a right hemisphere superiority on a task of sustained vigilant attention in normal subjects while Dimond (1980) reviews a series of experiments with visual, auditory and haptic tasks in patients with total commissurotomy showing superior right hemisphere vigilance.

MOTORIC SPEED AND INDIVIDUAL DIFFERENCES IN DYNAMIC PROCESS ASYMMETRIES

It follows from theories that differentiate left and right hemispheres on the grounds of positive versus negative affect (e.g., Flor-Henry, 1979b), approach versus withdrawal (Kinsbourne, 1982) and motoric versus sensory processors (Tucker & Williamson, 1984) that individuals with a left hemisphere disposition would have an advantage in motoric speed. We examined this with a haptic sorting task in which speed of sorting was compared for each hand and a 'processing' time obtained by subtracting the time taken to sort the objects without object identification, 'movement' time, from the time taken to sort them by class. In view of the contralateral mediation of active touch the right hand was used to index the left hemisphere and the left hand the right hemisphere.

In a pilot study (N=20), two tasks were developed which produced normal distributions in hemispheric advantages with the distribution on one task displaced to favour the left hemisphere and the other task to favour the right hemisphere. The left hemisphere task involved the discrimination of letters from numbers, and was termed the verbal task. The right hemisphere task involved the discrimination of straight from curved bordered digits and was termed the spatial task. Thirty-two dextral medical students were examined, subdivided by gender into two equal groups. The subjects were also subdivided on the basis of hemispheric dispositional asymmetries defined as the mean left-right processing difference over the first 4/8 trials of each sorting condition. A comparison of the extreme 25% of each lateral disposition and gender indicated that movement times were faster in those with a left hemispheric disposition ($p<0.01$). In other words, subjects whose processing times were initially better with the right hand, irrespective of whether the task was verbal or spatial, showed faster movement times bilaterally when objects were handled without identification (Baxter & Gruzelier, 1984).

These results were replicated in a further experiment involving 64 subjects divided by gender and age (over or under 30 years). When they were subdivided further on the basis of hemispheric disposition, the left hemisphere group again showed a significant advantage for movement time ($p<0.02$). Hemispheric disposition also interacted with age and gender ($p<0.007$). As shown in Table 2 this was largely due to a marked slowing in movement time in women over 30 years who had a right hemisphere disposition. This result may have a bearing on the controversial evidence that ageing, while leading to bilateral impairments, has more deleterious effects on the right hemisphere (Botwinick, 1985). The predominant but by no means unequivocal view about brain laterality and gender is that in women the left hemisphere is superior to the right whereas in men the right hemisphere is superior to the left (Harris, 1980). If the view (Flor-Henry, 1983) that pathology affects the more vulnerable hemisphere - the left in men and the right in women - is added to this, then a disadvantaged right hemisphere may be more a characteristic of ageing women than ageing men. Thus, taking gender

Table 2. Movement time in seconds for subjects classified by age, gender and hemispheric disposition in haptic sorting.

Hemisphere	<30 years		>30 years	
Disposition	M	F	M	F
Left	11.7	13.05	13.95	12.95
Right	13.6	12.95	13.46	16.95

into account may resolve some of the controversy surrounding cerebral laterality and ageing.

HYPNOTIC SUSCEPTIBILITY AND INDIVIDUAL DIFFERENCES IN DYNAMIC PROCESS ASYMMETRIES

Individual differences in asymmetries in haptic processing times have also been found to relate to hypnotic susceptibility. In one experiment students (Fig. 11) and in another middle-aged subjects, all naive to hypnosis, were tested on the verbal haptic test before and while listening to a tape recording of an induction of hypnotic relaxation. Prior to hypnosis those who were later to prove to be susceptible to the induction showed faster left than right hemisphere processing times. Under hypnosis there was an increase in right but not left hand processing times, and in students the increase in processing time was positively correlated ($r=0.65$, $p<0.01$) with the degree of hypnotic depth (Gruzelier, Brow, Perry et al., 1984). This implies that the induction of hypnosis produces a progressive slowing of left hemisphere processes.

In a previous experiment, lateral differences in electrodermal responses to tones in a session prior to hypnosis also proved to differentiate subjects who were to be susceptible to hypnosis (Gruzelier & Brow, 1985). Consistent with the haptic processing asymmetries, for whom susceptible subjects prior to hypnosis showed a left hemisphere advantage, the electrodermal responses of susceptible subjects were larger on the left hand (Fig. 12). Furthermore, under hypnosis susceptible subjects showed a reversal in asymmetries and an increase in habituation to tones occurring incidentally midway through the induction procedure. Habituation in unsusceptible subjects was retarded relative to the pre-hypnosis session. When considered in the context of cerebral asymmetry in attention and the dynamics of an initial left hemisphere advantage, its inhibition and a shift to the right hemisphere provides a neuropsychological framework for the familiar conventions of hypnotic induction. Hypnosis typically begins with the requirement to focus attention by fixating on a small object, which we interpret as engagement of a left hemispheric process. This is followed by suggestions of sleepiness and fatigue leading to a suspension of critical attitudes and an abdication or letting go of planning functions - an inhibition of left hemispheric functions. Coincidentally there is an increasing engagement of right hemisphere processes such as sensory memory, passivity, visual imagery, etc.

Unsusceptible subjects as a group do not show a left hemispheric activational bias prior to hypnosis in either haptic processing or electrodermal orientating responses. Nor do they undergo an inhibition of left hemispheric processes. Their tonic levels of skin conductance activity

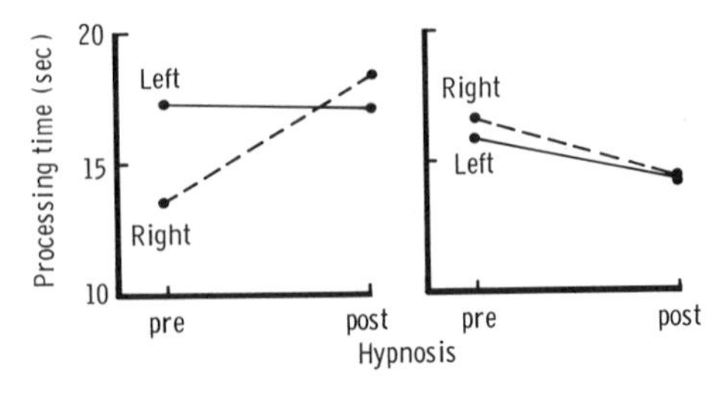

Fig. 11. Haptic processing time for left and right hands before hypnosis and after hypnotic induction while hypnotised in susceptible subjects (left) and unsusceptible subjects (right).

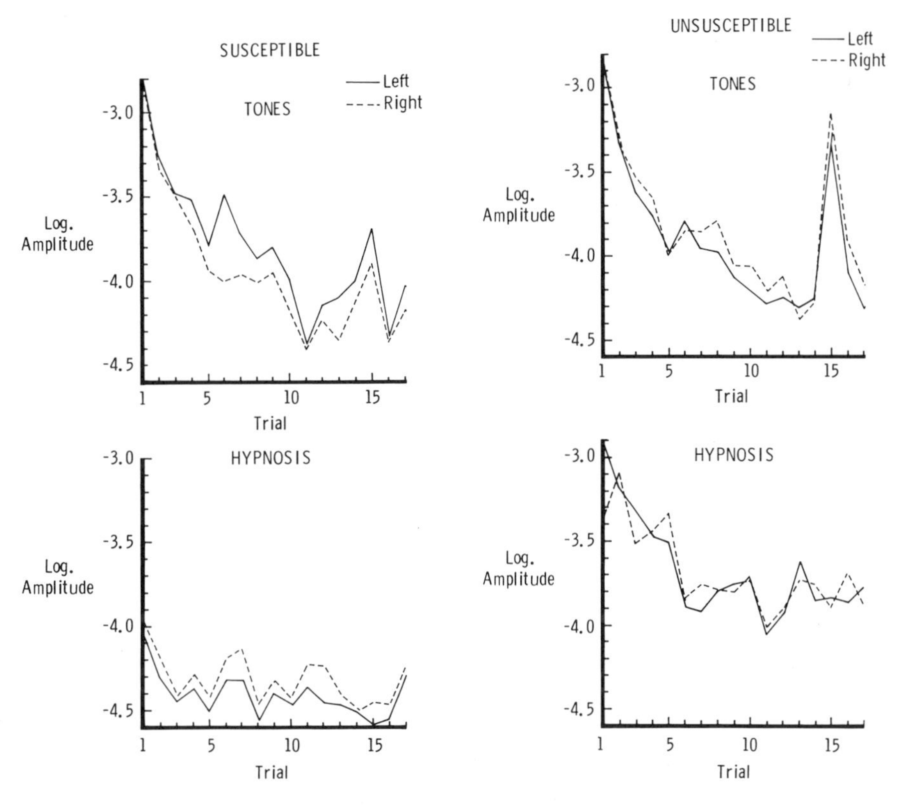

Fig. 12. Bilateral electrodermal response amplitudes to tones in susceptibles (left) and unsusceptibles (right) in session I (TONES) and in a later session under hypnosis (HYPNOSIS).

during the initial stages of the induction do not show the same decline or adaptation as those of susceptible subjects, non-specific responses are initially more frequent and haptic processing times show a small bilateral improvement. These results suggest a more defensive, anxious attitude and a failure to comply with the instructions for relaxation. In the light of these experiments hypnotic susceptibility is seen to consist of two components. The first is the ability to engage left hemisphere focused attention and the second to undergo an inhibition of left hemisphere planning functions. Unsusceptible subjects either do not engage left hemisphere focused attention, retaining a state of right hemisphere broad attention, or fail to let go of left hemisphere planning functions, i.e., undergo left hemisphere inhibition. Thus, individual differences in susceptibility to the induction of hypnotic relaxation reveal an interplay between dynamic and fixed process cerebral asymmetries.

ANXIETY AND INDIVIDUAL DIFFERENCES IN DYNAMIC PROCESS ASYMMETRIES

In our study of examination anxiety in students, a right visual field advantage for the processing of verbal stimuli gave way to a left visual field advantage, see Fig. 5. Tucker, Stenislie & Barnhardt (1978) had earlier found evidence for what was described as a processing load on the left hemisphere for high trait anxious students. This was inferred from two results. High trait anxious students compared with low had an asymmetry in favour of the right ear in loudness judgements, indicative of greater left

Table 3. Reaction time in msecs to verbal and spatial stimuli presented in the left and right visual fields in anxiety patients and dextral and sinistral controls.

Group	Task			
	Verbal		Spatial	
	LVF	RVF	LVF	RVF
Patients	582	608	697	734
Dextral	464	436	557	581
Sinistral	509	545	656	636

hemispheric activation, and at the same time displayed poorer right than left visual-field hemisphere processing of verbal and spatial stimuli. Our results were somewhat similar showing poorer left than right hemispheres processing in a verbal task when state anxiety was raised but this was shown to be due to an improvement in right hemisphere performance, Fig. 5.

A left visual field advantage in the same task was subsequently found in a group of eleven, dextral psychiatric patients (seven female), whose primary complaint was anxiety (Gruzelier & Bennett, in preparation). They were compared with seven dextral controls (four female) of similar age. In addition to the verbal task there was a spatial task which in two studies have shown a left visual field advantage in normal controls and psychotic patients (Connolly et al., 1979, 1983). The task consisted of a 3 x 3 dot matrix in which one dot was missing on target trials. Manual reaction time was measured with task and hand order counterbalanced in blocks of 40 trials with 240 trials per task. The results are shown in Table 3. There was a highly significant interaction between Group by Visual Field by Stimulus ($p<0.0001$). The controls showed the expected left hemisphere advantage for the verbal task and right hemisphere advantage for the spatial task ($p<0.0001$). Patients showed a right hemisphere advantage for both tasks ($p<0.0001$). The patients scored more highly on questionnaire measures of anxiety such as the IPAT and EPQ Neuroticism ($p<0.001$). In a test of hemisphericity in cognitive style (Zenhausern, 1978) the patients showed more left hemisphere reliance than controls ($p<0.05$). Thus, while patients revealed a left hemisphere disposition in cognitive style, it was their right hemisphere which showed a speed of processing advantage for verbal material as was seen in anxious students. Support for the role of left hemispheric activation in anxiety has been clarified somewhat using electrodermal measures. In the experiments reported above which revealed the relation between rate of habituation and the lateral asymmetry of responses (Gruzelier, Eves & Connolly, 1981), a subsequent analysis of questionnaire measures in the 31 hospital staff and 62 students revealed a relationship suggestive of a switch to left hemisphere control under highest levels of anxiety. For students, the larger sample permitted a comparison of those who were fast or moderate habituators (N=22) with those who were slow habituators. It is noteworthy that slow habituators were in the majority, contributing to an impression gained from a series of experiments that levels of stress in medical students are high. Given the large number of slow habituators it was possible to subdivide them further into those with larger responses on the right hand (N=18) and those with larger responses on the left hand (N=22). The latter could be regarded as an anomalous group with respect to the model of cerebral asymmetry and habituation examined earlier where fast habituation was under left

Table 4. Anxiety scales for students (N 62) classified according to rate of habituation and lateral asymmetry in orienting responses.

SCALE	GROUPS			
	Habituators (22)	Slow Habituators (40)	R/L Slow Habituators (18)	V/R Slow Habituators (22)
IPAT Total	27.15	31.16	30.09	32.27
Subscales				
Low self-control	5.29	7.00	7.00	7.00
Instability	3.50	3.00	2.68	3.38
Suspicion	2.46	3.13	3.25	3.00
Apprehension	7.38	8.22	6.50	9.94
Tension-frustration	6.46	8.01	7.38	8.63
EPI Neuroticism	6.75	8.43	8.22	8.64
Speilberger State	41.65	43.35	42.31	44.47
Trait	40.49	42.68	42.88	42.47

hemisphere control and slow habituation under right hemisphere control.

It can be seen in Table 4 that when comparing all subjects classified as either habituators or slow habituators a relationship between anxiety and retarded habituation was supported. All anxiety related scales, namely the total IPAT score, EPI Neuroticism, and Speilberger State/Trait were all higher in slow habituators. A novel finding was the evidence among slow habituators of higher anxiety in those with larger left hand responses, which, in terms of the model of cerebral influences on electrodermal activity is indicative of a dominance of left hemispheric influences. Comparisons between scales indicated that left hemisphere anxiety was characterised by the IPAT subscales of frustration-tension on which high scorers are described as "tense, frustrated, driven, overwrought", and more so by the apprehension subscale where high scorers are described as "apprehensive, self-reproaching, insecure, worrying and troubled." These scales indicate a verbal cognitive component to anxiety.

In the smaller sample of hospital staff the relationships were examined with correlations. Measures included the IPAT and EPI scales. Here only one significant correlation was found with an anxiety scale. This was between the lateral asymmetry in the amplitude of non-specific responses and the IPAT suspicion subscale, in the direction of higher scores of suspicion in subjects with larger left hand responses, see Table 5. As the hospital staff were a more heterogeneous group demographically, and in terms of life situation, the correlations were re-examined in another study of medical students (15 men and 15 women) aged between 20 and 23 years (Holland, 1984). They were dextral, all but for one whose exclusion did not alter the

Table 5. Correlations between electrodermal lateral asymmetries and anxiety

Subjects	Electrodermal Measure	Anxiety Scale	Correlation Rho
Hospital Staff (N 31)	non-specific responses	suspicion	=0.43, p<0.05
Students (N 30)	orienting responses		
		frustration-tension	=0.41, p<0.03
	dishabituating tones	Anxiety total	=0.44, p<0.03
		Emotional instability	=0.46, p<0.02
		Frustration-tension	=0.49, p<0.02
	skin conductance level	Anxiety total	=0.59, p<0.0001
		Low self-control	=0.49, p<0.003
		Apprehension	=0.32, p<0.04
		Frustration-tension	=0.49, p<0.02
		Neuroticism	=0.52, p<0.002

results. The same moderate intensity tone sequence was used as in the previous studies. A range of significant correlations were obtained. These are shown in Table 5. All showed an association of higher electrodermal reactivity on the left hand with high anxiety. The correlations were more numerous with asymmetries in tonic levels of activity. Correlations included IPAT anxiety and EPQ Neuroticism. Of the IPAT subscales frustration-tension again showed significant relationships and in this study included orienting and non-specific responses as well as skin conductance levels.

The relationship between lateral asymmetry in electrodermal activity, rate of habituation and individual differences in anxiety can be thought of as having the following implications for the influence of anxiety on dynamic process asymmetries (Fig. 13). Fast habituation and low anxiety reflect a dominance of left hemispheric inhibition. The shift in excitation-inhibition balance seen in decreasing rates of habituation reflects a reversal in hemispheric influences resulting in a dominance of right hemispheric activation. Under states of high anxiety, control may revert to the left hemisphere without influencing the excitation-inhibition balance which remains in a state of excitation, as reflected in slow habituation. The dominance of excitation, out of keeping with left-sided inhibition, may overload the operation of fixed structures of the left hemisphere. The schema in Fig. 13 is the familiar inverted - U relationship between arousal-anxiety and performance in a new guise. The behavioural disorganisation which characterises over-arousal and the down-swing of the inverted - U is depicted as a consequence of the shift in hemispheric control to the left hemisphere that occurs with high states of arousal-excitation.

Predictions should follow as to individual differences in the quality

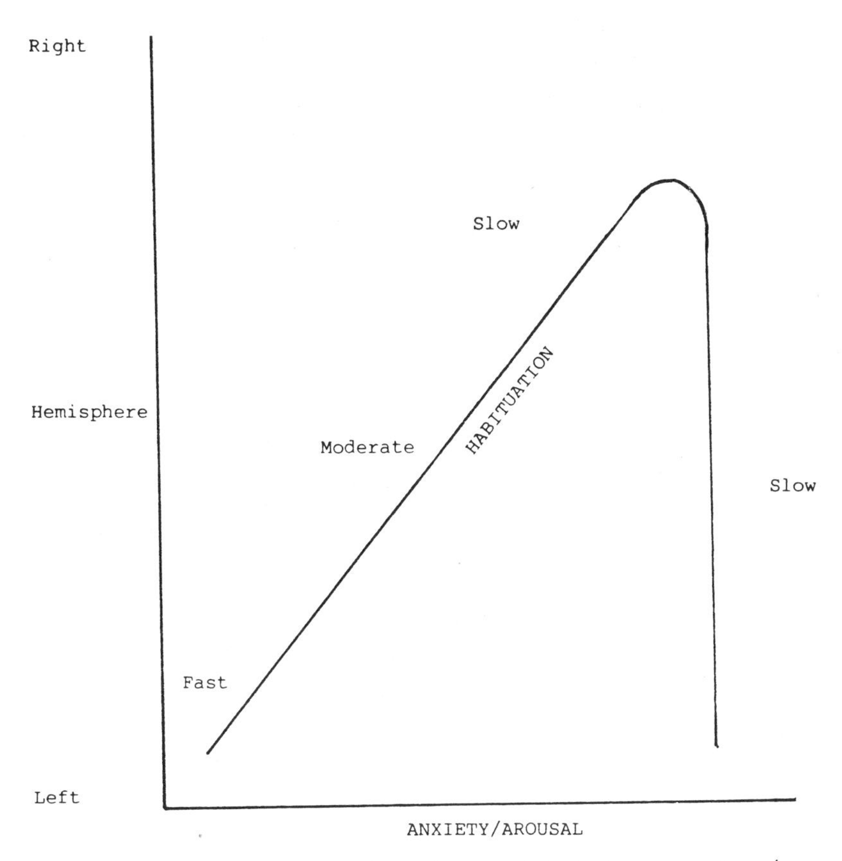

Fig. 13. Conceptualisation of the relationship between anxiety/arousal and lateral cerebral activation.

of anxiety according to the asymmetry in activational dominance. It might be anticipated that a left hemisphere dominance will underly cognitive, ruminative forms of anxiety whereas a right hemisphere dominance will underly generalised, free-floating forms of anxiety. There is some hints in support of this from the IPAT subscales. Those that showed an association with the left hemisphere activation dominance were the three scales with strong cognitive associations:- the 'frustration-tension' scale, the 'apprehension-worrying' scale and the 'suspicion' scale on which a high scorer is described as 'suspicious, jealour, hard-to-fool.' Whereas no relationships were found with the scale of 'emotional instability' - 'emotionally unstable, easily upset, changeable', or the scale of 'low self-control' - 'uncontrolled, follows own urges, careless of social rules'. Bearing in mind the inadequacies of self-report scales ratings obtained on clinical samples should provide a useful test of this hypothesis as exemplified by the approach of Rabavilas et al. (1979) who with unilateral recording compared rates of electrodermal habituation to very loud (100 dB SPL) tones in ten obsessive-compulsive patients with ruminative thoughts and in ten exhibiting ritualistic behaviour. The former group showed both higher reactivity and slower habituation to the tones. A systematic attempt combining ratings with bilateral skin conductance would be of interest. One such attempt in schizophrenia has produced theoretically interesting results and will now be reviewed.

SCHIZOPHRENIA AND INDIVIDUAL DIFFERENCES IN DYNAMIC PROCESS ASYMMETRIES

In psychiatry the need for sub-classification of schizophrenia is part of the contemporary Zeitgeist (Kety, 1980). Historically, since the time Kraepelin (1855-1926) first described catatonia, hebephrenia and paranoia as the single entity, dementia praecox, attempts have been made to account for the heterogeneity of the disorder with various principles of sub-classification such as acute/chronic, process/reactive, good/poor prognosis, good/poor pre-morbid, paranoid/non-paranoid. The contemporary need for sub-classification has come about for a number of reasons. It is clear that patients differ in response to neuroleptics some undergoing complete recovery and others a mild amelioration of symptoms. Some disorders spontaneously remit whereas others lead to chronic deterioration. Constellations of symptoms show wide variation between and within patients, at one extreme depicting florid acute symptoms and at the other socially and emotionally withdrawn, autistic, behaviour. While a neurological basis of schizophrenia has always had its advocates, it is the evidence from X-ray computed tomography that first gave this widespread credibility. However, the view that only some patients show CT signs and that these are the ones with cognitive deficits (Crow, 1980), has been based on the detection of fairly gross forms of pathology. It is likely that as scan resolution improves computed tomography together with techniques that explore the functional activation of the brain through measures of blood flow and metabolism (e.g., Sheppard et al., 1983; Gur et al., 1983) will reveal focal signs of abnormal structure and function more in keeping with evidence that all forms of schizophrenia possess at least transiently, central nervous system signs (Gruzelier, 1985). The most developed neuropsychophysiological theory of brain dysfunction in schizophrenia is currently one of lateralisation, either taking the form of a unilateral disorder, with most evidence implicating the left hemisphere, or because of faulty interhemispheric communication - evidence for which is reviewed in Gruzelier, this volume. There have been numerous reviews of these hypotheses such as Flor-Henry (1974; 1979,a,b; 1983,a,b), Gruzelier (1979a; 1981a,b; 1983), Newlin, Carpenter & Golden (1981), Walker & McGuire (1982) & Wexler (1980), to name a few. Most researchers of lateralised function in schizophrenia have taken a unitary view of the disorder, however evidence will be briefly outlined showing that individual differences in the form and course of the disorder may be unravelled through consideration of individual differences in the balance of hemispheric influences on dynamic processes (Gruzelier, 1981, 1983, 1984; Gruzelier & Manchanda, 1982). Undrugged new hospital admissions with a diagnosis of schizophrenia made with the help of the Present State Examination and CATEGO analysis (Wing, Cooper & Sartorius, 1974) were examined for rate of habituation with the moderate intensity tone series described above. If they were non-responsive to the tones, as is not uncommon in schizophrenia (Gruzelier & Venables, 1972; Bernstein et al., 1982), they were given a 90 dB SPL sequence of 1000 Hz tones of 5 second duration (see Gruzelier et al., 1981 b). Patients were sub-divided according to the direction of the lateral asymmetry in orienting responses and compared for syndromes (CATEGO) and ratings on the Brief Psychiatric Rating Scale (BPRS) (Overall and Gorham, 1962). Two experiments were run. The first involved 23 undrugged patients selected for a drug trial in which florid symptoms were one criterion for selection, and the second involved 25 consecutive undrugged admissions.

The two groups classified on the basis of a dynamic process lateral asymmetry differed on 13/38 CATEGO syndromes and 8/19 BPRS ratings. To sharpen the clinical description, the syndromes were examined for those features applicable to the present patients as each syndrome is usually made

Table 6. Symptoms of the two schizophrenia syndromes

Left hemisphere activation

Heightened self consciousness (simple ideas of reference)
Exaggerated self opinion and conviction of unusual ability (grandiosity)
Exaggerated concern about bodily welfare (hypochondriasis)

Euphoria (hypomania)
Situational anxiety; avoids situations; phobias

Pressure of speech
Flight of ideas (hypomania)

Right hemisphere activation

Lack of self confidence
Social withdrawal

Uncooperative (resistance, resentment, unfriendliness)
Irritable to hostile

Emotional withdrawal
Reduced emotional tone (blunted affect)

Motor underactivity

Slow speech
Muteness, restricted quantity of speech
Inefficient thinking (muddled, slow, not goal directed)

up of more than one symptom or sign. A discriminant function analysis was also undertaken to determine those syndromes and ratings which most economically distinguished the groups. This enlarged the profile to include a further six variables. A synthesis of these analyses is shown in Table 6. The groups were positioned at two extremes of an activation dimension involving self concepts, emotion, cognition, and motoric behavior.

The syndromes had an affinity with those described by Kety (1980). Consistent with Bleulerian schizophrenia patients with larger right hand responses had more blunted affect, reduced emotional tone, emotional withdrawal, slowness, restricted quantity and content of speech, reduced energy level, uncooperativeness, social withdrawal and conceptual disorganization. Consistent with the clinical profile of drug responsive schizophrenia those patients with larger left hand responses had many more florid features including hypomania, pressure of speech, flight of ideas, simple ideas of reference, delusions of grandeur, sexual and fantastic delusions, depressive delusions and hallucinations, and hypochondriacal delusions.

The syndromes overlap but are not to be confused with the distinction between acute, reactive, paranoid schizophrenia and chronic, non-paranoid schizophrenia, a classification which has revealed differences in cognitive functions (Magaro, 1980; Lang & Buss, 1965). What appears essential to draw out the difference is the coupling of paranoid features with a reactive, emotional component. This is the combination of features seen here, where, for example ideas of reference and hypomanic features such as pressure of speech were ranked first in the step-wise discriminant function analysis (Gruzelier, 1981b). The same components, this time with emotional reactivity defined psychophysiologically, distinguished paranoid and nonparanoid schizophrenic patients in an earlier dichotic listening study of hemispheric functions (Gruzelier & Hammond, 1979; 1980; see Gruzelier, this volume, Figs. 1 and 2).

Accordingly in schizophrenia activational dominance determines the form the psychosis will take, producing a florid psychosis when left hemisphere activation is dominant and a non-florid psychosis when right hemisphere activation is dominant. Schizophrenic patients whose responses were slow to habituate and who possessed heightened emotional reactivity showed the same lateral asymmetry as anxious normals, namely larger left hand responses. The group of patients with the opposite asymmetry possess right hemispheric activation, as was found in slow habituating normals whose levels of anxiety may be higher than average yet not as high as those slow habituators with the opposite asymmetry.

The model of hemispheric imbalance in the two syndromes is consistent with a neuropsychological interpretation of symptoms. The pressure of speech and flight of ideas of the florid syndrome implies an activation of left hemisphere verbal processes whereas the sluggish thinking, muteness and restricted quality of speech of the non-florid syndrome implies an underactivation of the left hemisphere. The association of euphoria with the left hemisphere and depression with the right hemisphere is in keeping with much evidence about the association of different polarities of mood with the two hemispheres (Sackheim et al., 1982).

As diagnosis of schizophrenia on the basis of hallucinations and delusions is widespread and not restricted to the diagnostic procedure employed in our studies, it follows that other studies of schizophrenia measuring lateralized functions should similarly reveal asymmetries of two kinds in different patients according to syndrome type. A review (Gruzelier, 1983) was undertaken and evidence was sought for two forms of asymmetry. Dimensions which have a bearing on the electrodermal syndrome may be age, chronicity, prognosis, paranoid versus residual or catatonic schizophrenia, anhedonic schizophrenia and in a recent report 'dominant' compared with 'non-dominant' hemisphere syndromes. The review did provide evidence of two syndromes in schizophrenia provided the measure reflected imbalances in hemispheric activation or attention and so belonged to the class of dynamic process asymmetries. The syndromes were consistent in character with those delineated by electrodermal responses. Furthermore in each case the nature of the asymmetry, perhaps with one exception, showed left hemisphere over-activation coupled with right hemisphere underactivation in acute, reactive, paranoid, good-prognosis schizophrenia, and left hemisphere underactivation coupled with right hemisphere overactivation in chronic, residual, poor-prognosis schizophrenia. A wide range of measures revealed bidirectional asymmetries in schizophrenia. These included levels of electrocortical activity as seen in the electroencephalogram (EEG), cortical evoked potentials to sensory stimuli, lateral eye movements to questions designed to stimulate the hemispheres independently, the recovery curve of the Hoffman reflex to somatosensory stimulation, measures of auditory processing including the shadowing of dichotically presented textual passages, the recall of digits presented dichotically, and thresholds to very brief stimuli, a test of somatosensory extinction whereby competing textures are presented to the palmar surfaces simultaneously and finally left-handedness (references in Gruzelier, 1983). It is noteworthy that all but the last measure reflect processes that are dynamic and reversible. Some vary with the activation levels of the hemispheres: electrodermal responsiveness, electrocortical activity, the Hoffman reflex and lateral eye movements; others with the direction of attention in the lateral plane: eye movements, auditory processing and somatosensory extinction. Handedness apart, in measures which belonged to the fixed process class, such as the recognition of verbal or spatial stimuli presented tachistoscopically, or the more traditional methods of neuropsychological test batteries, a left-sided abnormality proved to be ubiquitous. Thus the need to distinguish between the two classes of hemispheric asymmetry was critical.

More recent reports are compatible with the hemisphere-syndrome model. Rabavilas, Liappas & Stefanis (1986) have examined bilateral electrodermal activity in positive symptom paranoid schizophrenic patients. Correlations between symptoms and lateral asymmetries in specific and non-specific responses showed that the larger the asymmetry in favour of the left hand the larger were ratings of conceptual disorganisation, tension, grandiosity, suspiciousness and hallucinatory activity. Here, the ratings of tension and suspiciousness were consistent both with the schizophrenia model and with the results of the IPAT anxiety subscales in anxious normals outlined above. Further consistency was found in the relationship between symptoms and lateral asymmetries in tonic levels of activity. Comparisons were made between the six patients with the dextral preponderance in levels and the remaining 22 patients. Those with the rightward asymmetry in levels had higher ratings on negative symptoms such as emotional withdrawal and blunted affect and were rated as less tense. There were two ratings that were inconsistent with the model. Conceptual disorganisation was associated with both asymmetries, however, this is an ambiguous rating and only where it reflects anergia in thinking is it clearly a negative symptom. There was also less evidence of depression in the patients with the dextral preponderance in levels, which is out of keeping with other ratings of negative affect.

White, Svali & Charles (1986) has also found results consistent with the model with telemetric recording of bilateral skin conductance in chronic negative symptom schizophrenics. When compared with controls, patients were differentiated by higher right hand levels in tonic skin conductance under conditions of social stress when required to mime. Prolonged reaction time latency was also positively correlated with right hand levels of skin conductance.

Andrews et al., (1986), who divided patients on the basis of the CATEGO analysis of Gruzelier & Manchanda (1982) into active and withdrawn symptoms, found that this delineated a group of patients without the normal asymmetry in somatosensory evoked potentials to unimanual stimulation. Those with an abnormal symmetry in responses, which it is posited may reflect abnormal interhemispheric transfer or the development of abnormal ipsilateral pathways, belonged to the negative syndrome and had higher ratings of social withdrawal and slowness.

The relevance of the syndromes has also been extended to visuo-motor performance in tasks belonging to both structural and dynamic asymmetry classes (Gaebel, Ulrich & Frick, 1986). They compared 20 schizophrenic outpatients with 20 normal controls measuring lateral deviations in visual fixation of a stationary spot of light, and eye movements during a letter matrix search task. Individual differences were examined with the BPRS. Rightward deviations (left hemisphere activation) were associated with the rating of excitement and the factor score of activation. Leftward deviations (right hemisphere activation) were associated with emotional withdrawal. A long search time and search route, as in serial processing, was associated with the rightward deviation in eye fixation and less emotional withdrawal. Patients with the rightward deviation also had poorer search performances than controls. Remarkably, the patients with leftward eye deviations, who had the more severe ratings of psychopathology overall, had performance as good as controls. Their short search times and search routes were characteristic of parallel processing, a right hemispheric process. Right hemispheric activation was also consistent with their leftward eye movements and negative symptoms.

CONCLUSION

A small contribution has been made towards specifying dynamic process asymmetries with implications for individual differences. The independent variables have included asymmetries in rate of habituation of electrodermal orienting responses, non-specific electrodermal responses, and speed of haptic sorting. The process of 'attention', though integral to habituation of the orienting response, and involved in the cognitive tasks and hypnotic induction, does not have the applicability that 'activation' does in its unifying potential as a substrate for the behaviour under study, including the broad repertoire covered by the psychiatric ratings. Here 'attention' is subordinate to 'activation'.

In optimal circumstances hemisphere activation may be in consort with structural asymmetries. Levy et al. (1983) provide compelling evidence that the extent of right visual-field advantages in a syllable identification task was associated with asymetries on a free-vision, face-processing task such that as leftward asymmetries became larger, indicating increasing right hemispheric activation, right visual-field advantages on the verbal task diminished. Thus the extent of a structurally based asymmetry in one task was associated with an activation based asymmetry in another. This is in contrast to the dissociation between activation and structure in states of anxiety found by Tucker et al. (1978), shown here in the divided visual-field experiments in students under stress and in anxious patients, and implicit in the electrodermal asymmetries in highly anxious normals. Evidence points to an explanation of this dissociation in anxiety in terms of an overactivation of the left hemisphere giving rise to an overload of structurally based processing resources with a consequent shift in processing to the right hemisphere. However, this explanation is inferential and requires both concurrent monitoring of activation asymmetries under anxiety and manipulation of activation in order to provide validity for the interpretation of left hemispheric overactivation and overload.

Levy et al. (1983) also found that the subjects with strong right visual-field advantages rated their performances pessimistically and those with reduced or no asymmetries rated their performance optimistically. Research by Bear & Fedio (1976) into personality in unilateral temporal lobe epilepsy has shown a similar association of pessimism with the left hemisphere and optimism with the right from evaluations based on discrepancies between how patients rated themselves and how they were rated by their relative. The association of negative affect and the left hemisphere and positive affect and the right hemisphere is at odds with the hemisphere-syndrome model outlined above in schizophrenia. Aside from considerations of the adequacy of self-rating this conflict is mentioned here to acknowledge that the cerebral asymmetry of mood is but one of the controversial issues that surrounds the localisation of dynamic processes. An adequate discussion of this is outside the scope of the Chapter, suffice it to say that the fact that the controversy exists is probably not that one side of the argument is wrong, but that current models are oversimplifications. What is true of psychosis is unlikely to be true in all aspects for the more differentiated emotions of the normal brain nor is a unitary model of all the psychoses likely to be viable. Nevertheless, unifying principles have been drawn here between the normal and pathological brain on the basis of activational asymmetries. But the range of pathology covered has included only morbid anxiety and schizophrenia, and in the normal personality individual differences have considered only anxiety.

Apart from the one cognitive activation measure in which cerebral asymmetry in haptic processing was predictive of bilateral motoric handling speed, electrodermal responsiveness was the measure of activation. The literature on cerebral asymmetry and bilateral electrodermal responses is complex and notable for conflicting results. However, the central centres influencing electrodermal activity are also complex (Wang, 1964) and this complexity has seldom been recognised in the design of the experiments to date. Cognitive requirements in most have usually been multi-faceted with no converging operations to delineate discreet processes. Here the empirical consistency between replications and the theoretical consistency across studies may have arisen for two reasons. The first is that a standardised paradigm has been used throughout with only minor variations, usually in the number of trials. Secondly, the task is a passive one with minimal cognitive demands and no motor requirements. Holloway & Parsons (1969) have shown that when comparing brain damaged patients with controls, differences relating to the unilaterality of lesions were actually enhanced during passive stimulation compared with both resting conditions and motor and perceptual tasks. Asymmetries in controls also appeared to be larger under conditions of passive stimulation.

In this chapter, individual differences in asymmetric cerebral activation have been found predictive of rate of habituation to non-signal tones and flashes, the laterality of pain in migraine, the induction of relaxation under hypnosis, normal and pathological anxiety, syndromes in schizophrenia and individual variation in processes dependent on structural asymmetry such as verbal and spatial processing in both visual and somatosensory modalities. Just as handedness and gender are acknowledged as important variables to control for in studies of fixed process asymmetries, recognition should also be given to the need to control for individual differences in hemispheric activation.

REFERENCES

Andrews, H.B., Haus, A.D., Cooper, J.E., & Barber, C. (1986). The symptom pattern in schizophrenic patients with stable abnormalities of laterality of somatosensory cortical evoked responses. British Journal of Psychiatry. In Press.

Baxter, R., & Gruzelier, J. (1984). Lateral bias and individual differences in speed of haptic sorting. International Journal of Psychophysiology, 2, 310-311.

Bear, D., & Fedio, P. (1976). Quantitative analysis of interictal behavior in temporal lobe epilepsy. Archives of Neurology, 34, 454-467.

Bernstein, A., Frith, C.D., Gruzelier, J.H., Patterson, T., Straube, E., Venables, P.H., & Zahn, T.P. (1982). An analysis of the skin conductance orienting responses in samples of American, British and German schizophrenics. Biological Psychology, 14, 155-211.

Blakeslee, P. (1979). Attention and vigilance: performance and skin conductance response changes. Psychophysiology, 16, 413-419.

Boller, F., Cole, N., Vrtunski, T.B., Patterson, M., & Kim, Y. (1979). Power linguistic aspects of auditory comprehension in aphasia. Brain and Language, 7, 164-174.

Botwinick, J. (1985). The Neuropsychology of Ageing. In S.B. Filskov & T.J. Boll (Eds.) Handbook of Clinical Neuropsychology, Wylie: Chichester, 135-171.

Cohen, G. (1982). Theoretical interpretations of lateral asymmetries. In J.G. Beaumont (Ed.) Divided Visual Field Studies of Cerebral Organisation. New York: Academic Press, 87-115.

Coles, M.G. & Gale, A. (1971). Physiological reactivity as a predictor of performance in a vigilance task. Psychophysiology, 8: 594-599.

Connolly, J.F., Gruzelier, J.H., Kleinman, K.M., Hirsch, S.R. (1979). Lateralised abnormalities in hemisphere-specific tachistoscopic tasks in psychiatric patients and controls. In J.H. Gruzelier & P. Flor-Henry (Eds.) Hemisphere Asymmetries of Function in Psychopathology, Amsterdam, Elsevier/North Holland, 491-510.

Connolly, J.F., Gruzelier, J.H., Manchanda, R. (1983). Electrocortical and perceptual asymmetries in schizophrenia. In P. Flor-Henry & J.H. Gruzelier (Eds.) Laterality and Psychopathology. Amsterdam, Elsevier/North Holland, 363-378.

Crider, A., & Augenbraun, C.B. (1975). Auditory vigilance correlates of electrodermal response habituation speed. Psychophysiology, 12, 36-40.

Crow, T.J. (1980). Molecular pathology of schizophrenia: more than one disease process? British Medical Journal, 280, 66-68.

Dimond, S.J. (1980). Neuropsychology. London: Butterworths.

Dimond, S.J. & Beaumont, J.G. (1973). Difference in vigilance performance of the right and left hemisphere. Cortex, 9, 259-265.

Flor-Henry, P. (1974). Psychoses, neuroses & epilepsy. British Journal of Psychiatry, 124, 144-150.

Flor-Henry, P. (1979a). Commentary on theoretical issues and neuropsychological and electro-encephalographic findings. In J. Gruzelier & P. Flor-Henry (Eds.) Hemisphere Asymmetries of Function in Psychopathology. Amsterdam: Elsevier/North Holland, 639-646.

Flor-Henry, P. (1979b). On certain aspects of the localisation of the cerebral systems regulating and determining emotion. Biological Psychiatry, 14, 677-698.

Flor-Henry, P. (1983a). Cerebral Basis of Psychopathology. London: John Wright.

Flor-Henry, P. (1983b). Commentary and synthesis. In P. Flor-Henry & J. Gruzelier (Eds.) Laterality in Psychopathology, Amsterdam: Elsevier, 1-18.

Flor-Henry, P. & Gruzelier, J. (1983). Laterality and Psychopathology. Amsterdam: Elsevier.

Gaebel, W., Ulrich, G. & Frick, K. (1986). Eye-movement research with schizophrenic patients and normal controls using corneal reflection-pupil centre measurement. European Archives of Psychiatry and Neurological Sciences. In Press.

Gruzelier, J.H. (1979a). Synthesis and critical review of the evidence for hemisphere asymmetries of function in psychopathology. In J.H. Gruzelier & P. Flor-Henry (Eds.) Hemisphere Asymmetries of Function in Psychopathology, Amsterdam: Elsevier/North Holland, 647-672.

Gruzelier, J.H. (1979b). Lateral asymmetries in electrodermal activity and psychosis. In J.H. Gruzelier & P. Flor-Henry (Eds.) Hemisphere Asymmetries of Function in Psychopathology, Amsterdam: Elsevier /North Holland, 149-168.

Gruzelier, J.H. (1981a). Cerebral laterality and psychopathology: fact and fiction. Psychological Medicine, 11, 219-227.

Gruzelier, J.H. (1981b). Hemispheric imbalances masquerading as paranoid and non-paranoid syndromes. Schizophrenia Bulletin, 7, 662-673.

Gruzelier, J.H. (1983). A critical assessment and integration of lateral asymmetries in schizophrenia. In M. Myslobodsky (Eds.) Hemisyndromes: Psychobiology, Neurology and Psychiatry, New York: Academic Press, 265-326.

Gruzelier, J.H. (1984). Hemispheric imbalances in schizophrenia. International Journal of Psychophysiology, 1, 227-240.

Gruzelier, J.H. (1985). Schizophrenia: central nervous system signs in schizophrenia. In J.A.M. Fredericks (Ed). Handbook of Clinical Neurology, 2. Neurobehavioural Disorders. Amsterdam: Elsevier Science Publishers (B.V.), 481-521.

Gruzelier, J., Brow, T., Perry, A., Rhonder, J., Thomas, M. (1984). Hypnotic susceptibility: a lateral predisposition and altered cerebral asymmetry under hypnosis. International Journal of Psychophysiology, 2, 131-139.

Gruzelier, J.H. & Brow, T.D. (1985). Psychophysiological evidence for a state theory of hypnosis and susceptibility. Journal of Psychosomatic Research, 29, 287-302.

Gruzelier, J.H. & Bennett, C. (1986). Altered cerebral asymmetry for verbal processing in anxious patients. In Preparation.

Gruzelier, J.H., Nikolau, T., Conolly, J.F., Peretfield, R.C., Davies, A.T.G., Clifford-Rose, F. (1986). Laterality of pain in migraine distinguished by interlectal rates of publication of electrochemical responses to visual and auditory stimuli. Journal of Neurology, Neurosurgery and Psychiatry. In Press.

Gruzelier, J.H., Eves, F.F. & Connolly, J.F. (1981a). Habituation and phasic reactivity in the electrodermal system: reciprocal hemispheric influences. Physiological Psychology, 9, 313-317.

Gruzelier, J.H., Eves, F.F., Connolly, J.F. and Hirsch, S.R. (1981b). Orienting, habituation, sensitization and dishabituation in the electrodermal system of consecutive, drug-free admissions for schizophrenia. Biological Psychology, 12, 187-209.

Gruzelier, J.H. & Flor-Henry, P. (1979). Hemisphere Asymmetries of Function in Psychopathology. Amsterdam: Elsevier/North Holland.

Gruzelier, J.H. & Hammond, N.V. (1979). Lateralised auditory processing in medicated and unmedicated schizophrenic patients. In J.H. Gruzelier & P. Flor-Henry (Eds.) Hemisphere Asymmetries of Function in Psychopathology. Amsterdam: Elsevier/North Holland, 603-638.

Gruzelier, J.H. & Hammond, N.V. (1980). Lateralised deficits and drug influences on the dichotic listening of schizophrenic patients. Biological Psychiatry, 15, 759-779.

Gruzelier, J.H., Jutai, J.W., Connolly, J.F. & Hirsch, S.R. (1985). Cerebral asymmetries in unmedicated schizophrenic patients in EEG spectra and their relation to clinical and autonomic parameters. Advances in Biological Psychiatry, 15, 12-19.

Gruzelier, J.H. & Manchanda, R. (1982). The syndrome of schizophrenia: relations between electrodermal response, lateral asymmetries and clinical ratings. British Journal of Psychiatry, 141, 488-495.

Gruzelier, J.H. & Phelan, M. (1986). Lateral brain function and examination of stress: divided visual-field and electrodermal measures. In Preparation.

Gruzelier, J.H. & Venables, P.H. (1972). Skin conductance orienting activity in a heterogeneous sample of schizophrenics: possible evidence of limbic dysfunction. Journal of Nervous and Mental Disease, 55, 277-287.

Gur, R.E., Skolnick, B.E., Gur, R.C., Kaross, S., Rieger, W., Obrist, W.D., Younkin, D., Reivich, M. (1983). Brain function in psychiatric disorders. Archives General Psychiatry, 40, 1250-1254.

Harris, L.J. (1978). Sex differences in spatial ability: possible environmental, genetic and neurological factors. In M. Kinsbourne (Ed.) Asymmetrical Function of the Brain. London: Cambridge University Press, 405-522.

Heilman, K.M., Schwartz, H.D., Watson, R.T. (1978). Hypoarousal in patients with the neglect syndrome and emotional indifference. Neurology, 28, 229-232.

Heilman, K.M., Watson, R.T. & Valenstein, E. (1985). Neglect and related disorders. In K.H. Heilman & E. Valenstein (Eds.) Clinical Neuropsychology, 2nd edition, Oxford: Oxford University Press.

Holland, C. (1984). Personality differences and lateral asymmetries in electrodermal activity. Unpublished dissertation.

Holloway, F.A., and Parsons, O.A. (1969). Unilateral brain damages and bilateral skin conductance levels in humans, Psychophysiology, 6, 138-148.

Kety, S.S. (1980). The syndromes of schizophrenia: unresolved questions and opportunities for research. British Journal of Psychiatry, 136, 421-436.
Kinsbourne, M. (1970). The cerebral basis of lateral asymmetries in attention. Acta Psychologica, 33, 193-201.
Kinsbourne, M. (1975). The mechanism of hemispheric control of the lateral gradient of attention. In P.M.A. Rabbitt & S. Dornic (Eds.) Attention and Performance, London: Academic Press. 5, 81-97.
Kinsbourne, M. (1982). Hemispheric specialisation and the growth of human understanding. American Psychologist, 37, 411-420.
Krug, S.E., Scheier, I.H. & Cattell, R.B. (1976). Handbook for the IPAT Anxiety Scale. Champagne: Institute for Personality and Ability Testing.
Krupski, A., Ruskin, D.C. & Bakan, P. (1971). Physiological and personality correlates of commission errors in an auditory vigilance task. Psychophysiology, 8, 304-311.
Lang, P.J. & Buss, A.H. (1965). Psychological deficit in schizophrenia: II interference and activation. Journal of Abnormal Psychology, 70, 77-106.
Lashley, K.S. (1929). Brain Mechanisms and Intelligence, Chicago, University of Chicago Press.
Levy, J., Heller, W., Banich, M.T. & Burton, L.A. (1983). Are variations amongst right-handed individuals in perceptual asymmetries caused by characteristic arousal differences between hemispheres. Journal of Experimental Psychology: Human Perception and Performance, 9, No. 3.
Luria, A.R. (1973). The Working Brain, Harmondsworth: Penguin.
Magaro, P.A. (1980). Cognition in Schizophrenia and Paranoia. Hillsdale: Erlbaum.
Maltzman, I., Smith, N.J., Cantor, W. (1971). Effects of stress on habituation of the orienting reflex, Journal Experimental Psychology, 87, 207.
Morrow, L., Vrtunski, P.B., Kim, Y. & Boller, S. (1981). Arousal responses to emotional stimuli and laterality of lesion. Neuropsychologia, 19, 65-71.
Newlin, D.B., Carpenter, B., Golden, C.V.J. (1981). Hemisphere asymmetries in schizophrenia. Biological Psychiatry, 16, 561-582.
Overall, J.E. & Gorham, D.R. (1962). The brief psychiatric rating scale. Psychological Reports, 10, 799-812.
Rabavilas, A.D., Boulougouris, J.A.C., Perissaki, C. & Stefanis, C. (1979). Psychological differences between ruminations and compulsive acts in obsessive-compulsive neurotics. In J. Obiols, C. Ballus, E. Gonzalez, Monclus, & J. Pujol, (Eds.) Biological Psychiatry Today, Amsterdam: Elsevier/North Holland, 581-585.
Rabavilas, A.D., Liappas, J.A. & Stefanis, C.N. (1986). Electrodermal laterality indices in paranoid schizophrenics. In C.N. Stefanis (Ed.) Schizophrenia: Recent Biosocial Developments, New York: Human Sciences Press. In press.
Ross, S., Dardano, J.F. & Hackman, R.C. (1959). Conductance levels during vigilance task performance. Journal of Applied Psychology, 43, 65-69.
Sackheim, H.A., Greenberg, M.S., Wynman, A.L., Gur, R.C., Hungerbuhler, J.P. & Geschwind, N. (1982). Hemispheric asymmetry in the expression of positive and negative emotions: neurologic evidence. Archives of Neurology, 39, 210-218.
Satz, Z.P. (1980). Incidence of aphasia in left handers: a test of some hypothetical models of cerebral speech organisation. In J. Herron (Ed.) The Neuropsychology of Left-Handedness. New York: Academic Press, 189-198.
Sheppard, G., Gruzelier, J., Manchanda, R. & Hirsch, S.R. (1983). O Positron emission tomographic scanning in predominantly never-treated acute

schizophrenic patients. Lancet, 2, 118-120.
Siddle, D.A.T. (1972). Vigilance decrement and the speed of habituation of the GSR component of the orienting reflex. British Journal of Psychology, 63, 191-194.
Stern, R.M. (1966). Performance and physiological arousal during two vigilance tasks varying in signal presentation rate, Perceptual and Motor Skills, 23, 691-700.
Stern, J.A. & Janes, C.L. (1973). Personality and psychopathology. In W.F. Prokasy and D.C. Ruskin (eds). Electrodermal Activity in Psychological Research. London, Academic Press. 284-346.
Surwillo, W.W. & Quilter, R.E. (1965). The relation of frequency of spontaneous skin potential responses to vigilance and to age. Psychophysiology, 1, 272-276.
Tucker, D.M. (1981). Lateral brain function, emotion and conceptualization. Psychological Bulletin, 89, 19-46.
Tucker, D.M. & Williamson, P.A. (1984). Asymmetric neural control systems in human self-regulations. Psychological Review. 91, 185-215.
Tucker, D.M., Stenislie, C.E. & Barnhardt, T.N. (1978). Anxiety and lateral cerebral function. Journal of Abnormal Psychology, 87, 380-383.
Vygotsky, L.F. (1965). Psychology and localisation of functions. Neuropsychologia, 3, 381-386.
Walker, E. & McGuire, S. (1982). Intra and inter-hemispheric information processing in schizophrenia. Psychological Bulletin, 92, 701-725.
Wang, G.H. (1964). The Neural Control of Sweating. Madison, University of Wisconsin Press.
Waters, W.F., MacDonald, D.G. & Korenko, L. (1977). Habituation of the orienting responses: a gating mechanism subserving selective attention. Psychophysiology, 14, 228-236.
Wexler, B.E. (1980). Cerebral laterality in psychiatry: a review of the literature. American Journal Psychiatry, 137, 279-291.
White, C., Svali, J. & Charles, P. (1986). Psychophysiological responses, laterality, social stress and chronic schizophrenic disorder. British Journal of Psychiatry. In Press.
Wing. J.K., Cooper, J.E. & Sartorius, N. The Measurement and Classification of Psychiatric Symptoms. London, Cambridge University Press.
Wilkins, A., Nimmo-Smith, I., Tait, A., McManus, C., Della Sala, S., Tilley, A., Arnold, K.,Barrie, M., & Scott, S. (1984). A neurological basis for visual discomfort. Brain, 107, 989-1017.
Zenhausern, R. (1978). Imagery, cerebral dominance and style of thinking: A unified field model. Bulletin of the Psychonomic Society, 12, 381-384.

THE EVOLVING OF THE HOMEOSTATIC BRAIN: NEUROPSYCHOLOGICAL EVIDENCE

Michael Miran and Esta Miran

University of Rochester
Rochester, N.Y.
New York

Abstract

The purpose of this paper is to explore brain functioning as an integrated homeostatic system with lateralization and localization of brain function. The central question is what is the nature of the relationship between hemispheres as the brain processes information using specialized areas, neural pathways and whole brain interactions.

This paper discusses the neuropsychological, evolutionary, and developmental literature supporting the existence of homeostatic functioning within the brain. It cites the perceptual systems as a model and then reviews examples of lateralized and homeostatic functioning in motor and frontal systems. The inter- and intra- species evolutionary evidence for lateralization and homeostatic functions are reviewed. The role of developmental processes in adaptation to brain damage are discussed as exemplifying the role of homeostatic functions in the brain. The need for future research on the homeostatic process of the brain and the information flow from micro- to macro- systems within the brain are discussed.

INTRODUCTION

Left, right, right, left, like a patient being examined by a neuropsychologist, neuroscience's attention has shifted from hemisphere to hemisphere. Each new finding identifies a relative localization of one attribute or another on the right or left hemisphere. Broca (1865) proposed and Sperry (1968) expanded our understanding of cerebral asymmetries. Localization of function describes the operation of two subsystems i.e. the left and right hemisphere. This localized approach, however, leaves many loose ends and unresolved questions. One characteristic of studies of hemispheric asymmetries is that they show a broad range of individual differences, especially in populations other than right-handed males. Studying one hemisphere or the other can provide valuable data, yet the researcher risks neglecting the impact on the total brain of a specific stimulus.

The brain functions as an integrated homeostatic system in which information is transferred on anterior-posterior, lateralized, and sub-cortical levels (Miran & Miran, 1984a, b & c). As we have reviewed the literature on cerebral asymmetries, we have become aware that the brain is organized as a hierarchy of homeostatic systems. This hierarchical system codes and processes information. The smallest point in this hierarchy is the neuron and immediate neural networks. These microsystems are organized into mesosystems involving localized areas such as the occipital cortex. These mesosystems include specialized cell groups which may be dedicated to a specific function. The brain is organized into macrosystems including the anterior-posterior, right-left, and whole brain systems.

The potential interconnections of localized functions allows for the diversity, spontaneity and complexity of human behavior. While speech is "localized" in Broca's and Wernicke's areas and visuospatial imaging and affective recognition are "localized" on the right side of the brain, the important fact about human communication is that both affect and speech information are processed simultaneously to let us communicate with each other. As well as emphasizing the bi-cameral and divided nature of the brain, we emphasize the similarity, reciprocity, and coordination that is characteristic of human brain function.

The compelling question for further research is how does the brain process information both in the integrative mode and the localized mode i.e. to what extent do areas of the brain function independently and to what degree to they interact during any one cognitive event. In this interactive system how are the diverse areas of the brain involved in any one response?

Specifically, the identical sensory stimuli can trigger a variety of behavior. In order to select one of several potential behavioral patterns, the whole brain may function. It is quite possible that the brain selects one localized area to respond from, but in selecting, it is likely that the whole brain is scanned. Research using PET (Positron Emission Tomography), rCBF (Regional Cerebral Flow), and EEG has demonstrated activity throughout the brain during cognitive tasks, with relatively greater activity in specialized areas. Depending on the task, specific areas of the brain are activated in different orders. (Gur, 1983; Gur, Packer, Reivich and Weinberger, 1978; and Flor-Henry, 1983).

FROM PERCEPTION TO COGNITION: THE WORKING OF THE HOMEOSTATIC BRAIN

The Perceptual Mesosystems

The perceptual systems show very little lateralization, yet they are a good model of how the brain functions as an integrated homeostatic system. At a micro level, the perceptual systems are organized on a point by point homeotopic basis with bilateral representation (Hartline & Ratcliff, 1957; Hubel & Wiesel, 1962; and Hubel & Wiesel, 1967). Virtually all of our sensory experience reaches both hemispheres simultaneously. This bilateral representation establishes a stored data base for both hemispheres.

Evidence for the integrated brain model can be found in the perceptual systems of seeing, hearing, smelling and body sensing. For example Von Bekesy's (1968) model of the auditory system goes beyond the simple input-output connections to describe the more complex feedback interconnection, i.e., homeostatic processes. He cites the role of lateral inhibition at the sense organs and mid-brain levels which sharpens new "images" and inhibits "ground", thus providing a clear figure/ground relationship. He points out that these same organizing principles hold for higher order stimulus coding as well as basic stimulus impulse transductions.

Von Bekesy's (1968) article highlights, that from the moment a stimulus is turned on, that simultaneous homeostatic inhibitory and excitory trends occur. Thus, hearing a sound or seeing a light begins an active stimulus sharpening and coding process in the auditory or occipital corticles. At the same time that the stimulus is being registered on the display of the sensory cortices, it is also impacting the reticular activating system and arousing many other cortical areas in preparation for possible activity.

The stimulus is registered on the sensory cortex, and then it is contrasted to other available sensory images for recognition involving primary and secondary cortical areas. It is also mediated and shared via areas of multimodal sensation i.e., posterior parietal, anterior occipital lobe and association areas (Roberts, 1984). The stimuli can then be transferred in three directions: forward in each hemisphere, across the corpus callosum, and to subcortical areas (Allen, 1983; and Andreasen, 1984). The brain is now prepared for a number of responses; from the fight or flight responses of the affective limbic system to the sophisticated symbolic and conceptual analyses of speech and non-verbal prosody (Thompson, 1967). As the stimulus progresses from the perceptual system to the motor strip, temporal lobes and frontal lobes, more complex coding and response options are generated.

A possible transitional area between perceptual systems and the motor systems is the somatosensory cortex. This area is organized on a pattern of site specific representation based on innervation and activity of the sites involved. Our hearing and sight are closely interconnected with our bodily perception, sense of self, and the perception of our bodies in external space.

The somatosensory system also serves as one of the systems that completes a feedback loop for the motor system by making the individual aware of movement that has occurred and the somatosensory status of the body. We can relax and concentrate on the experience of our body at rest, or we can "do" something active and feel its impact on other areas of our body. The joint functions of perceiving changes in the state of our bodies and serving as a feedback loop are significant in a homeostatic model of the brain. The somatosensory system is bilaterally represented, and is adjacent to somatomotor areas.

The current models of homeostatic brain functioning provide new ways of looking at dichotic listening tasks and their results. Von Bekesy´s (1968) model, of lateral inhibition in the auditory system, suggests that many of the processes described in the dichotic listening literature may be a result of processes involving the cochlear nucleus and basilar membrane. He provides an example of localization of the sound of a click via complete inhibition of other sounds including a delayed click. This inhibition occurs within one millisecond of the first click. The right versus left preferences, observed in dichotic listening tasks, may reflect the preattentive processes of reticular activation or the executorial and inhibitory processes in the sub-cortical parts of the auditory system.

Dichotic listening studies demonstrate the interaction of perceptual and motor sets in shifting attention from hemisphere to hemisphere (Broadbent, 1970; Cherry, 1953; Moray, 1970; and Swets and Sewall, 1964). They note that switching sets is related to the time it takes from a subject to alternate between two attentional stimuli. This type of information exchange at neuronal and network levels underlies many of the processes by which homeostatic regulatin of behavior occurs.

Evarts, Shinoda, and Wise (1984) study higher brain functions in terms of "neural switching". They specifically discuss the relationship between perceptual and motor "sets" in preparing for responses. In terms of laterality the authors present data describing speech as an interaction with the motor activity of the mouth that depends on open-loop control in multiarticular coordination.

In summary, the neural input sets are linked to a variety of behaviors through the flexibility of switching of neural pathways. This process involves the rapid and efficient use of stored information or memory. Inhibition and activation are the key processes in neural switching. Patterns of inhibition and activation relate to localization of brain function. Many of the lateralized individual differences in PET and EEG studies appear to be examples of neural switching within a homeostatic brain system. The exact nature of these patterns is yet to be completely delineated.

The conclusions regarding the perceptual system are: 1. It provides a bilaterally represented data base for the brain, 2. Through the processes of lateral inhibition and excitation, stimuli are coded at sense organ, mid-brain, and cerebral levels, and 3. Neural networks are developed which connect perceptual and motor systems facilitating response patterns.

Motoric Mesosystems

As Von Bekesy (1968) describes a homeostatic process of inhibition and activation in the perceptual system, similar homeostatic networks are functioning within the motor and somatosensory systems. Netter (1980 p. 75) provides a thorough visual illustration of motoric system as an interactive feedback loop. The motor cortex is reciporically linked to the cerebellum and the basal ganglia. These structures may act as a servomechanism, with the subcortical areas acting as comparators (Heilman, 1979). Every muscle group, like the biceps and triceps, has reciprocal inhibition and activation which is necessary both to maintain resting tonus and do work. The cerebellum is involved with rapid movement (Ballistic); while the basal ganglia is involved with slow movement (Tonic) (Heilman, 1979).

While localized in opposite hemispheres, movement and verbal areas of the brain are integrally and reciprocally connected. Geschwind (1965) proposes the connection linking language to the elicitation of motor behavior by using neural substrates. Specifically, stimuli travel along the auditory pathways to Heschl's gyrus (primary auditory cortex), and then are relayed to the posterior superior portion of the temporal lobe (auditory association cortex). From there the stimuli travel by the arcuate fasciculus to the premotor areas (motor association cortex), and then to the primary motor area. Through these pathways, a behavior is initiated and monitored (Geschwind, 1965).

There is an interesting contrast between the functional and anatomical differences in lateralization of the motor and speech systems. In the speech system there is an apparent correspondence between the areas functionally associated with speech production and interpretation; and a specific set of brain locations i.e. Broca's area, Wernike's area, and the enlarged planum temporalis. The speech system is highly lateralized. In contrast, the motor system is highly lateralized i.e. one hand is usually preferred; yet, the neuroanatomical and cytoarchitectonic basis for this lateralization of motor behaviour is not as clear cut as it is for speech. Although the "dominant" hand may be associated with a slightly longer motor strip on the contralateral side of the brain, there is no clear gross anatomical differences between the two sides of the brain which are associated with a preference or greater dexterity of one hand over the other.

Motor systems do not function in isolation, but involve perceptual and verbal systems as a means of effecting complex human behavior. Behavior is a stream of reciprocal and homeostatic events involving many systems and subsystems of the brain. Particular subsystems continue to operate after a behavior has been initiated. The perceptual system continues to process incoming information while other cognitive and behavioral events are occurring in other systems of the brain. Hence, the individual can walk down the street, chew gum, and listen to a "Walkman" at the same time.

<u>Lateralization of specialized Abilities: Speech and Visuospatial</u>

Research over the last two decades has focused on the anatomical and functional specialization of each hemisphere (Gazzaniga, 1979; Geschwind, 1984; Kinsbourne, 1978; Milner Taylor & Sperry, 1968; Sperry, 1958; Sperry, Gazzaniga & Bogen, 1969; and Zaidel & Sperry, 1974). These studies strengthen the concept of relative specialization of the hemispheres. Each cerebral hemisphere is relatively specalized in functioning: the left hemisphere is specialized for verbal identification and speech production, and the right hemisphere is specialized for affective and visuospatial functions.

Scheibel & Scheibel (1954 & 1955) relate extended dendritic systems, increase in size of neuron somata, and in number of neuroglial cells to enriched functioning of gifted individuals. Scheibel (1984) studies dendritic growth and hemispheric asymmetry. He finds that in the non-verbal period of life, there is greater activity on the non-dominant side (the right hemisphere); while with the beginning of speech and conceptualization, there is a marked increase in the number of higher order dendritic branches on the dominant side (the left hemisphere). This is a shift of dendritic growth from the right to the left hemisphere.

In terms of development of specialized abilities, Scheibel (1984) suggests that skills are related to enriched dendritic growth. One feature of the brain, it may be hypothesised, is that it can "grow", and it can generate more information processing ability. In a dendritic tree each dendritic bifurcation represents a "go--no go" or "on-off" decision point. In terms of information generating or processing, more branches may be associated in a general way with "greater degrees of freedom" and "complexity" in thinking and behavior. Scheibel (1984) is careful to point out that this is a chicken and egg issue.

We cannot say whether there is a genetic program determining dendritic growth; or alternatively, whether environmental stimulation influences dendritic growth patterns of the hemispheres. But it appears that there are individual differences in patterns of dendritic growth. There is a mathematical model called fractals that describes the symmetric and asymmetric branching in naturally ocurring systems. According to Mandelbort (1982), this model describes dendritic systems.

<u>Speech</u>. Strong support for specialization of function in each hemisphere comes from research demonstrating that speech is located in the left temporal lobe above the planum temporale -- Wernicke's area (Springer & Deutsch, 1981). Porac & Coren (1981) present data on the architectonic site of the enlarged planum temporale and the greater length of the Sylvan fissure in the left hemisphere. Yet Porac & Coren (1981) further state that speech may not be completely located in the left hemisphere for most individuals; but rather, it may be located to a greater or lesser degree in the left hemisphere. Language may be bilaterally located for most individuals, with individual differences as to the extent to which language is located in the left hemisphere.

For the majority of the population speech is located in the left hemisphere, yet for 1% to 4% of right-handed people and 25% of left-handed people, speech is located in the right hemisphere (Porac & Coren, 1981). Galaburda (1984, p. 20) cites evidence that anatomical differences are the basis for lateralization for language: 1. There is a larger Broca's areas and associated Tpt and PC areas in 66% of the population. 2. Postmortem analysis of brains found asymmetries in proportion of pyramidal fibers crossing from one side to the other in the decussations of the medulla. 3. In the lower medulla, analysis of brains finds an aberrant circumolivary bundle deriving from the pyramidal tract on the left side more frequently than on the right.

While evidence points to a left hemisphere dominance, yet more careful examination of the anatomical data on lateralization reflects a wide range of individual differences. "In Broca's area architectonic asymmetries fluctuate between 15% and 259% in favor of the left side (Galaburda, 1984, p. 19). These individual differences are observed in research on architechtonic and behavioral studies of localization of speech.

Recent findings suggest that some speech may be located in the right hemisphere. Gur et. al. (1984) show that in a left-handed callosotomy patient, both left and right hemispheres were involved in language i.e. writing. Gazzaniga (1983) describes a patient with language in the right hemisphere (although it occurs infrequently). Gardner (1982) discusses evidence that the right hemisphere can think verbally. Based on aphasia studies, Geschwind & Galaburda (1984) suggest that language is bilaterally represented to some extent in most of the population.

Cappa and Vignolo (1979) extend the localization of language from a left-right hemisphere task to a task involving the thalamus. They found aphasia after left thalamic lesions. Language may in fact involve the participation of subcortical areas of the brain.

The two hemispheres of the brain together produce streams of word-behavior interactions. Thus, the first event in a speech sequence may be a facial or tonal recognition,i.e. a right hemisphere task. Next, there is the analysis of content, organization of a response, and finally the response. The redundant and homeostatic quality of this system allows for various solutions, i.e. subcortical-affective, visual-motor, and verbal cognitive solutions to the same problems.

In conclusion, speech is a dynamic phenomena in which perceptual images and words are combined via a variety of decision rules to create symbols which are useful in communication to others, and to serve as an internal communication system within the brain. As Piaget (1954) aptly points out, speech is an extension of sensori-motor concepts to include verbal stimuli. The speech system, as we now understand it, can include both right and left hemispheres, the thalamus, and other areas. The functioning of this marvelously human symbol system requires multiple representations and events on both sides of the brain to produce words and to coordinate action.

Visuospatial. Research has yet to arrive at a clear, specific, consistent, definitive, right hemisphere task i.e., affect, facial recognition, visual patterns, musical patterns, and prosody. What has been designated a right hemisphere task for the purpose of research, may actually involve more brain systems than was originally anticipated.

More recent research findings are contradicting earlier results on right hemisphere functioning. Parsons (1984) replicated Kimura's (1969) visual dot-location task. In contrast to Kimura's results of right hemisphere superiority. Parsons (1984, p. 31) found "a significant overall left hemisphere superiority".

The path that information takes in being processed may be other than task specific. In fact, the processing of information in the brain may depend on other factors such as speed of task. Marquis, Glass and Corlett

(1984) describe EEG patterns of the same task, varying speed of task. When the task is performed at a slower speed (25%) than their self-chosen pace, the individual processes information in the left hemisphere. In contrast, as the pace increases (25%) the individual processes the same information in the right hemisphere. Marquis et al. (1984, p. 206) suggest: "A sequential processing strategy is being used at the slow pace whilst a parallel or Gestalt processing strategy is being used at the fast pace". In any behavior there may be many combinations of inter- and intra-hemispheric contributions.

In conclusion, communication is the integration of speech and visuospatial systems. Language is the product of a system involving multiple locations in the brain. Language involves inputs and outputs from and to perceptual, motor, frontal, and subcortical areas. Thus, even in the most classic case where anatomical evidence for lateralization is strongest for speech located in the left hemisphere, the functioning of the lateralized hemisphere is constantly interacting with other areas of the brain via the corpus callosum and anterior posterior fibers within each hemisphere.

Actual communication between individuals and within the individual is a whole brain process. For example, hearing the words involves auditory cortex, reading and writing involves the visual cortex, and the verbal content involves intellectual and affective interpretations. Andreasen (1984) proposes that the language function works properly when special cortical centers and the "writing" that connects them function properly.

An example of whole brain function occurs when an individual picks up a pencil. He/she sees the color, feels the shape, interprets the prospect of having to work, and initiates the behavior of writing. Recognizing the pencil and initiating appropriate behavior of writing words involves both multiple brain areas (speech and visuospatial), and interactive brain processes.

The Frontal Lobes

As we have progressed through the brain from the back to the front, we have examined the brain's structure from hardwired sensory systems to hard wired motor systems. What then do the frontal lobes do in this system? They provide the internal monitoring feedback and modulating systems that synchronizes and sequences activity. The frontal lobes can coordinate the complex motoric sequencing and suppressing of subcortical systems; thereby completing a feedback loop via activating or damping the motor potential, subcortical response, and the reticular activation system. The frontal lobe's role in the relative excitation and inhibition of other behaviors is relatively well understood. A great deal has been learned since Phineas Gage's unfortunate accident!

While the literature on hemispheric asymmetries has emphasized the role of the corpus callosum in exchanging information between the left and the right hemisphere within the brain, our review of the literature has pointed to numerous other ipsilateral tracts. Notably these include the frontal-occipital, frontal thalamic, and frontal limbic tracts, which provide input and output for feedback loops involving the frontal lobes. The frontal lobes play a modulating role in complex cognition by inhibiting subcortical structures and motor structures. The anterior-posterior, right-left, and cortical-subcortical tracts complete the feedback system and orchestrate the activities of the homeostatic brain.

Evidence for and against asymmetry in the frontal lobes is hotly debated. Porac and Coren (1981) present new evidence showing the right frontal lobe is generally larger, and blood pressure and volume are greater in the right hemisphere.

In conclusion, the frontal lobes increasingly have the capacity for the type of covert behavioral rehearsals that we call "thinking". The issue of consciousness is less problematic if we assume that we are dealing with a homeostatic system that is capable of monitoring its own behavior. Much of the work of personality theorists has described the process of "insight". The nervous system is organized to not only have the right hand know what the left hand is doing in an on-line fashion; but more importantly, to allow them to work in a coordinated fashion to accomplish a task, while the mouth may be talking or the feet running -- or emotionally, the heart may be "breaking".

Brain Functions during Cognition

PET studies show patterns of activation i.e. what happens when the brain performs a cognitive task. Studies of regional cerebral blood flow demonstrate that for different populations of subjects, there are varying patterns of brain activation. Gur, Skolnick, Gur, Caroff, Rieger, Obrist, Younkin and Reivich (1985) examine patient and non-patient subjects and find differences: 1. probe by probe, 2. resting vs. cognitive task, 3. male vs female, 4. patient vs. non-patient, 5. spatial vs. verbal task, and 6, left vs. right hemisphere. Variance in blood flow can be interpreted as variance in brain activation. While one part of the brain may be more active for a specific task, all parts of the brain are receiving a continual flow of blood and are active in varying degrees.

Using (rCBF), Sheppard, Gruzelier, Manchanda, & Hirsch (1983) record patterns of symmetrical and asymmetrical responding in normal and schizophrenic subjects. Subject's responding appear to be variable depending on the specific probe studies i.e. which brain area in which subject population is being studied.

EEG studies support the concept of a homeostatic integrated brain: there are patterns of individual and group differences in hemispheric brain activation and inhibition (Glass, Butler & Carter, 1984; Etevenon, 1984; Flor-Henry and Koles, 1984; and Rebert and Lowe, 1984). These studies show, that when the brain performs a cognitive, spatial, or affective task in varying subject populations, then different areas of the brain are stimulated and activated.

THE ROLE OF EVOLUTION IN BRAIN DEVELOPMENT

In our study of human brain behavior relationships and the development of laterality, let us start at the beginning. We take a change of scene from the modern computerized neuroscience laboratories that we associate with brain research and return to the African veldt of 3 million years ago. What could such a trip teach us about human brain-behavior relationships? Such a trip provides information on the origins of human brain functions and the brain's evolutionary development of complexity and laterality.

The proposed homeostatic model delineates patterns of brain organization existing prior to differentiation of the human species, and then describes the interactive neuropsychological patterns of brain behavior relationships seen today. Evolutionary theory traces the development of specific systems from general ones and complex systems from simple ones, i.e., the nervous system from hormonal signalling mechanisms (Horridge, 1968). The evolutionary data expand and clarify the question of lateralized cerebral asymmetries and integrated brain functioning in terms of: 1. Co-evaluation of brain and behavior. 2. The evolution of improved neural sequencing and homeostatic regulation of complex behavior. 3. Evolution of size and convultions. 4. The evolution of laterality.

Presented below are examples of other species that reflect the evolutionary processes at work in the development of human lateralized speech, and complex behavior. The echidna is an example of an animal that evolved increased intelligence via increased brain size and convultions without specific lateralization of functions. In contrast, the song bird is an animal that evolved lateralization of singing without increased size and complexity of convolutions. Non-human primates exhibit complex behavior similar to humans; yet they do not consistently display laterality. Australopithecines have a brain which is smaller than the human brain and has many ape-like characteristics. Yet the Australopithecine's brain may offer clues to the beginnings of hemispheric asymmetry.

Echidna

In exploring the evolution of the human brain, an examination of the echidna provides valuable information. Fossil evidence indicates that this species' brain increased in size and convolutions just as the human brain evolved. Echidnas are monotremes. They lack a corpus callosum their nervous system has commissures and tracts in a relatively symmetrical brain, but they do have two hemispheres. Yet, in the evolutionary process, they demonstrate the importance of brain size and convolutions. Their neocortex occupies 43% to 48% of brain weight; almost as much as primates brains where the neocortex occupies 54%. Their neocortex is expanded and nearly spherical; while its surface is richly convoluted in a series of deep folds and bumps. Although the echidna has many "reptilian" features such as laying eggs, it functions behaviorally at the level of a cat or primate, and demonstrates rapid learning on certain tasks (Gould, 1985).

Song Bird

Song birds and canaries are vertebrates that demonstrate lateralized sound production. Nottebohm (1984) points out that such birds are the only vertebrates in which there is a relation between a naturally occurring learned behavior and the brain pathways that control it. Although singing is lateralized and occurs without ever hearing the song of other birds, such a song is simple and lacks other social cues. For a bird's song, to reflect the unique song patterns of particular species, it requires hearing, imitation, and learning. This study points out, that for a relatively "hardwired" neurosystem to act, requires interaction with a specific social environment.

Primates

Evidence for the co-evaluation of brain structures and complex human behaviors can be found in primate research. Recent research highlights the issue of inter- and intrahemispheric information processing in macaques and humans (Doty, Overman and Negrao, 1979; and Overman & Doty, 1982). While we cannot study neurosystems and synaptic processes of prehominids and early hominids, and compare it to humans, we can compare and contrast non-human primates and humans. The above mentioned studies of non-human primates and humans demonstrate similar complex brain processes of interhemispheric transfer and coordination; yet in laboratory studies, the primates (Macaques) do not display consistent lateralization of task and behavior.

Terrace (1985) compares speeh in apes with speech in humans. He suggests that language has two levels: words and sentences. While apes may be able to put two words together to make a simple sentence, only humans can use language to express complex ideas and concepts. Language is processed in a range from simple labelling to conceptual expression.

Australopithecines

Falk (1984) made endocasts (casts of the inside of the braincase) of our earliest known human ancestors called "australopithecines". These endocasts are dated approximately 2.5 - 3.5 million years ago. The major distinction found between australopithecines and humans are in brain convolutions, density, and size; as well as organization. Falk (1984) cites the evidence: "The organization of the australopithecines is ´ape-like´ in that it identifies a particular groove, known as the fronto-orbital sulcus, that courses from the side of the frontal lobe to its surface. In place of this feature, human brains display a particular pattern of convolutions" (Falk, 1984, p. 38).

LeMay & Culebras (1972) examined the endocasts of the Neanderthal man who lived 30,000 to 50,000 years ago. The endocasts had typical sylvan asymmetry of a longer, straighter left fissure on the right that turned up. One aspect of this evolving pattern of convolutions is the establishing of patterns of cerebral asymmetries in the human brain.

The above authors´ research further provides evidence showing that the human brain evolved as a complex integrative structure. The human brain: (a) is considerably larger than australopithecines´ (1450 cc instead of 450 cc´s), (b) has clearly identifiable frontal lobes which are involved in human higher-order cognitive and behavioral control functions, (c) has identifiable fissures and gyri which help provide more information processing area in a more compact volume, and (d) has an identifiable Broca´s and Wernicke´s areas, which are associated with hemispheric asymmetry, specifically the human characteristic of speech.

The evolution of the human lateralized verbal system required the following changes from "ape like" structures: 1. a Broca´s area or equivalent structure, 2. a voice box that is low enough in the throat to permit human speech sounds, 3. a structure of facial muscles that can be used for form human sounds. 4. the necessary innervation connecting the brain and facial muscles, and 5. learning of socially relevant and significant sounds.

Based on the limited cases known so far, it appears that these changes did not always take place in a synchronized fashion. In some cases, such as the evolution from *Homo habilis* to *Homo erectus* to *Homo sapiens (neanderthal)* to *Homo sapiens (modern)*, the brain size appears to have preceded the thinning of the skull and reducing of brow ridges (Holloway, 1974; Laitman, 1984). In other situations such as the transition from *Australopithicus afarensis* to *robustus* to *zinjantrophus*, the advent of bidpedalism provided a basis for change in the angle of the head, and eventually evolution of the voice box. The progress of human evolution has been a result of change in many interacting systems. These changes resulted in an enlarged brain with a subsystem that permits the acquisition of speech in a social environment that provides the necessary models and reinforcement.

Johanson and Edey (1982) and Jolly (1972) highlight the following behaviors evolved from australopithecines to humans: bipedalism, tool making and use, speech and use of language, changes in reproductive behavior that allows for genetic penetrance, and cooperative food gathering. Miran (1983) cites the intergenerational storage and transmission of information as a characteristic of human communities and the socialization of "creative" behaviors. The increased brain size and complexity, as well as hemispheric asymmetry, extends human mental abilities.

There is a question as to the relationship between the degree of laterality and the evolution of expanded and convoluted structures. Most of the efforts to explain this relationship have oversimplified the complexity of both phylogenetic and ontogentic processes. We can identify at least three interrelated processes: (a) the establishing of layers of neuronal tissue in phylogentic and developmental sequences. These cell layers are the basis for the microsystems and cellular specialization, (b) the folding and refolding of these sheets of cells, and (c) the asymmetrical arrangements of convolutions and cell layers into asymmetrical structures associated with the behavior of speech.

We believe that these processes may reflect some underlying evolutionary and developmental patterns that permit increasing complexity of neural tissues. Although they are still at an early stage of use in neurosciences, there are mathematical models called fractals which describe the symmetric and asymmetric distributions of a phenomena around an axis. These models have been used to describe dendritic trees and neural structures. They appear to be a good fit. What we observe as lateralization may be a result of mathematical functions that govern the distribution of neural structures around a central axes (Mandelbrot, 1982), may be speculated.

Geschwind (1984) makes the point that the evolution of language development includes internal cognitive information processing. The use of internal language is a means by which an organism can self-regulate. He suggests that internal language processing systems that exist in humans may also exist in other species.

In conclusion the human brain system evolved in a broad co-evolutionary context. First the human brain systems and subsystems are distinct in terms of increased brain size, density, and convolutions. Next the human brain developed laterality, each hemisphere is specialized for a function i.e. the left hemisphere is specialized for verbal identification while the right is specialized for visuospatial functions.

DEVELOPMENTAL ISSUES IN THE NORMAL AND ABNORMAL BRAIN

In the preceding section we discussed the role of evolution in the establishment of systems and subsystems of the brain. In particular, we examined the development of lateralized functioning. In this section we consider the developmental process and the influences of various types of injury. By comparing child with adult brains, we can see the differences as unique systems which have unique patterns of resonse to injury. Many previous conceptualizations, such as "plasticity", can be evaluated in terms of the interaction of systems and subsystems in a homeostatic brain.

Some parts of the brain can compensate for damage to other areas (Roberts, 1984). This compensation is often a result of shared equivalent functions and the "plasticity" of behavior that is both fundamental to human beings and capable of being developed through socialization (Miran, 1983). It is possible that adaptation, learning, and recovery all reflect an underlying inherent characteristic of neural tissue.

Previous research comparing brain function and dysfunction of children with adults and senior adults has yielded problematic findings (Reitan, 1979; Telzrow, 1984; Towbin, 1978). Problems can be expected when you compare one dynamic system with a different dynamic system; and essentially we have two distinct brain systems i.e. infant and adult. These systems may share many functions in common, yet these systems also may involve different mechanisms. Miller (1984) suggests that, in comparing children to adults, what we are seeing are different disease etiology, processes, and compensation. This makes statistical comparisons of normal and brain damaged infants and adults problematic. St. James-Roberts (1981, p. 47) states: "Valid comparisons are prevented by failure to control these procedures, the status of residual system, diaschisic variables, recovery periods and experimental variables ... and differences in surgical techniques". What we can discuss is the brain mechanisms and related diseases as the brain develops. The homeostatic brain model offers a paradigm for examining brain damage in terms of age, location, and severity.

Infants-and-Children

At birth the human brain includes genetic encoding for "hard wired" or "dedicated" areas, i.e. medullary and mid-brain regulation of biological processes perceptual and motor systems. Frontal, temporal and association areas are less clearly "dedicated" and to a greater degree are dependent on environmental "software" i.e. socialization for their functioning.

At its earliest stage much of brain function can be compared to a "computer core" which has the potential for memory but which has not been "programmed" to any specific pattern or filled with data. In later development this early system is progressively developed both neuroanatomically and behaviorally, including myelination of the corpus callosum. The child acquires behavior and increases synaptic and dendritic connections. As this system develops neuroanatomically, it develops functionally (Bigelow, Nasrallah and Rausche, 1983; Miller, 1984).

Miller (1984) describes the maturation of the brain and concludes that in early stages of brain maturation (childhood) the brain may not be able to carry out the functions of a mature brain (adulthood). Based on a primarily sensori-motor input-output system, a process of development and interaction is initiated. Piaget (1954) describes the process of development as primary circular reactions, the maturing of the perceptual-motor coding systems, and the development of cognition and organized sensori-motoric response patterns. As each system develops, it enhances the devlepment of other brain subsytems and systems.

At birth the brain is hardwired for perceptual motor activities. The frontal areas of the brain are relatively plastic i.e., the part of the brain which later moderates speech and higher level cortical functioning. Towbin (1978) discusses damage to the cerebral system in pre- and perinatals. He describes two phenomena: (1) reduction in cerebral function -- mental retardation; and, (2) distortion of cerebral function -- the uncontrolled motor patterns such as cerebral palsy, epilepsy, and related pathology. The most common cause of cerebral damage in perinatals is hypoxia or lack of oxygen. Damage leads to a system wide dysfunction. The greater the damage, then the more severe the dysfuntion (Towbin, 1978). Brain damage in infants and young children is most likely to impact the development of the total system. Miller (1984) cites malnutrition as a similar system wide problem. A smaller less differentiated system is affected in a relatively equal and generalized fashion than a larger more specialized system.

The diseases that attack children are different from those that attack adults. Children are vulnerable to residual impairment from infectious diseases, malnutrition, and cranial irradiation (Levin, Eisenberg, Wigg and Kobayashi, 1982). These conditions do occur in adults but are more frequently identified in children. In terms of flexibility children appear to have an advantage over adults. There is a great deal of still relatively undifferentiated area in the brain. If there is damage to the left hemisphere, then other brain subsystems i.e. the right hemisphere can take over its functions. If unfilled memory banks are damaged, the child experiences no specific losses, but an overall delay in development. Kertesz (1979) attributes plasticity in the young (before the age of 10-12 years) to the adaptability of Golgi Type II cells. The flexibility of these neurons man terminate in the teens by hormonal changes.

While evidence strongly supports developmental changes in brain organization and function, individual differences remain a key concept. Within this developmental sequence, the actual rate of growth or recovery from damage differs for individual children (Roberts, 1984).

Adults and Senior Adults

In the adult, the brain functions efficiently and effectively in an interactive mode. Moscovitch (1979) finds that the right hemisphere plays a supportive role in normal communication and memory. Deptula and Yozawitz (1984) examine depression as a right hemisphere deficit. Springer and Deutch (1981) examine normal subjects' hemispheric functioning. Attention and task interference influence information processing. Cognitive tasks increase the activity in the brain as measured by EEG. This increase in activity is related to specialization of hemispheric function (Glass, 1984).

In the adult the "core" has been filled with data and programmed for specific functions. Damage to the "core" results in specific deficits (Hellige, 1983); yet malnutrition and starvation in adulthood do not have such permanent devastating effects as seen in children. Wiesel (1970) describes starvation conditions in the Nazi Concentration Camps; yet when adequate nutrition was restored, these camp survivors have gone on to function cognitively on a high level. They have recorded their intellectual functions.

Another group of disorders in adults are system wide breakdowns. Specifically, they are Huntington's Chorea, and similar disorders. These diseases involve the breakdown of information processing and flow within the brain.

These diseases exemplify the problems of studying brain function and dysfunction. Cain (1985) discusses Huntington's Chorea, a genetically determined neurological degenerative disease. Yet the course of the disease, the nature of the symptoms, and the speed of deterioration differs between individuals. This type of disorder describes the breakdown of sub-systems and eventually multiple sub-systems within the brain.

Senior adults have problems of: strokes, progressive dementia, toxic psychosis, anoxia, loss of blood supply to the brain, and related blood flow and systems diseases. There is a progressive atrophy and loss of function of cerebral tissue. Joynt and Shoulson (1979) and Scheff (1984) discuss Dementia as a possible localized deficit, yet they caution that whole brain functions are disturbed such as cognition, orientation, memory, abstraction and reaction to stress.

The brain diseases of adults and senior adults include tumors that may or may not be cancerous; and strokes which involve the decrease of blood flow to the brain. These cases are localized lesions of the brain and affect specific subsystems, as well as specific microsystems.

In conclusion, the integrated brain model provides a paradigm for studying brain function and dysfunction. This is the study of multiple areas of the brain as they interact in either a successful manner or a damaged manner. Any head injury or deteriorative process affects the homeostatic systems and sub-systems of the brain. The patterns of recovery and deficits provide information as to the development of localized brain function at varying ages. The advantage of the homeostatic brain model is that it helps us understand the development of the brain and the effects of damage at different ages.

CONCLUSIONS

Previous models of brain function have been based on lesion and deficit models, split brain studies, and other relatively rare conditions. While these models describe aspects of brain function, they do not completely describe the complex and dynamic interactions taking place within the human brain on a routine basis. We need to understand how information is transmitted from micro to meso to macrosystems. While micro- and "molar" processes are important, the next step in brain research is to develop the means to study complex interactions of dynamic structures and functions. Our model of integrated homeostatic brain function takes into consideration the complex and dynamic nature of brain behavior relationships.

Data from more recent studies suggest that, while there is relative specialization of hemispheres, the brain is more flexible in lateralization of function than was originally believed. What is emerging is a concept of brain function as a complex interactive systems process. One hemisphere may be dedicated to a specific function; yet while the hemisphere is performing the designated function, it is in interaction with the "whole" brain, both simultaneously receiving and sending information. This is an internal dynamic network in search of a homeostatic balance.

One of the key features of the homeostatic brain model is that it takes into consideration hemispheric asymmetry in providing a model for understanding individual differences. The individual differences in size and organization of the hemispheres of the brain underlies many of the observed differences in hemispheric asymmetries. In a particular area of a hemisphere which is "specialized for a function", the dendritic enrichment appears to be associated with the development of specialized abilities such as speech or visuo-motor skills. This model also permits a better understanding of the role of the environment and socialization as necessary conditions for the development of a complex skill such as speech.

The homeostatic brain model has a significant heuristic value. The concepts of development, plasticity, and compensation may all be different words describing similar processes of adaptation to environmental influences or injuries. Many fundamental issues in psychology such as cognitive dissonance and repression are being formulated in terms of the right and left brain model or more recently the idea of brain modules and microprocessors.

One of the contributions of the model of the homeostatic brain is that it makes us begin to rethink our definitions of wellness, psychopathology, and brain damage. The distinction between different types of psychopathology and brain damage from normals may be more than a matter of site-specific deficits: it may be the disruption of homeostatic systems.

The next task is to evaluate empirically and operationally the homeostatic brain model. We anticipate further research focusing not on localized processes but on the functioning of systems and subsystems within the homeostatic brain.

REFERENCES

Allen, M. (1983). Models of hemispheric specialization. Psychological Bulletin, 93, 73-104.

Andreasen, N.C. (1984). The Broken Brain. New York: Harper & Row.
Bigelow, L., Nasrallah, H. and Rauscher, F. (1983). Corpus callosum thickness in schizophrenia. British Journal of Psychiatry, 142, 284-287.
Broadbent, D. (1970). Stimulus set and response set: two kinds of selective attention. In D. Mostofsky (Ed.) Attention: Contemporary Theory and Analysis. New York: Appleton-Century-Crofts.
Broca, P. (1865) Sur la faculte de language articule. Bulletin Societe Anthropolgia, 6, 493-494. In J. Hellige (Ed.) Cerebral Hemisphere Asymmetry: Method, Theory, and Application. New York: Praeger. [1983].
Cain, E. (1985). Grand Rounds on Neuropsychiatry. Strong Memorial Hospital, University of Rochester. February 27, 1985.
Cappa, S. and Vignolo, L. (1979). Transcortical features of aphasia following left thalamic hemorrhage. Cortex, 15, 121-130.
Cherry, E. (1953). Some experiments on the recognition of speech, with one and with two ears. Journal of Acoustic Society of America, 25, 975.
Deptula, D. and Yozawitz, A. (1984). Lateralized brain dysfunction in depression: Analysis of memory. Paper presented at the 12th annual meeting of the International Neuropsychological Society. February, 1984.
Doty, R.W., Overman, W.H., and Negrao, N. (1979). Role of forebrain commisures in hemispheric specialization and memory in macaques. In I. Steele Russel, M. W. van Hof and G. Berlucchi (Eds.) Structure and Function of Cerebral Commissures. Baltimore: University Park Press, 333-342.
Etevenon, P. (1984). Intra and inter-hemispheric changes in alpha intensities in EEGs of schizophrenic patients versus matched controls. Biological Psychology, 19, 147-256.
Evarts, E.V., Shinoda, Y., Wise, S. (1984). Neurophysiological Approaches to Higher Brain Functions. New York: Wiley & Son.
Falk, D. (1984). The petrified brain. Natural History, 93, 36-39.
Flor-Henry, P. (1983). Functional hemispheric asymmetry and psychopathology. Integrative Psychiatry 1, 46-52.
Flor-Henry, P., and Koles (1984). Statistical quantitative EEG studies of depression, mania, schizophrenia and normals. Biological Psychology, 19, 257-279.
Galaburda, A.M. (1984). Anatomical asymmetries. In N. Geschwind and A.M. Galaburda (Eds.) Cerebral Dominance: The Biological Foundations. Cambridge: Harvard University Press.
Gardner, H. (1982). Art, Mind & Brain: A Cognitive Approach to Creativity. New York: Basic Books.
Gazzaniga, M., (1979). Handbook of Behavioural Neurobiology. Vol. 2. New York: Plenum Press.
Gazzaniga, M. (1983). Right hemisphere language: A twenty year perspective. American Psychologist, 525-549.
Geschwind, N. (1965). Disconnection syndromes in animals and man. Brain, 88, 237-294, 585-644.
Geschwind, N. (1984). Historical introduction. In N. Geschwind and A.M. Galaburda (Eds.), Cerebral Dominance: The Biological Foundations. Cambridge: Harvard University Press.
Geschwind, N., and Galaburda, A.M. (1984). Cerebral Dominance: The Biological Foundations. Cambridge: Harvard University Press.
Glass, A. (1984). Cognitive and EEG asymmetry. Biological Psychology, 19, 213-217.
Glass, A., Butler, S., and Carter, J. (1984). Hemispheric asymmetry of EEG alpha activity: Effects of gender and familial handedness. Biological Psychology, 19, 169-187.
Gould, S.J. (1985). Bligh's bounty. Natural History, 94, 2-10.
Gur, R. (1983). A cognitive-motor network demonstrated by positron emission tomography. Neuropsychologia, 21, 601-606.
Gur, R.E., Gur, R.C, Sussman, N., O'Connor, M., and Vey, M. (1984).

Hemispheric control of the writing hand: The effect of callostomy in a left-hander. Neurology, 34, 904-908.
Gur, R., Packer, I., and Reivich, M. and Weinberger (1978). Cognitive task effects on hemispheric blood flow in humans. Paper presented APA Annual Convention, Toronto, Canada.
Gur, R.E., Skolnick, B.E., Gur, R.C., Caroff, S., Rieger, W., Obrist, W., Younkin, D. & Reivich, M. (1984). Brain function in psychiatric disorders. Archives of General Psychiatry, 41, 695-699.
Hartline, H.K., and Ratcliff, F. (1957). Inhibitory interaction of receptor units in the eye of limulus. Journal of General Physiology, 40, 357-376.
Heilman, K. (1979). The neuropsychological basis of skilled movement in man. In M. Gazzaniga (Ed.) Handbook of Behavioral Neurobiology. New York: Plenum Press.
Hellige, J. (1983). The study of cerebral hemisphere differences: Introduction and overview. In J. Hellige (Ed.) Cerebral Hemisphere Asymmetry: Method, Theory and Application. New York: Praeger.
Holloway, R.L. (1974). The casts of fossil hominid brains. Scientific American. 106-115.
Horridge, G.A. (1968). Interneurons. San Francisco: W. H. Freeman.
Hubel, D.H., and Wiesel, T.N. (1962). Receptive fields, binocular interaction and functional architecture in the cat's visual cortex. Journal of Physiology, 160, 106-154.
Hubel, D.H., and Wiesel, T.N. (1967). Cortical and callosal connections concerned with the vertical meridian of visual fields in the cat. Journal of Neurophysiology, 30, 1561-1573.
Johanson, D., and Edey, M. (1982). Lucy: The Beginning of Humankind. New York: Warner Books, Inc.
Jolly, A. (1972). The Evolution of Primate Behavior. New York: Macmillan Co.
Joynt, R., and Shoulson, I. (1979). Dementia. In K. Heilman and E. Valenstein (Eds.) Clinical Neuropsychology. New York: Oxford University Press.
Kertesz, A. (1979). Recovery and treatment. In K. Heilman and E. Valenstein (Eds.) Clinical Neuropsychology. New York: Oxford University Press.
Kimura, D. (1969). Spatial localization in left and right visual fields. Canadian Journal of Psychology, 23, 445-458.
Kinsbourne, M. (1978). Asymmetrical Function of the Brain. Cambridge: Cambridge University Press.
LeMay, M., and Culebras, A. (1972). Human brain: Morphologic differences in the hemispheres demonstrable by carotid arteriography. New England Journal of Medicine, 287, 168-170.
Laitman, J.T. (1984). The anatomy of human speech. Natural History, 93, 20-27.
Levin, H.S., Eisenberg, H.M., Wigg, N.R., and Kobayashi, K. (1982). Memory and intellectual ability after head injury in children and adolescents. Neurosurgery, 11, 668-673.
Mandelbrot, B. (1982). The Fractal Geometry of Nature. New York: W.H. Freeman & Co.
Marquis, F., Glass, A., and Corlet, E.N. (1984). Speed of work and EEG asymmetry. Biological Psychology, 19, 205-211.
Miller, E. (1984). Recovery and Management of Neuropsychological Impairments. New York: Wiley.
Milner, B., Taylor, L., and Sperry, R.W. (1968). Lateralized suppression of dichotically presented digits after commissural section in man. Science, 161, 184-185.
Miran, E. (1983). The Ecology of Creativity. Dissertation. Teachers College, Columbia.
Miran, M.D. and Miran, E.R. (1984a). Interhemispheric communication in schizophrenia: Individual differences. Chair symposium at NATO

Conference on Individual Differences in Hemispheric Specialization (this volume).
Miran, M.D., and Miran, E.R. (1984b). Cerebral asymmetries: Neuropsychological measurement and theoretical issues. Biological Psychology, 19, 295-304.
Miran, M.D., and Miran, E. R. (1984c). Cerebral asymmetries: Theoretical and conceptual issues in models of the integrated brain. Paper presented at the NATO Conference on Individual Differences in Hemispheric Specialization. Maratea, Italy, October 1984. (see this volume).
Moray, N. (1970). Attention: Selective Processes in Vision and Hearing. New York: Academic Press.
Moscovitch, M. (1979). Information processing and the cerebral hemispheres. In M. Gazzaniga (Ed.) Handbook of Behavioral Neurobiology. Vol. 2. New York: Plenum Press.
Netter, F.H. (1980). The CIBA Collection of Medical Illustrations: Nervous System. New Jersey: CIBA.
Nottebohm, F.H. (1984). Learning, forgetting, and brain repair. In N. Geschwind & A.M. Galaburda (Eds.) Cerebral Dominance: The Biological Foundations. Cambridge: Harvard Press.
Overman, W., and Doty, R. (1982). Hemispheric specialization displayed by man but not macaques for analysis of faces. Neuropsychologia, 20, 113-128.
Parsons, O. (1984). Recent developments in clinical neuropsychology. In G. Goldstein (Ed.) Advances in Clinical Neuropsychology, Vol. 1 New York: Plenum Press.
Piaget, J. (1954). The Construction of Reality in the Child. New York: Basic Books.
Porac, P., and Coren, S. (1981). Lateral Preferences and Human Behavior. New York: Springer-Verlag.
Rebert, C., and Lowe, R. (1984). Hemispheric lateralization of event-related potentials in a cued reaction-time task. Biological Psychology, 19, 189-204.
Reitan, R.M. (1971). Trail making test results for normal and brain-damaged children. Perceptual and Motor skills, 33, 575-581.
Roberts, F. (1984). Differential Diagnosis in Neuropsychiatry. New York: John Wiley.
Scheff, S.W. (1984). Aging and Recovery of Function in the Central Nervous System. New York: Plenum Press.
Scheibel, M.E., and Scheibel, A.B. (1954). Observations on the intracortical relations of the climbing fibers of the cerebellum: a Golgi study. Journal of Comparative Neurology, 101, 733-764.
Scheibel, M.E., and Scheibel, A.B. (1955). The inferior olive: A Golgi study. Journal of Comparative Neurology, 102, 77-132.
Scheibel, A.B. (1984). A dendritic correlate of human speech. In N. Geschwind and A. Galaburda (Eds.), Cerebral Dominance: The Biological Foundations. Cambridge: Harvard University Press.
Sheppard, G., Gruzelier, J., Manchanda, R., & Hirsch, S.R.(1983). 0 positron emission tomographic scanning in predominantly never-treated acute schizophrenic patients. Lancet, 1148-1152.
Sperry, R.W. (1958). The corpus callosum and interhemispheric transfer in the monkey. Anatomical Records, 131, 297.
Sperry, R.W. (1968). Hemispheric disconnection and unity in conscious awareness. American Psychologist, 23, 723-733.
Sperry, R.W., Gazzaniga, M., and Bogen, J. (1969). Interhemispheric relationships: The neocortical commissures; syndromes of hemisphere disconnection. In P. Vinken, & P. Bruyn (Eds.) Handbook of Clinical Neurology. Amsterdam: North-Holland.
Springer, S., and Deutsch, G. (1981). Left Brain, Right Brain. San Francisco: Freeman & Co.
St. James-Roberts, I. (1981). A reinterpretation of hemispherectomy data

without functional plasticity of the brain. Brain and Language, 13, 31-53.

Swets, F., & Sewall, A. (1964). Stimulus versus response uncertainty in recognition. In J. Swets (Ed.) Signal Detection and Recognition by Human Observers. New York: Wiley.

Telzrow, C.F. (1984). Applying neuropsychological framework to the assessment of preschool children. Presented at the Meeting of the American Psychological Association, Toronto.

Terrace, H. (1985). In the beginning was the "Name". American Psychologist, 40, 1011-1028.

Thompson, R. (1967). Foundations of Physiological Psychology. New York: Harper & Row.

Towbin, A. (1978). Cerebral dysfunctions related to perinatal organic damage: Clinical-neuropathological correlations. Journal of Abnormal Psychology, 87, 617-635.

Von Bekesy, G. (1968). Similarities of Inhibition in the Different Sense Organs. Paper presented at the 75th meeting of the American Psychological Association, San Francisco, California.

Wiesel, E. (1970). A Beggar in Jerusalem. New York: Schocken Books.

Zaidel, E., and Sperry, R. (1974). Memory impairment after commisurotomy in man. Brain, 97, 263-272.

AN ARGUMENT CONCERNING SCHIZOPHRENIA: THE LEFT HEMISPHERE DRAINS THE SWAMP

Rue L. Cromwell

Department of Psychology
University of Kansas
Lawrence
Kansas, 66045-2462

Two major traditions exist in the psychological study of schizophrenia. One, stemming from cognitive theory, has focused upon attentional and information-processing factors with minimal concern for brain-behavior relations. The other has been neuropsychological, which focuses upon brain-behavior relations and topography of brain function with minimal concern for cognitive formulation. The purpose of this chapter will be to recount some problems in schizophrenia research and, in somewhat iterative fashion, discuss what might be learned from the interplay of cognitive and psychoneurological concepts in understanding schizophrenia.

TWO PROBLEMS IN SCHIZOPHRENIA RESEARCH

Two of the many problems in current schizophrenia research are the classification of individuals with only one data domain and the "merry-go-round effect".

Classifying people as schizophrenic patients with only one data domain (Cromwell, 1984, p.16) means that individuals are sorted out primarily on the basis of what they say. Reported delusions, reported hallucinations, and/or inference from verbal report about thought disorder are the defining symptom features. Based upon this limited domain an arbitrary group is designated who present problems to self and others and who are received by health institutions for care and custody.

On the other hand, it is indeed hazardous to assume that the necessary antecedent conditions of this disorder all fall within that same verbal domain. To the extent other factors, non-verbal in nature, are involved, effective progress in classification (for purposes of intervention, prevention and prognosis) may be curtailed.

The "merry-go-round effect" (Cromwell, 1972) refers to how investigators so often tend to view the antecedent-consequent relationships in schizophrenia. Many tend to "ride only one horse in the carousel". They attend to variables on only one level of description (e.g., biochemical, structural-anatomical, electrophysiological, information-processing, interpersonal communication, stress and expressed emotion, conceptual uncertainty, or such), and they assume that their horse is "leading the pack". Variables on other levels of description are viewed as secondary effects.

STRATEGIES TO SOLVE THESE PROBLEMS

Of the strategies (Cromwell, 1984, p.18-28) to circumvent these problems, two deserve mention here: (a) the search for schizophrenia-related variants (SRVs), and, (b) the search for the earliest information-processing event in which something goes wrong among schizophrenic patients.

The SRV strategy

A schizophrenia-related variant (SRV) is a variable which
(a) is associated with schizophrenia (though not necessarily exclusively, since the boundaries for clinical definition of schizophrenia must remain suspect),

(b) can be identified in patients before and after, as well as during, any episode of illness, and,

(c) is familial, e.g., can be identified in healthy first degree relatives.

The SRV strategy is directed toward variables in any domain of data. Since they, by definition, must be antecedent to the manifest disorder, it is hypothesized that they will serve more clearly as phenotypes for genetic study than the later features which accompany or define the manifest disorder.

An analogy to the SRV strategy can be seen in the work of Annett (see this volume; 1978; Annett & Kilshaw, 1983), wherein handedness is assumed to be the final result of both genetic and environmental forces and, therefore, not an appropriate phenotype for genetic study. Through her assumption that right vs. left verbal dominance is indeed more clearly a phenotype, the distribution of handedness as a combined result of sundry environmental and underlying constitutional determinants becomes more clear. In similar fashion, the SRV strategy takes seriously the commonly accepted notion that schizophrenia is a product of hereditary and environmental factors. Then, without the contradictory assumption that schizophrenia itself should be pursued as a phenotypic target, the search is directed toward SRVs as appropriate phenotypes.

Since schizophrenic symptoms emerge later in life than handedness, its genetic and other constitutional determinants are more elusive. SRVs which precede psychotic episodes may be useful not only for genetic study but as potential risk predictors. However, since they also occur in relatives who never become ill, they may help isolate and clarify the genetic and environmental factors necessary for the precipitation of illness. Thus, an SRV is not a subclinical form or a one-to-one correlate of schizophrenia which follows along later. It is a precisely defined variable which may or may not be sufficient to presage a schizophrenic episode.

Since the SRV strategy is a new one, few SRVs have been identified. Some candidates for study are (a) reaction time cross-over, (b) span of apprehension, (c) visual evoked response augmenting-reducing, (d) blood platelet monoamine oxidase, (e) plasma dopamine-beta-hydroxylase, (f) dichotic listening intrusions, and (g) smooth pursuit eye tracking disturbance (Cromwell 1984, p.33-34). In addition, Iacono (e.g. Iacono et al., 1983; personal communication) has also studied (h) skin conductance recovery and (i) capillary nailbed structure. Zubin and Steinauer (Reference Note 2) are also studying pupillary dilation, pupillary light reaction, vigilance, heart rate and blink rate.

The implications of these variables for the present paper is that the ones which currently appear to be most strong in meeting the SRV criteria are those which are involved in the early stages of information processing. The implications of this for the understanding of the psychoneurology of schizophrenia and of normal brain functioning will be examined.

"Earliest Processing Event" Strategy

Another strategy for studying schizophrenic (or SRV-deviant) individuals involves the search for the earliest event of breakdown in the sequence of information reduction. Little attention had been given to this area of cognition until recent years, and certainly no standard nomenclature to describe these information-processing events has yet been agreed upon. Yet, some convergent findings are worth mentioning.

The first set of events ordinarily examined are those involving sense organ input. With these receptive and peripheral afferent events essentially no important impairments have been found to be specific to schizophrenic patients (e.g., see review by Shakow, 1963).

Once information reaches the central nervous sytem, iconic integrity and decay represent the next candidacy in the search for schizophrenic breakdown. Spaulding, Rosenzweig, Huntzinger, Cromwell, Briggs and Hayes (1980) examined visual decay (and integration) features by presenting two dot matrices separated by a variable interval without visual presentation. If the second dot matrix entered the icon before the first one had decayed then the subject could integrate the two dot patterns and report a two-digit number. No number or other recognizable image was detectable when each dot pattern was viewed separately. Schizophrenic patients were shown to be remarkably similar to other comparison groups in this decay and integration function.

Similar results were found earlier when Knight, Sherer and Shapiro (1977) presented successive line fragment images separated by a variable interval. With smaller intervals the subjects were more able to recognize a line drawing which resulted from the integration of the two fragment images. Although schizophrenics again did not differ from control subjects, the differences in the decay function suggested that more than one mechanism may be involved in this stage.

The next stage concerns processing from the icon. One aspect of this has been studied through partial report span of apprehension. In this procedure an array of stimuli (usually letters) are displayed tachistoscopically for 50 msec. The subject is preinstructed to look for either of two targets (such as a T or F). The subject then either reports or guesses which of the two targets occurred. From the percentage correct performance, the mean number of elements processed is calculated. This technique, along with vigilance measures likely related to it, has revealed a wide range of findings of schizophrenic deviance. Schizophrenics (Neale, McIntyre, Fox & Cromwell, 1969), remitted schizophrenics (Asarnow, Steffy, MacCrimmon & Cleghorn, 1978), children at biological risk for schizophrenia (Asarnow, Steffy, MacCrimmon & Cleghorn, 1977), and childhood schizophrenics (Sherman & Asarnow, 1984) are all deviant on this task. Moroever, little or no impairment is found when subjects are asked for "full report" of all they have seen in the array (Cash, Neale & Cromwell, 1972). Thus, the impairment appears to be related to the scanning and disengagement from irrelevant stimuli rather than the scanning capacity and short term memory factors.

The next search for schizophrenia breakdown would be in the stages where information is transformed to representational forms. These representations would be the unit products which eventually lead to verbal emissions. Here many redundant systems are likely involved, and the current nomenclature for these stages of processing is even less clear. However, ample examples of schizophrenic breakdown appear to arise from this "represented" or "labeled" information. Often these breakdowns are referred to as distorted perceptions, such as in inkblot testing. The major point to be remembered here is that the deficit already described above in visual search inevitably carries forward faulty information for this representational processing stage.

In the stages after representation, where verbal concepts have been formed, numerous studies have shown schizophrenic impairment. In verbal or other effector response, in executive decision process, and in use (efficient retrieval) of long term memory, the impairments are so pervasive in schizophrenia that it is difficult (and theoretically more interesting) to identify where the schizophrenic is unimpaired. All of these deficit measures involve focal attention and subject awareness. Indeed, as indicated previously, the definition of the schizophrenic disorder itself lies in these latter phases of processing.

The conclusion to be drawn from the foregoing data and their interpretations is that the schizophrenic deficit is not so diffuse as to defy isolation forever. The deficit is not encountered in the very early stages of informational input, but, once it does occur, the stages which follow appear generally impaired.

Qualifying comment

Among the many statements of caution and qualifying assumptions when interpreting attentional and information processing data a few are certainly worthy of comment here.

In such a recent area of research, as already suggested, the useful divisions of processing stages and the nomenclature are certainly not clear. The methods are certainly subject to exploration and refinement. A best possible outcome would be that the present modes of description become more and more obsolete.

It is also important to remember that informational input does not come in discrete bits, even though this is attempted in the experimental laboratory. Instead, with the continuous flow of input from all modalities, the various stimuli and processing events do not occur in isolation.

In this respect some recent work is directed as to how the subject disattends (disengages processing) from immediately prior stimulation. Such disengagement mechanisms would appear important in order for the subject to be prepared for the continuing stream of input. Posner (1982; see also Posner, Cohen, Choate, Hockey & Maylor, Reference Note 1) has shown that after successive visual processing from Site A to Site B, the subject can more easily process information from a third Site C than return to process new information at Site A. This mechanism may be important in "clearing the slate" in order to deal with continually new input in real life situations.

PRELIMINARY COMMENTS CONCERNING HEMISPHERIC FUNCTION

Before discussing the implications of these cognitive findings in schizophrenia for hemispheric brain functioning, it is appropriate to make some preliminary statements and assumptions about the latter.

Proposed assumptions

The following proposed assumptions, all challengable but potentially useful, represent a starting point for the final argument of this paper:

1. The right hemisphere is primarily responsible for preattentional processing, especially in the visual domain.

Here, as elsewhere, it is recognized that one can speak only of hemispheric advantage, not of discrete functions.

2. This preattentional processing is inevitably a massive effort in brain function. Stated differently, the task to select, "group," and enhance the "relevant", and to inhibit the "irrelevant", from the vast array of input occurring at a given moment is massive as compared to the trivial requirements to transform greatly reduced information into pathways of verbal resolution, decision making, and action.

3. The sequence of processing on this preattentive level remains localized primarily in the right hemisphere, at least in the stages of visuospatial separations of figure from ground.

4. Information in these early preattentive stages must become sufficiently reduced to be transferred across the corpus callosum.

5. After crossing to the left hemisphere the information is received and further reduced through verbal coding and logical manipulation.

This stage is required to make the information storable in long term memory (with the possible exception of memory for images).

This stage is also required for performing output resolutions of conscious thought, decision making, verbal response, and other effector reaction.

Proposed hypotheses

If the foregoing crude picture of relative specialization is valid, then one could hypothesize the following:

1. The right hemisphere should have the relatively greater energy demand since it is dealing with massive amounts of unreduced information.

2. The right hemisphere might be expected to be structurally different in order to accommodate to this specialized massive task. That is, if the massive preattentional function (at least the visual one) is primarily and most frequently located in the right hemisphere, one would well expect the appropriate cell structure, vascular structure, and chemistry to differ on the right side in a manner which would accommodate to this greater preattentional and energy-demanding function.

3. The functions of the left hemisphere might more easily be simulated today by modern computers with "verbal" and logical algorithms which deal with information after it is already in reduced form.

Implications

1. Since the final stages of information processing are primarily in the left hemisphere, and these are the stages of which we are most immediately aware, they are the ones, historically, more likely to be studied and identified first. Thus, we might be expected to label this side as the "dominant" hemisphere. The present formulation would suggest that the left side of the brain is dominant only in the sense that the side of the moon which faces us is dominant. Likewise, one could call the surface portion of an iceberg the dominant portion.

The formulation of dominant vs. non-dominant hemisphere has probably retarded progress in the understanding of brain function. Better it would be, if one can speak of relative advantage, that the left hemisphere has the "later" and the right hemisphere has the "earlier" processing functions.

2. Although the right hemisphere is regarded as the locus of visual-spatial "ability", it might more appropriately be viewed as the site of relative advantage for certain visual processing stages.

Evidence

Gur (this volume) has reported greater blood flow in the right hemisphere. If blood flow is related to lateral energy demand, then the right side of the brain, as hypothesized above, should in general have the higher rates.

With respect to task-specific blood flow, tasks with strong visuo-spatial emphasis should be expected to put a greater demand upon the right hemisphere, and tasks with strong verbal emphasis should put a greater demand upon the left hemisphere. Again, Gur (this volume; Gur, Gur, Rosen, Warach, Alavi, Greenberg & Reivich, 1983) has shown that this relative difference in blood flow in the direction expected (verbal tasks elevating left hemisphere blood flow and visual tasks elevating right hemisphere blood flow). However, as might be expected, this difference is not as great as the overall difference of right over left side. The massive preattentional screening must be taking place regardless of the emphasis in the task being performed.

Relevance for schizophrenia

When one attempts to fit the data of schizophrenic deficit onto the topographic map of brain hemispheres there is ample reason for caution. To begin, several reasons may be cited for this caution.

1. Little work in schizophrenia has been done with split field techniques; therefore, inferences usually must be based upon bilateral information exposure plus some prior notion of how these are laterally specialized.

2. As has been stated so often elsewhere (see this volume) one can at best speak only of hemispheric advantage, not of unique hemispheric function. Perhaps because of redundant systems in the brain which are topographically separated, functions can be acquired in one hemisphere when the other has lost the function. Nevertheless, a picture can be constructed of the relative hemispheric advantage in the sequences of processing events earlier described in the search for schizophrenic deficit.

3. Contrary to the "staging events", abstracted and interpreted from the attentional and information-processing research on schizophrenia, both the input of stimulation and the monitoring and rhythm of the brain are ongoing. They are not activated (set in motion) by a particular input at a particular moment.

Given these and other appropriate cautions, the tentative formulation concerning schizophrenia is as follows:

1. Sensory input and iconic storage functions have no significant impairment in schizophrenia.

2. Preattentional functions are faulty in the schizophrenic patient. To the extent that the right hemisphere is at advantage for these functions, then the right hemisphere is faulty in this early phase of information processing.

3. The information, reduced in faulty form, is transmitted across the corpus callosum to the left hemisphere.

4. The left hemisphere, being the recipient of any fault arising from the right hemisphere (or from transport therefrom), is therefore compromised in performing the later stages of processing. It would not be unexpected that the left hemisphere is often found deviant in arousal and other psychophysiological indices (as reported by Flor-Henry, 1979; Gruzelier, Jutai, Connolly & Hirsch, 1984; and others).

The essence of the present assertion is that deviant left hemisphere functioning does not necessarily mean that the left hemisphere is the source of the problem of schizophrenia. For the "later" left hemispheric mechanisms to be operating smoothly and efficiently in the face of earlier impairments in the chain of processing would be difficult to argue. If the right hemisphere shows deviance (also reported by Gruzelier et al., 1984, and see this volume) its effects should become evident in informational products of the left hemisphere.

In short, the left hemisphere of the schizophrenic finds itself "up to its 'crotch' in crocodiles under conditions while its major purpose is to drain the swamp" i.e., to perform the final homeostatic stages of information resolution.

5. As suggested by Venables (1984), visuo-spatial processing functions, normally assumed by the right hemisphere, may have to be picked up by the left hemisphere. If so, one could reasonably assume, as he does, that an interference has resulted from the proximity of these earlier and later processing functions. This formulation also would support the notion of left hemisphere aberration as a secondary matter in schizophrenia.

However, it is important to note here that the proximity/interference hypothesis is not necessary in order to account for the deviant functioning in the left hemisphere as being a secondary feature. The mere transmission of faulty information from the right hemisphere is sufficient to argue this point.

CONCLUSIONS

1. Research on information processing stages might be useful for developing a sequential dynamic, rather than a static ability, formulation of brain function. Such a formulation would suggest that early (preattentional) processing stages are primarily in the right and later (verbal, decision-making, executive action) stages are primarily in the left hemisphere.

2. A formulation that the massive preattentional screening is primarily in the right hemisphere might explain some of the empirical findings of lateral differences in cell density, blood flow, and task-related blood flow.

3. Left hemisphere hyperarousal and other left brain deviations would be explained as secondary outcomes resulting from faulty information being transferred to that hemisphere from earlier preattentional processing in the right hemisphere.

REFERENCES

Annett, M. (1978). A single gene explanation of right and left handedness and brainedness. Coventry, England: Lanchester Polytechnic.

Annett, M. & Kilshaw, D. (1983). Right- and left-hand skill: II. Estimating the parameters of the distribution of L-R differences in males and females. British Journal of Psychology, 74, 269-283.

Asarnow, R.F. Steffy, R.A., MacCrimmon, D.J. and Cleghorn, J.M. (1977). An attentional assessment of foster children at risk for schizophrenia. Journal of Abnormal Psychology, 86, 267-275.

Asarnow, R.F., Steffy, R.A., MacCrimmon, D.J., and Cleghorn, J.M. (1978). An attentional assessment of foster children at risk for schizophrenia. In L. C. Wynne, R.L. Cromwell, and S. Mathysse, S. (Eds.) The Nature of Schizophrenia. New York: John Wiley and Sons.

Cash, T.F., Neale, J.M., and Cromwell, R.L. (1972). Span of apprehension in acute schizophrenics: a full report technique. Journal of Abnormal Psychology, 79, 322-326.

Cromwell, R.L. (1972). Strategies for studying schizophrenic behavior. Psychopharmacalogia, 24, 121-146.

Cromwell, R.L. (1984). Preemptive thinking and schizophrenia research. In W.D. Spaulding and J.K. Cole (Eds.) Nebraska Symposium on Motivation. Lincoln, NE: University of Nebraska Press, 1-46.

Flor-Henry, P. (1979). Commmentary on theoretical issues and neuropsychological and electroencephalographic findings. In J.H. Gruzelier and P. Flor-Henry, (Eds.) Hemisphere Asymmetries of Function in Psychopathology. Amsterdam: Elsevier/North Holland Co., 189-222.

Gruzelier, J.H., Jutai, J.W., Connolly, J.F. & Hirsch, S.R. (1984). Cerebral asymmetries in unmedicated schizophrenic patients in EEG spectra and their relation to clinical and autonomic parameters. In J. Mendelwicz and H.M. van Praag (Eds.) Advances in Biological Psychiatry. Basel: S. Karger.

Gur, R.C., Gur, R.E., Rosen, A.D., Warach, S., Alavi, A., Greenberg, J., & Reivich, M. (1983). A cognitive-motor network demonstrated by positron emission tomography. Neuropsychologia, 21, 601-606.

Iacono, W.G., Lykken, D.T., Peloquin, L.J., Lumry, A.E., Valentine, R.H., & Tuason, V.B. (1983). Electrodermal activity in euthymic unipolar and bipolar affective disorders. Archives of General Psychiatry, 40, 557-565.

Knight, R., Sherer, M., and Shapiro, J. (1977). Iconic imagery in overinclusive and nonoverinclusive schizophrenics. Journal of Abnormal Psychology, 86, 245-255.

Neale, J.M., McIntyre, C., Fox, R., and Cromwell, R.L. (1969). The span of apprehension in acute schizophrenics. Journal of Abnormal Psychology, 74, 593-596.

Posner, M.I. (1982). Neural systems control of spatial orienting. Philosophical Transactions of Royal Society, London, B 298, 187-198.

Shakow, D. (1963). Psychological deficit in schizophrenia. Behavior Science, 8, 275-305.

Sherman, T. & Asarnow, R.F. (1984). The cognitive disabilities of the schizophrenic child. In M. Sigman (Ed.), Children with Dual Disabilities: Mental Retardation and Emotional Disorders, Orlando, Florida: Grune & Stratton.

Spaulding, W.D., Rosenzweig, L.H., Huntzinger, R.S., Cromwell, R.L., Briggs, D., and Hayes, T. (1980). Visual pattern integration in psychiatric patients. Journal of Abnormal Psychology, 89, 635-643.

Venables, P. (1984). Cerebral mechanisms, autonomic responsiveness and attention in schizophrenia. In W.D. Spaulding and J.K. Cole (Eds.) Nebraska Symposium on Motivation. Lincoln: University of Nebraska Press, 47-92.

Reference

1. Posner, M.I., Cohen, Y., Choate, L., Hockey, R. & Maylor, E. Sustained concentration: Passive filtering or active orienting? Paper delivered to the meeting on Preparatory Processes, sponsored by NSF and CNRS, Ann Arbor, Michigan, August 1982.

2. Zubin, J. & Steinhauer, S. Psychobiological indicators in schizophrenics and their siblings. Unpublished manuscript.

CEREBRAL LATERALITY & SCHIZOPHRENIA: A REVIEW OF THE INTERHEMISPHERIC DISCONNECTION HYPOTHESIS

John Gruzelier

Department of Psychiatry
Charing Cross and Westminster Medical School
Fulham Palace Road, London W6 8RP. U.K.

INTRODUCTION

The disorders of interhemispheric communication revealed by the animal experiments of Sperry & colleagues in the 1960's (Sperry, 1964), and by the patients of Bogen & Vogel undergoing callosectomy (1962), gave impetus not only to basic research on hemispheric specialisation but also provided theoretical concepts of relevance to neurology and psychiatry (Geschwind, 1965; Galin, 1974). The splitting of psychic functions which gave schizophrenia its name was plausibly seen to stem from disconnection in neurological terms between a rational, verbally mediated left hemisphere and a holistic, nonverbally mediated right hemisphere. Conceivably, the observed dissociation between affect and cognition may have a similar origin to the dissociation between verbal and facial expression sometimes seen after callosectomy. Alternatively, a disorder of hemispheric specialisation which disrupted the normal processes of interhemispheric integration might be at fault; the popular presumption was that this was likely to be left-sided in view of the linguistic disturbances of schizophrenia. Acordingly, the split mind of the schizophrenic might arise from a split brain. Substance to these ideas was added by two influential reports. Flor-Henry (1969) surveyed the lateralisation of epileptic foci in patients with temporal lobe epilepsy combined with schizophrenia or manic-depressive features. Schizophrenic-like psychoses were associated with left-sided or bilateral foci and manic-depressive psychoses with right-sided foci. Rosenthal & Bigelow (1972) reported that the only pathology found in an examination at post-mortem, admittedly of the right hemisphere only, in ten schizophrenic patients was an enlargement of the corpus callosum. Subsequent research polarised initially around these competing ideas, though by the time of a conference called to review the findings (Gruzelier & Flor-Henry, 1979) they were seen to be mutually compatible.

Now there are a number of theories about the nature of lateralised and interhemispheric dysfunction in schizophrenia:- 1). There is a functional reversal of normal hemispheric specialisation for verbal and nonverbal processes. 2) Right hemisphere deficits are the forerunner of left hemisphere dysfunction. 3) The left hemisphere problem is primary, but opinions diverge as to whether this produces a loss or overactivation of function. 4) The interhemispheric pathways play a primary role. 5) Opposite states of hemispheric balance in activation lead to a relative dominance of the left hemisphere in acute, reactive, florid, remitting schizophrenia and right hemispheric dominance in chronic, retarded schizophrenia. These theories have been recently reviewed by the author (Gruzelier, 1986) and this chapter will be concerned only with the role of the corpus callosum in schizophrenia.

The origin of the theory that madness arises from incongruous and independent actions of the hemispheres has been traced by Harrington (1986) to the 18th century, when it was said of the insanity of Pascal 'madness and wisdom each had its compartment or its lobe, the two sides separated by a fissure' (La Mettrie, 1747). However it was Wigan (1844) who fully elaborated the thesis in a book 'A New View of Insanity: Duality of Mind'. He proposed that the normal exertion of control over one hemisphere by the other is lost and the resulting interhemispheric conflict is manifested by psychosis. At the time this view did not gain hold for several reasons, one of the more persuasive of which has been the puzzle of agenesis of the corpus callosum. William Ireland (1891) writing in the British Medical Journal 'On the discordant action of the double brain' expressed this puzzlement 'It is, however, surprising that the complete absence of this organ has been noted half a dozen times without entailing any apparent functional deficiency, for examination of the literature shows that where there has been imbecility there has always been some other grave defect. On the other hand .... where the brain is otherwise well developed there may be no disturbance of mobility, co-ordination, general or special sensibility, reflexes, speech, or intelligence, whether the defect of the corpus callosum be primary or secondary. This view has also been confirmed by observations of destruction of the corpus callosum from disease".

Research on callosal agenesis up to the present day has not resolved this problem. It is now thought that sparing of the anterior commissure in acallosals may permit interhemispheric transfer, or that compensation has occurred through duplication of functions bilaterally with or without dependency on highly developed ipsilateral fibres (Bogen, 1985). Nor is the callosal agenesis model appropriate for callosectomy; acallosals show few of the disconnection deficits seen in split-brain patients. Needless to say, this along with the existence of coincidental noncallosal pathology, does not bode well for the application of the acallosal model to schizophrenia.

The split-brain model also appears inappropriate for schizophrenia. After callosectomy neither acute nor chronic clinical signs have any resemblance to psychosis and in cases of organic pathology of the callosum psychotic symptoms occur only rarely. Notwithstanding the relevance of such symptoms to some of the accompanying features or subclinical disorders of schizophrenia, the origin of the symptoms appears outside of the callosum. These are what Bogen (1985) terms 'neighbourhood signs'. He concludes a discussion of the mental symptoms by saying "In my experience, patients with anterior callosal lesions often do have 'a certain apathy'. This 'imperviousness' occurs in patients with acute or progressive callosal lesions - especially the malignancy which is sometimes called a 'butterfly glioma' because it spreads its wings into both frontal lobes. The patient who is impervious to instructions will eventually respond, and often appropriately (but sometimes incompletely), but after repeated requests and considerable delay. We are now inclined to attribute this symptom not to involvement of the genu of the corpus callosum (which is, to be sure, involved) but rather to involvement of the medial aspects of the frontal lobes including the anterior cingulate gyri. And we suppose the imperviousness to be a milder form of akinesia, often approaching a mute immobility, of a patient who has what is sometimes called 'the subfrontal syndrome' consequent to bleeding from an anterior cerebral artery aneurysm, or with a third ventricle tumor".

"In any event, imperviousness can be a useful sign of anterior callosal lesions, although it is probably not a result of callosal interruption. This seems, in retrospect, a good example of anatomic relationships being important clinically, although misleading from the point of view of physiological theory".

"Neighborhood signs have also been noted with posterior callosal lesions, with involvement of the hippocampi. Translating Escourolle et al. (1975) (see Bogen, 1985):

A certain number of our tumors of the splenium (twice as common as genu gliomas) were accompanied by memory dysfunction, whereas the anterior tumors were more often manifested by akinetic states with mutism, probably because of bilateral anterior cingulate involvement."

Thus structural lesions in and around the callosum vary along an anterior-posterior axis in their behavioural effects, and these bear little resemblance to the fundamental deficits of schizophrenia.

Theory aside, the real impetus to current interest in the callosum in schizophrenia arose from the post-mortem study of Rosenthal & Bigelow (1972) which revealed callosal enlargement. However, this has been qualified in a recent replication (Bigelow et al., 1983) where enlargement was restricted to anterior and medial aspects and was confined to a subgroup of mostly nonparanoid patients with disorders of early onset. Another publication on the same series of patients which examined the callosum for gliotic cells (Nasrallah et al., 1983) cast yet a different light on the findings. Gliosis was associated with a thinner callosum in posterior regions and occurred in late onset paranoids as well as in manic-depressives. This may imply regional differences, both with respect to the nature of the pathology and the associated clinical syndromes, and possibly the functional implications of the interhemispheric transfer deficit. A cautious interpretation is probably warranted for Nasrallah et al. question the pathological significance of so-called ′enlargement′ in view of the fact that the dimensions of the callosal cross-section in schizophrenic patients were no larger than in medical and surgical controls.

The inappropriateness of models of defective interhemispheric transfer in schizophrenia on the basis of agenesis of the callosum, or callosectomy, or for that matter organic lesions of the callosum, is clearly apparent. Nevertheless, this does not rule out a more subtle problem of interhemispheric integration in schizophrenia. The nature of the problem may lie in faulty transmission such as a poor signal to noise ratio (Butler, 1979) or a loss of contralateral inhibition. Logically a disorder of interhemispheric transmission may not simply involve a problem in the interhemispheric pathway but in the transmission or reception of the signal in either hemisphere, thus a problem in interhemispheric integration may coexist with a lateralised disorder.

AUDITORY PROCESSING

Dichotic listening studies of schizophrenic patients have a bearing on callosal dysfunction. Split brain patients have shown an abnormally large right ear advantage for verbal material which may reflect an impairment of left ear input across the interhemispheric pathway. Comparisons between binaural and monaural hearing have provided the somewhat unusual finding that acute schizophrenic patients under some circumstances perform more poorly under the more natural binaural condition. Considering first the dichotic listening paradigm, split-brain patients have shown an abnormally large right ear advantage to verbal material due to a reduction in the recall of words heard in the left ear (Milner et al., 1968; Sparks and Geschwind, 1968; Springer and Gazzaniga, 1975). This evidence has been

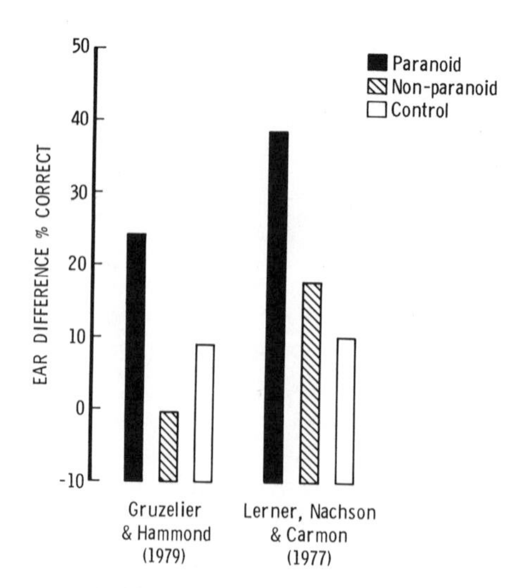

Fig. 1. Ear differences in free recall of digits.

used to validate dichotic listening as a method for investigation of hemispheric effects via contralateral auditory projections; the ipsilateral auditory projections having been functionally occluded, apparently by virtue of competition of input in either ear. The failure of split-brain patients to recall much of the verbal input to the left ear implies that the normal route of transmission extending from left ear to right hemisphere via the contralateral pathway and then across the callosum has been prevented by commissurotomy. Transmission does not occur via the ipsilateral pathway.

In schizophrenia, a callosal disconnection theory predicts larger than normal right ear advantages to verbal material analysed in the left hemisphere, and conversely larger left ear advantages to non-verbal material analysed in the right hemisphere. Of the verbal dichotic studies several have found abnormally large right ear advantage in subgroups of patients. Lerner et al. (1977) examined the recall of strings of three or four digits, comparing 30 paranoid with 30 non-paranoid patients and 20 normal controls. All groups showed the expected right ear advantage but this was largest in paranoids (x: 38.56), followed by non-paranoids (x: 17.53), and smallest in controls (x: 10.25), see Fig. 1. Two thirds of each group were at acute stages of illness but chronicity had no bearing on the data. Subsequently the results were interpreted by one of the authors as evidence of abnormal left hemispheric overactivation in paranoid patients (Nachson, 1980); a callosal disconnection interpretation has also been offered (Walker et al., 1981).

Larger than normal right ear advantages have also been reported in recalling consonant-vowel-consonant trigrams and common words, again in a subgroup of schizophrenic patients (Lishman et al., 1978). These were male patients with active symptoms, auditory hallucinations at some time of their clinical history, and no genetic association with schizophrenia. Colbourn and Lishman (1979) went on to examine these effects using consonant-vowel syllables and tone contours. The majority of schizophrenic patients showed normal left ear recall of the syllables but on this occasion the majority of male patients failed to show a right ear advantage. The authors concluded that the normal left ear recall indicated no support for a callosal transmission impairment while the reduction in right ear recall in male patients suggested a left hemisphere impairment. There was no impairment in recall of the tone contours in the seven patients tested.

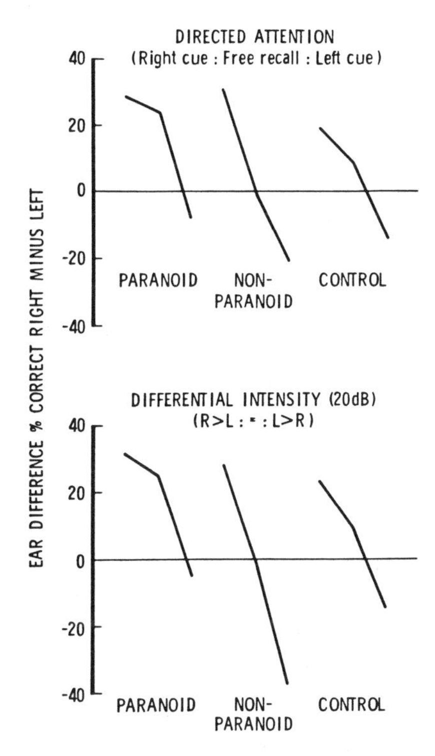

Fig. 2. Ear differences under conditions of directed attention and differential activity.

Gruzelier and Hammond (1979, 1980) carried out essentially the same experiment as Lerner et al. (1977) with two modifications. In addition to the free recall condition they included blocks of trials where subjects were instructed to recall all the numbers heard in one ear before the other, a conventional procedure for varying the direction of attention to either ear. Additional blocks involved intensity manipulations where digits were 20 decibels louder in one ear than the other. Eighteen chronic, male schizophrenic patients were tested, and as was found in the Israeli study, the extent of the right ear advantage depended upon a clinical history of paranoid symptoms; paranoid patients showed a larger than normal right ear bias. In this study non-paranoids showed less of a right ear advantage than controls. Psychophysiological recording revealed that the majority of paranoid patients were autonomically responsive, as shown by the frequency of electrodermal responses to moderate intensity tones, whereas non-paranoids for the most part were non-responsive. The ear differences are shown in Fig. 1.

The conditions in which directed attention and intensity were manipulated had a particular bearing on the callosal hypothesis. It can be seen from Fig. 2 that the paranoid patients in the conventional free recall condition showed a right ear bias similar to that where attention was directed to the right ear, or where stimulus intensities were 20 decibels louder in the right ear than the left. It was as though in ordinary circumstances their attention had been directed to the right ear. It is important to note that they could reverse the ear advantage when instructions or stimulus conditions favoured the left ear. The dynamic nature of the ear advantages would not support an interhemispheric disconnection theory based on a structural impairment of the callosum. The results were more in keeping with a dynamic imbalance in intra- and interhemispheric processing. Paranoid patients showed an imbalance advantaging the left hemisphere and disadvantaging the right. As a consequence they were not as competent at directing attention leftward as controls and non-paranoid patients, the latter of whom were particularly proficient at recalling left ear material. Non-paranoid patients revealed the opposite asymmetry, a dominance of the right hemisphere over the left. The effects were reliable across six experimental sessions given at fortnightly intervals and were not influenced by the withdrawal of medication for four weeks, or its reinstatement for a further four weeks.

Interhemispheric interactions were also implicit in the condition where subjects were required to withold the recall of louder digits until after the recall of quieter digits. Here 13/18 patients had greater difficulty when witholding recall of right ear digits whereas controls (10/16) had greater difficulty in withholding left ear digits. This would imply in patients greater excitatory influences from the left to the right hemisphere, an issue returned to in discussing evoked potential findings.

Two other dichotic listening studies involved longitudinal testing on a different type of patient - new admissions. Wexler & Heninger (1979) examined 8 schizophrenic, 6 schizo-affective and 12 patients with primary depression admitted while unmedicated to a research ward and then retested while undergoing treatment. Stimuli were nonsense syllables consisting of stop consonants preceded and followed by ′a′. Apart from five patients with reliable left ear preferences, ear differences were on the whole labile. There were no ear asymmetries when symptoms were severe but with recovery from symptoms the normal right ear advantage emerged. Symptoms in these patients were predominantly of the positive type:- paranoid behaviour, hallucinations, odd and unusual thoughts, thought disorder, verbal anxiety and expressed anger; while negative symptoms included psychomotor retardation and depressive mood.

The second dichotic study (Johnson & Crockett, 1982) also involved acute patients - 16 schizophrenics and 16 depressives - tested first on admission while unmedicated and again when symptoms remitted on drugs. They compared the recall of words and musical chords. For verbal material their results were consistent with the earlier study. Once again on admission the schizophrenic patients showed no ear advantage for words yet on recovery, when medicated, right ear advantages were clear. Curiously on admission they showed the expected left ear advantage for chords but on recovery the ear advantage reversed. It is conceivable, though this remains to be tested, that the reversal in ear advantage for chords from the normal right hemisphere advantage in admission to the abnormal right ear advantage on recovery was due to drugs. This is in view of the fact that a number of reports on the effects of antipsychotic neuroleptics in schizophrenia are consistent in showing enhanced left hemisphere functions on medication (Serafetinides, 1972, 1973; Hammond & Gruzelier, 1978; Gruzelier & Hammond 1979).

Reconciliation of the dichotic listening studies requires consideration of three factors:- syndromes, outcome, and severity of symptoms. Patients with acute disturbance and patients attending outpatient clinic (Colbourn & Lishman, 1979; Wexler & Henninger, 1979; Johnson & Crockett, 1982), both signs of a remitting form of schizophrenia, do not show the expected right ear advantage for verbal material, which in view of the normal right hemisphere specialisation for tone contours and musical chords (Colbourn & Lishman, 1979; Johnson & Crockett, 1982) suggests a left hemisphere impairment and a sparing of the right hemisphere rather than a hemispheric reversal of functions. Chronic patients on the other hand show abnormal right or left ear advantages for verbal material with the direction of ear advantage depending on the syndrome:- a right ear-left hemispheric bias in the reactive, florid, paranoid syndrome and a left ear-right hemisphere bias in the nonreactive, chronic, retarded syndrome. All experiments support the dynamic nature of the ear advantages. This rules out callosal impairment of the structural, disconnection type.

The relevance of chronicity is also becoming clear in results with another auditory procedure which compares the effects of monaural and binaural listening on story comprehension. P. Green & Kotenko (1980) examined twenty schizophrenic outpatients who had evidence in case histories of auditory hallucinations, ten normal controls and ten neurotic patients. After excluding four schizophrenic patients, two for poor performance and two because of left ear advantages, the remainder were found to exhibit poorer left ear performance and poorer binaural performance relative to the monaural conditions. The results were interpreted as evidence of impaired callosal transfer on two counts. The impaired left ear performance was in accord with the callosectomy model, while it was hypothesised that the binaural impairment may follow from faulty interhemispheric integration via the interhemispheric pathways.

Both results were replicated in a second study (Green et al., 1983) which involved 26 acute and 22 chronic patients, again compared with neurotic patients and normal controls. 18/26 acute and 7/22 chronic patients, a considerably smaller number than before, showed abnormally reduced left ear scores. As before the acute group showed binaural listening that was inferior to the preferred monaural ear. This time about half the patients showed a left ear preference. The authors also examined 13 children at genetic risk for schizophrenia and 13 control children. The high risk group had poorer comprehension overall and showed the binaural impairment seen in acute patients. Ear differences did not reach significance but the children were deficient in a test of speech sound perception suggesting left hemisphere involvement. Thus, the studies of P. Green and colleagues revealed a deficiency in binaural listening relative to the better monaural condition which was confined to acute schizophrenic patients and children at genetic risk for schizophrenia. In the acute patients the binaural deficit coexisted with poorer ear performance. In chronic patients on the other hand there was no sign of the impairment in binaural listening, yet there was a loss of right ear advantages, and in fact the opposite asymmetry was found in about half of them.

Replications using the same comprehension tests have been attempted by Kugler and colleagues (Kugler 1983) and by E. Green (1985). Examining chronic patients Kugler found that performance on both ears was reduced but there was no abnormal asymmetry in either direction. She also adapted the task for dichotic listening with competing stories in either ear and instructions to shadow one ear. While there was no evidence of attentional difficulties in shadowing, patients showed negligible right ear advantages and reduced performance overall. The lack of a right ear advantage was in keeping with evidence of impaired left hemisphere performance and was consistent with the absence of a right ear advantage in the chronic patients by P. Green and colleagues. Another replication by E. Green (1985) involved twelve acute and twelve chronic schizophrenics who were compared with normal controls and patients with borderline to mild mental handicap. The binaural impairment and right ear advantage found by P. Green in acute patients was replicated. The chronic patients also produced results consistent with the earlier studies namely no impairment in binaural listening and no ear advantage in monaural listening.

One other instance of poorer performance with binaural than monaural listening has been reported (Gruzelier & Hammond, 1979). This was found in comparing reaction times to monaural stimuli which were the words 'left' or 'right' presented in a response compatability paradigm (Simon, 1969). Schizophrenic patients at different stages of the illness and drug treatment were compared. Those with active, positive symptoms were alone in showing slower reactions when the words were heard binaurally (x: 700 msecs) than when heard in the left ear (x: 650 msecs) or right ear (x: 674 msecs). This provided evidence that the performance deficit with binaural listening was restricted to patients with active, positive symptoms of schizophrenia.

Thus, auditory studies in schizophrenia have revealed a performance decrement in acute patients when binaural listening was involved. This has been interpreted as an interhemispheric integration deficit and it coincided with a larger than normal right ear advantage in monaural listening for story comprehension. In chronic patients, the right ear advantage was lost and in one study was replaced in half the patients by a left ear advantage. In chronic patients there was no binaural hearing deficit and the results were consistent with a left hemispheric deficit. In acute patients the right ear advantage accompanying the binaural deficit may reflect either a right hemisphere and interhemispheric deficiency or a left hemisphere dysfunction, possibly of an overactivation type. In children at genetic risk for schizophrenia it was found to coincide with impairments in speech sound perception, presumably involving the left hemisphere. The association of opposite ear advantages with chronicity is compatible with the hemispheric imbalance syndrome model mentioned in connection with the dichotic listening studies. An overactive and dysfunctional left hemisphere could also underly a binaural, verbal integration deficit.

The dichotic and story comprehension results are compatible in showing relationships between ear advantages and syndromes having an affinity with acute or chronic schizophrenia. Yet they are incompatible with respect to the absence of the ear advantage in verbal dichotic listening in the acutely disturbed patients compared with the right ear advantage in story comprehension in acute patients. This may signify that level of processing is important such that the dependency on deeper levels of processing in story comprehension places a greater reliance on left hemispheric functions. Alternatively the descriptor 'acute' may be too vague and may disguise differences between the studies in severity of psychopathology.

HAPTIC PROCESSING

Haptic tasks involving intermanual transfer have provided another test of interhemispheric communication in schizophrenia. Up to the time of previous reviews (Gruzelier 1979, 1981, 1983) there was little in the way of conclusive evidence. However, recent evidence indicates transmission problems in a subgroup of patients co-existing with lateralised aspects. P. Green (1978) had studied 20 mainly outpatient, medicated patients and compared them with normal controls and neurotic patients. Green found that the schizophrenic patients showed poorer learning of tasks involving intermanual transfer, nevertheless there was no control for drugs or intelligence, and the tasks on which the schizophrenic patients experienced difficulty were the more complex ones and hence the effect could have simply reflected task difficulty. Subsequently Carr (1980), attempted to control for intelligence by equating patients and controls for socioeconomic status and employment. She compared 10 chronic medicated patients with age and sex matched normal controls. The task required subjects to remember an array of objects palpated without vision after an interval of up to three minutes. Unfamiliarity of objects and increasing delay made the task harder for both groups but only the patients experienced more difficulty than controls when tasks involved intermanual transfer. These results support a transfer deficit unless the effects could be attributed to medication.

Dimond et al. (1979, 1980) performed tests known to show impairment after commissurotomy. The 24 schizophrenic patients were institutionalised and middle-aged (mean age: 52 years). They were compared with six normal controls and six cases of severe anxiety or depressive illness requiring long term care. Unfortunately, errors in performance were summed over patients making the results difficult to interpret. Even so it was clear that impairments in intermanual transfer in hand tapping, touch localisation and object naming did occur in some patients, but these were often transient, for example on the second of the four monthly test sessions no patient showed anomia.

Weller and Kugler (1979) also selected tests on the basis of the callosectomy model. These included touch localisation, reproduction of imposed posture, and identification of inscribed digits. They managed to include some patients free of neuroleptics in their sample of 24 acute schizophrenics and 16 chronic schizophrenic patients. These were compared with schizoaffectives (N = 8), depressives (N = 11) and normal controls. Close inspection of the results (see Gruzelier 1979, 1983) suggests some evidence of a lateralised left hemisphere deficit but is ambiguous with respect to callosal transfer. Further experiments by Kugler and Henly (1979) involved intermanual matching of objects and measures of the time taken and errors made in discriminating shapes, letters and numbers. They found left-sided rather than callosal transfer deficits.

New evidence by Huret et al. (1984) using a collection of tests of parietal function found a variety of deficits. These implicated the left hemisphere in tests of right-left disorientation, finger agnosia and dyscalculia, all components of Gerstmann's left parietal syndrome. Less prevalent were problems in interhemispheric transfer which were seen in tactile dyslexia occurring bilaterally or on the left side alone, and from motor and praxic features of left dysgraphia. Some patients also displayed constructional apraxia suggesting right parietal involvement in keeping with a small body of evidence implicating right parietal dysfunction in schizophrenia. All-in-all, these results are exemplary of trends in the research on lateralisation and interhemispheric deficits in schizophrenia with respect to the relative incidence of left-sided, interhemispheric and right-sided deficits. Clinical correlates remain an unexplored issue in tests of haptic and motoric processing but should prove informative as it is apparent from the existing reports that patients vary in patterns of deficits.

VISUAL PROCESSING

In the visual modality, divided visual-field studies have provided measures of callosal crossing time. No evidence of interhemispheric transmission impairments in schizophrenia has been found. Yet most studies have provided evidence of left hemisphere deficits. The conventional finding in these reports is slower processing of verbal information presented in the right visual-field which passes directly to the left hemisphere for processing than when the same material is presented in the left visual-field and reaches the left hemisphere for analysis via the callosal pathway (e.g. Connolly et al., 1979, 1983). Gur (1978) found that this extended to inaccuracies in verbal performance, but so far this is an isolated finding - typically the problem is one of speed of processing rather than inaccuracy, indicating that the left hemisphere remains specialised for verbal analysis in schizophrenia as it does in the normal brain (see Walker & McGuire, 1982).

Estimates of callosal crossing time revealed no suggestion of a deficit (Connolly et al., 1979, 1983; Shelton & Knight, 1984). In Connolly et al's (1979) experiment the response involved the time taken to identify vocally numbers in a sequence of letters and numbers (verbal task) or the absence of a dot on a 3 x 3 matrix (spatial task). Absolute differences between visual fields were no different for schizophrenic patients and controls. This was despite the apparent need for two crossings of the callosum in the case of spatial processing of stimuli presented in the right visual field - the first crossing from left hemisphere to right for the spatial analysis and the second from right to left for the vocal report.

Bilateral hemi-field presentation requiring the matching of stimuli in opposite visual fields provides some evidence of matching difficulties (Beaumont & Dimond, 1973; Eaton, 1979). Nevertheless the possibility remains that these could be a function of the greater attentional demands of binocular presentation. These studies, together with a third which did not replicate the bilateral matching impairments (Magaro & Page, 1983), were all consistent in showing right visual-field impairments in letter matching (Beaumont & Dimond, 1973), name matching (Eaton, 1979), and in one, impairments in face, letter and shape matching in nonparanoids as distinct from paranoids (Magaro & Page, 1983).

A final visual study used the acallosal model of disconnection and accordingly requires a cautious interpretation of the role of the interhemispheric pathway. This entailed the moncular viewing of a rotating pattern and the interocular transfer of movement after-effects, the transfer of which is absent in acallosals. About two thirds of sixteen, medicated, chronic male patients reported reliable after-effects. Unexpectedly these were of shorter duration in patients (Tress et al., 1979).

SOMATOSENSORY EVOKED POTENTIALS

Another approach has been to examine somatosensory cortical evoked potentials bilaterally in response to unilateral stimulation, on the assumption that the ipsilateral potential is generated via callosal fibres. The results are confusing, because investigators have produced deficits unique to their study and are in conflict as to the putative role of interhemispheric and ipsilateral pathways. Nevertheless, all are invoked in support of interhemispheric integration problems in schizophrenia! The common assumption is that the ipsilateral response occurs in response to stimuli transmitted from the contralateral cortex (via the callosum) resulting in an attenuation of amplitudes and a delay in latencies. The effects have been found to be more pronounced in children, a result in keeping with the known maturational development of the cerebral commissures throughout childhood (Salamy, 1978). To date, in comparisons between schizophrenic patients and controls, the supposedly clear-cut picture of contralateral dominance is seldom seen in controls let alone patients. In accounting for this Gulmann et al. (1982) reveal the uncertainties that surround contribution of interhemispheric and intrahemispheric influences on the homolateral somatosensory potential. According to Regan (1972) bilateral representation occurs only for components later than 65 msecs, yet Gulmann et al. (1982) deduce that it is unlikely that all components later than 65 msecs depend on callosal transmission because of evidence from infarct cases in whom only early contralateral components were normal, yet who did have abnormal late ipsilateral components. This implies that later components were generated secondarily in the ipsilateral hemisphere from the earlier component dependent on callosal transfer (Gulmann et al., 1982). It is evident that more basic research in normal subjects and patients with well defined neurological lesions is imperative before results with psychiatric patients can be interpreted.

Problems of interpretation are further compounded by the results found in schizophrenic patients. These do not follow straightforward predictions from disconnection theory. Ipsilateral responses are not attenuated nor are latencies delayed. The opposite situation often prevails. Evoked potentials are either symmetrical in amplitude, latency, or both, or latency differences imply shorter transmission times in patients than controls. Perhaps not surprisingly authors tend to be coy about their hypotheses and invoke post hoc explanations invoking abnormal ipsilateral influences, or accelerated transmission across the cerebral hemispheres, or abnormal development of ipsilateral pathways secondary to a defective callosum.

Tress et al. (1979) using a tap to the forearm, which produced results similar to median nerve stimulation, found abnormal ipsilateral latencies. Twelve male, medicated, chronic patients were compared with seven normal controls. Schizophrenic patients shared an attenuation of ipsilateral amplitudes with controls but the ipsilateral latency of the component about 50 msecs was not delayed as it was in controls. This effect was subsequently re-examined in another group of patients and compared with twelve controls (Tress et al., 1983). The groups again differed in latencies with less evidence of longer ipsilateral than contralateral responses in patients than controls. The components showing the differences often differed from the previous study and ranged from 25 - 60 msecs. In a third study another 15 patients were tested, eleven on two occasions where on the second of which pindalol or placebo had been added to neuroleptic medication. On the first occasion the normal advantage to contralateral latencies was found at 35 msecs but not at 25 or 50 msecs. On retesting, the peak at 25 msecs failed to show the contralateral advantage in those patients on placebo.

Thus it can be seen that in the three experiments of Tress et al., latency rather than amplitude measures showed abnormal contralateral-ipsilateral relationships. Contrary to disconnection theory, ipsilateral latencies were shorter not abnormally long in patients. However, the components varied both across experiments and with retesting within the same experiment: experiment I, P50; experiment II, P25, N35 and P50; experiment II session 1 P25 and P50; session 2 P25. At the same time the other components which included N60 and P75 did show the expected asymmetry throughout. With respect to a neuropsychological interpretation, the authors noted that the contralateral-ipsilateral differences in controls were too small to meaningfully reflect a delay due to callosal transmission. The authors referred to ipsilateral influences and 'more rapid spread of activity, in the schizophrenic subjects from the ipsilateral primary receiving area to other areas'.

The report of Jones and Miller (1981) is the most controversial on this phenonomen. Using vibration ot the finger, the critical feature of which was a sudden extension of the finger (Jones and Miller, 1982), they reported responses that were symmetrical in amplitude and latency in both patients and controls.

An attempted replication had a different outcome. Andrews et al. (1984) compared 19 schizophrenic and 9 manic-depressive patients with 10 normal controls. They used a vibratory and kinesthetic stimulus modelled on the Jones and Miller procedure with electrodes placed over parietal cortex. Their procedures took account of various methodological criticisms advanced by Shagass et al. (1982) of the Jones and Miller work. In addition in a pilot study they compared the vibratory stimulus with median nerve stimulation. In general, amplitudes to both stimuli were larger over contralateral cortex, especially in the earlier peaks at P30 and P45 which were the most clearly identifiable. They concluded that ipsilateral responses in normals were too small to be reliably identifiable and this did not permit an accurate estimate of inter-hemispheric transfer time. Furthermore, peak identification was unreliable bilaterally in about a third of the patients who were discarded from the analysis. In the remaining subjects three patterns of response were observed, one of which was specific to patients. All three groups had a similar incidence of symmetrical responses whereby stimulation on one side produced appreciable ipsilateral activity in addition to the contralateral responses. The third pattern which was seen only in schizophrenic patients and one manic-depressive patient with a sister with schizophrenia consisted of symmetrical responses from stimulation of either forefinger.

The schizophrenic patients with the bilateral responses were distinguished by clinical symptomatology. A syndrome analysis was undertaken after that of Gruzelier & Manchanda (1982), (see Gruzelier, this volume) who found that patients with a predominance of negative symptoms of schizophrenia in combination with Schneiderian symptoms were homogeneous in the direction of the asymmetry of electrodermal responses compared with patients with predominantly positive symptoms who showed the opposite asymmetry. Evidence from both electrodermal and electrocortical recording (Gruzelier, 1983, 1984; Gruzelier et al., 1985) indicated that the group with the negative symptoms possessed an imbalance in hemispheric activity in the direction of higher right than left hemisphere activity. This would make a comparison of hemispheres in the Andrews et al. study of considerable interest. They were cautious in the interpretation of the abnormal bilateral response pattern seen in about half of the patients, concluding that in these patients there is 'a demonstrative abnormality of early cortical registration of somatosensory information which tends to an excessive cortical response ipsilateral to the stimulated finger'.

Median nerve stimulation has been used in one study of schizophrenia. Gulmann et al. (1982) examined 10 chronic medicated schizophrenic patients and 10 age and sex matched normal controls. Reliable ipsilateral responses were found after 65 msecs. Contralateral versus ipsilateral latency differences distinguished patients from controls when the right arm was stimulated. Transmission time was in fact shorter in patients, not longer, as a callosal disconnection theory would predict. When expressed as a ratio the effect was restricted to male patients and lost significance when female patients were added.

In summary, none of the studies found evidence in keeping with callosectomy or an impairment in callosal transmission akin to abnormally delayed maturation. When lateral differences arose these were only in early components up to about 65 msecs and varied with the stimulus methods involved. There is reason to question whether these could reflect callosal influences in view of their short latency or because the ipsilateral-contralateral differences in normals were too small to reflect inter-hemispheric transmission times. If the finding of Gulmann et al. (1982) relating to the P65 component is exempt from these qualifications then the abnormality in patients was restricted to information transmitted from left to right, and not _vice versa_, and indicated faster than normal transmissions in patients, not delayed transmission. If this is a reliable finding, even for a subgroup of patients, and if the interpretation is valid, it suggests an important lateralised anomaly and may imply a reduction in left hemisphere inhibitory influences over the right hemisphere. Finally, assuming that the ipsilateral responses recorded by Tress et al. (1979, 1983), Jones and Miller (1981) and Andrews et al. (1984) are not artefactual, then the results appear to reflect an abnormal spread of activity in the homolateral cortex to somatosensory stimulation. The findings of Andrews et al. suggest that this may be restricted to patients with a predominance of negative symptoms, patients who have been shown elsewhere to possess an imbalance in hemispheric activity in favour of the right hemisphere, (see Gruzelier, this volume).

CONCLUSION

There is now sufficient research on lateralisation and interhemispheric relationships in schizophrenia to reveal some consistent patterns of results across studies. These offer rays of hope for elucidating the nature of brain functional organisation in a disorder notorious for a lack of replicable findings about basic neural mechanisms. Though the implications from these results raise more questions than answers, some clear guidelines are offered for future research.

Firstly there is no evidence to suggest that a disorder of interhemispheric relationships is a unitary deficit in schizophrenia. Anatomical and histopathological studies of callosal structure, together with neuropsychological measures of function suggest that the type of dysfunction may relate to clinical syndromes which have a bearing on chronicity and symptoms. As has been discussed, Gruzelier, this volume, and see Gruzelier (1981; 1983; 1985; 1986) for more detailed accounts, there is reason to suggest an important distinction between symptoms associated with acute, reactive, florid, remitting schizophrenia and chronic, retarded, poor prognosis schizophrenia. These may be loosely termed positive and negative syndromes, provided Schneiderian symptoms are seen as a separate syndrome one which may coexist with the other two. In tests of the callosal hypothesis the need for this distinction has been shown in measures of dichotic listening, story comprehension and in somatosensory evoked potentials. These combine with a more extensive range of variables from studies of lateralisation:- electrodermal orientating responses, EEG, cortical evoked potentials, the Hoffman reflex, conjugate lateral eye movements, somatosensory extinction and handedness to provide support for the syndrome-asymmetry model (Gruzelier, 1983). Recent tests which explicity test the model find support from an EEG spectrum analysis of visual evoked potentials (Gruzelier et al., 1985) and from lateral deviations in visual fixation, visual search and pupillary dynamics (Gaebel and Ullrich, 1986). It appears that hemispheric balance is one neuropsychological dimension that distinguishes the syndromes:- greater left than right hemispheric activation in the positive syndrome and the opposite state of imbalance in the negative syndrome. A further refinement to the model is here suggested from comparisons of binaural and monaural listening; this relates to a chronicity dimension. The positive syndrome in acute patients may co-exist with a problem of interhemispheric integration, a coexistence not seen in chronic patients with the positive syndrome who do not show the integration deficit. However, further tests of this hypothesis are required and these should include ratings of symptom severity to give precision to the acute/chronic distinction.

Evidence to date suggests little in the way of the functional disconnection that would follow structural impairments in interhemispheric pathways akin to callosectomy or agenesis of the callosum. Only in the somatosensory modality in the evoked potential studies (Tress et al., 1979, 1983; Gulmann et al., 1982; Andrews et al., 1985) and the behavioural studies of Carr (1980) and Huret et al. (1984) were there implications for a nonfunctional callosum. In the Andrews et al. study this was restricted to the negative syndrome patients and in the Carr study the patients were institutionalised and therefore may also have been characterised by negative features. The interesting possibility arises that the negative symptom - right hemisphere syndrome, also co-exists with an as yet unspecified callosal dysfunction. This would provide an alternative account to the one of a left hemisphere loss of function for the poverty of speech, which may extend in some cases to mutism, and other features of cognitive reduction and motoric retardation seen in the negative syndrome.

Other results suggest that the nature of disordered interhemispheric relationships in some patients may stem from faulty transmission such as would occur through disruption of the fine tuning of patterns of excitation and inhibition that underly interhemispheric communication. Little is understood of these dynamics in the normal brain and to date the influence of contralateral inhibition has received the most consideration (Denenberg, 1979; Flor-Henry, 1983). Two strands of evidence have been mentioned which suggest that there may be an asymmetry in these relationships, such as excessive excitation, or conversely a failure of inhibition in a left to right direction (Gruzelier & Hammond, 1979, 1980; Gulmann et al., 1982). There is another study which has also provided tentative support for left-sided origin to a callosal transmission impairment in schizophrenia (Buchsbaum et al., 1979). Visual evoked potentials were measured to flashes of variable intensity presented to the nasal hemi-retinae while patients ignored or attended to the temporal sequence. Sixteen unmedicated schizophrenic patients were compared with patients with left or right-sided temporal lobectomy. The schizophrenic patients resembled the left temporal lobectomy patients in failing to show an increased N120 response to stimuli transmitted by the indirect callosal pathway to the left temporo-parietal region. This represented a failure in callosal transmission of a signal in a leftward direction. However, when this is seen in conjunction with a left-sided impairment (left lobectomy in the neurological cases), it would appear that the disruptive influence was transmitted from left to right and hence was in keeping with the other reports.

The most clear cut evidence of an impairment in interhemispheric integration arose from evidence of poorer reaction times and story comprehension with binaural than monaural listening. This occurred in the case of the reaction time impairment only in actively disturbed patients with a predominance of moderate-to-severe positive symptoms. The comprehension impairment occurred in acute patients and in those whose symptoms had abated to permit their treatment as outpatients. The impairment also occurred in asymptomatic children with a schizophrenic parent. Thus the binaural integration impairment is not a neuropsychological concomitant of acute, positive symptoms but may instead reflect a trait marker for the remitting form of schizophrenia.

While a problem of interhemispheric integration is a reasonable inference as to the nature of the neuropsychological deficit, it was mentioned at the outset that a problem of interhemispheric transfer could also stem from a lateralised deficit. There is little doubt that in acute patients, without resorting to the literature on lateralisation in schizophrenia which does not concern itself with interhemispheric deficits, the bulk of evidence implicates the left hemisphere:- the exaggerated right ear advantages in dichotic listening, the absence of impairments in nonverbal dichotic listening and spatial hemi-retinal processing, the verbal-right visual field deficit in tachistoscopy, and the greater prevalence of right than left hand deficits in haptic processing. Nevertheless, a smaller body of evidence implicates the right hemisphere in perhaps a subgroup of patients, seen here in the results of Huret et al. If the evoked potential experiments are correct in suggesting an abnormal development of ipsilateral pathways in some patients with schizophrenia then interpretation of the auditory studies in schizophrenia is placed on a different footing. The neuropsychological model of dichotic listening which provides the basis for the belief that the simultaneous presentation of competing monaural input causes a suppression of the ipsilateral auditory pathways may not hold in schizophrenia. However, before this theory warrants further consideration, basic research is needed on ipsilateral-contralateral relationships in normal subjects. Research in our laboratory has shown that a dominance of contralateral over ipsilateral evoked potential relationships to monaural stimuli occurs in only 50% of normals, and, while a significant group effect in favour of contralateral dominance is obtained with repeat sessions, the effect is not reliable within subjects (Connolly et al., 1985). Andrews et al. (1984) did find reliable responses to somatosensory stimulation at retest intervals of one year, but again only 50% of their normal subjects displayed the contralateral dominance phenomenon. Whether there is asymmetry in the instances of ipsilateral dominance would be a worthwhile question.

Finally, two other, possibly interrelated, factors require consideration. These are levels of processing and regional localisation within the interhemispheric commissures. The evidence of Bigelow et al. (1983) and Nasrallah et al. (1983), though inconsistent with one another, do raise the possibility of regional differences in structural impairments which appear to have clinical counterparts. This may account for the reason why studies of visual processing with tachistocopy, unlike auditory and somatosensory measures, have failed to support the interhemispheric relationship hypothesis in schizophrenia. Interhemispheric commissures which are devoted to the relatively low order processing required in the various divided-visual field studies may escape impairment in schizophrenia. Level of processing may also be relevant to interhemispheric and lateralised dysfunction as seen in the discrepancy between ear advantages in the different auditory procedures in acute patients.

Thus, regional localisation within the cerebral interhemispheric pathways, as well as levels of processing, are added to the many factors that require consideration in future attempts to test the model of impaired interhemispheric relationships in schizophrenia. The other factors described in this chapter included dynamic versus structural processes, positive versus negative syndromes, acute versus chronic stages of the disorder, lateralised versus interhemispheric pathway deficits, leftward versus rightward interhemispheric influences, and the concept of disconnection versus disordered communication. The callosal theory of schizophrenia holds promise of continuing to fulfil an heuristic role in delineating the nature of brain disorder in schizophrenia.

REFERENCES

Andrews, H.B., Cooper, J.E., & Barber, E. (1984). The symptom pattern in schizophrenic patients with stable abnormalities of laterality of somato-sensory cortical evoked responses. Paper presented at the Second International Meeting of the International Organisation of Psychophysiology, London. July.

Beaumont, J.G., & Dimond, S.J. (1973). Brain Disconnection and schizophrenia. British Journal of Psychiatry, 123, 661-662.

Bigelow, L.B., Nasrallah, H.A., Raauscher, F.P. (1983). Corpus callosum thickness in chronic schizophrenia. British Journal of Psychiatry, 142, 284-287.

Bogen, J.E. (1985). The Callosal Syndrome. In Oxford: K.N. Heilman and E. Valenstein. (Eds.) Clinical Neuropsychology, Oxford University Press, Oxford.

Bogen, J.E., and Vogel, P.J. (1962). Cerebral commissurotomy in man. Bulletin of the Los Angeles Neurological Society. 27, 169-172.

Buchsbaum, M.S., Carpenter, W.T., Fedio, P., Goodwin, F.M., Murphy, D.L., and Post, R.M. (1979). Hemispheric differences in evoked potential enhancement by selective attention to hemiretinally presented stimuli in schizophrenic, affective and post-temporal lobectomy patients. In J.H. Gruzelier & P. Flor-Henry (Eds). Hemisphere Asymmetries of Function in Psychopathology. Amsterdam: Elsevier/North Holland, Amsterdam. 317-328.

Butler, S.R. (1979). Interhemispheric relations in schizophrenia. In J. Gruzelier and P. Flor-Henry (Eds.) Hemisphere Asymmetries of Function in

Psychopathology, Amsterdam: Elsevier/North Holland Biomedial Press, 47-63.

Carr, S.A. (1980). Interhemispheric transfer of stereognostic information in chronic schizophrenics. British Journal of Psychiatry, 136, 53-58.

Colbourn, C.J., & Lishman, W.A. (1979). Lateralisation of function and psychotic illness: a left hemisphere deficit? In J. Gruzelier and P. Flor-Henry (Eds.) Hemisphere Asymmetries of Function in Psychopathology. Amsterdam: Elsevier/North Holland, 647-672.

Connolly, J.F., Gruzelier, J.H., Kleinman, K.M., and Hirsch, S.R. (1979). Lateralised abnormalities in hemisphere-specific tachistoscopic tasks in psychiatric patients and controls. In J. Gruzelier and P. Flor-Henry (Eds.) Hemisphere Asymmetries of Function in Psychopathology, Amsterdam: Elsevier/North Holland, 491-510.

Connolly, J.F., Gruzelier, J.H. and Manchanda, R. (1983). Electrocortical and perceptual asymmetries in schizophrenia. In P. Flor-Henry and J. Gruzelier (Eds.) Laterality and Psychopathology, Amsterdam: Elsevier/North Holland.

Connolly, J.F., Manchanda, R., Gruzelier, J.H., and Hirsch, S.R. (1985). Hemispheric differences in the event-related potential (ERP) to monaural stimulation: A comparison of schizophrenic patients with normal controls. Biological Psychiatry, 20, 293-303.

Denenberg, C.H. (1979). Brain laterality and behavioural asymmetry in the rat. In: P. Flor-Henry and J. Gruzelier (Eds.) Laterality and Psychopathology. Amsterdam: Elsevier Science Publishers.

Dimond, S.J., Scammell, R.E., Pryce, J.Q., Huss, D. and Gray, C. (1979). Callosal transfer and left-hand anomia in schizophrenia. Biological Psychiatry, 14, 735-739.

Dimond, S.J., Scammell, R., Pryce, I.J., Haws, D. and Gray, C. (1980). Some failures of intermanual and cross-lateral transfer in chronic schizophrenia. Journal of Abnormal Psychology, 89, 505-509.

Eaton, E.M. (1979). Hemisphere-related visual information processing in acute schizophrenia before and after neuroleptic treatment. In J. Gruzelier and P. Flor-Henry (Eds.) Laterality and Psychopathology, Amsterdam: Elsevier Science Publishers, 511-526.

Flor-Henry, P. (1969). Psychoses and temporal lobe epilepsy: a controlled investigation. Epilepsia, 19, 363-395.

Flor-Henry, P. (1983). Cerebral Basis of Psychopathology. London: John Wright.

Gaebel, W., Ulrich, G. & Frich, K. (1986). Eye movement research with schizophrenic patients and normal controls using corneal reflection pupil centre measurement. European Archives of Psychiatry and Neurological Sciences. In press.

Galin, D. (1974). Implications for psychiatry of left and right cerebral specialisations. A neuropsychological context for unconscious processes. Archives of General Psychiatry, 31, 572-583.

Geschwind, N. (1965). Disconnection syndromes in animals and men. Brain, 88, 237-294.

Green, E. (1985). Interhemispheric coordination and focussed attention in chronic and acute schizophrenia. British Journal of Clinical Psychology, in press.

Green, P. (1978). Defective interhemispheric transfer in schizophrenia. Journal of Abnormal Psychology, 87, 472-480.

Green, P., and Kotenko, V. (1980). Superior speech comprehension in schizophrenics under monaural versus binaural listening conditions. Journal of Abnormal Psychology, 89, 399-408.

Green, P., Hallett, S., and Hunter, H. (1983). Abnormal interhemispheric integration and hemispheric specialisation in schizophrenics and high-risk children. In P. Flor-Henry and J. Gruzelier (Eds.) Laterality and Psychopathology, Amsterdam: Elsevier Biomedical Press, 433-470.

Gruzelier, J.H. (1979). Synthesis and critical review of the evidence for hemispheric asymmetries of function in psychopathology. In J.H.

Gruzelier and P. Flor-Henry (Eds.) Hemisphere Asymmetry of Function and Psychopathology, Amsterdam: Elsevier/North Holland Biomedial Press. 647-672.

Gruzelier, J.H. (1981). Cerebral laterality and psychopathology: fact and fiction. Psychological Medicine, 11, 93-108.

Gruzelier, J.H. (1983). A critical assessment and integration of lateral asymmetries in schizophrenia. In M.S. Myslobodsky (Ed.). Hemisyndromes, Psychobiology, Neurology and Psychiatry. New York: Academic Press, 265-326.

Gruzelier, J.H. (1984). Hemispheric imbalances in schizophrenia. International Journal of Psychophysiology, 1, 227-240.

Gruzelier, J.H. (1985). Schizophrenia: Central nervous system signs in schizophrenia. In J.A.M. Frederiks (Ed.). Handbook of Clinical Neurology, 2 Neurobehavioural Disorders. Amsterdam: Elsevier Science Publishers, B.V., 481-521.

Gruzelier, J.H. (1986). Theories of lateralised and interhemispheric dysfunction in syndromes of schizophrenia. (Eds.) G. Burrows, T. Norman and N. Rubenstein. Handbook of Studies on Schizophrenia, Part 2, Amsterdam: Elsevier Science Publishers B.V. (Biomedical Division). In press.

Gruzelier, J.H., & Flor-Henry, P. (Eds.). (1979). Hemisphere Asymmetries of Function in Psychopathology Elsevier/North Holland, Amsterdam.

Gruzelier, J.H., & Hammond, N.V. (1979). Lateralised auditory processing in medicated and unmedicated schizophrenic patients. In J. Gruzelier & P. Flor-Henry (Eds.). Hemisphere Asymmetries of Function in Psychopathology. Amsterdam: Elsevier/North Holland, 603-638.

Gruzelier, J.H., & Hammond, N.V. (1980). Lateralised deficits and drug influences on the dichotic listening of schizophrenic patients. Biological Psychiatry, 15, 759-779.

Gruzelier, J.H., Jutai, J.W., Connolly, J.F., & Hirsch, S.R. (1985). Cerebral asymmetries in unmedicated schizophrenic patients in EEG spectra and their relation to clinical and autonomic parameters. Advances in Biological Psychiatry, 15, 12-19.

Gruzelier, J.H., & Manchanda, R. (1982). The syndrome of schizophrenia: Relations between electrodermal response lateral asymmetries and clinical ratings. British Journal of Psychiatry, 141, 488-495.

Gulmann, N.C., Wildschiodtz, G., & Orback, K. (1982). Alteration of interhemisphere conduction through corpus callosum in chronic schizophrenia. Biological Psychiatry, 17, 585-594.

Gur, R.E. (1978). Left hemisphere dysfunction and left hemisphere overactivation in schizophrenia. Journal of Abnormal Psychology, 87, 226-238.

Hammond, N.V., & Gruzelier, J.H. (1978). Laterality, attention and rate effects in the auditory temporal discrimination of chronic schizophrenics: the efffect of treatment with chlorpromazine. Quarterly Journal of Experimental Psychology, 30, 91-103.

Harrington, A. (1986). Ninteenth century ideas on hemisphere differences and ′duality of mind′. Behavioural Brain Sciences, in press.

Huret, J.D., Kabbas, Y., Le Trait, S., & Lemperiere, T. (1984). Gerstmann′s Syndrome and defective interhemispheric transfer in schizophrenia. Paper presented to Second Conference of International Organisation of Psychophysiology, London, July.

Ireland, W.W. May 30th (1891). On the discordant action of the double brain. British Medical Journal, 1167-1169.

Johnson, O., & Crockett, H.D. (1982). Changes in perceptual asymmetries with clinical improvement of depression & schizophrenia. Journal of Abnormal Psychology, 91, 45-54.

Jones, G.M., & Miller, J.J. (1981). Functional tests of the corpus callosum in schizophrenia. British Journal of Psychiatry, 139, 553-557.

Jones, G.M., & Miller, J. (1982). The corpus callosum and brain function in schizophrenia. British Journal of Psychiatry, 141, 535-537.
Kugler, B.T. (1983). Auditory processing in schizophrenia patients. In P. Flor-Henry & J.H. Gruzelier (eds.). Laterality and Psychopathology. Amsterdam: Elsevier/North Holland, 471-506.
Kugler, B.T. & Henley, S.H.A. (1979). Laterality effects in the tactile modality in schizophrenia. In J. Gruzelier & P. Flor-Henry (Eds.). Hemisphere Asymmetries of Function in Psychopathology. Amsterdam: Elsevier/North Holland, Biomedical Press. 475-490.
La Mettrie, J.D. (1747). Man a machine. (French-English) Ed. La Salle, Open Court, 1912.
Lerner, J., Nachson, I., & Carmon, A. (1977). Responses of paranoid and non-paranoid schizophrenics in a dichotic listening task. Journal of Nervous and Mental Diseases, 164, 247-252.
Lishman, W.A., Toone, B.K., Colbourn, C.T., McMeekan, E.R.L., & Mance, R.M. (1978). Dichotic listening in psychiatric patients, British Journal of Psychiatry, 132, 333-341.
Magaro, P.A. & Page, J. (1983). Brain disconnection, schizophrenia and paranoia. Journal of Nervous and Mental Disease, 169, 546-557.
Milner, B., Taylor, L., & Sperry, R.W. (1968). Lateralised suppression of dichotically presented digit after commissural section in man. Science, 161, 184-186.
Nachson, G. (1980). Hemispheric dysfunction in schizophrenia. Journal of Nervous and Mental Disease, 168, 241-242.
Nasrallah, H.A., Mcalley-Whitters, M., Bigelow, L.B., & Rauscher, F.P. (1983). A histological study of the corpus callosum in chronic schizophrenia. Psychiatry Research, 10, 251-260.
Regan, D. (1972). Evoked Potentials in Psychology, Survey Physiology and Clinical Medicine. London: Chapman & Hall, 114-117.
Rosenthal, R., & Bigelow, L.B. (1972). Quantitative brain measurement in chronic schizophrenia. British Journal of Psychiatry, 121, 259-264.
Salamy, A. (1978). Commissural transmission: maturational changes in humans. Science, 200, 1409-1411.
Serafetinides, E.A. (1972). Laterality and voltage in the EEG of psychiatric patients. Diseases of the Nervous System, 33, 622-623
Serafetinides, E.A. (1973). Voltage Laterality in the EEG of psychiatric patients. Diseases of the Nervous System, 34, 190-191.
Shagass, C., Josiassen, R.C., Roemer, R.A., Straumanis, J.J., & Stepner, S.M. (1983). Failure to replicate evoked potential observation suggesting corpus callosum dysfunction in schizophrenia. British Journal of Psychiatry, 142, 471-476.
Shelton, E.J. & Knight, R.G. (1984). Inter-hemispheric transmission times in schizophrenics. British Journal of Clinical Psychology. 23, 227-228.
Simon, J.R. (1969). Reactions toward the source of stimulation. Journal Experimental Psychology, 81, 174-176.
Sparks, R.E., Geschwind, M. (1968). Dichotic listening in man after section of neurocortical commissures. Cortex, 4, 3-16.
Sperry, R.W. (1964). The great cerebral commissure. Scientific American, 210, 42-52.
Springer, S.P., & Gazzaniga, M.S. (1975). Dichotic listening in partial and complete split-brain patients. Neuropsychologia, 13, 341-346.
Tress, K.H., Kugler, B.T., & Caudrey, D.J. (1979). Interhemispheric integration in schizophrenia. In J.H. Gruzelier and P. Flor-Henry (Eds.) Hemisphere Asymmetries of Function in Psychopathology. Amsterdam: Elsevier/North Holland, 449-462.
Tress, K.H., Caudrey, D.J., & Mehta, B. (1983). Tactile-evoked potentials in schizophrenia. British Journal of Psychiatry, 143, 156-164.
Walker, E., Hopper, E., & Emory, A. (1981). A reinterpretation of findings on hemispheric dysfunction in schizophrenia. Journal of Nervous and Mental Diseases, 169, 378-380.

Walker, E., & McGuire, S. (1982). Intra- and interhemispheric information processing in schizophrenia. *Psychological Bulletin*, *92*, 701-725.

Weller, M., & Kugler, B.T. (1979). Tactile discrimination in schizophrenia and affective psychoses. In J. Gruzelier and P. Flor-Henry (Eds.) *Hemisphere Asymmetries of Function in Psychopathology*. Amsterdam: Elsevier/North Holland, 463-474

Wexler, B.F., & Heninger, G.R. (1979). Alterations in cerebral laterality in acute psychotic illness. *Archives of General Psychiatry*, *137*, 279-284.

Wigan, A.L. (1844). *A New View of Insanity: The Duality of the Mind*. London: Longmann, Brown, Green and Longman.

INDIVIDUAL DIFFERENCES IN CEREBRAL LATERALIZATION: HOMEOSTATIC BRAIN FUNCTIONS OF SCHIZOPHRENICS

Michael Miran and Esta Miran

Strong Memorial Hospital
University of Rochester
Medical School
Rochester, New York

INTRODUCTION

One of the typical experiences of a psychiatrist or psychologist interviewing a schizophrenic patient is the feeling that if they asked the right question, they would get a coherent answer. Yet each new question produces a tangent, a circumstantial response, and a fragmented piece of fictious perception with the gestalt never completed. After a while the clinician finds elements of an underlying theme. This elusive and inconclusive search for the right question and answer is typical of our scientific study of schizophrenia. Each new study and theory gives us yet another fragment in our search for understanding and ultimately for better treatment methods.

Let us consider a more specific example of the communication process with schizophrenic patients. The following is an excerpt from a therapy group described by Rokeach (1964), p.53.

> "Sir," Leon began, "I told them my sincere belief but these gentlemen also stated their sincere belief. I don't care to lead their life, and they have a right to live their own."
>
> Then he turned to Joseph: "Captain Davy Jones, will you get up there and talk about your subconscious institution pertaining to your character? Therefore, do you have any part subconscious reflections that you wondered about pertaining to? Do you have any dreams?
>
> "I'm just simply God and I work for the cause of the English," Joseph answered.
>
> "Sir, Jesus Christ, man!" Leon exclaimed. "I have to disagree with you on that because England..."

This interaction between two of the three Christs of Ypsilanti highlights the difficulties schizophrenics have in communicating with each other as well as other normal individuals. Schizophrenics' language, as in this example, is not just grammatically faulty, but it is organized around different principles and premises than is the case with normal communication (Abse, 1971).

This chapter focuses on the role of intrahemispheric and interhemispheric communication in schizophrenia. We propose that the deficits in cognition, perception, and communication that are typical of schizophrenics can be understood as breakdowns in the internal communication processes within a homeostatic brain. This chapter examines the literature on models of schizophrenia.

THEORIES OF LEFT AND RIGHT HEMISPHERIC DYSFUNCTION

Flor-Henry (1969) began studying the role of the left hemisphere in schizophrenia when he noticed that there were many similarities between the behavior of temporal lobe epileptics and schizophrenics. Further studies Flor-Henry, Koles, Howarth, & Burton (1979); Flor-Henry & Koles (1980); and Flor-Henry, Koles, & Tucker (1982) provide evidence for the hypothesis of a left temporal lobe overactivation in schizophrenics.

In their latest work Flor-Henry, Koles & Sussman (1984) and Flor-Henry & Koles (1984) use sophisticated statistical techniques to compare the performance of the two hemispheres in different populations on verbal and visual tasks. Flor-Henry finds changes in the interhemispheric organization of schizophrenics and other psychotics. The interhemispheric coherence of schizophrenics appears to be lowest of all psychotic groups. He also highlights that in normals, during verbal and block design tasks, there are partial shifts of laterality. He finds that there are anterior posterior shifts in psychoses.

Based on current theories of left- and right-hemisphere organization, Flor-Henry (1983), Gruzelier and Hammond (1976), Gruzelier and Manchanda (1982) and Gur (1979), have examined schizophrenia in terms of dominant (usually left) temporal lobe overactivation, frontal lobe deficits, and right hemisphere deficits. Gur (1979) used a tachistoscope and reported that schizophrenics tended to "overactivate their left dysfunctional hemisphere." Using regional cerebral blood flow (rCBF), Gur, Packer, Reivich and Weinberger (1978) further supported the hypothesis of left hemispheric overactivation. In their most recent research Gur, Gur, Skolnick, Caroff, Obrist, Resnick, and Reivich (1985), using (rCBF), find: "Schizophrenics, however, had no change in hemispheric asymmetry of flow for the verbal task and the greater left hemispheric increase for the spatial task." Gur's research reflects the role of individual differences and inconsistent variance in the functioning of both hemispheres during verbal and visual task performance in schizophrenics. While intra- and interhemispheric coordination appear to be significant factors in schizophrenic dysfunction, the exact nature of this lack of coordination needs further investigation.

Based on data from electrodermal lateral asymmetries, Gruzelier (1984) presents evidence for two syndromes in schizophrenia: "A florid syndrome is coincidental with left-hemispheric overactivation, and a nonflorid syndrome with right-hemispheric overactivation." Using electroencephalographic measurement, Gruzelier (1983b) found: "...asymmetrical electro-cortical activity in both directions in schizophrenia, with right-hemisphere abnormalities coincidental with the retarded syndrome and left-hemisphere abnormalities with the more florid one." Gruzelier's evidence suggests that in schizophrenia there are individual differences in hemispheric functioning. Gruzelier interprets the differences as showing that the florid schizophrenics that respond to treatment are more like normals, and that the more chronic schizophrenics appear more like brain damaged patients.

A series of papers presented at the European Winter Conferences On Brain Research in Chamonix, France highlight many of the issues in EEG studies of cerebral asymmetries in schizophrenics. Tucker (1984) suggests that anxiety is a function of dopamine systems. Serafetinides (1984) distinguishes left hemispheric dysfunction as associated with thought disorder symptoms and right hemispheric dysfunction with symptoms of anxiety. He highlights that there is a change in the EEG functioning of schizophrenics in the course of treatment. He also cites several references which reflect changes in right left patterns in schizophrenia. Etevenon (1984) makes the point that there are major intrahemispheric as well as interhemispheric changes in EEG potentials in schizophrenics.

Another study by Etevenon, Pidoux, Rioux, Peron-Magnan, Verdeaux and Deniker (1979) compare the spectral analysis of the EEG's of two groups of schizophrenics and a control group. They find both intrahemispheric and interhemispheric difference between the schizophrenics and the control groups. Roemer, Shagass, Straumanis and Amadeo (1979) examine auditory and somatosensory evoked potentials in schizophrenics, nonpsychotic psychiatric patients and normals. They find that only the auditory evoked potential results augmented the previous visual evoked potential evidence of left hemisphere involvement in schizophrenia. A comparison is made by Toone, Cooke & Lader (1979) of electrodermal activity in patients who had temporal lobe surgery to determine if the same type of responses would be identified in temporal lobe patients as are seen in schizophrenics. They found no differences between the operated and non-operated sides of the body and between control and temporal lobe patients on electrodermal responding.

The evidence from these researchers fails to establish a single consistent localized deficit in brain structure. However, their results highlight a breakdown in feedback, synchronization, and effective co-processing between the hemispheres in schizophrenics. In addition to the individual differences discussed above, the literature has a number of instances of failures to identify left temporal lobe overactivation. Schneider's (1983) findings contradict this hypothesis: "Unlike normals, schizophrenics' left-hemisphere function appeared to be inferior to the right-hemisphere function in the perceptual task". These disparate findings suggest a process which unpredictably oscillates in its impact on the right and left hemispheres. The cause of this oscillating is in the process of being delineated through ongoing research in this area.

THEORIES OF CORPUS CALLOSUM DYSFUNCTION

Other researchers suggest that schizophrenia is the result of a corpus callosal dysfunction and enlarged ventricles. Beaumont and Dimond (1973) used a four screen tachistoscope to project stimuli into the separate hemispheres of the brain by presenting them into a particular visual field. Their result showed that schizophrenics performed significantly poorer than controls in conditions requiring cross-matching of stimuli between hemispheres; while schizophrenics showed no deficit in separate hemispheric identifications.

Henn & Nasrallah (1982) discuss schizophrenia in terms of brain disease. Bigelow, Nasrallah & Rauscher (1983) found significantly thicker corpus callosums in 21 early onset chronic schizophrenics; as compared to 8 late onset schizophrenics, 13 neurological patients and 14 with other psychiatric diagnosis. Nasrallah, Jacoby, McCalley-Whitters & Kuperman (1982) used computed tomography to show that: "55 young men with chronic schizophrenia and 27 age- and sex-matched control subjects showed a significantly higher ventricle-brain ratio (VBR) in the patients with chronic schizophrenia." Nasrallah, McCalley-Whitters, & Jacoby (1982) found significantly enlarged ventricles in manic and schizophrenic groups, as compared to normal controls. "Cerebral atrophy was more frequent in schizophrenia, while cerebellar atrophy was more frequent in mania ...

Cerebral atrophy was not associated with ventricular enlargement in either disorder." Studies of corpus callosal dysfunction in schizophrenia first suggested that there were enlarged ventricles and a thickened corpus callosum. Further investigation is providing more specific evidence supporting or refuting structural variance between schizophrenics and normal controls.

Dimond, Scammell, Pryce, Huws & Gray (1980) used intermanual and cross-lateral transfer to study split-brain symptoms in chronic schizophrenia. Patients were compared to a mixed group of depressed and anxious patients. The results showed a failure of interhemispheric transfer in performance in schizophrenic patients. Researchers suggest that the deficit is a "noisy channel" that produces the failure in the transmission of information. The researchers stress that performance is inconsistent, i.e., that information transfer is "inefficient" rather than "disconnected." Jaynes (1976) makes a similar point highlighting that normal people change hemispheres during a task at a rate of once per minute while schizophrenics appear to change at a rate of once in 4 minutes. Many studies suggest a difficulty in interhemipheric information processing in schizophrenia.

Other studies have examined right and left hemisphere functioning in schizophrenics. Tress, Caudrey & Mehta (1983) did an evoked potential study using tactile evoked potentials. They find differences between schizophrenics and controls. The schizophrenics did not show a "lateralization effect" suggesting that schizophrenics were not receiving tactile evoked potentials transmitted across the corpus callosum.

Walker, Hoppes & Emory (1981) have reinterpreted earlier studies as reflecting a deficit interhemispheric transfer rather than left hemisphere over-activation. Magaro and Page (1983) specifically postulated an interhemispheric deficit in schizophrenia. The findings of their study supported a left hemisphere deficit not an interhemispheric deficit.

Gazzaniga (1983) presented data on cerebral asymmetries in split brain patient with right-hemisphere language. The patient had his corpus callosum severed to control epileptic seizures. The patient functioned normally in his everyday life. As he functioned in his daily routine, information was being fed separately and simultaneously to both hemispheres. In the experimental condition, stimuli were simultaneously presented to the left and then the right hemisphere of the brain. Specifically, the stimuli consisted of a story: "MARY ANN MAY COME VISIT INTO THE TOWNSHIP TODAY". Stimuli to the left hemisphere sequenced alternately: "MARY MAY VISIT THE SHIP." Stimuli to the right hemisphere sequenced alternately: "ANN COME INTO TOWN TODAY." The brain then had to integrate this information without the help of the corpus callosum. The patient's response was: "Ma ought to come into town today to visit Mary Ann on the boat." In an attempt to integrate the stimuli, the patient became confused and confabulated a story similar, but not necessarily identical, to what we might interpret as schizophrenic responses.

Considering the evidence for callosal dysfunction from a homeostatic brain model, it appears that the corpus callosum plays a role in generation of schizophrenic symptoms. However, the evidence suggests a dynamic i.e. information flow deficit. Further investigations are needed in order to draw specific conclusions as to the relationship between structural and biochemical variations in the corpus callosum and deviant behavior.

FRONTAL LOBE DYSFUNCTION

A cerebral blood flow study by Ingvar and Franzen (1974) identified schizophrenics as having hypofrontality and lower levels of frontal lobe arousal. A later study by Gur, Skolnick, Gur, Caroff, Reiger, Obrist, Younkin & Reivich (1983) has shown that when a verbal or visual task is involved these effects diminish and are replaced by lateralized differences.

Sheppard, Gruzelier, Manchanda, Hirsch, Wise, Franckowiak & Jones (1983) find no hypoactivity in the frontal lobes of schizophrenics. They find evidence that metabolism within the basal ganglia is reduced in schizophrenics. They state: "suppport was obtained for the hypothesis that an abnormality in hemisphere laterality may underlie schizophrenic illness."

Although (rCBF) studies have produced negative findings on frontal lobe dysfunction, the homeostatic brain model suggests that the frontal lobes may be involved in producing schizophrenic symptoms. First, the dopamine system which is influenced by psychotropic medication is heavily represented in frontal lobe areas. Secondly, the frontal lobes have interconnections with many other areas of the brain via ascending and descending tracts. We suspect that at least some of these anterior posterior tracts are involved in the generation of schizophrenic symptoms such as slowed and inaccurate responses. The use of leucotomy and lobotomy may have achieved results because they interrupted dysfunctional circuits involved in psychotic behavior. However, the surgical technique is so unselective that it cuts many functional channels and disrupts many behaviors which are not "psychotic".

COMPARISONS OF SCHIZOPHRENIC AND BRAIN INJURED PATIENTS

Davison and Bagley (1969) review many schizophrenic-like organic conditions. They provide a model of organic brain disorders in the aetiology of schizophrenic-like symptoms. The model they use to discuss the aetiology of schizophrenia includes physical stress, social stress, psychological stress, genetic predisposition, alcohol, amphetamines, Parkinsionianism, and narcolepsy. These factors contribute to biochemical changes (along with organic brain disorder) which lead to specific brain dysfunction and to schizophrenic experience and behavior.

They cite parallels between schizophrenics' behavior and various types of brain damage from different aetiological entitles. These studies do not provide evidence of any one specific site in the brain which produces schizophrenia. Rather they suggest many types of brain damage result in schizophrenic-like symptoms. In terms of schizophrenic symptoms, areas like the temporal lobes and diencephalon seem particularly important.

Gruzelier (1983a, p.276) discusses Taylor's (1975) study of temporal lobe epileptic patients. He cites evidence for and against "alien tissue" i.e. small tumors, hematomas, and focal dysphasias, as a basis for psychosis. His conclusion is: "Given that many alien tissue lesions occur at early stages of brain development, the relatively late occurrence of psychosis may involve the breakdown of compensatory mechanisms." Flor-Henry (1983) makes the connection between the symptoms of schizophrenia and left temporal lobe epilepsy.

Many schizophrenics have behavioral deficits typical of brain damaged people. However, their deficits are not reliably associated with a specific location in the brain or aetiology. Further, neuropsychological investigation into schizophrenia may provide more definitive brain behavior relationships.

Many schizophrenics have behavioral deficits typical of brain damaged people. However, their deficits are not reliably associated with a specific location in the brain or aetiology. Further, neuropsychological investigation into schizophrenia may provide more definitive brain behavior relationships.

DEFICITS IN PARIETAL AND OCCIPITAL FUNCTIONING

Given that many of the key symptoms of schizophrenia involve gross perceptual defects such as hallucinations and body schema distortions, it is reasonable to examine the evidence of rear brain involvement. Erwin and Rosenbaum (1979) did a study comparing schizophrenics and patients with known parietal lobe lesions. They found that schizophrenics and parietal lobe damaged patients had similar impairments in weight discrimination when compared with controls. The parietal lobe damaged patients showed a deficit in tactile sensory perception. Schizophrenics showed disturbances in body image. They interpret these findings as reflecting different types of neurological dysfunctions.

Despite difficulties in controlling for attentional and cognitive artifacts, most of the studies of visual and auditory functions in schizophrenics do not find a difference between schizophrenics and normals in the functioning of their visual and auditory systems. However, when tasks become more complex and generalization or inhibition of responses becomes part of the task, then schizophrenics have significant problems.

Polyakov (1969) wrote an article examining the problem of perceptual and cognitive deficits in schizophrenia. He describes the breakdown in schizophrenics' cognitive functioning in terms of a series of links with a fundamental (biochemical) process being the initial link in the series. He compared schizophrenics to a control group of sixty normals on a series of experimental tasks. He studied generalization and comparison using an experimental method based on the work of Meleshko (1965). The first series of problems consisted of asking the patients to make comparisons of 12 pairs of objects. In the second experiment 39 objects pictured on cards were shown to the subjects, and they were asked to divide the cards into groups. Polyakov (1969) finds: "...if their logical, operational schema of activity is preserved, schizophrenics are able to abstract any attribute and on the basis of the attribute make a generalization as well as the normal person."

Schizophrenics auditory perception was also studied by Polyakov (1969). A "muffled word" was used as the target stimulus. They used "strong," "medium," and "weak" noise levels; and measured the percentage of correct identifications. Under the strong noise condition there was little difference between normals and schizophrenics, because the stimulus was overwhelmed by the noise. In the weak noise group the schizophrenics scored only slightly worse than normals. On the intermediate level of difficulty schizophrenics were best at perceiving unlikely endings and normals were best at perceiving likely endings.

A similar study was conducted of visual perception among schizophrenics (Bogandov, 1965). Subjects were presented with blurred pictures of objects which they were asked to identify. In this study schizophrenics had the highest thresholds for identifications when common standard images were presented. On unusual images schizophrenics had lower thresholds i.e. faster recognition than normals.

Investigations by Cornblatt & Erlenmeyer-Kimling (1985) show global attentional deviance is a marker in children at risk of schizophrenia. Knight, Elliott, & Freedman (1985) study backward masking of information in schizophrenics. They find "the perceptual-organization-deficit hypothesis best accounts for the apparent disruptions in poor premorbids' short term visual memory."

Two recent studies illustrate the difficulties doing neuropsychological research with schizophrenics. Knight et al. (1985) examined short term visual memory functioning with cognitive and backward masks. They find significant differences between good and poor premorbid schizophrenics. The poor premorbid schizophrenics treated both cognitive and backward masks equivalently and showed deficits. The results are discussed in terms of a theory of visual-input dysfunction in schizophrenia. In a second study, again using a visual information processing task, Knight, Youard and Wools (1985) report a general difference between schizophrenics and normals but no specific differences in various conditions. The results are discussed as reflecting a generalized cognitive disturbance. They relate the level of difficulty on particular tasks to the amount of capacity required to complete processing. Knight et al. (1985) discuss the difference between their study and others in the field.

The point that we would emphasize is the contradiction between findings on one hand of a specific visual information processing deficit; and on the other hand, broader cognitive processing deficits. The schizophrenia literature has many of these contradictions. The finding of both specific deficits and general information processing difficulties may reflect the multiple levels of dysfunction characteristic of schizophrenics. We believe this occurs because of the malfunctioning of homeostatic processes within schizophrenics′ brains.

DEFICITS IN NEUROPSYCHOLOGICAL FUNCTIONING

Current studies discuss neuropsychological test results of schizophrenics (Green & Walker, 1985; Gruzelier, 1983a; Lewis, 1979). These studies find that schizophrenics do not have the specific deficits typical of lateralized stroke or head injury patients; yet their test responses are different than "normals". Lewis (1979) found that schizophrenics tend to be slower than normals on many tasks. As discussed by Gruzelier (1983a), schizophrenics have problems with both right and left hemisphere tasks which may reflect dynamic processing problems rather than the type of structural neuropsychological problems usually identified via neuropsychological test batteries.

Lewis (1979) compares various levels of chronicity and length of hospitalization. Lewis finds that these variables do not, in and of themselves, account for most of the variance in the deficits of schizophrenics′ behaviors. In her study, schizophrenics had difficulties on Reaction Time, Receptive Speech, Expressive Speech, Writing, Arithmetic, Pathognomic and Intelligence Subscales of the Luria Nebraska Neuropsychological Battery. She also discusses the role of deficits on the Motor Rhythm, Receptive Speech, Arithmetic, and Memory Subtests. Another study (Purisch, Golden & Hammeke, 1983) used the Luria Nebraska to discriminate between schizophrenic and brain injured patients. A similar neuropsychological study by Selin and Gottschalk (1983) compared schizophrenics, depressives and conduct disorders. They found that schizophrenics have deficits in rhythm, perception, abstraction, and on EEG measures.

Green and Walker (1985) compare schizophrenics with positive and negative symptoms on a battery of neuropsychological tests. They summarize their findings as follows: "The central finding from this study is that positive and negative symptoms are associated with different patterns of performance deficit on neuropsychological tests. Negative symptoms are associated with poorer performance on tests that measure visual-motor and visual-spatial skills. The positive symptoms, on the other hand, are related to deficits on tests that involve short term verbal memory. These findings are not consistent with the notion that negative symptoms are uniquely associated with generalized cognitive impairment. Instead, it appears that the situation is more complex and that both types of symptoms are associated with characteristic performance deficits (Green and Walker, 1985 p.466)." Gruzelier (1983) finds electrodermal differences in patients with positive and negative symptoms. Their symptom patterns are consistent with cognitive deficits discussed by Green and Walker (1985).

Holden, Stock & Itil (1985) study neuropsychological deficits in chronic treatment resistant schizophrenics. They find that therapy resistance was positively correlated with scores for organic impairment on the Modified Word Learning Test. They find positive correlations between their neuropsychological measures. They also collect EEG data. They find that their treatment resistant patients had a greater incidence of low frequency bands (3-8 Hz bands). They find that the Modified Word Learning Test was the most useful tool for measuring organic deterioration in their sample.

SCHIZOPHRENIA: THE BREAKDOWN OF THE HOMEOSTATIC BRAIN

Brain Systems: Micro-, Meso- , and Macrosystems

In the human brain, information is coded and processed as a hierarchy of homeostatic systems from neurons to whole brain functions: from micro- to meso- to macrosystems. Evidence suggests that there are microsystems breakdowns in schizophrenic brain functioning. Microsystems consist of neurons and their immediate neural networks. Schizophrenic behavior patterns include generating stimuli which do not correspond to external stimuli i.e. hallucinations. This may reflect random and unsynchronized firing of neurons or groups of cells in the brain.

Mesosystems consist of specialized cell groups organized into localized areas such as the occipital cortex, perceptual, and motor systems, and each hemisphere. These localized areas perform a specific function. Evidence from EEG studies (Flor-Henry, Koles & Sussman, 1984; Flor-Henry & Koles, 1984; and Koukkou, 1985); (rCBF) studies (Gur, 1979); and behavioral studies (Lewis, 1979) suggest that in schizophrenia there may be a breakdown in information processing in communication between these localized areas.

In a normal brain, a stimulus is going to be attenuated as a result of lateral inhibition. In a schizophrenic's brain the stimulus is not modulated. Repetitive cycles are established which lead to repeating behavior without change. This can be seen, for example, when a patient rocks and rocks all day. This involves the malfunctioning of information feedback loops and is seen in perceptual, motor, and cognitive systems. Processing time is slow and there are many errors. Schizophrenics show individual variability in their functioning.

The whole brain functions as a macrosystem. This includes the anterior-posterior, right-left, and whole brain systems. The normal brain has a relative synchronicity of EEG which varies depending on the task and reciprocal inhibition of right and left sides of the brain. This helps the individual remain focused on a specific task and utilizes the brain effectively to perform that task. In schizophrenia, the individual does not stay focused. Tangential information is processed as if it were the main thought.

Schizophrenic behavior and laterality tasks

The homeostatic brain model can be used to evaluate the behavior of schizophrenics on laterality tasks. The search for a specific locus of schizophrenic deficit is confounded with the tasks used to study it, and more importantly with the endogenous "tasks" that schizophrenics themselves use routinely. In many studies, Morstyn, Duffy & McCarley (1983); Shaw, Colter & Resek (1983); Bernstein, Taylor, Starkey, Juni, Lubowsky & Paley (1981); Abrams & Taylor (1980); and Hiramatsu, Kameyana, Saitoh, Niwa, Rymar & Itoh (1984) there is a conflict between the task demands of the experiment, and covert verbal and visual activities of schizophrenic patients i.e. the patients are generating their own stimuli. For example, in resting EEG and (rCBF) studies schizophrenics often show left hemisphere arousal. Such a phenomena could occur if schizophrenics were actively involved in repetitive ideas and covert speech. The problem with laterality tests of schizophrenics is that the internal stimuli to themselves, conflicts with the external stimuli of the experiment. The internal information processing of schizophrenics interferes with the experimental tasks, and produce anomalous results.

One study explores the cortical and subcortical arousal processes in schizophrenia, and they (Pfefferbaum, Horvath, Roth, Tinklenburg & Kopell, 1980) find deficits in cortical but not subcortical arousal in schizophrenics. Other studies such as Siegel, Waldo, Mizner, Adler & Freedman (1984) have studied sensory gate keeping and inhibition in schizophrenics. Schizophrenics were impaired in their ability to inhibit and had abnormalities in wave length and power. An EEG telemetry study is particularly interesting. Stevens and Livermore (1982) find slow activity and less alpha activity in schizophrenics than normal controls. During periods of abnormal behavior, they showed the "Ramp Spectra" often associated with epileptic spikes. No epileptic spikes were observed.

There is a dichotomy between what schizophrenics think and do, and external reality. In addition to interhemispheric dysfunction, Etevenon (1984) cites the role of intrahemispheric dysfunction. Thus, a stimulus presented to one hemisphere is not effectively coded i.e. a stimulus may be inhibited, "lost," transferred, or greatly amplified. The inter- and intrahemispheric communication problems led to some of the more bizarre behavior of schizophrenics. Stimuli generated within the brain are treated as if they are external realities, and external realities are not responded to as such. The breakdown contributes further to changes in body perception, confusion, and disorganization of thinking.

Individual differences in the course of the schizophrenia process

To gain a better understanding of the progression of schizophrenia from acute to chronic states and lateralization of function, we can consider a sample case and the major changes in his life. In the classic case of Dr. Daniel Paul Schreber, we see an individual who had a relatively good premorbid adjustment, some family history of mental illness, and a previous hypochondriacal episode from which he recovered successfully. During a period of stress, Schreber becomes highly anxious, hypersensitive to stimuli, withdraws socially, and develops increasingly bizarre and idiosyncratic ideas. The symptoms include sleeplessness and hypochondriacal behavior. Following this initial stage of high anxiety and perceptual distortion, his bizarre experiences and idiosyncratic ideas become organized into a delusional system. After many years of hospitalization, his symptoms subside to a point where he can again be released to the community (Freud, 1968).

Serafetinides (1984) discusses the advantages of examining EEG findings as a "temporary disequilibrium, with an onset, course and resolution, either towards a normative pattern or towards a pathological one..." If we translate Dr. Schreber's experiences into patterns of hemispheric activation, we would see the following changes in terms of onset, course, and resolution. In terms of onset we would see very high levels of activation in one or both hemispheres. During the course of his psychosis, we would expect to find poor interhemispheric coordination. In the residual stage we would expect lowered arousal and a predominance of right hemisphere oriented behaviors.

Several studies find behavioral and EEG changes during the course of a schizophrenic process (Frith, Stevens, Johnstone, & Crow, 1982; Iacono, 1982; Koukkou, 1985; Rizzo, Albani, Spadaro 2, Moroculti, 1983; Rizzo, Spadaro, Albani, & Moroculti, 1984). The findings in these studies are that acute schizophrenics who remit show more normalized EEG's than schizophrenics who remain symptomatic. "Chronic" schizophrenics' EEG's continue to show a variety of differences from normal and remitted schizophrenics.

The main point about this analysis is that it reflects a dynamic process of brain behaviour relationships as the individual experiences different levels of dysfunctional adaptation. The course of the illness, as seen in both EEG (Serafetinides, 1984) and factor analytic studies (Carr, 1983), reveals that it is difficult to isolate a specific localized deficit related to schizophrenic behavior. What you have is a progressive loss of the person's overall ability to function, and then a regaining of functions. Schizophrenia is a dynamic process with levels of dysfunction, and with an onset, course and resolution. The dynamic processes should be seen to a lesser extent in individuals at risk for schizophrenia than in schizophrenics. Children of schizophrenics, who are at risk of schizophrenia, show some of the same EEG and neuropsychological functions that are seen in schizophrenics (Prentky, Salzman, & Klein, 1981). Likewise, schizotypal people show some of the same characteristics but to a lesser degree. These people have some characteristics of schizophrenics, but do not develop schizophrenia (Braff, 1981). Without the intense stressers and subsequent stages of deterioration, these people do not exhibit schizophrenic symptoms. Since children at risk show some similarities to schizophrenics, they provide the beginning and important stages in the development of schizophrenia. Further study of these children will add to the body of knowledge on predictors of schizophrenia. This then can be applied to early intervention and treatment.

Implications for assessment of schizophrenics

Excellent research is being done using EEG and brain imaging, yet this research has not provided the field with practical methods to assess schizophrenic patients. Our high technology imaging techniques provide new information about hemispheric functioning, but the data is not easily translatable into categories of schizophrenia associated with specific brain-behavior dysfunction. There is a schism between factorial validity or other forms of validity, and clinical judgment (Cromwell, 1983). To increase the validity of clinical assessment, what is needed is a way to apply the results of the EEG and brain imaging studies to assessment of schizophrenia.

One area where clinical assessment and research findings have converged is the study of positive and negative symptoms of schizophrenia. Cromwell (1983) and Holden et al. (1985) point out the significance of positive and negative symtoms for understanding and treating schizophrenia. Type I, positive symptoms, appear to be identified with left hemisphere overactivation. The patients have hallucinations, delusions, and thought disorders. Their behavior is bizarre, disturbed, and sometimes dangerous. Type II, negative symptoms, appear to be associated with right hemisphere or bilateral activation. The patients become passive, and their affect flattens. Behaviorally they stay at home, fail at school or their jobs, and take on virtually no household responsibilities. This diagnostic categorization has direct implications for treatment, management, medication, and prognosis.

Based on a homeostatic model, clinical assessment of schizophrenia can be made more accurate and applicable to treatment. The way to do this is to develop assessment instruments which examine feedback loops within the brain. It is possible to identify subtypes of schizophrenia in terms of systems functions and dysfunctions: 1. Arousal homeostasis; 2. Motor and perceptual systems, especially proprioceptive feedback; 3. Interhemispheric balance and reciprocity; 4. Anterior-posterior feedback; 5. Cortical-subcortical feedback systems; and, 6. Whole brain functioning.

The evidence of slowing of responses in schizophrenia suggests another assessment method. Schizophrenia can be examined in terms of the speed (reaction time) and accuracy of information processing. Neuropsychological tasks to assess the roles of inhibition/excitation and feedback within brain areas need to be developed. Much of the future focus in assessment should be on dynamic processes. This can be done by improving our existing: (a) neuropsychological techniques;(b) EEG techniques; (c) brain imaging techniques; (d) clinical interview and assessment techniques.

Implications for treatment of schizophrenia

In light of the above evidence let us reconsider our treatment of schizophrenics. Holden et al. (1985) found that: "some chronic schizophrenic patients do have neuropsychological and neurophysiological profiles consistent with organic brain dysfunction." If schizophrenics, and especially treatment resistant schizophrenics, have organic brain dysfunction, then it makes sense to treat them like brain damaged people.

Yozawitz, Charters, Iskander & Reiter (1985) describe a system of assessment and rehabilitation of psychiatric patients with neuropsychological deficits. They combined cognitive habilitation techniques with traditional activities of daily living (ADL) skills. The patients showed post treatment improvement on Performance I.Q., the Similarities, Digit Span, and Block Design Wechsler subtests, and the Arithmetic Subtest of the Wide Range Achievement test. Control subjects, only receiving the ADL program, improved on the Wechsler Information Subtest only. Their results can be interpreted as indicating that neuropsychological interventions will be effective with schizophrenic patients.

There is an example of an intensive treatment program for brain injured adults which may provide a model of possible treatment programs for schizophrenics. Hoofien & Ben-Yishay (1982) and Ben-Yishay, Rattok, Lakin, Piasetsky, Ross, Silver, Zide & Ezrachi (1985) present a description of a treatment program for brain injured adults. Treatment includes: " (a) intensity (daily treatment), (b) communality (emphasis on group-therapy) and (c) cognitive remediation (improvement of cognitive functions by special techniques of learning). Treatment is evolved in three stages: assessment (two weeks); intensive treatment (nine months); vocational experience (unlimited duration). According to follow-up data, about 80% of the patients achieve positive rehabilitational results." Considering that schizophrenics appear to be brain injured, this type of intensive program may help them recover their functioning. We are not ignoring nor minimising the substantial differences between brain damaged and schizophrenic patients in premorbidity, history, course of illness, life experience, and behavior. Rather, we are emphasizing the potential utility of cognitive retraining and psychosocial rehabilitation.

A second treatment technique derived from the homeostatic brain model is to train schizophrenic patients to improve the functioning of feedback loops within their own brains; and thereby increase their ability to modulate, inhibit, and coordinate their behavior. Several experimental treatments are possible. Using biofeedback techniques such as those used in studies of meditation and EEG (Kamiya, 1969), schizophrenics could be reinforced for increasing the inter- and intrahemispheric coherence of their EEG's.

Schneider and Pope (1982) did a study conditioning EEG changes in schizophrenics through biofeedback. Using biofeedback, they trained schizophrenics to "mimic the EEG changes that have been shown to occur with neuroleptic induced clinical improvement." A related approach is to help reduce the levels of psychotropic medication with side effects such as Tardive Dyskinesia. Biofeedback techniques could be used to train the patients to use other techniques to obtain the effect of the drug. The introduction of biofeedback should produce the therapeutic effect, while decreasing the level of drug.

There is evidence from innovative programs that schizophrenic are able to revive feedback loops and regain competent daily functioning and a sense of self-worth (Katz and Katz, 1983). They use art to help recreate or revive feedback systems in the brain. Their treatment program has focused on the self-esteem and living skills of the artist/patients. The Katzs' suggest that the patients are developing new brain behavior relationships by expressing themselves through their art work.

CONCLUSION

Our conclusions are as follows

1. The two hemispheres play a role in production of schizophrenic symptoms and states. However, the situation is not as simple as that a single dominant hemisphere is over or under active. The lateralization problems in schizophrenia reflect a breakdown in reciprocity, synchronicity, inhibition and coherence between the two hemispheres.

2. Schizophrenia is a multileveled breakdown in the functioning of the brain. Specific neural functions and neural transmitter mediated functions are disrupted (best understood in the dopamine circuits). There is a breakdown in what Walsh (1981) calls "arousal homesotasis" involving both cortical and mid-brain areas. Multiple perceptual, motor, and cognitive systems are impaired. The particular systems involved show wide individual differences. Interhemispheric and whole brain cognitive functions are impaired.

3. Schizophrenia is a breakdown in intra- and interhemispheric feedback systems. This breakdown has periods of synchronicity and functioning, and periods of dissynchronicity and dysfunction. The process is similar to the electrical disorganization which occurs in an epileptic seizure. Yet schizophrenics do not have as dramatic behavioral and EEG abnormalities as do epileptics. Their behavioral and EEG functions are disturbed compared to normals, but do not reflect the specific spike and wave and convulsions of epilepsy.

4. The acute phases of schizophrenia have the potential for patients to return to normal or near normal states. Continuation of schizophrenic states results in functional and neuropsychological deterioration similar to that seen in other types of brain damage.

5. The assessment of schizophrenia is undergoing a transition. Neurophysiological, neuropsychological and familial studies suggest revaluation of the symptom oriented classification of schizophrenics. It is

recommended that the classification of schizophrenics be based on the feedback systems within the brain which are impaired.

6. The outlook for treatment of schizophrenics may be improving. In addition to biochemical methods, intensive neuropsychological intervention, EEG, and Biofeedback techniques can be used to at least reduce some of the severity of schizophrenic symptoms. The groups of schizophrenics most resistant to current treatments appear to represent the groups most in need of neuropsychological intervention.

7. Research on the neuropsychology and neurophysiology of schizophrenia should focus on homeostatic processes and feedback processes within the brain.

REFERENCES

Abrams, R. and Taylor, M. (1980). Psychopathology and the electroencephalogram. Biological Psychiatry, 15, 871-8.

Abse, D.W. (1971). Speech and Reason: Language Disorder in Mental Disease. Translation of P. Wegener. The Life of Speech. Charlottesville: The University Press of Virginia.

Ben-Yishay, Y., Rattok, J., Lakin, P., Piasetsky, E., Ross, B., Silver, S., Zide, E., and Ezrachi, O. (1985). Neuropsychologic rehabilitation: Quest for a holistic approach. Seminars in Neurology, 5, 252-259.

Beaumont, R.G., and Dimond, S.F. (1973). Brain disconnection and schizophrenia. British Journal of Psychiatry, 123, 661-662.

Bernstein, A., Taylor, D., Starkey, P., Juni, S., Lubowsky, J., & Paley, H. (1981). Bilateral skin conductance, finger pulse volume, and EEG orienting response to tones of differing intensities in chronic schizophrenics and controls. Journal of Nervous and Mental Disorders, 169, 513-28.

Bigelow, L., Nasrallah, H., and Rauscher, F. (1983). Corpus callosum thickness in schizophrenia. British Journal of Psychiatry, 142, 284-287.

Bogdanov, Y.I. (1965). The study of visual perception under conditions of incomplete information. In B.V. Zeigarnik (Ed.) Problems of experimental psychopathology. Moscow.

Braff, D. (1981). Information processing in schizotypal patients. Paper presented at APA Convention.

Carr, V. (1983). Recovery from schizophrenia: A review of patterns of psychosis. Schizophrenia Bulletin, 9, 95-121.

Cornblatt, B. and Erlenmeyer-Kimling, L. (1985). Global attentional deviance as a marker of risk for schizophrenia: Specificity and predictive validity. Journal of Abnormal Psychology, 94, 470-486.

Cromwell, R. (1983). Preemptive thinking and schizophrenia research. Nebraska Symposium on Motivation.

Davison, K. and Bagley, C. (1969). Schizophrenia-like psychoses associated with organic disorders of the central nervous system: A review of the literature. British Journal of Psychiatry, 4, 114-184.

Dimond, S.F., Scammell, R., Pryce, I.F., Huws, D., and Gray, C. (1980). Some failure of intermanual and cross-lateral transfer in chronic schizophrenia. Journal of Abnormal Psychology, 89, 505-509.

Erwin, B.J., and Rosenbuam, G. (1979). Parietal lobe syndrome and schizophrenia: Comparison of neuropsychological deficits. Journal of Abnormal Psychology, 88, 234-241.

Etevenon, P. (1984). Intra and inter-hemispheric changes in alpha intensities in EEGs of schizophrenic patients versus matched controls. Biological Psychology, 19, 147-256.

Flor-Henry, P. (1969). Psychosis and temporal lobe epilepsy - a controlled investigation. Epilepsia, 10, 363-395.

Flor-Henry, P. (1983). Functional hemispheric asymmetry and psychopathology. Integrative Psychiatry, 1, 46-52.
Flor-Henry, P. and Koles, A.J. (1980). EEG studies in depression, mania and normals: Evidence for partial shifts of laterality in the affective psychoses. Advances in Biological Psychiatry, 4, 21-43.
Flor-Henry, P., and Koles (1984). Statistical quantitative EEG studies of depression, mania, schizophrenia and normals. Biological Psychology, 19, 257-279.
Flor-Henry, P., Koles, Z.J., Howarth, B.G., and Burton, L. (1979). Neurophysiological studies of schizophrenia, mania and depression. In J. Gruzelier & P. Flor-Henry, (Eds). Hemisphere Asymmetries of Function in Psychopathology. Amsterdam: Elsevier/North Holland.
Flor-Henry, P., Koles, A., and Sussman, P. (1984). Further observations on right/left hemispheric energy oscillations in the endogenous psychoses Advances in Biological Psychiatry, 15, 1-11.
Flor-Henry, P., Koles, Z.J., and Tucker, D.M. (1982). Studies in EEG power and coherence (8-13 Hz) in depression, mania and schizophrenia compared to controls. Advances in Biological Psychiatry, 9, 1-7.
Freud, S. (1968). Three Case Histories. New York: Collier Books.
Firth, C., Stevens, M., Johnstone, E., and Crow, T. (1982). Skin conductance habituation during acute episodes of schizophrenia: qualitative differences from anxious and depressed patients. Psychological Medicine, 12, 575-83.
Gazzaniga, M.S. (1983). Right hemisphere language: A twenty year perspective American Psychologist, 38, 525-549.
Green, M, and Walker, E. (1985). Neuropsychological performance and positive and negative symptoms in schizophrenia. Journal of Abnormal Psychology, 94, 460-69.
Gruzelier, J. (1983). Critical assessment and integration of lateral asymmetries in schizophrenia. In M.S. Myslobodsky (Ed.) Hemisyndromes: Psychobiology, Neurology, Psychiatry. New York: Academic Press.
Gruzelier, J. (1983). Left- and right-sided dysfunction in psychosis. Implications for electroencephalographic measurement. Advances in Biological Psychiatry, 13, 192-195.
Gruzelier, J. (1984). Hemispheric imbalances in schizophrenia. International Journal of Psychophysiology, 19, 227-240.
Gruzelier, J. (1984). Schizophrenia and spectral analysis of the visual evoked potential. British Journal of Psychiatry, 145, 496-501.
Gruzelier, J., and Hammond, N. (1976). Schizophrenia - A dominant hemisphere temporal lobe disorder? Research Communications in Psychology, Psychiatry, and Behavior, 1, 33-72.
Gruzelier, J., and Manchanda, R. (1982). The syndrome of schizophrenia: Relations between electrodermal response, lateral asymmetries and clinical ratings. British Journal of Psychiatry, 141, 488-495.
Gur, R. (1979). Cognitive concomitants of hemispheric dysfunction in schizophrenia. Archives of General Psychiatry, 36, 269-274.
Gur, R., Packer, I., Reivich, M. and Weinberger, J. (1978). Cognitive task effects on hemispheric blood flow in humans. Paper presented APA Annual Convention, Toronto, Canada.
Gur, R., Skolnick, B., Gur, R., Caroff, S., Reiger, W., Obrist, W., Younkin, D., and Reivich, M. (1983). Brain function in psychiatric disorders. Archives of General Psychiatry, 40, 1250-1254.
Gur, R., Gur, R., Skolnick, B., Caroff, S., Obrist, W., Resnick, S., and Reivich, M. (1985). Brain function in psychiatric disorders. Archives of General Pysychiatry, 42, 329-334.
Henn, F.A., and Nasrallah, J.A. (1982). Schizophrenia as a Brain Disease. New York: Oxford University Press.
Hiramatsu, K., Kameyama, T., Saitoh, O., Niwa, S., Rymar, K., and Itoh, K. (1984). Correlations of event-related potentials with schizophrenic deficits in information processing and hemispheric dysfunction. Biological Psychology, 19, 281-294.

Holden, F.M., Stock, W.J., and Itil, T.M. (1985). Neuropsychological deficit in chronic treatment resistant schizophrenia. Prepublication Draft.

Hoofien, D. and Ben-Yishay, Y. (1982). Neuropsychological therapeutic community rehabilitation of severely brain-injured adults. Neuropsychological Community Rehabilitation. In Psychological Research inRehabilitation. Israel: Ministry of Defense, Department of Rehabilitation.

Iacono, W. (1982). Bilateral electrodermal habituation - dishabituation and resting EEG in remitted schizophrenics. Journal of Nervous and Mental Disorders, 170, 91-101.

Ingvar, D. and Franzen, G. (1974). Distribution of cerebral activity in chronic schizophrenia. Lancet, 2, 1484-1486.

Jaynes, J. (1976). The Origin of Consciousness in the Breakdown of the Bicameral Mind. Boston: Houghton Mifflin.

Kamiya, J. (1969). Operant control of the EEG alpha rhythm and some of its reported effects on consciousness. In C.T. Tart (Ed.), Altered States of Consciousness. New York: Wiley.

Katz, F. and Katz, E. (1983). Art and Disabilities. Berkeley: Institute of Art and Disabilities.

Knight, R., Elliott, D. and Freedman, E. (1985). Short-term visual memory in schizophrenics. Journal of Abnormal Psychology, 94, 427-224.

Knight, R., Youard, P., and Wools, I. (1985). Visual information-processing deficits in chronic schizophrenic subjects using tasks matched for discriminating power. Journal of Abnormal Psychology, 94, 545-459.

Koukkou, M. (1985). EEG correlates of information processing in acute and remittent schizophrenic patients. Archives Suisses de Neurologie, Neurochirurgie et de Psychiatrie, 136, 37-43.

Lewis, G. (1979). Effects of chronicity and hospitalizaton on neuropsychological performance in schizophrenics. Paper presented APA Convention.

Magaro, P. and Page, J. (1983). Brain disconnection, schizophrenia, and paranoia. Journal of Nervous Mental Disorders, 171, 133-140.

Meleshko, T. (1965). Variation on the method of studying the process of comparison in schizophrenics. In B.V. Zeigarnik (Ed.), Problems of Experimental Psychopathology. Moskow.

Morstyn, R., Duffy, F., & McLarley, R. (1983). Altered topography of EEG spectra content in schizophrenia. Electroencephalography & Clinical Neurophysiology, 56, 263-71.

Moscovitch, M. (1983). The linguistic and emotional functions of the normal right hemisphere. In E. Perecman (Ed.), Cognitive Processing in the Right Hemisphere. New York: Academic Press.

Nasrallah, J., Jacoby, C., McCalley-Whitters, M., and Kuperman, S. (1982). Cerebral ventricular enlargement in subtypes of chronic schizophrenia. Archives of General Psychiatry, 39, 774-777.

Pfefferbaum, A., Horvath, T., Roth, W., Tinklenberg., & Kopell, B. (1980). Auditory brain stem and cortical evoked potentials in schizophrenia. Biological Psychiatry, 15, 209-23.

Polyakov, U.F. (1969). The experimental investigation of cognitive functioning in schizophrenia. In M. Cole and I. Maltzman (Eds.), Handbook of Contemporary Soviet Psychology. New York: Basic Books.

Prentky, R., Salzman, L., & Klein, R. (1981). Habituation and conditioning of skin conductants responses in children at risk. Schizophrenia Bulletin, 7, 281-91.

Purisch, A.D., Golden, C.J., and Hammeke, T.A. (1983). Discrimination of schizophrenic and brain-injured patients by a standardized version of Luria's Neuropsychological tests. Journal of Consulting & Clinical Psychology, 46, 1266-1273.

Rizzo, P., Albani, G., Spadaro, M., and Moroculti, C. (1983). Brain slow potentials (CNV), prolactin, and schizophrenia. Biological Psychiatry, 18, 175-83.

Rizzo, P., Spadaro, M., Albani, G., & Moroculti, C. (1984). Contingent negative variation and schizophrenia: A long-term follow-up study. Biological Psychiatry, 19, 1719-24.

Roemer, R., Shagass, C., Straumanis, J., & Amadeo, M. (1979). Somatosensory and auditory evoked potential studies of functional differences between the cerebral hemispheres in psychosis. Biological Psychiatry, 14(2), 357-73.

Rokeach, M. (1964). The Three Christs of Ypsilanti. New York: Alfred A. Knopf.

Schneider, S. (1983). Multiple measures of hemispheric dysfunction in schizophrenia and depression. Psychological Medicine, 13, 287-297.

Schneider, S.J. and Pope, A.T. (1982). Neuroleptic-like electroencephaloraphic changes in schizophrenics through biofeedback. Biofeedback & Self-Regulation, 7, 479-491.

Selin, C.L. and Gottschalk, L.A. (1983). Schizophrenia, conduct disorder and depressive disorder: Neuropsychological, speech sample and EEG results. Perceptual & Motor Skills, 57, 427-444.

Serafetinides, E. (1984). EEG lateral asymmetries in psychiatric disorders. Biological Psychology, 19, 137-246.

Shaw, J., Colter, N., and Resek, G. (1983). EEG coherence, lateral preference and schizophrenia. Psychological Medicine, 13, 299-306.

Sheppard, G., Gruzelier, J., Manchanda, R., Hirsch, S.R., Wise, R., Franckowiak, R., & Jones, T. (1983). O positron emission tomographic scanning in predominantly never-treated acute schizophrenic patients. Lancet, 1148-1152.

Siegel, C., Waldo, M., Mizner, G., Adler, L., and Freedman, R. (1984). Deficits in sensory gating in schizophrenic patients and their relatives. Evidence obtained with auditory evoked responses. Archives of General Psychiatry, 41, 607-12.

Stevens, J. and Livermore, A. (1982). Telemetered EEG in schizophrenia: Spectral analysis during abnormal behavior episodes. Journal of Neurology, Neuropsychology and Psychiatry, 45, 385-95.

Taylor, D.C. (1975). Factors influencing the occurrence of schizophrenia-like psychoses in patients with temporal lobe epilepsy. Psychological Medicine, 5, 249-256.

Toone, B., Cook, E., and Lader, M. (1979). The effect of temporal lobe surgery on electrodermal activity: implications for an organic hypothesis in the aetiology of schizophrenia. Psychological Medicine, 9, 281-5.

Tress, K., Caudrey, D., and Mehta, B. (1983). Tactile-evoked potentials in schizophrenia. Interhemispheric transfer and drug effects. British Journal of Psychiatry, 143, 156-64.

Tucker, D. (1984). Lateral brain function in normal and disorderd emotions: Interpreting electroencephalographic evidence. Biological Psychology, 19, 219-235.

Walker, E., Hoppes, E., and Emory, E. (1981). A reinterpretation of findings on hemispheric dysfunction in schizophrenia. Journal of Nervous and Mental Disorders, 169, 378-380.

Walsh, R. (1981). Towards an Ecology of Brain. New York: SP Medical & Scientific Books.

Yozawitz, A., Charters, M.B., Iskander, T., & Reiter, S. (1985). Cognitive habilitation of neuropsychological deficit in psychiatric patients. Paper presented at the International Neuropsychological Society, February, 1985.

CEREBRAL LATERALITY AND PSYCHOPATHOLOGICAL DISORDERS

E. A. Serafetinides

Veterans Administration Medical Center
West Los Angeles, (Brentwood Division)
and The Department of Psychiatry and
The Brain Research Institute
UCLA School of Medicine
Los Angeles, California

INTRODUCTION

In a previous report (Serafetinides, 1965), I presented evidence suggesting not only a marked presence of aggressive behavior in, mostly young, male temporal lobe epileptics (TLE), but also a left dominant hemisphere focus for the majority of them. Indeed, looking at the results again, it can be seen that out of 100 TLE cases, 36 were diagnosed as aggressive, 25 of which had a left (L) focus vs. only 11 with a right (R) focus. The 64 non-aggressive TLE cases were evenly split between the two hemispheres (33 R and 31 LTL focus respectively). This difference is significant at the $p < .05$ level. Discussing the implications, then, I stressed the significance of the dominant hemisphere for learning, linking defect of the latter with frustration in coping, and aggression as a maladjusted form thereof.

THE EVIDENCE

Indirect supportive evidence for this argument is provided by Camp (1977) and Camp, Zimet, van Dornick and Dahlem (1977) who found indeed that verbal abilities are impaired in aggressive boys. The children studied had an IQ of 90 and above. A few verbal tests were found to be significantly inferior in aggressive boys as compared to normals, but much more importantly, what differentiated the two groups was what the authors termed verbal mediational ability. The hypothesis is made that "both learning and behavior problems in aggressive boys may be symptomatic of an ineffective linguistic control system". These boys are characterized by a rapid response style and when speculating about control, and thus dyscontrol mechanisms, one ought to consider that, as will be touched upon later, the dominant hemisphere for speech might be involved in "pacing" response styles through mechanisms similar to those it employs in "pacing" production of speech.

Additional such evidence is presented by Krynicki (1978). In two of three male adolescent groups that he studied, 7 with a history of multiple assaultive incidents and 6 with an organic brain syndrome, the results of neuropsychological tests and of EEG examination were indistinguishable, showing poorer performance on the tests and paroxysmal type EEG abnormalities compared to a third group of 8 adolescents with an antisocial behavior pattern but without assaultive incidents. The nonassaultive patients had also a better established hand dominance, fewer perserveration errors and a better verbal memory. It is interesting that the EEG abnormalities were of a frontal distribution which the author considered as evidence of fronto-temporal limbic connections.

A negative report was published by Hare (1979). Tachistoscopic letter recognition tasks administered to the right and left visual fields of 55 prison inmates, divided according to psychopathy ratings (high, medium, low) failed to differentiate among the groups thus tested. Sandel and Alcorn (1980) also failed to discern any lateralization in 25 antisocial patients by employing the conjugate lateral eye movement index. In contrast, however, to these reports, other investigators have reported on the following lateralizing evidence. Pritchard, Lombroso and McIntyre (1980), reported on 56 temporal lobe epileptic patients, among whom 36% were found to have developed "psychological complications". Psychopathology, including antisocial manifestations, was seen more often in patients with left temporal lobe spike foci (43%) and in males (42%) but the authors stated that these trends did not attain statistical significance. Yeudall, Fromm-Auch and Davies (1982), examined the possibility of neuropsychological impairment in 99 consecutively admitted juvenile delinquents (64 males and 35 females). By using the Halstead-Reitan Battery and 12 additional tests and comparing the results to a control group's findings (47 Ss) they concluded that delinquents presented more abnormal profiles and their pattern of deficits implicated the right or non-dominant hemisphere. However, further analysis disclosed that the delinquent group in question included a significantly lower than average number of violent adolescents and/or a correspondingly higher percentage of delinquents with depression. The neuropsychological state of adolescent delinquent boys was also investigated by Wolff, Waber, Bauermeister, Cohen and Ferber (1984). In contrast to controls the delinquent boys showed impairment on all language measures and, by implication, impairment of the dominant or left hemisphere. It is interesting to juxtapose to this report the findings by Weintraub and Mesulam (1983) on 14 patients with neurologic and neuropsychologic signs consistent with right-hemisphere dysfunction. These patients were characterized by shyness, visuospatial disturbances, and inadequate paralinguistic communicative abilities which impaired their interpersonal skills. Although of average intelligence, they were poor in arithmetic and finally despite their difficulty in conveying their feelings (avoiding eye contact and lacking the gestures and prosody that normally accompany speech) there was no evidence that they were unable to experience affect. Hare and McPherson (1984) employing a verbal dichotic listening task on 146 male prison inmates and control group of 159 male non-criminals found that psychopathic prisoners (according to DSM-III and other criteria) exhibited a less lateralised performance than the other groups. The authors raised the possibility that language processes are not strongly lateralized in psychopaths.

Returning to epileptic patients, in an attempt to determine the relationship between unilateral temporal lobe epileptic focus and behavior, Bear and Fedio (1977) analyzed quantitatively 18 interictal traits selected on the basis of prior reports and pilot tests. The samples consisted of 27 epileptics (15 RTL and 12 LTL) and 2 control groups, one of 12 normal adults and one of 9 patients with neuromuscular disorders. In terms of laterality, RTL epileptics reported more elation, LTL epileptics more anger, paranoia, and dependence. Observers rated the RTL group as higher in terms of affect and the LTL group as higher in terms of paranoia. In reinterpreting these findings in the light of my own experience I would formulate the differences between R and LTL epileptics thus: in LTL epileptics, aggression can be seen as either a variation of the catastrophic reaction or an expression of the paranoid trait; in RTL epileptics, aggression can be seen as an occasional manifestation of a basically unstable affect.

It seems, thus, on the evidence presented so far, that psychopathological disorders of an assaultive type are associated with language or communication difficulties and thus by implication with dysfunction of the left or dominant hemisphere. The employment of alternative coping strategies, relying more on non-verbal as well as "illogical" mechanisms, are hypothetical explanations which might be worth considering in discussing the origins of violent behavior in psychopathic disorders.

SPECULATIONS AND HYPOTHESES

The evidence for the role that the dominant hemisphere for speech plays in the control of aggressive impulses cannot be overlooked. Whether this role depends largely on the neocortical elements involved in learning, especially verbal learning, or is drawn from the related subcortical elements, is an area needing further research. Certainly the importance of the amygdala in rage reactions cannot be denied nor the role of hippocampus in learned behavior. The question that could be asked is whether the limbic system of the dominant hemisphere for speech has a dominant role in the generation or control of aggression, either in connection with other aspects of learning, or, independently.

In previous publications I discussed the three principles characterising cerebral dysfunction, namely those of deficit, disinhibition and adaptation (Serafetinides, 1980; 1981; 1984). I also alluded to various hemispheric specific assumed impairments implicit in such conditions, namely, impairment not only of speech but also of time-sequencing and pacing of mental events in relation to dominant hemisphere dysfunction, paralleled, in non-dominant hemisphere dysfunction, with impairment of the spatial "framing" of mental events. The resulting behavioral pattern, in the case of dominant hemisphere disorders, is thus postulated to be characterized by an explosive simultaneity of mental output, stemming from the impaired ability to "hold" or "defer" the expression of such output, as would normally be the case. In non-dominant hemisphere disorders an analogous pattern is postulated, i.e., that of mood implosion, resulting from the inability to maintain a stable perspective of the self *vis á vis* the internal or external environment, as perceived.

In epilepsy (Serafetinides, 1980) such a formulation has to accomodate the phenomena associated with seizure discharges and their significance in terms of stimulation or hyperconnection and inhibition or disconnection. Thus, somatosensory symptoms and signs, such as the ones present in convulsions, can be considered as phenomena of the first type (stimulation-hyperconnection) whereas aphasia and cognitive or amnestic symptoms, can be considered as phenomena of the second type (inhibition or disconnection). Furthermore, the organism might show in the end a host of other phenomena as well, which as they tend to represent its overall response to functions perceived as having been interfered with, can be understood as (mal) adaptation or compensation phenomena depending on the case and the circumstances. These can include aggression, paranoid thinking, catastrophic reaction, anxiety, tension, grandiosity and depression. As the terms imply, such (mal) adaptation or compensation phenomena can follow impaired function of either hemisphere, at least theoretically; in practice, however, some empirical correlations seem to surface again and again, such as those of aggression, paranoid thinking and catastrophic reactions with lesions of the dominant hemisphere, and depression, anxiety and grandiosity with non-dominant ones.

Proceeding from the specific case of epilepsy to the more general one of psychiatric disorders, a further hypothetical formulation (Serafetinides 1981; 1984) runs as follows: dominant hemisphere dysfunctions can manifest themselves as perturbations of consciousness of either a quantitative nature (e.g. excitement or withdrawal) or of a qualitative one (e.g. paranoid delusions). Associated phenomena such as inappropriate affect or obscure somatic symptoms might represent in this respect (mal) adaptation or compensation phenomena arising from a deregulated non-dominant hemisphere. Symptoms directly attributable, on the other hand, to non-dominant hemisphere dysfunction can be considered as those arising from perturbations of awareness (visuo-spatial and self) of either a quantitative nature (e.g. mania or depression) or of a qualitative one (e.g. bodily delusions). Here the associated ruminative or verbal phenomena, such as stereotype thoughts or flight of ideas can be considered as (mal) adaptation or compensation phenomena arising from a deregulated dominant hemisphere.

Thus, as can be seen from the above attempts at formulating testable schemes of brain-behavior relationships various complexities of such relationships have to be constantly redefined and notions such as the ones presented here (e.g. direct vs indirect symptoms, excitation vs inhibition, (mal) adaptation vs compensation) might prove useful in these redefinitions. In the end, however, only further research can determine their -at present only heuristic - value.

Acknowledgement

The author wishes to acknowledge the support of the VA Medical Research Service.

REFERENCES

Bear, D.M., and Fedio, P. (1977). Quantitative analysis of interictal behavior in temporal lobe epilepsy. Archives of Neurology, 34, 454-467.

Camp, B.W. (1977). Verbal mediation in young aggressive boys. Journal of Abnormal Psychology, 86, 145-153.

Camp, B.W., Zimet, S.G., van Dornick, W.J., and Dahlem, N.W., (1977). Verbal abilities in young aggressive boys. Journal of Educational Psychology, 69, 129-135.

Hare, R.D., and McPherson, M. (1984). Psychopathy and perceptual asymetry during verbal dichotic listening. Journal of Abnormal Psychology, 93, 141-149.

Hare, R.C. (1979). Psychopathy and laterality of cerebral function. Journal of Abnomal Psychology, 88, 605-610.

Krynicki, V.E., (1978). Cerebral Dysfunction in repetitively assaultive adolescents. Journal of Nervous and Mental Diseases, 166, 59-67.

Pritchard, P.B., Lombroso C.T. and McIntyre, M. (1980). Psychological complications of temporal lobe epilepsy. Neurology, 30, 227-232.

Sandel, A. and Alcorn, J.D. (1980). Individual hemispherity and maladaptive behaviors, Journal of Abnormal Psychology, 89, 514-517.

Serafetinides, E.A. (1965). Aggressiveness in temporal lobe epileptics and its relation to cerebral dysfunction and environmental factors. Epilepsia, 6, 33-42.

Serafetinides, E.A. (1980). Epilepsy, Cerebral dominance and behavior. In M. Girgis, and L.G. Kilsh, (Eds.). Limbic Epilepsy and the Dyscontrol Syndrome. Elsevier-North Holland, New York.

Serafetinides, E.A. (1981). Psychopathology of the cerebral hemispheres. In E.A. Serafetinides (Ed.) Psychiatric Research in Practice. Grune and Stratton, New York.

Serafetinides, E.A. (1984). EEG Lateral asymmetries in psychiatric disorders. Biological Psychology, 19, 237-246.

Weintraub, S. and Mesulam, M.M. (1983) Developmental learning disabilities in the right hemisphere. _Archives of Neurology_, _40_, 463-468.

Wolff, P.H., Waber, D., Bauermeister, H., Cohen, C., Ferber, R. (1982). The neuropsychological status of adolescent delinquent boys. _Journal of Child Psychology and Psychiatry_, _23_, 267-279.

Yeudall, L.T., Fromm-Auch, D. and Davies, P. (1982). Neuropsychological impairment of persistent delinquency. _Journal of Nervous and Mental Disorders_, _170_, 257-265.

CONTRIBUTORS

MARIAN ANNETT, Department of Applied Social Studies, Coventry (Lanchester) Polytechnic, Coventry, CV1 5FB, U.K.

PAMELA, J. BION, Psychology Department, Lancaster University, Lancaster, LA1 4YF, U.K.

NIELS BIRBAUMER, Department of Clinical and Physiological Psychology, Gartenstrasse 29, University of Tubingen, Tubingen D4700, F.R. Germany.

JOHN L. BRADSHAW, Department of Psychology, Monash University, Clayton Victoria, Australia 3168.

STUART R. BUTLER, Department of Anatomy, Medical School, University of Birmingham, Birmingham B15 2TJ, U.K.

RUE L. CROMWELL, Department of Psychology, University of Kansas, Lawrence, Kansas, 66045-2462, U.S.A.

THOMAS ELBERT, Department of Clinical and Physiological Psychology, Gartenstrasse 29, University of Tubingen, Tubingen D7400, F.R. Germany.

P. FLOR-HENRY, Alberta Hospital Edmonton, Box 307, Edmonton, Alberta, T6G 2G3, Canada.

ALAN GLASS, Department of Anatomy, Medical School, University of Birmingham, Birmingham B15 2TJ, U.K.

JOHN GRUZELIER, Department of Psychiatry, Charing Cross and Westminster Medical School, 22/24 St. Dunstans Road, London.

RAQUEL E. GUR, Brain Behavior Laboratory, Departments of Psychiatry and Neurology, University of Pennsylvania, Philadelphia, Pennsylvania 19014, U.S.A.

RUBEN C. GUR, Brain Behavior Laboratory, Departments of Psychiatry and Neurology, University of Pennsylvania, Philadelphia, Pennsylvania, 19014, U.S.A.

DEBRA L. KIGAR, Department of Psychiatry, McMaster University, Hamilton, Ontario, Canada.

Z.J. KOLES, Department of Applied Sciences in Medicine, Edmonton, Alberta, T5J 2J7, Canada.

WERNER LUTZENBERGER, Arbeitsbereich Klinische und Physiologische Psychologie, Gartenstr. 29, D-4700 Tubingen, F.R. Germany.

JOHN C. MARSHALL, Neuropsychology Unit, The Radcliffe Infirmary, Woodstock Road, Oxford, OX2 6HF, U.K.

KATHRYN H. MCWEENY, Psychology Department, Lancaster University, Lancaster, LA1 4YF, U.K.

MICHAEL MIRAN, Department of Psychiatry, University of Rochester Medical Centre, Rochester, New York, 14642, U.S.A.

ESTA MIRAN, University of Rochester, Rochester, New York, 14642, U.S.A.

DENNIS L. MOLFESE, Department of Psychology and School of Medicine, Southern Illinois University, Carbondale, IL 62901, U.S.A.

VICTORIA J. MOLFESE, Department of Psychology and School of Medicine, Southern Illinois University, Carbondale IL 62901, U.S.A.

THOMAS R. O'CONNELL, Counselling Department, University of North Dakota, U.S.A.

LELON J. PEACOCK, Department of Psychology, University of Georgia, Athens, Georgia 30602, U.S.A.

ANTONIO E. PUENTE, Department of Psychology, University of North Carolina at Wilimington, North Carolina, 28403, U.S.A.

WILLIAM J. RAY, Department of Psychology, Penn State University, University Park, Pa. 16802, U.S.A.

J.R. REDDON, Alberta Hospital Edmonton, Box 307, Edmonton, Alberta, T6G 2G3, Canada.

BRIGITTE ROCKSTROH, Arbeitsbereich Klinische und Physiologische Psychologie, Gartenstr. 29, D-4700 Tubingen, F.R. Germany.

S.J. SEGALOWITZ, Department of Psychology, Brock University, St. Catharines, Ontario, L2S 3A1, Canada.

E.A. SERAFETINIDES, Veterans Administration Medical Center West Los Angeles, (Brentwood Division), and The Department of Psychiatry and The Brain Research Institute, UCLA School of Medicine, Los Angeles, California, U.S.A.

W.N. SCHOFIELD, Department of Experimental Psychology, University of Cambridge, Downing Street, Cambridge, CB2 3EB. U.K.

THOMAS B. SCOTT, Counselling Department, University of North Dakota, U.S.A.

DON M. TUCKER, Department of Psychology, University of Oregon, Eugene, OR97403, U.S.A.

A.M. WEBER, Department of Psychology, University of Victoria, P.O. Box 1700, Victoria, B.C., V8W 2Y2D, Canada.

SANDRA F. WITELSON, Department of Psychiatry, Faculty of Health Sciences, McMaster University, Hamilton, Ontario, L8N 3Z5, Canada.

ANDREW W. YOUNG, Psychology Department, Lancaster University, Lancaster, LA1 4YF. U.K.

INDEX